Modern Birkhäuser Classics

Many of the original research and survey monographs in pure and applied mathematics published by Birkhäuser in recent decades have been groundbreaking and have come to be regarded as foundational to the subject. Through the MBC Series, a select number of these modern classics, entirely uncorrected, are being re-released in paperback (and as eBooks) to ensure that these treasures remain accessible to new generations of students, scholars, and researchers.

Visions in Mathematics

GAFA 2000 Special Volume, Part I

N. Alon
J. Bourgain
A. Connes
M. Gromov
V. Milman
Editors

Reprint of the 2000 Edition

Birkhäuser

Editors:

N. Alon
School of Mathematical Sciences
University of Tel Aviv
Tel Aviv 69978
Israel
e-mail: nogaa@math.tau.ac.il

A. Connes
Collège de France
3, rue d'Ulm
75231 Paris cedex 05
France
e-mail: alain@connes.org

V. Milman
Department of Mathematics
University of Tel Aviv
Tel Aviv 69978
Israel
e-mail: milman@post.tau.ac.il

J. Bourgain
School of Mathematics
Institute for Advanced Study
Princeton University
Princeton, NJ 08540
USA
e-mail: bourgain@math.ias.edu

M. Gromov
Institut des Hautes Études Scientifiques
35, Route de Chartres
91440 Bures-sur-Yvette
France
e-mail: gromov@ihes.fr

Originally published under the title "Visions in Mathematics. Towards 2000"
as Special Volume, Part I, pp. 1-453 of the Journal "GAFA Geometric And
Functional Analysis" by Birkhäuser Verlag, Switzerland, ISSN 1016-443X
© 2000 Birkhäuser Verlag, P.O. Box 133, CH-4010 Basel, Switzerland

2000 Mathematics Subject Classification 35-02, 37-02, 00A05, 00A79, 01A67,
17B37, 42B20, 53C23, 76F55, 82C03

Library of Congress Control Number: 2010920640

Bibliographic information published by Die Deutsche Bibliothek
Die Deutsche Bibliothek lists this publication in the Deutsche Nationalbibliografie;
detailed bibliographic data is available in the Internet at <http://dnb.ddb.de>.

ISBN 978-3-0346-0421-5

© 2010 Birkhäuser, Springer Basel AG
P.O. Box 133, CH-4010 Basel, Switzerland
Part of Springer Science+Business Media
Printed on acid-free paper produced of chlorine-free pulp. TCF ∞

ISBN 978-3-0346-0421-5   e-ISBN 978-3-0346-0422-2

9 8 7 6 5 4 3 2 1   www.birkhauser.ch

Table of Contents

Foreword

The meeting "Visions in Mathematics – Towards 2000" took place mainly at Tel Aviv University in August 25-September 3, 1999, with a few days at the Sheraton-Moriah Hotel at the Dead Sea Health Resort. The meeting included about 45 lectures by some of the leading researchers in the world, in most areas of mathematics and a number of discussions in different directions, organized in various forms.

The goals of the conference, as defined by the scientific committee, consisting of N. Alon, J. Bourgain, A. Connes, M. Gromov and V. Milman, were to discuss the importance, methods, future and unity/diversity of mathematics as we enter the 21st Century, to consider the relation between mathematics and related areas and to discuss the past and future of mathematics as well as its interaction with Science.

A new format of mathematical discussions developed by the end of the Conference into an interesting addition to the more standard form of lectures and questions.

We believe that the meeting succeeded in giving a wide panorama of mathematics and mathematical physics, but we did not touch upon the interaction of mathematics with the experimental sciences.

This is the first part of the proceedings of the meeting. The second part will appear later this year.

It is a pleasure to thank Mrs. Miriam Hercberg for her great technical help in the preparation of this manuscript.

N. Alon   J. Bourgain

A. Connes   M. Gromov

V. Milman

GAFA2000

Tel Aviv, Israel

W. T. Gowers

S. Bloch

M. Gromov

J. Bourgain

V. Kac

J. Fröhlich

D. Kazhdan

S. Klainerman

Y. Ne'eman

A. Kupiainen

S. Novikov

E. H. Lieb

Ya. G. Sinai

L. Lovász

E. M. Stein

Visions in Mathematics – Towards 2000
Scientific Program

Wednesday, August 25

Lev Auditorium, Tel Aviv University

09:15-09:30 OPENING REMARKS

Geometry   *Chair: L. Polterovich*

09:30-10:15 *M. Gromov*: Geometry as the art of asking questions

10:30-11:15 *H. Hofer*: Holomorphic curves and real three-dimensional dynamics

11:45-12:30 *Y. Eliashberg*: Symplectic field theory

Ergodic Theory, Dynamic Systems   *Chair: S. Mozes*

14:00-14:45 *H. Furstenberg*: Dynamical methods in diophantine problems

15:00-15:45 *G. Margulis*: Diophantine approximation, lattices and flows on homogeneous spaces

16:15-17:00 *Y. Sinai*: On some problems in the theory of dynamical systems and mathematical physics

17:15-   Discussion with introduction by *Y. Sinai*

Thursday, August 26

Lev Auditorium, Tel Aviv University

Mathematical Physics   *Chair: A. Jaffe*

09:00-09:45 *J. Fröhlich*: Large quantum systems

10:00-10:45 *Y. Ne'eman*: Physics as geometry - Plato vindicated

11:15-12:00 *A. Connes*: Non-commutative geometry

Computer Science *Chair: N. Alon*

14:00-14:45 *P. Shor*: Mathematical problems in quantum information theory

15:00-15:45 *A. Razborov*: Complexity of proofs and computation

16:15-17:00 *A. Wigderson*: Some fundamental insights of computational complexity

17:15-18:00 *M. Rabin*: The mathematics of trust and adversity

Friday, August 27

Moriah Hotel, Dead Sea

Chair: A. Connes

14:00-14:45 *A. Jaffe*: Mathematics of quantum fields

15:00-15:45 *S. Novikov*: Topological phenomena in real physics

16:00-16:45 Discussion on Mathematical Physics with introduction by *A. Connes*

17:00-18:00 Discussion on Geometry with introduction by *M. Gromov*

Sunday, August 29

Moriah Hotel, Dead Sea

10:00- Discussion on Mathematics in the Real World (image, applications etc.), with introduction by *R. Coifman*

14:00- Discussion on Computer Science and Discrete Mathematics with introduction by *M. Rabin*

Monday, August 30

Lev Auditorium, Tel Aviv University

Analysis   *Chair: A. Olevskiĭ*

09:00-09:45   *R. Coifman*: Challenges in analysis 1

10:00-10:45   *P. Jones*: Challenges in analysis 2

11:15-12:00   *E. Stein*: Some geometrical concepts arising in harmonic analysis

12:10-12:40   Discussion

Number Theory   *Chair: Z. Rudnick*

14:30-15:15   *H. Iwaniec*: Automorphic forms in recent developments of analytic number theory

15:30-16:15   *P. Sarnak*: Some problems in number theory and analysis

16:45-17:30   *D. Zagier*: On "q" (or "Connections between modular forms, combinatorics and topology")

17:45-   Discussion: "The unreasonable effectiveness of modular forms" introduced by *P. Sarnak*

Tuesday, August 31

Lev Auditorium, Tel Aviv University

Discrete Mathematics   *Chair: A. Wigderson*

09:00-09:45   *L. Lovász*: Discrete and continuous: two sides of phenomena

10:00-10:45   *N. Alon*: Probabilistic and algebraic methods in discrete mathematics

11:15-12:00   *G. Kalai*: An invitation to Tverberg's theorem

12:10-12:40   Discussion

Analysis   *Chair: E. Stein*

14:30-15:15 *T. Gowers*: Rough structure and crude classification

15:30-16:15 *J. Bourgain*: Some problems in Hamiltonian PDE's

16:45-17:30 *V. Milman*: Topics in geometric analysis

17:45-   Discussion

Wednesday, September 1

Lev Auditorium, Tel Aviv University

Algebra   *Chair: V. Kac*

09:00-09:45 *S. Bloch*: Characteristic classes for linear differential equations

10:00-10:45 *V. Voevodsky*: Motivic homotopy types

11:15-12:00 *D. Kazhdan*: The lifting problems and crystal base

12:10-12:40 Discussion

Algebra   *Chair: D. Kazhdan*

14:30-15:15 *A. Beilinson*: Around geometric Langlands

15:30-16:15 *J. Bernstein*: Equivariant derived categories

16:45-17:30 *V. Kac*: Classification of infinite-dimensional simple groups of supersymmetries and quantum field theory

17:45-   Discussion with introduction by *D. Kazhdan*

Thursday, September 2

Lev Auditorium, Tel Aviv University

Chair: M. Gromov

09:00-09:45 *R. MacPherson*: On the applications of topology

10:00-10:45 *M. Kontsevich*: Smooth and compact

11:15-12:00 *D. Sullivan*: String interactions in topology

12:10-12:40 Discussion

Chair: G. Kalai

14:30-15:15 *I. Aumann*: Mathematical game theory: Looking backward and forward

15:30-16:15 *E. Hrushovski*: Logic and geometry

16:45- Discussion: The role of homotopical algebra in physics, with introductions by *D. Sullivan* and *M. Kontsevich*

Friday, September 3

Lev Auditorium, Tel Aviv University

Mathematical Physics and PDE   *Chair: V. Zakharov*

09:00-09:45 *T. Spencer*: Universality and statistical mechanics

09:50-10:35 *E. Lieb*: The mathematics of the second law of thermodynamics

11:00-11:45 *A. Kupiainen*: Lessons for turbulence

11:50-12:35 *S. Klainerman*: Some general remarks concerning nonlinear PDE's

12:40-13:10 Discussion

13:10-13:30 CLOSING REMARKS

13:30- Continuation of Discussion

GAFA, Geom. funct. anal.
Special Volume – GAFA2000, 1 – 31
1016-443X/00/S10001-31 $ 1.50+0.20/0

❙GAFA Geometric And Functional Analysis

GAUß–MANIN DETERMINANT CONNECTIONS AND PERIODS FOR IRREGULAR CONNECTIONS

SPENCER BLOCH AND HÉLÈNE ESNAULT

Abstract

Gauß–Manin determinant connections associated to irregular connections on a curve are studied. The determinant of the Fourier transform of an irregular connection is calculated. The determinant of cohomology of the standard rank 2 Kloosterman sheaf is computed modulo 2 torsion. Periods associated to irregular connections are studied in the very basic $\exp(f)$ case, and analogies with the Gauß–Manin determinant are discussed.

Everything's so awful reg'lar a body can't stand it.

The Adventures of Tom Sawyer
Mark Twain

1 Introduction

A very classical area of mathematics, at the borderline between applied mathematics, algebraic geometry, analysis, mathematical physics and number theory is the theory of systems of linear differential equations (connections). There is a vast literature focusing on regular singular points, Picard-Fuchs differential equations, Deligne's Riemann-Hilbert correspondence and its extension to $\mathcal{D}$-modules, and more recently various index theorems in geometry.

The classification of irregular singular points in terms of Stokes structures is explained in [M, chap. IV]. It involves the Riemann-Hilbert correspondence for holonomic $\mathcal{D}$-modules with regular singular points, together with a filtration related to the rate of growth of local solutions.

Nonetheless, there are some very modern themes in the subject which remain virtually untouched. On the arithmetic side, one may ask, for example, how irregular connections can be incorporated in the modern theory of motives? How deep are the apparent analogies between wild ramification in characteristic p and irregularity? Can one define periods for irregular connections? If so, do the resulting period matrices have anything to do with ϵ-factors for ℓ-adic representations? On the geometric side, one can ask for a theory of characteristic classes $c_i(E, \nabla)$ for irregular connections such that c_0 is the rank of the connection and c_1 is the isomorphism class of the determinant. With these, one can try to attack the Riemann-Roch problem.

In this note, we describe some conjectures and examples concerning what might be called a families index theorem for irregular connections. Let

$$f : X \to S \qquad\qquad (1.1)$$

be a smooth, projective family of curves over a smooth base S. Let $\mathcal{D}$ be an effective relative divisor on X. Let E be a vector bundle on X, $\nabla :$ $E \to E \otimes \Omega^1_X(\mathcal{D})$ an integrable, absolute connection with poles along $\mathcal{D}$. The relative de Rham cohomology $\mathbb{R}f_*(\Omega^*_{X-\mathcal{D}} \otimes E)$ inherits a connection (Gauß–Manin connection), and the families index or Riemann-Roch problem is to describe the isomorphism class of the line bundle with connection

$$\det \mathbb{R}f_*(\Omega^*_{X-\mathcal{D}} \otimes E). \qquad\qquad (1.2)$$

The Gauß–Manin construction is fairly standard and we do not recall it in detail. By way of example, we cite two classical formulas (Gauß hypergeometric and Bessel functions, respectively):

$$\frac{\Gamma(b)\Gamma(c-b)}{\Gamma(c)} F(a, b; c; z) = \int_0^1 u^{b-1}(1-u)^{c-b-1}(1-uz)^{-a} du$$

$$J_n(z) = \frac{1}{2\pi i} \int_{S_0} u^{-n} \exp \frac{z}{2}\left(u - \frac{1}{u}\right) \frac{du}{u} \qquad (S_0 = \text{circle about } 0).$$

In both cases, the integrand is a product of a solution of a rather simple degree 1 differential equation in u, the solution being either

$$u^{b-1}(1-u)^{c-b-1}(1-uz)^{-a} \quad \text{or} \quad u^{-n} \exp \frac{z}{2}\left(u + \frac{1}{u}\right),$$

with an algebraic 1-form (du or du/u). The integral is taken over a chain in the u-plane. The resulting functions $F(a, b; c; z)$ and $J_n(z)$ satisfy Gauß–Manin equations, which are much more interesting degree 2 equations in z.

It is not our purpose to go further into the classical theory, but, to understand the role of the determinant, we remark that in each of the above cases, there is a second path and a second algebraic 1-form such that the two integrals, say $f_1(z)$ and $f_2(z)$, satisfy the same second order equation. The Wronskian determinant

$$\begin{vmatrix} f_1 & f_2 \\ \frac{df_1}{dz} & \frac{df_2}{dz} \end{vmatrix}$$

satisfies the degree 1 equation given by the determinant of Gauß–Manin.

We are working here with algebraic de Rham cohomology. Analytically (i.e. permitting coordinate and basis transformations with essential singularities on $\mathcal{D}$), the bundle $E^{\mathrm{an}}|_{X-\mathcal{D}}$ can be transformed (locally on S) to have regular singular points along $\mathcal{D}$, but the algebraic problem we pose is more subtle.

If $\mathcal{D} = \emptyset$, or, more generally, if ∇ has regular singular points, the answer is known:

$$\det \mathbb{R}f_*(\Omega^*_{X-\mathcal{D}} \otimes E) = f_*\big((\det E^{\vee}, -\det \nabla) \cdot c_1(\omega_{X/S})\big). \quad (1.3)$$

(In the case of regular singular points, the c_1 has to be taken as a relative class [ST], [BE1].)

In a recent article [BE2], we proved an analogous formula in the case when E was irregular and rank 1. For a suitable $\mathcal{D}$, the relative connection induces an isomorphism

$$\nabla_{X/S,\mathcal{D}} : E|_{\mathcal{D}} \cong E|_{\mathcal{D}} \otimes \big(\omega(\mathcal{D})/\omega\big),$$

i.e. a trivialization of $\omega(\mathcal{D})|_{\mathcal{D}}$. The connection pulls back from a rank 1 connection $(\mathcal{E}, \nabla_{\mathcal{E}})$ on the relative Picard scheme $\mathrm{Pic}(X, \mathcal{D})$, and the Gauß–Manin determinant connection is obtained by evaluating $(\mathcal{E}, \nabla_{\mathcal{E}})$ at the priviledged point $(\omega(\mathcal{D}), \nabla_{X/S,\mathcal{D}}) \in \mathrm{Pic}(X, \mathcal{D})$.

We want now to consider two sorts of generalizations. First, we formulate an analogous conjecture for higher rank connections which are admissible in a suitable sense. We prove two special cases of this conjecture, computing the determinant of the Fourier transform of an arbitrary connection and, up to 2-torsion, the determinant of cohomology of the basic rank 2 Kloosterman sheaf.

Second, we initiate in the very simplest of cases $E = \mathcal{O}$, $\nabla(1) = df$, the study of periods for irregular connections. Let $m = \deg f + 1$. We are led to a stationary phase integral calculation over the subvariety of $\mathrm{Pic}(\mathbb{P}^1, m \cdot \infty)$ corresponding to trivializations of $\omega(m \cdot \infty)$ at ∞. The subtle part of the integral is concentrated at the same point

$$\big(\omega(m \cdot \infty), \nabla_{X/S, m \cdot \infty}\big) \in \mathrm{Pic}(\mathbb{P}^1, m \cdot \infty)$$

mentioned above.

Our hope is that the geometric methods discussed here will carry over to the sort of arithmetic questions mentioned above. We expect that the distinguished point given by the trivialization of $\omega(\mathcal{D})$ plays some role in the calculation of ϵ-factors for rank 1 ℓ-adic sheaves and that the higher rank conjectures have some ℓ-adic interpretation as well.

We would like to thank Pierre Deligne for sharing his unpublished letters with us. The basic idea of using the relative Jacobian to study ϵ-factors for rank 1 sheaves we learned from him. We also have gotten considerable inspiration from the works of T. Saito and T. Terasoma cited in the bibliography. The monograph [K] is an excellent reference for Kloosterman sheaves, and the subject of periods for exponential integrals is discussed briefly at the end of [Ko].

2   The Conjecture

Let S be a smooth scheme over a field k of characteristic 0. We consider a smooth family of curves $f : X \to S$ and a vector bundle E with an absolute, integrable connection $\nabla : E \to E \otimes \Omega^1_X(*D)$. Here $D \subset X$ is a divisor which is smooth over S. We are interested in the determinant of the Gauß–Manin connection

$$\det \mathbb{R}f_*(\Omega^*_{(X-D)/S} \otimes E) \in \text{Pic}^\nabla(S) := \mathbb{H}^1(S, \mathcal{O}_S^\times \xrightarrow{d\log} \Omega^1_S)$$

$$\cong \Gamma(S, \Omega^1_S/d\log \mathcal{O}_S^\times). \quad (2.1)$$

PROPOSITION 2.1. *Let $K = k(S)$ be the function field of S. Then the restriction map $\text{Pic}^\nabla(S) \to \text{Pic}^\nabla(\text{Spec}(K))$ is an injection.*

Proof. In view of the interpretation of Pic^∇ as $\Gamma(\Omega^1/d\log \mathcal{O}^\times)$, the proposition follows from the fact that for a meromorphic function g on S, we have g regular at point $s \in S$ if and only if dg/g is regular at s. □

Thus, we do not lose information by taking the base S to be the spectrum of a function field, $S = \text{Spec}(K)$. We shall restrict ourselves to that case.

Let

$$\mathcal{D} = \sum_{x \in D(\overline{K})} m_x x \quad (2.2)$$

be an effective divisor supported on D. Suppose that the relative connection has poles of order bounded by $\mathcal{D}$, i.e.

$$\nabla_{X/S} : E \to E \otimes \omega(\mathcal{D}), \quad (2.3)$$

and that $\mathcal{D}$ is minimal with this property. (We frequently write ω in place of $\Omega_{X/S}$.) We view $\mathcal{D}$ as an artinian scheme with sheaf of functions $\mathcal{O}_\mathcal{D}$ and relative dualizing sheaf $\omega_\mathcal{D} := \omega(\mathcal{D})/\omega$. (To simplify, we do not use the notation $\omega_{\mathcal{D}/K}$).

PROPOSITION 2.2. *Define* $E_{\mathcal{D}} := E \otimes \mathcal{O}_{\mathcal{D}}$. *Then* $\nabla_{X/S}$ *induces a function-linear map*

$$\nabla_{X/S,\mathcal{D}} : E_{\mathcal{D}} \to E_{\mathcal{D}} \otimes \omega_{\mathcal{D}}.$$

Proof. Straightforward.   □

Let $j : X - D \hookrightarrow X$. There are now two relative de Rham complexes we might wish to study

$$j_* j^* E \to j_* j^* (E \otimes \omega)$$
$$E \to E \otimes \omega(D).$$

The first is clearly the correct one. For example, its relative cohomology carries the Gauß–Manin connection. On the other hand, the second complex is sometimes easier to study, as the relative cohomology of the sheaves involved has finite dimension over K. Since we are primarily interested in the irregular case, that is when points of $\mathcal{D}$ have multiplicity ≥ 2, the following proposition clarifies the situation.

PROPOSITION 2.3. *Let notation be as above, and assume every point of* $\mathcal{D}$ *has multiplicity* ≥ 2. *Then the natural inclusion of complexes*

$$\iota : \{E \to E \otimes \omega(D)\} \hookrightarrow \{j_* j^* E \to j_* j^* (E \otimes \omega)\}$$

is a quasiisomorphism if and only if $\nabla_{X/S,\mathcal{D}} : E_{\mathcal{D}} \to E_{\mathcal{D}} \otimes \omega_{\mathcal{D}}$ *in Proposition 2.2 is an isomorphism.*

Proof. Let $\mathcal{D} : h = 0$ be a local defining equation for $\mathcal{D}$. Write $\mathcal{D} = \mathcal{D}' + D$ where D is the reduced divisor with support $= \mathrm{supp}(\mathcal{D})$. We claim first that the map ι is a quasiisomorphism if and only if for all $n \geq 1$ the map G defined by the commutative diagram

$$E/E(-D) \xrightarrow{\ \nabla_{X/S,\mathcal{D}} - n \cdot id \otimes \frac{dh}{h}\ } (E(\mathcal{D}')/E(-D)) \otimes \omega(D)$$

$$\cong \downarrow {}^{``}{\cdot}h^{-n\text{''}} \qquad\qquad \cong \downarrow {}^{``}{\cdot}h^{-n\text{''}}$$

$$E(n\mathcal{D})/E((n-1)\mathcal{D}) \xrightarrow{\ G\ } (E(n\mathcal{D}+\mathcal{D}')/E((n-1)\mathcal{D}+\mathcal{D}')) \otimes \omega(D)$$

given by $h^{-n} e \mapsto h^{-n} \nabla_{X/S,\mathcal{D}}(e) - n h^{-n-1} e \otimes dh$ is a quasi-isomorphism. This follows by considering the cokernel of ι

$$j_* j^* E/E \to j_* j^* (E \otimes \omega)/E \otimes \omega(\mathcal{D})$$

and filtering by order of pole. The assertion of the proposition follows because $n h^{-n-1} e \otimes dh$ has a pole of order strictly smaller than the multiplicity of $(n+1)\mathcal{D}$ at every point of $\mathcal{D}$.   □

We will consider only the case

$$\nabla_{X/S,\mathcal{D}} : E_{\mathcal{D}} \cong E_{\mathcal{D}} \otimes \omega_{\mathcal{D}}. \tag{2.4}$$

We now consider the sheaf $j_* j^* \Omega^1_X$ of absolute (i.e. relative to k) 1-forms on X with poles on D. Let $\mathcal{D} = \sum m_x x$ be an effective divisor supported on D as above, and write $\mathcal{D}' = \mathcal{D} - D$.

DEFINITION 2.4. *The sheaf $\Omega^p_X\{\mathcal{D}\} \subset j_* j^* \Omega^p_X$ is defined locally around a point $x \in D$ with local coordinate z by*

$$\Omega^p_X\{\mathcal{D}\}_x = \Omega^p_X(\mathcal{D}')_x + \Omega^{p-1}_{X,x} \wedge \tfrac{dz}{z^{m_x}}.$$

The graded sheaf $\bigoplus_p \Omega^p_X\{\mathcal{D}\}$ is stable under the exterior derivative and independent of the choice of local coordinates at the points of D. One has exact sequences

$$0 \to f^* \Omega^p_S(\mathcal{D}') \to \Omega^p_X\{\mathcal{D}\} \to \Omega^{p-1}_S(\mathcal{D}') \otimes \omega(D) \to 0. \tag{2.5}$$

DEFINITION 2.5. *An integrable absolute connection on E will be called admissible if there exists a divisor $\mathcal{D}$ such that $\nabla : E \to E \otimes \Omega^1_X\{\mathcal{D}\}$ and such that $\nabla_{X/S,\mathcal{D}} : E_{\mathcal{D}} \cong E_{\mathcal{D}} \otimes \omega_{\mathcal{D}}$.*

REMARK 2.6. *When E has rank 1, there always exists a $\mathcal{D}$ such that ∇ is admissible for $\mathcal{D}$ ([BE2, Lemma 3.1]). In higher rank, this need not be true, even if $\nabla_{X/S,\mathcal{D}}$ is an isomorphism for some $\mathcal{D}$. For example, let $\eta \in \Omega^1_K$ be a closed 1-form. Let $n \geq 1$ be an integer and let $c \in k$, $c \neq 0$. The connection matrix*

$$A = \begin{pmatrix} \frac{cdz}{z^m} & \frac{\eta}{z^n} \\ 0 & \frac{cdz}{z^m} - \frac{ndz}{z} \end{pmatrix}$$

satisfies $dA + A \wedge A = 0$ for all $m, n \in \mathbb{N}$, but the resulting integrable connection is not admissible for $n > m$, although $\nabla_{X/S,(0)}$ is an isomorphism for $c \neq n$ if $m = 1$. Note in this case it is possible to change basis to get an admissible connection. We don't know what to expect in general. There do exist connections for which $\nabla_{X/S,\mathcal{D}}$ is not an isomorphism for any $\mathcal{D}$, for example, if one takes a sum of rank 1 connections with different m_x as above (see notations (2.2)) and a local basis adapted to this direct sum decomposition.

Henceforth, $S = \operatorname{Spec}(K)$ is the spectrum of a function field, and we consider only integrable, admissible connections $\nabla : E \to E \otimes \Omega^1_X\{\mathcal{D}\}$ with $\mathcal{D} \neq \emptyset$. In sections 3 and 4 we will see many importnat examples (Fourier transforms, Kloosterman sheaves) of admissible connections. By abuse of notation, we write

$$H^*_{DR/S}(E) := \mathbb{H}^* \big(X, E \to E \otimes \omega(\mathcal{D}) \big). \tag{2.6}$$

We assume (Prop. 2.3) this group coincides with $H^*_{DR}(E|_{X-D})$. The isomorphism class of the Gauß–Manin connection on the K-line $\det H^*_{DR/S}(E)$ is determined by an element

$$\det H^*_{DR/S}(E) \in \Omega^1_{K/k}/d\log(K^\times)$$

which we would like to calculate.

Suppose first that E is a line bundle. Twisting E by $\mathcal{O}(\delta)$ for some divisor δ supported on the irregular part of the divisor D, we may assume $\deg E = 0$. In this case, the result (the main theorem in [BE2]) is the following. Since E has rank 1, $\nabla_{X/S,\mathcal{D}} : E_\mathcal{D} \cong E_\mathcal{D} \otimes \omega_\mathcal{D}$ can be interpreted as a section of $\omega_\mathcal{D}$ which generates this sheaf as an $\mathcal{O}_\mathcal{D}$-module. The exact sequence

$$0 \to \omega \to \omega(\mathcal{D}) \to \omega_\mathcal{D} \to 0 \tag{2.7}$$

yields an element $\partial\nabla_{X/S,\mathcal{D}} \in H^1(X,\omega) \cong K$ which is known to equal $\deg E = 0$. Thus, we can find some $s \in H^0(X,\omega(\mathcal{D}))$ lifting $\nabla_{X/S,\mathcal{D}}$. We write (s) for the divisor of s as a section of $\omega(\mathcal{D})$ (so (s) is disjoint from D). Then the result is

$$\det H^*_{DR/S}(E) \cong -f_*\big((s) \cdot E\big). \tag{2.8}$$

(When (s) is a disjoint union of K-points, the notation on the right simply means to restrict E with its absolute connection to each of the points and then tensor the resulting K-lines with connection together.) Notice that unlike the classical Riemann-Roch situation (e.g. (1.3)) the divisor (s) depends on $(E, \nabla_{X/S})$.

Another way of thinking about (2.8) will be important when we consider periods. It turns out that the connection (E, ∇) pulls back from a rank 1 connection $(\mathcal{E}, \nabla_\mathcal{E})$ on the relative Picard scheme $\mathrm{Pic}(X, \mathcal{D})$ whose points are isomorphism classes of line bundles on X with trivializations along $\mathcal{D}$. The pair $(\omega(\mathcal{D}), \nabla_{X/D,\mathcal{D}})$ determine a point $t \in \mathrm{Pic}(X, \mathcal{D})(K)$, and (2.8) is equivalent to

$$\det H^*_{DR/S}(E) \cong -(\mathcal{E}, \nabla_\mathcal{E})|_t. \tag{2.9}$$

Let $\omega^\times_\mathcal{D} \subset \omega_\mathcal{D}$ be the subset of elements generating $\omega_\mathcal{D}$ as an $\mathcal{O}_\mathcal{D}$-module. Let $\partial : \omega_\mathcal{D} \to H^1(X,\omega) = K$. Define $\tilde{B} = \omega^\times_\mathcal{D} \cap \partial^{-1}(0)$. One has a natural action of $K^\times$, and the quotient $\omega^\times_\mathcal{D}/K^\times$ is identified with isomorphism classes of trivializations of $\omega(\mathcal{D})|_\mathcal{D}$, and hence with a subvariety of $\mathrm{Pic}(X, \mathcal{D})$. One has

$$t \in B := \tilde{B}/K^\times \subset \omega^\times_\mathcal{D}/K^\times \subset \mathrm{Pic}(X, \mathcal{D}). \tag{2.10}$$

The relation between $t, B, \mathcal{E}$ is the following. $\mathcal{E}|_B \cong \mathcal{O}_B$, so the connection $\nabla_{\mathcal{D}}|_B$ is determined by a global 1-form Ξ. Then

$$\Xi(t) = 0 \in \Omega^1_{B/K} \otimes K(t). \tag{2.11}$$

Indeed, t is the unique point on B where the relative 1-form Ξ/K vanishes (cf. [BE2, Lemma 3.10]).

Now suppose the rank of E is > 1. We will see when we consider examples in the next section that $\det H^*_{DR/S}(E)$ depends on more than just the connection on $\det E$ (Remark 3.3). Thus, it is hard to imagine a simple formula like (2.8). Indeed, there is no obvious way other than by taking the determinant to get rank 1 connections on X from E. The truly surprising thing is that if we rewrite (2.8) algebraically we find a formula which does admit a plausible generalization. We summarize the results, omitting proofs (which are given in detail in [BE2]). For each $x_i \in D$, choose a local section s_i of $\Omega^1_X\{\mathcal{D}\}$ whose image in $\omega(\mathcal{D})$ generates at x_i. Write the local connection matrix in the form

$$A_i = g_i s_i + \frac{\eta_i}{z_i^{m_i-1}} \tag{2.12}$$

with z_i a local coordinate and $\eta_i \in f^*\Omega^1_S$. Let s be a meromorphic section of $\omega(\mathcal{D})$ which is congruent to the $\{s_i\}$ modulo $\mathcal{D}$. Define

$$c_1\big(\omega(\mathcal{D}), \{s_i\}\big) := (s) \in \mathrm{Pic}(X, \mathcal{D}). \tag{2.13}$$

As in (2.8), we can define

$$f_*\big(c_1(\omega(\mathcal{D}), \{s_i\}) \cdot \det(E, \nabla)\big) \in \Omega^1_K/d\log K^\times. \tag{2.14}$$

We further define

$$\big\{c_1(\omega(\mathcal{D})), \nabla\big\} := f_*\big(c_1(\omega(\mathcal{D}), \{s_i\}) \cdot \det(E, \nabla)\big) - \sum_i \mathrm{res}\, \mathrm{Tr}(dg_i g_i^{-1} A_i). \tag{2.15}$$

Here res refers to the map

$$\Omega^2_X\{\mathcal{D}\} \to \Omega^1_K \otimes \omega(\mathcal{D}) \to \Omega^1_K \otimes \omega_{\mathcal{D}} \xrightarrow{\text{transfer}} \Omega^1_K. \tag{2.16}$$

CONJECTURE 2.7.   Let $\nabla : E \to E \otimes \Omega^1_X\{\mathcal{D}\}$ be an admissible connection as in Definition 2.5. Then

$$\det H^*_{DR/S}(X - D, E) = -\big\{c_1(\omega(\mathcal{D})), \nabla\big\} \in \Omega^1_K/d\log(K^\times) \otimes_{\mathbb{Z}} \mathbb{Q}.$$

Our main objective here is to provide evidence for this conjecture. Of course, one surprising fact is that the right-hand side is independent of choice of gauge, etc. Again, the proof is given in detail in [BE2] but we reproduce two basic lemmas. There are function linear maps

$$\nabla_{X,\mathcal{D}} : E_{\mathcal{D}} \to E_{\mathcal{D}} \otimes \big(\Omega^1_X\{\mathcal{D}\}/\Omega^1_X\big); \quad \nabla_{X/S,\mathcal{D}} : E_{\mathcal{D}} \to E_{\mathcal{D}} \otimes \omega_{\mathcal{D}}, \tag{2.17}$$

and it makes sense to consider the commutator

$$[\nabla_{X,\mathcal{D}}, \nabla_{X/S,\mathcal{D}}] : E_{\mathcal{D}} \to E_{\mathcal{D}} \otimes \left(\Omega^1_X\{\mathcal{D}\}/\Omega^1_X\right) \otimes \omega_{\mathcal{D}}.$$

LEMMA 2.8. $[\nabla_{X,\mathcal{D}}, \nabla_{X/S,\mathcal{D}}] = 0.$

Proof. With notation as in (2.12), we take $s_i = dz_i/z_i^{m_i}$, where z_i is a local coordinate. Integrability implies

$$dA_i = dg_i \wedge \frac{dz_i}{z_i^{m_i}} + d\left(\frac{\eta_i}{z_i^{m_i-1}}\right) = A_i^2 = [\eta_i, g_i]\frac{dz_i}{z_i^{2m_i-1}} + \epsilon \quad (2.18)$$

with $\epsilon \in \Omega^2_K \otimes K(X)$. Multiplying through by $z_i^{m_i}$, we conclude that $[\eta_i/z_i^{m_i-1}, g_i]$ is regular on D, which is equivalent to the assertion of the lemma.   □

The other lemma which will be useful in evaluating the right-hand term in (2.15) is

LEMMA 2.9. *We consider the situation from (2.12) and (2.15) at a fixed $x_i \in D$. For simplicity, we drop the i from the notation. Assume $ds = 0$. Then*

$$\operatorname{res}\operatorname{Tr}(dgg^{-1}A) = \operatorname{res}\operatorname{Tr}\left(dgg^{-1}\frac{\eta}{z^{m-1}}\right).$$

Proof. We must show $\operatorname{res}\operatorname{Tr}(dgs) = 0$. Using (2.18) and $\operatorname{Tr}[g,\eta] = 0$ we reduce to showing $0 = \operatorname{res}\operatorname{Tr}(d(\eta z^{1-m})) \in \Omega^1_K$. Since $\eta \in f^*\Omega^1_K$, we may do the computation formally locally and replace d by d_z. The desired vanishing follows because an exact form has no residues.   □

With Lemma 2.8, we can formulate the conjecture in a more invariant way in terms of an AD-cocycle on X. Recall [E]

$$AD^2(X) := \mathbb{H}^2(X, \mathcal{K}_2 \xrightarrow{d\log} \Omega^2_X) \cong H^1(X, \Omega^2_X/d\log\mathcal{K}_2). \quad (2.19)$$

The AD-groups are the cones of cycle maps from Chow groups to Hodge cohomology, and as such they carry classes for bundles with connections. There is a general trace formalism for the AD-groups, but in this simple case the reader can easily deduce from the right-hand isomorphism in (2.19) a trace map

$$f_* : AD^2(X) \to AD^1(S) = \Omega^1_K/d\log K^\times. \quad (2.20)$$

When the connection ∇ has no poles (or more generally, when it has regular singular points) it is possible to define a class

$$\epsilon = c_1(\omega) \cdot c_1(E, \nabla) \in AD^2(X) \quad (2.21)$$

with $f_*(\epsilon) = [\det H^1_{DR/S}(X, E)]$. Remarkably, though it no longer has the product description (2.21), one can associate such a class to any admissible connection.

Fix the divisor $\mathcal{D}$ and consider tuples $\{E, \nabla, \mathcal{L}, \mu\}$ where (E, ∇) is an admissible, absolute connection, $\mathcal{L}$ is a line bundle on X, and $\mu : E_{\mathcal{D}} \cong E_{\mathcal{D}} \otimes \mathcal{L}_{\mathcal{D}}$. We require

$$0 = [\mu, \nabla_{X,\mathcal{D}}] : E_{\mathcal{D}} \to E_{\mathcal{D}} \otimes \mathcal{L}_{\mathcal{D}} \otimes \left(\Omega_X^1\{\mathcal{D}\}/\Omega_X^1 \right). \qquad (2.22)$$

Of course, the example we have in mind, using Lemma 2.8, is

$$\{E, \nabla\} := \left\{ E, \nabla, \omega(\mathcal{D}), \nabla_{X/S,\mathcal{D}} \right\}. \qquad (2.23)$$

To such a tuple satisfying (2.22), we associate a class $\epsilon(E, \nabla, \mathcal{L}, \mu) \in AD^2(X)$ as follows. Choose cochains $c_{ij} \in GL(r, \mathcal{O}_X)$ for E, $\lambda_{ij} \in \mathcal{O}_X^\times$ for $\mathcal{L}$, $\mu_i \in GL(r, \mathcal{O}_{\mathcal{D}})$ for μ, and $\omega_i \in M(r \times r, \Omega_X^1\{\mathcal{D}\})$ for ∇. Choose local liftings $\tilde{\mu}_i \in GL(r, \mathcal{O}_X)$ for the μ_i.

PROPOSITION 2.10.   *The Čech hypercochain*

$$\left(\{\lambda_{ij}, \det(c_{jk})\}, d\log\lambda_{ij} \wedge \mathrm{Tr}(\omega_j), \mathrm{Tr}(-d\tilde{\mu}_i\tilde{\mu}_i^{-1} \wedge \omega_i) \right)$$

represents a class

$$\epsilon(E, \nabla, \mathcal{L}, \mu) \in \mathbb{H}^2\left(X, \mathcal{K}_2 \to \Omega_X^2\{\mathcal{D}\} \to \Omega_X^2\{\mathcal{D}\}/\Omega_X^2 \right) \cong AD^2(X).$$

This class is well defined independent of the various choices. Writing $\epsilon(E, \nabla) = \epsilon(E, \nabla, \omega(\mathcal{D}), \nabla_{X/S,\mathcal{D}})$, *we have*

$$f_*\epsilon(E, \nabla) = \left\{ c_1(\omega(\mathcal{D})), \nabla \right\}$$

where the right-hand side is defined in (2.15).

Proof. Again the proof is given in detail in [BE2] and we omit it.   □

As a consequence, we can restate the main conjecture:

CONJECTURE 2.11.   *Let* ∇ *be an integrable, admissible, absolute connection as above. Then*

$$\det H_{DR/S}^*(E, \nabla) = -f_*\epsilon(E, \nabla).$$

To finish this section, we would like to show that behind the quite technical cocyle written in Proposition 2.10, there is an algebraic group playing a role similar to $\mathrm{Pic}(X, \mathcal{D})$ in the rank 1 case. Let G be the algebraic group whose K-points are isomorphism classes $(\mathcal{L}, \mu)$, where $\mathcal{L}$ is an invertible sheaf, and $\mu : E_{\mathcal{D}} \to E_{\mathcal{D}} \otimes \mathcal{L}_{\mathcal{D}}$ is an isomorphism commuting with $\nabla_{X,\mathcal{D}}$. It is endowed with a surjective map $q : G \to \mathrm{Pic}(X)$. As noted, G contains the special point $(\omega(\mathcal{D}), \nabla_{X/S,\mathcal{D}})$. The cocyle of Proposition 2.10 defines a class in $\mathbb{H}^2(X \times_K G, \mathcal{K}_2 \to \Omega_{X \times G}^2\{\mathcal{D} \times G\} \to \Omega_{X \times G}^2\{\mathcal{D} \times G\}/\Omega_{X \times G}^2)$. Taking its trace (2.20), one obtains a class in $(\mathcal{L}(E), \nabla(E)) \in AD^1(G)$, that is a rank one connection on G. Then $f_*\epsilon(E, \nabla)$ is simply the restriction of $(\mathcal{L}(E), \nabla(E))$ to the special point $(\omega(\mathcal{D}), \nabla_{X/S,\mathcal{D}})$.

Now we want to show that this special point, as in the rank 1 case, has a very special meaning. By analogy with (2.10), we define

$$\tilde{B} = \left(\mathrm{Ker}(\mathrm{Hom}(E_{\mathcal{D}}, E_{\mathcal{D}} \otimes \omega_{\mathcal{D}}) \xrightarrow{\mathrm{res\,Tr}} K)\right) \cap \mathrm{Isom}(E_{\mathcal{D}}, E_{\mathcal{D}} \otimes \omega_{\mathcal{D}}) \tag{2.24}$$

$$B = \tilde{B}/K^{\times} \subset G. \tag{2.25}$$

We observe that Lemma 2.9 shows that $g \in B$. Choosing a local trivialization $\omega_{\mathcal{D}} \cong \mathcal{O}_{\mathcal{D}} \frac{dz}{z^m}$ and a local trivialization of $E_{\mathcal{D}}$, we write $\nabla_{X/S}|_{\mathcal{D}}$ as a matrix $g \frac{dz}{z^m}$, with $g \in GL_r(\mathcal{O}_{\mathcal{D}})$. Θ is then identified with a translation invariant form on the restriction of scalars $\mathrm{Res}_{\mathcal{D}/K} GL_r$

$$\Theta := \left(\mathcal{L}(E), \nabla_{G/K}(E)\right) = \mathrm{res\,Tr}\left(d\mu \mu^{-1} g \frac{dz}{z^m}\right). \tag{2.26}$$

The assumption that $g \in B$ implies that Θ descends to an invariant form on $\mathrm{Res}_{\mathcal{D}/K} GL_r/\mathbb{G}_m \supset G_0$, where $G_0 := \{(\mathcal{O}, \mu) \in G\}$. By invariance, it gives rise to a form on the G_0-torsor $G_{\omega(\mathcal{D})} := \{(\omega(\mathcal{D}), \mu) \in G\}$. We have

$$B \subset G_{\omega(\mathcal{D})} \subset G.$$

Let $S \subset G_0$ be the subgroup of points stabilizing B.

PROPOSITION 2.12. $\Theta|_B$ *vanishes at a point* $t \in B$ *if and only if* t *lies in the orbit* $g \cdot S$.

Proof. Write $th = g$. Write the universal element in $\mathrm{Res}_{\mathcal{D}/K} GL_r$ as a matrix $X = \sum_{k=0}^{m-1} (X_{ij}^{(k)})_{ij} z^k$. The assertion that $\Theta|_B$ vanishes in the fibre at t means

$$\mathrm{res\,Tr}\left(\sum_k \left(d(X_{ij}^{(k)})_{ij} z^k\right) h \frac{dz}{z^m}\right)(t) = a\left(\sum_i dX_{ii}^{(m-1)}\right)(t)$$

for some $a \in K$. Note this is an identity of the form

$$0 = \sum c_{ij}^{(k)} dX_{ij}^{(k)} \in \Omega^1_{G/K} \otimes K(t); \quad c_{ij}^{(k)} \in K. \tag{2.27}$$

We first claim that in fact this identity holds already in $\Omega^1_{G/K}$. To see this, write $\mathcal{G} = \mathrm{Res}_{\mathcal{D}/K} GL_r$. Note $\Omega^1_{\mathcal{G}}$ is a free module on generators $dX_{ij}^{(k)}$. Also, $G \subset \mathcal{G}$ is defined by the equations

$$\left[\sum_{k=0}^{m-1} (X_{ij}^{(k)})_{ij} z^k, \sum_{k=0}^{m-1} (g_{ij}^{(k)})_{ij} z^k\right] = 0,$$

$$\left[\sum_{k=0}^{m-1} (X_{ij}^{(k)})_{ij} z^k, \sum_{k=0}^{m-2} (\eta_{ij}^{(k)})_{ij} z^k\right] = 0,$$

which are of the form

$$\sum_{i,j,k} b^{(k)}_{ijp} X^{(k)}_{ij} = 0, \qquad p = 1, 2, \ldots, M; \ b^{(k)}_{ijp} \in K,$$

that is are linear equations in the $X^{(k)}_{ij}$ with K-coefficients. Thus, we have an exact sequence

$$0 \to N^\vee \to \Omega^1_{G/K} \otimes \mathcal{O}_G \to \Omega^1_{G/K} \to 0, \qquad (2.28)$$

where $N^\vee$ is generated by K-linear combinations of the $dX^{(k)}_{ij}$. We have, therefore, a reduction of structure of the sequence (2.28) from $\mathcal{O}_G$ to K, and therefore $\Omega^1_{G/K} \cong \Omega^1_0 \otimes_K \mathcal{O}_G$, where $\Omega^1_0 \subset \Omega^1_{G/K}$ is the K-span of the $dX^{(k)}_{ij}$. Hence, any K-linear identity among the $dX^{(k)}_{ij}$ which holds at a point on G holds everywhere on G. As a consequence, we can integrate to an identity

$$\operatorname{res} \operatorname{Tr} \left(X h \tfrac{dz}{z^m} \right) = a \Big(\sum_i X^{(m-1)}_{ii} \Big) + \kappa \qquad (2.29)$$

with $\kappa \in K$. If we specialize $X \to t$ we find

$$0 = \operatorname{res} \operatorname{Tr} \left(g \tfrac{dz}{z^m} \right) = a \cdot \operatorname{res} \operatorname{Tr} \left(t \tfrac{dz}{z^m} \right) + \kappa = \kappa. \qquad (2.30)$$

We conclude from (2.29) and (2.30) that $h \in S$. □

3 The Fourier Transform

In this section we calculate the Gauß–Manin determinant line for the Fourier transform of a connection on $\mathbb{P}^1 - D$ and show that it satisfies the Conjecture 2.7. Let $\mathcal{D} = \sum m_\alpha \alpha$ be an effective k-divisor on $\mathbb{P}^1_k$. Let $\mathcal{E} = \oplus_r \mathcal{O}$ be a rank r free bundle on $\mathbb{P}^1_k$, and let $\Psi : \mathcal{E} \to \mathcal{E} \otimes \omega(\mathcal{D})$ be a k-connection on $\mathcal{E}$. Let $(\mathcal{L}, \Xi)$ denote the rank 1 connection on $\mathbb{P}^1 \times \mathbb{P}^1$ with poles on $\{0, \infty\} \times \{0, \infty\}$ given by $\mathcal{L} = \mathcal{O}_{\mathbb{P}^1 \times \mathbb{P}^1}$ and $\Xi(1) = d(z/t)$. Here z, t are the coordinates on the two copies of the projective line. Let $K = k(t)$. We have a diagram

$$\begin{array}{ccc} (\mathbb{P}^1_z - \mathcal{D}) \times \mathbb{P}^1_t & \hookleftarrow & (\mathbb{P}^1_z - \mathcal{D}) \times \operatorname{Spec}(K) \\ \downarrow p_1 & & \downarrow p_2 \\ \mathbb{P}^1_z - \mathcal{D} & & \operatorname{Spec}(K) \end{array} \qquad (3.1)$$

The Gauß–Manin determinant of the Fourier transform is given at the generic point by

$$\det H^*_{DR/K}\big((\mathbb{P}^1_z - \mathcal{D})_K, \, p^*_1(\mathcal{E}, \Psi) \otimes (\mathcal{L}, \Xi)\big)$$

$$= \det H^*_{DR/K}\big((\mathbb{P}^1_z - \mathcal{D})_K, (E, \nabla)\big) \quad (3.2)$$

with $E := p^*_1 \mathcal{E} \otimes \mathcal{L}|_{(\mathbb{P}^1_z - \mathcal{D})_K}$ and $\nabla = \Psi \otimes 1 + 1 \otimes \Xi$. We have the following easy

REMARK 3.1. *Write*

$$\Psi = \sum_\alpha \sum_{i=1}^{m_\alpha} \frac{g^\alpha_i dz}{(z - \alpha)^i} + d(g^\infty_1 z + \ldots + g^\infty_{m_\infty - 1} z^{m_\infty - 1}) \quad (3.3)$$

where $g^\alpha_i \in M(r \times r, k)$. *Then*

$$\nabla = \Psi + \frac{dz}{t} - \frac{zdt}{t^2} \quad (3.4)$$

is admissible if and only if either

(i) $m_\infty \leq 2$ *and* $g^\alpha_{m_\alpha}$ *is invertible for all* $\alpha \neq \infty$, *or*
(ii) $m_\infty \geq 3$, $g^\alpha_{m_\alpha}$ *is invertible for all* $\alpha \neq \infty$, *and* $g^\infty_{m_\infty - 1}$ *is invertible.*

Theorem 3.2. *The connection* (E, ∇) *satisfies Conjecture 2.7.*

Proof. We first consider the case when Ψ has a pole of order ≤ 1 at infinity, so the $g^\infty_i = 0$ in (3.3). A basis for

$$H^0\Big(\mathbb{P}^1_K, E \otimes \omega\Big(\sum m_\alpha \alpha + 2\infty\Big)\Big)$$

is given by

$$e_j \otimes dz; \quad e_j \otimes \frac{dz}{(z - \alpha)^i}, \quad 1 \leq i \leq m_\alpha, \ 1 \leq j \leq r. \quad (3.5)$$

$H^0_{DR/K} = (0)$ and $H^1_{DR/K} = \operatorname{coker}(H^0(E) \to H^0(E \otimes \omega(\sum m_\alpha \alpha + 2\infty)))$ has basis

$$e_j \otimes \frac{dz}{(z - \alpha)^i}; \quad 1 \leq i \leq m_\alpha, \ 1 \leq j \leq r. \quad (3.6)$$

To compute the Gauß–Manin connection, we consider the diagram (here $\mathcal{D} = \sum m_\alpha \alpha + 2\infty$ and $\mathcal{D}' = \mathcal{D} - D = \sum(m_\alpha - 1)\alpha + \infty$)

$$
\begin{array}{ccc}
H^0(E) & = & H^0(E) \\
\downarrow{\scriptstyle \nabla_X} & & \downarrow{\scriptstyle \nabla_{X/S}} \\
\end{array}
$$

$$0 \to \quad H^0(E(\mathcal{D}')) \otimes \Omega^1_K \quad \to \quad H^0(E \otimes \Omega^1_{\mathbb{P}^1}\{\mathcal{D}\}) \xrightarrow{a} H^0(E \otimes \omega(\mathcal{D})) \to 0$$

$$
\begin{array}{ccc}
\downarrow{\scriptstyle \nabla_{X/S} \otimes 1} & & \downarrow{\scriptstyle \nabla_X} \\
\end{array}
$$

$$H^0(E(\mathcal{D}') \otimes \omega(\mathcal{D})) \otimes \Omega^1_K \xrightarrow{\cong} H^0(E \otimes \Omega^2_{\mathbb{P}^1}\{\mathcal{D} + \mathcal{D}'\})$$

$$(3.7)$$

One deduces from this diagram the Gauß–Manin connection

$$H^1_{DR/K}(E) \cong \operatorname{coker}(\nabla_{X/S}) \xrightarrow{\nabla_{GM}} H^1_{DR/K}(E) \otimes \Omega^1_K; \quad w \mapsto \nabla_X(a^{-1}(w)). \tag{3.8}$$

We may choose $a^{-1}\!\left(e_j \otimes \frac{dz}{(z-\alpha)^i}\right) = e_j \otimes \frac{dz}{(z-\alpha)^i}$, so by (3.4)

$$\nabla_{GM}\left(e_j \otimes \frac{dz}{(z-\alpha)^i}\right) = \nabla_X\left(e_j \otimes \frac{dz}{(z-\alpha)^i}\right) = e_j \otimes \frac{z\,dz \wedge dt}{(z-\alpha)^i t^2}. \tag{3.9}$$

In $H^1_{DR/K} \cong \operatorname{coker}(\nabla_{X/S})$ we have the identity

$$e_j \otimes dz = -t\Psi e_j. \tag{3.10}$$

We conclude

$$\nabla_{GM}\left(e_j \otimes \frac{dz}{(z-\alpha)^i}\right)$$
$$= \begin{cases} \left(e_j \otimes \frac{dz}{(z-\alpha)^{i-1}} + \alpha e_j \otimes \frac{dz}{(z-\alpha)^i}\right) \wedge \frac{dt}{t^2} & 2 \le i \le m_\alpha \\ \left(-t\Psi e_j + \alpha e_j \otimes \frac{dz}{z-\alpha}\right) \wedge \frac{dt}{t^2} & i = 1. \end{cases} \tag{3.11}$$

In particular, the determinant connection, which is given by $\operatorname{Tr}\nabla_{GM}$, can now be calculated:

$$\operatorname{Tr}\nabla_{GM} = \sum_\alpha \frac{rm_\alpha \alpha\, dt}{t^2} - \operatorname{Tr}\sum_\alpha \frac{g_1^\alpha dt}{t}. \tag{3.12}$$

We compare this with the conjectured value which is the negative of (2.15). Define

$$F(z) := \sum_\alpha \frac{1}{(z-\alpha)^{m_\alpha}} - 1 = \frac{G(z)}{(z-\alpha)^{m_\alpha}}; \quad s := F(z)dz. \tag{3.13}$$

One has

$$c_1\left(\omega(\mathcal{D}), \left\{\frac{dz}{(z-\alpha)^{m_\alpha}}, dz\right\}\right) = (G), \tag{3.14}$$

the divisor of zeroes of G. We need to compute $(\det E, \det \nabla)|_{(G)}$. We have

$$G(z) = \sum_\alpha \prod_{\beta \neq \alpha} (z-\beta)^{m_\beta} - \prod_\alpha (z-\alpha)^{m_\alpha}$$
$$= -z^{\sum m_\alpha} + \left(\sum m_\alpha \alpha + \#\{\alpha \mid m_\alpha = 1\}\right) z^{(\sum m_\alpha)-1} + \dots \tag{3.15}$$

Note that the coefficients of G do not involve t, so the dz part of the connection dies on (G) and we get

$$\mathrm{Tr}\nabla|_{(G)} = -\tfrac{rz\,dt}{t^2}\big|_{(G)} = -\tfrac{r\,dt}{t^2} \sum_{\substack{\beta \\ G(\beta)=0}} \beta$$

$$= -\tfrac{r\,dt}{t^2}\Big(\sum m_\alpha \alpha + \#\{\alpha \mid m_\alpha = 1\}\Big). \tag{3.16}$$

It remains to evaluate the correction terms $\mathrm{res}\,\mathrm{Tr}(dgg^{-1}A)$ occurring in (2.15). In the notation of (2.12), $\eta/z^{m-1} = -z\,dt/t^2$, and by Lemma 2.9 we have $\mathrm{res}\,\mathrm{Tr}(dgg^{-1}A) = -\mathrm{res}\,\mathrm{Tr}\big(dgg^{-1}\tfrac{z\,dt}{t^2}\big)$. Clearly, the only contribution comes at $z = \infty$. Take $u = z^{-1}$. At ∞ the connection is

$$A = -\Big(\sum_\alpha \sum_i \frac{g_i^\alpha u^i}{(1 - u\alpha)^i} + \frac{1}{t}\Big)\frac{du}{u^2} - \frac{dt}{ut^2}. \tag{3.17}$$

We rewrite this in the form $A = gs + \tfrac{\eta}{u}$ as in (2.12) with s as in (3.13) and $\eta = -dt/t^2$. We find

$$g = \frac{\sum_{i,\alpha} \frac{g_i^\alpha u^i}{(1-u\alpha)^i} + t^{-1}}{\sum_\alpha \frac{u^{m_\alpha}}{(1-u\alpha)^{m_\alpha}} - 1} = \frac{\kappa}{v}, \tag{3.18}$$

(defining κ and v to be the numerator and denominator, respectively). Then

$$\mathrm{res}\,\mathrm{Tr}(dgg^{-1}A) = -\mathrm{res}\,\mathrm{Tr}\big(dgg^{-1}\tfrac{dt}{ut^2}\big)$$

$$= \big(-\mathrm{res}\,\mathrm{Tr}(d\kappa\kappa^{-1}u^{-1}) + r\cdot\mathrm{res}\,\mathrm{Tr}(dvv^{-1}u^{-1})\big)\tfrac{dt}{t^2}$$

$$= \Big(-t\sum \mathrm{Tr}(g_1^\alpha) - r\#\{\alpha \mid m_\alpha = 1\}\Big)\tfrac{dt}{t^2}. \tag{3.19}$$

Combining (3.19), (3.16), and (3.12) we conclude

$$\mathrm{Tr}\nabla_{GM} = -\big(\mathrm{Tr}\nabla|_{(G)} - \mathrm{res}\,\mathrm{Tr}(dgg^{-1}A)\big), \tag{3.20}$$

which is the desired formula.

We turn now to the case where Ψ has a pole of order ≥ 2 at infinity. We write

$$\Psi = \sum_\alpha \sum_{i=1}^{m_\alpha} \frac{g_i^\alpha\,dz}{(z - \alpha)^i} + g^\infty\,dz; \quad g^\infty = g_2^\infty + \ldots + g_{m_\infty}^\infty z^{m_\infty - 2} \tag{3.21}$$

$$\nabla = \Psi + \tfrac{dz}{t} - \tfrac{z\,dt}{t^2}. \tag{3.22}$$

A basis for $\Gamma(\mathbb{P}^1, \omega(\sum m_\alpha \alpha + m_\infty \infty))$ is given by

$$e_j \otimes \frac{dz}{(z - \alpha)^i}; \quad 1 \leq i \leq m_\alpha; \quad e_j \otimes z^i dz; \quad 0 \leq i \leq m_\infty - 2. \tag{3.23}$$

A basis for the Gauß–Manin bundle is given by omitting $e_j \otimes z^{m_\infty - 2} dz$. As in (3.7)–(3.9), the Gauß–Manin connection is

$$w \mapsto zw \wedge \frac{dt}{t^2} . \tag{3.24}$$

To compute the trace, note that in $H^1_{DR/K}$, we have if $m_\infty \geq 3$

$$e_j \otimes z^{m_\infty - 2} dz = -(g^\infty_{m_\infty})^{-1} \left(\sum_{i,\alpha} g^\alpha_i(e_j) \otimes \frac{dz}{(z-\alpha)^i} \right.$$
$$\left. + (g^\infty_2(e_j) + \ldots + g^\infty_{m_\infty - 1}(e_j) z^{m_\infty - 3} + t^{-1} e_j) \otimes dz \right). \tag{3.25}$$

If $m_\infty = 2$,

$$e_j \otimes dz = -(g^\infty_2 + t^{-1})^{-1} \left(\sum_{i,\alpha} g^\alpha_i(e_j) \otimes \frac{dz}{(z-\alpha)^i} \right). \tag{3.26}$$

It follows that

$$\mathrm{Tr}\nabla_{GM} = \begin{cases} \left(\sum_\alpha r m_\alpha \alpha - \mathrm{Tr}((g^\infty_{m_\infty})^{-1} g^\infty_{m_\infty - 1}) \right) \frac{dt}{t^2} & m_\infty \geq 3 \\ \left(\sum_\alpha r m_\alpha \alpha - \sum_\alpha \mathrm{Tr}((g^\infty_2 + t^{-1})^{-1} g^\alpha_1) \right) \frac{dt}{t^2} & m_\infty = 2. \end{cases} \tag{3.27}$$

To compute the right-hand side in Conjecture 2.7, we take as trivializing section

$$s = \sum_\alpha \frac{dz}{(z-\alpha)^{m_\alpha}} - z^{m_\infty - 2} dz = \frac{G(z) dz}{\prod_\alpha (z-\alpha)^{m_\alpha}} \tag{3.28}$$

where

$$G(z) = \sum_\alpha \prod_{\beta \neq \alpha} (z-\beta)^{m_\beta} - z^{m_\infty - 2} \prod_\alpha (z-\alpha)^{m_\alpha} . \tag{3.29}$$

We have $(s) = (G)$, the divisor of zeroes of the polynomial G. Again, G does not involve t, so if $G = -\prod(z - a_k)$, we have

$$\mathrm{Tr}\nabla|_{z=a_p} = -\frac{r a_p dt}{t^2}. \tag{3.30}$$

Thus

$$\mathrm{Tr}\nabla|_{(s)} = \begin{cases} -\frac{(\sum r m_\alpha \alpha + r \cdot \#\{\alpha \mid m_\alpha = 1\}) dt}{t^2}, & m_\infty = 2 \\ -\frac{\sum r m_\alpha \alpha dt}{t^2}, & m_\infty \geq 3. \end{cases} \tag{3.31}$$

Finally we have to deal with the correction term $\mathrm{res}\,\mathrm{Tr}(dg g^{-1}(-z dt/t^2))$. There is no contribution except at ∞. We put $u = z^{-1}$ as before, and

$g = \kappa/v$ with

$$\kappa = \sum_\alpha \sum_{i=1}^{m_\alpha} \frac{u^{m_\infty - 2 + i} g_i^\alpha}{(1 - \alpha u)^i} + (g_2^\infty + t^{-1}) u^{m_\infty - 2} + \ldots + g_{m_\infty}^\infty \qquad (3.32)$$

$$v = \sum_\alpha \frac{u^{m_\infty - 2 + m_\alpha}}{(1 - \alpha u)^{m_\alpha}} - 1 . \qquad (3.33)$$

Assume first $m_\infty = 2$. Then

$$\operatorname{res} \operatorname{Tr}(dg g^{-1} A) = -\operatorname{res} \operatorname{Tr}\left(dg g^{-1} \frac{dt}{ut^2}\right)$$

$$= -\operatorname{res} \operatorname{Tr}(d_z \kappa \kappa^{-1} u^{-1}) \frac{dt}{t^2} + r \cdot \operatorname{res} \operatorname{Tr}(d_z v v^{-1} u^{-1}) \frac{dt}{t^2}$$

$$= -\operatorname{Tr}\left((g_2^\infty + t^{-1})^{-1} \sum g_1^\alpha\right) \frac{dt}{t^2} - r \cdot \#\{\alpha \mid m_\alpha = 1\} \frac{dt}{t^2} . \quad (3.34)$$

In the case $m_\infty \geq 3$ we find

$$\operatorname{res} \operatorname{Tr}(dg g^{-1} A) = -\operatorname{Tr}\left((g_{m_\infty}^\infty)^{-1} g_{m_\infty - 1}^\infty\right) \frac{dt}{t^2} . \qquad (3.35)$$

The theorem follows by comparing (3.27), (3.31), (3.34), and (3.35).   □

REMARK 3.3.   *The presence of nonlinear terms in the g_i^α in (3.27) means that the connection on $\det H_{DR/K}^*(E)$ is not determined by $\det E$ alone.*

4   Kloosterman Sheaves

In this section we show that the main conjecture holds at least up to 2-torsion for the basic rank 2 Kloosterman sheaf [K]. The base field k is $\mathbb{C}$. Fix $a, b \in K^\times$ (in fact one can work over $K = \mathbb{C}[a, b, a^{-1}, b^{-1}]$). Let $\alpha, \beta \in \mathbb{C} - \mathbb{Z}$ and assume also $\alpha - \beta \in \mathbb{C} - \mathbb{Z}$. Consider two connections $(\mathcal{L}_i, \nabla_i)$ on the trivial bundle on $\mathbb{G}_m$ given by $1 \mapsto \alpha d\log(t) + d(at)$ and $1 \mapsto \beta d\log(u) + d(bu)$ where t, u are the standard parameters on two copies of $\mathbb{G}_m$. Let $X := \mathbb{G}_m \times \mathbb{G}_m$ and consider the exterior tensor product connection on X

$$\mathcal{L} := \mathcal{L}_1 \boxtimes \mathcal{L}_2 = (\mathcal{O}_X, \nabla); \ \nabla(1) = \alpha d\log(t) + \beta d\log(u) + d(at + bu) . \quad (4.1)$$

Note all the above are integrable, absolute connections.

PROPOSITION 4.1.

1. $H_{DR}^i(\mathcal{L}_i/K) \cong \begin{cases} K & i = 1, \\ 0 & i \neq 1, \end{cases}$

2. $H_{DR}^p(\mathcal{L}/K) \cong \begin{cases} H_{DR}^1(\mathcal{L}_1/K) \otimes H_{DR}^1(\mathcal{L}_2/K) & p = 2, \\ 0 & p \neq 2. \end{cases}$

Proof. The de Rham complex (of global sections on X) for $(\mathcal{L}, \nabla)$ is the tensor product of the corresponding complexes for the ∇_i, so (2) follows from (1). For (1) we have, e.g., the complex of global sections $\mathcal{O} \xrightarrow{\nabla_{1,K}} \omega$, $1 \mapsto \alpha d\log(t) + a dt$. In H^1_{DR} this gives for all $n \in \mathbb{Z}$

$$a t^n dt \equiv -(\alpha + n) t^{n-1} dt.$$

Assertion (1) follows easily.   □

We now compute Gauß–Manin. We will (abusively) use sheaf notation when we mean global sections over $\mathbb{G}_m$ or X. Also we write ∇ for either one of the ∇_i or the exterior tensor connection. ∇_K is the corresponding relative connection. One has the diagram

$$
\begin{array}{ccc}
\mathcal{O} & === & \mathcal{O} \\
\downarrow{\scriptstyle \nabla} & & \downarrow{\scriptstyle \nabla_K} \\
\mathcal{O} \otimes \Omega^1_K \longrightarrow & \Omega^1 \xrightarrow{\ \sigma\ } & \omega \\
\downarrow{\scriptstyle \nabla_K \otimes 1} & \downarrow{\scriptstyle \nabla} & \\
\omega \otimes \Omega^1_K \xrightarrow[\cong]{\ \iota\ } & \Omega^2/F^2. &
\end{array}
$$

Here σ is the obvious function linear map (e.g. for $\nabla = \nabla_1$, $\sigma(t^n dt) = t^n dt$), and $F^2 \subset \Omega^2$ is the subgroup of 2 forms coming from the base. This leads to the Gauß–Manin diagram

$$
\begin{array}{ccc}
\mathcal{O} & \xrightarrow{\nabla - \sigma \nabla_K} & \mathcal{O} \otimes \Omega^1_K \\
\downarrow{\scriptstyle \nabla_K} & & \downarrow{\scriptstyle \nabla_K \otimes 1} \\
\omega & \xrightarrow{-\iota^{-1} \nabla \sigma} & \omega \otimes \Omega^1_K.
\end{array}
$$

For example when $\nabla = \nabla_1$ we get on H^1_{DR}

$$\nabla_{GM}(t^n dt) = -\iota^{-1}\big(\nabla(1) \wedge t^n dt\big) = -\iota^{-1}(t da \wedge t^n dt) = t^{n+1} dt \otimes da.$$

Since $t dt \equiv \frac{-(\alpha+1)}{a} dt$, we get

$$\nabla_{GM}(dt) = -(\alpha + 1) dt \otimes d\log(a) \equiv -\alpha dt \otimes d\log(a) \mod d\log(K^\times).$$

On the (rank 1) tensor product connection $H^2_{DR}(\mathcal{L}) = H^1_{DR}(\mathcal{L}_1) \otimes H^1_{DR}(\mathcal{L}_2)$, the Gauß–Manin determinant connection is therefore

$$- \alpha d\log(a) - \beta d\log(b). \tag{4.2}$$

Note that we computed this determinant here by hand, but we could have as well applied directly Theorem 4.6 of [BE2]: the determinant of

$H^1_{DR}(\mathcal{L}_1)$ is just the restriction of ∇_1 to the divisor of $\mathbb{P}^1$ defined by the trivializing section $\alpha d \log(t) + a dt$ of $\omega(0 + 2\infty)$, that is by $\frac{a}{t} + \alpha = 0$. Thus the determinant is $-\alpha d \log(a) \in \Omega^1_K / d \log K^\times$, and similarly for $H^1_{DR}(\mathcal{L}_2)$, the determinant is $-\beta d \log(b) \in \Omega^1_K / d \log K^\times$.

The idea now is to recalculate that determinant connection using the Leray spectral sequence for the map $\pi : X \to \mathbb{G}_m$, $\pi(t, u) = tu$. Write v for the coordinate on the base, so $\pi^*(v) = tu$.

PROPOSITION 4.2. *We have*

$$R^i \pi_{*, DR}(\mathcal{L}) = \begin{cases} 0 & i \neq 1 \\ \text{rank 2 bundle on } \mathbb{G}_m & i = 1. \end{cases}$$

Proof. Let ∇_π be the relative connection on $\mathcal{L}$ with respect to the map π, and take t to be the fibre coordinate for π. Write $u = v/t$. Then

$$\nabla_\pi(1) = \alpha d \log(t) + \beta d \log(u) + a dt + b du = (\alpha - \beta) d \log(t) + a dt - b v \frac{dt}{t^2}$$

so in $R^1 \pi_{*, DR}$ we have

$$0 \equiv \nabla_\pi(t^n) = (\alpha - \beta + n) t^{n-1} dt + a t^n dt - b v t^{n-2} dt .$$

It follows that $R^1 \pi_{*, DR}(\mathcal{L})$ has rank 2 (generated, e.g., by dt and dt/t), and the other $R^i = (0)$ as claimed. $\qquad\square$

Define

$$\mathcal{E} := R^1 \pi_{*, DR}(\mathcal{L}), \quad \nabla = \nabla_{GM} : \mathcal{E} \to \mathcal{E} \otimes \Omega^1_{\mathbb{G}_m} .$$

Theorem 4.3. *The Gauß–Manin connection on $H^*_{DR/K}(\mathbb{G}_m, \mathcal{E})$ satisfies Conjecture 2.7 up to 2-torsion.*

REMARK 4.4. *In fact, we will see that ∇ is not admissible in the sense of Definition 2.5, but its inverse image via a degree 2 covering is. Since the new determinant of de Rham cohomology obtained in this way is twice the old one, we lose control of the 2-torsion. We do not know whether the conjecture holds exactly in this case or not.*

Proof. We can now calculate the connection $\nabla := \nabla_{GM}$ on $\mathcal{E}$ just as before. We have the Gauß–Manin diagram $(\sigma(t^n dt) = t^n dt)$.

$$\begin{array}{ccc} \mathcal{O}_X & \xrightarrow{\nabla - \sigma \nabla_\pi} & \mathcal{O}_X \otimes \Omega^1_{\mathbb{G}_m} \\ \downarrow{\scriptstyle \nabla_\pi} & & \downarrow{\scriptstyle \nabla_\pi \otimes 1} \\ \omega_\pi & \xrightarrow{-\iota^{-1} \nabla \sigma} & \omega_\pi \otimes \Omega^1_{\mathbb{G}_m}. \end{array}$$

Here

$$\nabla(1) = (\alpha - \beta) \frac{dt}{t} + \beta \frac{dv}{v} + a dt + t da + t^{-1} d(bv) + bv d(t^{-1}) .$$

We get

$$\nabla_{GM}\left(\tfrac{dt}{t}\right) = -\iota\left(\nabla(1)\wedge\tfrac{dt}{t}\right) = \tfrac{dt}{t}\otimes\beta\tfrac{dv}{v} + dt\otimes da + \tfrac{dt}{t^2}\otimes d(bv)$$
$$\nabla_{GM}(dt) = -\iota^{-1}\left(\left(\beta\tfrac{dv}{v} + tda + t^{-1}d(bv)\right)\wedge dt\right)$$
$$= dt\otimes\beta\tfrac{dv}{v} + tdt\otimes da + \tfrac{dt}{t}\otimes d(bv)\,.$$

We can now substitute

$$\tfrac{dt}{t^2} \equiv (bv)^{-1}\left((\alpha-\beta)\tfrac{dt}{t} + adt\right)$$
$$tdt \equiv \frac{bv}{a}\frac{dt}{t} - \frac{\alpha-\beta+1}{a}dt$$

getting finally

$$\nabla_{GM}\left(\tfrac{dt}{t}\right) = \tfrac{dt}{t}\otimes\left((\alpha-\beta)\tfrac{db}{b} + \alpha\tfrac{dv}{v}\right) + dt\otimes\left(da + a\tfrac{db}{b} + a\tfrac{dv}{v}\right)$$
$$\nabla_{GM}(dt) = dt\otimes\left(\beta\tfrac{dv}{v} - (\alpha-\beta+1)\tfrac{da}{a}\right) + \tfrac{dt}{t}\otimes\left(bv\tfrac{da}{a} + d(bv)\right)\,.$$

For convenience define

$$\theta = \tfrac{da}{a} + \tfrac{db}{b} + \tfrac{dv}{v}\,.$$

Representing an element in our rank two bundle as a column vector

$$\left(\begin{smallmatrix} r \\ s \end{smallmatrix}\right) = rdt + s\tfrac{dt}{t}$$

the matrix for the connection on $\mathcal{E}$ becomes

$$A := \begin{pmatrix} \beta\theta - (\alpha+1)\tfrac{da}{a} - \beta\tfrac{db}{b} & a\theta \\ bv\theta & a\theta - \alpha\tfrac{da}{a} - \beta\tfrac{db}{b} \end{pmatrix}\,.$$

The corresponding connection has a regular singular point at $v = 0$ and an irregular one at $v = \infty$. Extending $\mathcal{E}$ to $\mathcal{O}^2$ on $\mathbb{P}^1_K$, we can take $\mathcal{D} = (0) + 2(\infty)$ but the matrix g is not invertible at ∞. In order to remedy this, make the base change $z^{-2} = v$, and adjoin to K the element $\sqrt{ab}$. Notice the base change modifies the Gauß–Manin determinant computation. Let us ignore this for a while and continue with the determinant calculation.

Define $\gamma := \sqrt{ab}/z$. Make the change of basis

$A_{\text{new}} =$

$$\begin{pmatrix} 1 & \tfrac{-z^2}{b}(\beta-\gamma) \\ 0 & \tfrac{z^2}{2b} \end{pmatrix}\begin{pmatrix} \beta\theta - (\alpha+1)\tfrac{da}{a} - \beta\tfrac{db}{b} & a\theta \\ \tfrac{b}{z^2}\theta & a\theta - \alpha\tfrac{da}{a} - \beta\tfrac{db}{b} \end{pmatrix}$$

$$\times\begin{pmatrix} 1 & 2(\beta-\gamma) \\ 0 & \tfrac{2b}{z^2} \end{pmatrix} + \begin{pmatrix} 1 & \tfrac{-z^2}{b}(\beta-\gamma) \\ 0 & \tfrac{z^2}{2b} \end{pmatrix}\begin{pmatrix} 0 & -2d\gamma \\ 0 & d(\tfrac{2b}{z^2}) \end{pmatrix}\,.$$

This works out to

$$A_{\text{new}} = \begin{pmatrix} \gamma\theta - (\alpha+1)\tfrac{da}{a} - \beta\tfrac{db}{b} & (2\alpha+2\beta+1)\gamma\theta - 2\beta(\alpha+1)\theta \\ \tfrac{\theta}{2} & (\alpha+\beta+2)\theta - \gamma\theta - (\alpha+1)\tfrac{da}{a} - \beta\tfrac{db}{b} \end{pmatrix}\,.$$

Here of course $\theta = \frac{da}{a} + \frac{db}{b} - 2\frac{dz}{z}$.

Note

$$\text{Tr}(A_{\text{new}}) = (\alpha + \beta + 2)\theta - 2(\alpha + 1)\frac{da}{a} - 2\beta\frac{db}{b}$$
$$\equiv (\beta - \alpha)\frac{da}{a} + (\alpha - \beta)\frac{db}{b} - 2(\alpha + \beta)\frac{dz}{z} \quad \text{mod } d\log(K^\times).$$

At $z = 0$ the polar part of A_{new} looks like

$$\begin{pmatrix} \frac{\sqrt{ab}}{z}(\frac{da}{a} + \frac{db}{b} - 2\frac{dz}{z}) & (2\beta + 2\alpha + 1)\frac{\sqrt{ab}}{z}(\frac{da}{a} + \frac{db}{b} - 2\frac{dz}{z}) + 4\beta(\alpha + 1)\frac{dz}{z} \\ \frac{-dz}{z} & \frac{-\sqrt{ab}}{z}(\frac{da}{a} + \frac{db}{b} - 2\frac{dz}{z}) - 2(\alpha + \beta + 2)\frac{dz}{z} \end{pmatrix}.$$

Writing $A_{\text{pol},0} = g_0\frac{dz}{z^2} + \frac{\eta_0}{z}$, the matrix for g_0 with coefficients in $\mathbb{C}[z]/(z^2)$ is

$$g_0 = \begin{pmatrix} -2\sqrt{ab} & -2(2\alpha + 2\beta + 1)\sqrt{ab} + 4\beta(\alpha + 1)z \\ -z & 2\sqrt{ab} - 2(\alpha + \beta + 2)z \end{pmatrix}.$$

Also

$$\eta_0 = \begin{pmatrix} \sqrt{ab}(\frac{da}{a} + \frac{db}{b}) & (2\alpha + 2\beta + 1)\sqrt{ab}(\frac{da}{a} + \frac{db}{b}) \\ 0 & -\sqrt{ab}(\frac{da}{a} + \frac{db}{b}) \end{pmatrix}.$$

With respect to the trivialization $d\log(z^{-1})$ the matrix g at $z = \infty$ is

$$g_\infty = \begin{pmatrix} 0 & 4\beta(\alpha + 1) \\ -1 & -2(\alpha + \beta + 2) \end{pmatrix}.$$

Notice that the matrices for g are invertible both at 0 and ∞. Writing $A_{\text{pol},\infty} = g_\infty\frac{dz}{z} + \eta_\infty$ the contribution $\text{res Tr } dg_\infty g_\infty^{-1}\eta_\infty$ is of course vanishing, as well as the contribution at ∞ obtained by changing the trivialization dz/z to unit $\cdot\, dz/z$. At 0 we get ($\bar{g} := g \mod (z)$)

$$\text{res Tr}\left(dgg^{-1}\frac{\eta}{z}\right) = \text{res Tr}\left(d\bar{g}\bar{g}^{-1}\frac{\eta}{z}\right) = \frac{1}{2}\left(\frac{da}{a} + \frac{db}{b}\right) \times \qquad (4.3)$$

$$\text{Tr}\left[\begin{pmatrix} 0 & 4\beta(\alpha + 1) \\ -1 & -2(\alpha + \beta + 2) \end{pmatrix}\begin{pmatrix} -1 & -(2\beta + 2\alpha + 1) \\ 0 & 1 \end{pmatrix}^{-1}\begin{pmatrix} 1 & 2\beta + 2\alpha + 1 \\ 0 & 1 \end{pmatrix}\right]$$

$$= \frac{1}{2}\left(\frac{da}{a} + \frac{db}{b}\right)(2\beta + 2\alpha + 4) \equiv (\alpha + \beta)\left(\frac{da}{a} + \frac{db}{b}\right) \quad \text{mod } d\log(K^\times).$$

Now we compare with the conjectural formula 2.7

$$\det(H^*_{DR}(\mathcal{E}))^{-1} = c_1(\omega(\mathcal{D}), s) \cdot \det(\nabla) - \text{res Tr}\left(dgg^{-1}\frac{\eta}{z}\right).$$

Here s can be taken to be the divisor defined by the trivializing section $\frac{dz}{z^2} - \frac{dz}{z}$ of the sheaf $\omega(2\cdot 0 + \infty)$, that is $z = 1$. Thus one has

$$\text{Tr}\,A_{\text{new}} = (\beta - \alpha)\frac{da}{a} + (\alpha - \beta)\frac{db}{b}.$$

Further, we have to write

$$s = \frac{dz}{z^2}w,$$

where $w = 1 - z \in \mathcal{O}_{X,(20)}^{\times}$. Since $\mathrm{Tr}\,\eta_0 = 0$, $\mathrm{res}\,\mathrm{Tr}\,\frac{dw}{w}\frac{\eta_0}{z} = 0$ as well, thus the local contribution at 0 is given by (4.3).

The conjecture gives (writing $\delta_2 : \mathbb{P}^1 \to \mathbb{P}^1$, $x \mapsto x^2$)

$$\det(H_{DR}^*(\delta_2^*\mathcal{E}))^{-1} \overset{?}{=} (\beta - \alpha)\tfrac{da}{a} + (\alpha - \beta)\tfrac{db}{b} - (\alpha + \beta)\left(\tfrac{da}{a} + \tfrac{db}{b}\right)$$
$$= -2\alpha\tfrac{da}{a} - 2\beta\tfrac{db}{b}.$$

Notice we have adjoined $\sqrt{ab}$ to K so we have lost some 2-torsion. Bearing in mind that $\mathcal{E} = R^1\pi_{*,DR}$ which introduces a minus sign in the determinant calculations and comparing with our earlier calculation (4.2) above, we find that what we need to finish is

PROPOSITION 4.5. *The Gauß–Manin determinant for de Rham cohomology of $\delta_2^*\mathcal{E}$ is twice the corresponding determinant for $\mathcal{E}$.*

Proof. Again we use sheaf notation for working with modules. Recall for the pullback we substituted $z^2 = w = v^{-1}$. We can write $\delta_2^*\mathcal{E} = \mathcal{E} \oplus z\mathcal{E}$. We have

$$\delta_2^*\nabla(ze) = z\nabla(e) + ze \otimes \tfrac{dv}{-2v},$$

so with respect to the above decomposition we can write

$$(\delta_2^*\mathcal{E}, \delta_2^*\nabla) = (\mathcal{E}, \nabla) \oplus \left(\mathcal{E}, \nabla - \tfrac{1}{2}\tfrac{dv}{v}\right)$$

The second term on the right is the connection obtained by tensoring $\mathcal{E} = R^1\pi_{*,DR}(\mathcal{L}_1 \boxtimes \mathcal{L}_2)$ with $(\mathcal{O}, -\tfrac{1}{2}\tfrac{dv}{v})$. Using the projection formula and the invariance of the latter connection, this is the same as the connection on $R^1\pi_{*,DR}((\mathcal{L}_1 - \tfrac{1}{2}\tfrac{dt}{t}) \boxtimes (\mathcal{L}_2 - \tfrac{1}{2}\tfrac{du}{u}))$, i.e. it amounts to replacing α, β by $\alpha - \tfrac{1}{2}, \beta - \tfrac{1}{2}$. Using (1), this changes the Gauß–Manin determinant by $d\log(\sqrt{ab})$ which is trivial. It follows that the Gauß–Manin determinant of $\delta_2^*\mathcal{E}$ is twice that of $\mathcal{E}$, which is what we want. $\square$

This concludes the proof of Theorem 4.3. $\square$

5 Periods

Let $X/\mathbb{C}$ be a smooth, complete curve. We consider a connection (relative to $\mathbb{C}$) $\nabla : E \to E \otimes \omega_X(\mathcal{D})$. Let $\mathcal{E}$ be the corresponding local system on $X(\mathbb{C}) - \mathcal{D}$. Notice that we do not assume ∇ has regular singular points, so $\mathcal{E}$ does not determine (E, ∇). For example, it can happen that $\mathcal{E}$ is a trivial local system even though ∇ is highly nontrivial. In this section, we consider the question of associating periods to $\det H_{DR}^*(X - \mathcal{D}, E)$. We work with algebraic de Rham cohomology in order to capture the irregular structure.

The first remark is that it should be possible using Stokes structures [M] to write down a homological dual group $H_1(X^*, \mathcal{E})$ and perfect pairings ($\mathcal{E}^\vee$ is the dual local system)

$$H_1(X^*, \mathcal{E}^\vee) \times H^1_{DR}(X - \mathcal{D}, E) \to \mathbb{C}. \qquad (5.1)$$

Here X^* is some modification of the Riemann surface X. (The point is that, e.g., in the example we give below the de Rham group can be large while the local system $\mathcal{E}$ is trivial and $X - D = \mathbb{A}^1$.) Let $F \subset \mathbb{C}$ be a subfield, and assume we are given (i) an F-structure on $\mathcal{E}$, i.e. an F-local system $\mathcal{E}_F$ and an identification $\mathcal{E}_F \otimes \mathbb{C} \cong \mathcal{E}$. (ii) A triple $(X_0, \mathcal{D}_0, E_0)$ defined over F and an identification of the extension to $\mathbb{C}$ of these data with $(X, \mathcal{D}, E)$. When, e.g., (E, ∇) satisfies the condition of Proposition 2.3, one has

$$\det H^*_{DR}(X - \mathcal{D}, E)$$
$$\cong \mathbb{C} \otimes_F \det H^*(X_0, E_0) \otimes \det H^*(X_0, E_0 \otimes \omega(\mathcal{D}_0))^{-1} \quad (5.2)$$

so the determinant of de Rham cohomology gets an F-structure, even if ∇ is not necessarily itself defined over F. Of course, (i) determines an F-structure on $H_*(X^*, \mathcal{E})$. Choosing bases $\{p_j\}, \{\eta_k\}$ compatible with the F-structure and taking the determinant of the matrix of periods $\int_{p_j} \eta_k$, (5.1) yields an invariant

$$\mathrm{Per}(E_0, \nabla, \mathcal{E}_F) \in \mathbb{C}^\times / F^\times. \qquad (5.3)$$

(More generally, one can consider two subfields $k, F \subset \mathbb{C}$ with a reduction of $\mathcal{E}$ to F and a reduction of E to k. The resulting determinant lies in $F^\times \backslash \mathbb{C}^\times / k^\times$.) In the case of regular singular points these determinants have been studied in [ST].

Notice that the period invariant depends on the choice of an F-structure on the local system $\mathcal{E} = \ker(\nabla^{\mathrm{an}})$. When (E, ∇) are "motivic", i.e. come from the de Rham cohomology of a family of varieties over X, the corresponding local system of Betti cohomology gives a natural $\mathbb{Q}$-structure on $\mathcal{E}$. By a general theorem of Griffiths, the connection ∇ in such a case necessarily has regular singular points. In a non-geometric situation, or even worse, in the irregular case, there doesn't seem to be any canonical such $\mathbb{Q}$ or F-structure. For example, the equation $f' - f = 0$ has solution space $\mathbb{C} \cdot e^x$. Is the $\mathbb{Q}$-reduction $\mathbb{Q} \cdot e^x$ more natural than $\mathbb{Q} \cdot e^{x+1}$? Of course, in cases like this where the monodromy is trivial, the choice of $\mathcal{E}_F$ is determined by choosing an F-point $x_0 \in X_0 - \mathcal{D}_0$ and taking $\mathcal{E}_F = E_{x_0}$.

Even if there is no canonical F-structure on $\mathcal{E}$, one may still ask for a formula analogous to Conjecture 2.7 for $\mathrm{Per}(E_0, \nabla, \mathcal{E}_F)$. In this final section we discuss the very simplest case

$$X = \mathbb{P}^1, \ \mathcal{D} = m \cdot \infty, \ E = \mathcal{O}_X, \ \nabla(1) = df = d(a_{m-1}x^{m-1} + \ldots + a_1 x). \tag{5.4}$$

Period determinants in this case (and more general confluent hypergeometric cases) were computed by a different argument in [T]. We stress that our objective here is not just to compute the integral, but to exhibit the analogy with formula (2.9). We would like ultimately to find a formula for periods of higher rank irregular connections which bears some relation to Conjectures 2.7 and 2.11. We consider the situation (5.4) with $a_{m-1} \neq 0$. Then $H_{DR}^0 = (0)$ and $H_{DR}^1 \cong \mathbb{H}^1(\mathbb{P}^1, \mathcal{O} \to \omega(m \cdot \infty))$ has as basis the classes of $z^i dz$, $0 \leq i \leq m - 3$. One has $\mathcal{E}^\vee = \mathbb{C} \cdot \exp(f(z))$ (trivial local system) so we take the obvious $\mathbb{Q}$-structure with basis $\exp(f)$. We want to compute the determinant of the period matrix

$$\left(\int_{\sigma_i} \exp(f(z)) z^{j-1} dz \right)_{i,j=1,\ldots,m-2} \tag{5.5}$$

for certain chains σ_i on some X^*. Let S be a union of open sectors about infinity on $\mathbb{P}^1$ where $Re(f)$ is positive (i.e. S is a union of sectors of the form (here $N \gg 1$ and $\epsilon \ll 1$ are fixed)

$$S_k := \left\{ re^{i\theta} \mid N < r < \infty, \ \frac{-\arg(a_{m-1}) + (2k - \frac{1}{2} - \epsilon)\pi}{m-1} < \theta_k \right.$$
$$\left. < \frac{-\arg(a_{m-1}) + (2k + \frac{1}{2} + \epsilon)\pi}{m-1} \right\}$$

so $X^* := \mathbb{P}^1 - S \sim \mathbb{P}^1 - \{p_1, \ldots, p_{m-1}\}$ where the p_k are distinct points. In particular, $H_1(X^*) = \mathbb{Z}^{m-2}$. Define $\sigma_k := \gamma_k - \gamma_0$, where

$$\gamma_k := \left\{ r \exp(i\theta) \mid 0 \leq r < \infty; \ \theta = \frac{-\arg(a_{m-1}) + (2k+1)\pi}{m-1} \right\}. \tag{5.6}$$

The σ_k, $1 \leq k \leq m - 2$ form a basis for $H_1(X^*, \mathbb{Z})$.

Write $P_{ij} = \int_{\sigma_i} \exp(f(z)) z^{j-1} dz$.

LEMMA 5.1. We have

$$\det(P_{ij})_{1 \leq i,j \leq m-2}$$

$$= \int_{\sigma_1 \times \cdots \times \sigma_{m-2}} \exp(f(z_1) + \ldots + f(z_{m-2})) \prod_{i<j} (z_j - z_i) dz_1 \wedge \ldots \wedge dz_{m-2} \tag{5.7}$$

Proof. The essential point is the expansion

$$\prod_{i<j} (z_j - z_i) = \sum_a (-1)^{\text{sgn}(a)} z_1^{a(1)-1} z_2^{a(2)-1} \cdots z_{m-2}^{a(m-2)-1}, \tag{5.8}$$

where a runs through permutations of $\{1, \ldots, m - 2\}$.   □

We will evaluate (5.7) by stationary phase considerations precisely parallel to the techniques described in section 2 and [BE2]. Indeed, the degree $m - 2$ part $J^{m-2}(\mathbb{P}^1, m \cdot \infty) \subset J(\mathbb{P}^1, m \cdot \infty)$ of the generalized jacobian is simply the $\mathcal{O}_{m \cdot \infty}^\times$-torsor $\omega_{m \cdot \infty}^\times$ of trivializations of $\omega_{m \cdot \infty} = \omega(m \cdot \infty)/\omega$ modulo multiplication by a constant in $\mathbb{C}^\times$. Writing $u = z^{-1}$, we may identify this torsor with

$$\left\{ b_0 \tfrac{du}{u} + \ldots + b_{m-1} \tfrac{du}{u^m} \mid b_{m-1} \neq 0 \right\}. \tag{5.9}$$

The quotient of such trivializations up to global isomorphism is

$$\omega_{m \cdot \infty}^\times / \mathbb{C}^\times = \left\{ s_{m-1} \tfrac{du}{u} + \ldots + s_1 \tfrac{du}{u^{m-1}} + \tfrac{du}{u^m} \right\} = \left\{ (s_{m-1}, \ldots, s_1) \right\}. \tag{5.10}$$

Let $B \subset \omega_{m \cdot \infty}^\times / \mathbb{C}^\times$ be defined by $s_{m-1} = 0$. Let $\Gamma(\mathbb{P}^1, \omega(m \cdot \infty))^\times$ denote the space of sections which generate $\omega(m \cdot \infty)$ at ∞. We have

$$\mathbb{A}^{m-2} \twoheadrightarrow \mathrm{Sym}^{m-2}(\mathbb{A}^1) \overset{div}{\cong} \Gamma(\mathbb{P}^1, \omega(m \cdot \infty))^\times / \mathbb{C}^\times \cong B \subset \omega_{m \cdot \infty}^\times / \mathbb{C}^\times. \tag{5.11}$$

Let $z_1, \ldots, z_{m-2}$ be as in (5.7), and add an extra variable z_{m-1}. Take $s_k(z_1, \ldots, z_{m-1})$ to be the k-th elementary symmetric function, so, e.g., $s_{m-1} = z_1 z_2 \cdots z_{m-1}$. We have a commutative diagram

$$
\begin{array}{ccc}
\mathbb{A}^{m-2} & \longrightarrow & B \\
\downarrow{\scriptstyle z_m - 1 = 0} & & \downarrow \\
\mathbb{A}^{m-1} & \xrightarrow{z \mapsto (s_{m-1}(z), \ldots, s_1(z))} & \omega_{m \cdot \infty}^\times / \mathbb{C}^\times.
\end{array}
\tag{5.12}
$$

Notice that

$$\prod_{i < j} (z_j - z_i) dz_1 \wedge \ldots \wedge dz_{m-2} = ds_1 \wedge \ldots \wedge ds_{m-2}. \tag{5.13}$$

Let $p_k(z_1, \ldots, z_{m-2}) = z_1^k + \ldots + z_{m-2}^k$ be the k-th power sum (or k-th Newton class). Define

$$F(s_1, \ldots, s_{m-2}) := f(z_1) + \ldots + f(z_{m-2}) = a_1 p_1 + \ldots + a_{m-1} p_{m-1}. \tag{5.14}$$

Notice that, although the right-hand expression makes sense on all of $\omega_{m \cdot \infty}^\times / \mathbb{C}^\times$, we think of F as defined only on $B : s_{m-1} = 0$. Let Ψ be the direct image on B of the chain $\sigma_1 \times \cdots \times \sigma_{m-2}$ on $\mathbb{A}^{m-2}$. The integral (5.7) becomes

$$\int_\Psi \exp\left(F(s_1, \ldots, s_{m-2})\right) ds_1 \wedge \ldots \wedge ds_{m-2}. \tag{5.15}$$

LEMMA 5.2. *Let $b \in B = Sym^{m-2}(\mathbb{A}^1)$ correspond to the divisor of zeroes of $df = f'dz = (a_1 + 2a_2 z + \ldots + (m-1)a_{m-1}z^{m-2})dz$. Then dF vanishes at b and at no other point of B.*

Proof. The differential form

$$\eta := a_1 dp_1(z_1, \ldots, dz_{m-1}) + \ldots + a_{m-1}dp_{m-1}(z_1, \ldots, z_{m-1})$$

on $\omega^\times_{m \cdot \infty}/\mathbb{C}^\times$ is translation invariant. Indeed, to see this we may trivialize the torsor and take the point $s_1 = \ldots = s_{m-1} = 0$ to be the identity. Introducing a formal variable T with $T^m = 0$, the group structure is then given by $s \oplus s' =: s''$ with

$$\left(1 - s_1 T + \ldots + (-1)^{m-1}s_{m-1}T^{m-1}\right)\left(1 - s'_1 T + \ldots + (-1)^{m-1}s'_{m-1}T^{m-1}\right)$$
$$= \left(1 - s''_1 T + \ldots + (-1)^{m-1}s''_{m-1}T^{m-1}\right). \quad (5.16)$$

Since $-\log(1 - s_1 T + \ldots + (-1)^{m-1}s_{m-1}T^{m-1}) = p_1 T + \ldots + p_{m-1}T^{m-1}$ it follows that the p_i are additive, whence η is translation invariant.

Note that $dF = \eta|_B$. Define $\pi : \mathbb{A}^1 \to B \subset \omega^\times_{m \cdot \infty}/\mathbb{C}^\times$ by $\pi^* p_k = z^k$, $k \le m - 2$, $\pi^* s_{m-1} = 0$. Then $\pi^* \eta = df$. In particular, $\pi^* \eta$ vanishes at the zeroes of $df = f'dz$. It follows that since $b = (f') \in Sym^{m-2}(\mathbb{A}^1)$, we have $DF|_b = 0$ as well. The proof that b is the unique point where $\eta|_B$ vanishes is given in [BE2, Lemma 3.10]. We shall omit it here. $\qquad\square$

Note that

$$b = (b_1, \ldots, b_{m-2}) = \left(\frac{a_1}{(m-1)a_{m-1}}, \frac{2a_2}{(m-1)a_{m-1}}, \ldots, \frac{(m-2)a_{m-2}}{(m-1)a_{m-1}}\right)$$

in the s-coordinate system on B. Set $t_i = s_i - b_i$, and write $F(s) = F(b) + G(t)$, so $G(t_1, \ldots, t_{m-2})$ has no constant or linear terms.

LEMMA 5.3. *There exists a non-linear polynomial change of variables of the form*

$$t'_j = t_j + B_j(t_1, \ldots, t_{j-1})$$

such that $B(0, \ldots, 0) = 0$ and $G(t) = Q(t')$ where Q is homogeneous of degree 2.

Proof. The proof is close to [BE2, Lemma 3.10]. We write (abusively) $p_k(s)$ for the power sum $z_1^k + \ldots + z_{m-1}^k$, taken as a function of the elementary symmetric functions $s_1, \ldots, s_k$. The quadratic monomials $s_i s_{m-1-i}$ all occur with nonzero coefficient in p_{m-1}. By construction, $F(s) = a_1 p_1(s) + \ldots + a_{m-1}p_{m-1}(s)$ with $a_{m-1} \ne 0$. If we think of s_i as having weight i, $p_k(s)$ is pure of weight k, so $s_i s_{m-1-i}$ occurs with nonzero coefficient in $F(s)$. Since the weight $m - 1$ is maximal, $t_i t_{m-1-i}$ will occur with nonzero coefficient in $G(t)$ as well. Thus, we have

$$G(t) = Q(t_1, \ldots, t_{m-1}) + H(t)$$

where Q is quadratic and contains $t_i t_{m-1-i}$ with nonzero coefficient, and H has no terms of degree < 3. Further, H has no terms of weight $> m - 1$. In particular, the variable t_{m-2} does not occur in H. If we replace t_{m-2} by $t'_{m-2} := t_{m-2} + A_{m-2}(t_1, \ldots, t_{m-3})$ for a suitable polynomial A_{m-2}, we can eliminate t_1 from H completely, $G(t) = Q(t_1, \ldots, t_{m-3}, t'_{m-2}) + \tilde{H}(t_2, \ldots, t_{m-3})$. The weight and degree conditions on $\tilde{H}$ are the same as those on H, so we conclude that $\tilde{H}$ does not involve t_{m-3}. Also, this change does not affect the monomials $t_i t_{m-1-i}$ in Q for $i \geq 2$. Thus we may write

$$G = (*)t_1 t'_{m-2} + (**)t_2 t_{m-3} + \tilde{Q}(t_3, \ldots, t_{m-4}) + \tilde{H}(t_2, \ldots, t_{m-4})$$

since $(**) \neq 0$, we may continue in this fashion, writing $t'_{m-3} = t_{m-3} + A_{m-3}(t_1, \ldots, t_{m-4})$, etc. □

The constant term can be written

$$F(b) = \sum_{\substack{\beta \\ f'(\beta)=0}} f(\beta). \tag{5.17}$$

The nonlinear change of variables $t \mapsto t'$ has jacobian 1. Also, the quadratic form above is necessarily nondegenerate (otherwise F would have more that one critical point). One has

PROPOSITION 5.4.

$$\det(P_{ij})_{1 \leq i,j \leq m-2} = \prod_{\substack{\beta \\ f'(\beta)=0}} \exp(f(\beta)) \int_\Theta \exp(Q(t)) dt_1 \wedge \ldots \wedge dt_{m-2}, \tag{5.18}$$

where

$$Q = t_1^2 + \ldots + t_{m-2}^2 \tag{5.19}$$

is the standard, nondegenerate quadric on B, and Θ is some $n - 2$-chain on B. Moreover, $\int_\Theta$ is determined up to $\mathbb{Q}^\times$-multiple on purely geometric grounds.

Since the shape of the integral is obviously coming from the change of coordinates $t \mapsto t'$, we have to understand the meaning of $\int_\Theta$.

Let $W \subset \mathbb{P}^n \times \mathbb{P}^1$ be the family of quadrics over $\mathbb{P}^1$ defined by

$$UQ(S_1, \ldots, S_{m-2}) - VT^2 = 0. \tag{5.20}$$

Here $S_1, \ldots, S_{m-2}, T$ are homogeneous coordinates on $\mathbb{P}^{m-2}$, U, V are homogeneous coordinates on $\mathbb{P}^1$, and $Q(S_1, \ldots, S_{m-2})$ is a nondegenerate quadric. We have Weil divisors $Y : U = T = 0$; $Z : Q = T = 0$ in W. Note Y and Z are smooth, and $W_{\text{sing}} = Y \cap Z$. Let $\pi : W' \to W$ be the blowup of W along the Weil divisor Y. Let $Y' \subset W'$ be the exceptional divisor.

LEMMA 5.5. i. W' is smooth.

ii. The strict transform Z' of Z in W' is isomorphic to Z and
$$Y' \cap Z' \subset Y'_{\text{smooth}}.$$

Proof. Let $P' = \mathrm{BL}(Y \subset \mathbb{P}^n \times \mathbb{P}^1)$ be the blowup. Then W' is the strict transform of W in P'. Since $Y \cap Z$ is the Cartier divisor $U = 0$ in Z, it follows that the strict transform of Z in P' or W' is isomorphic to Z.

We consider the structure of W' locally around the exceptional divisor. We may assume some S_i is invertible and write $s_j = S_j/S_i$, $t = T/S_i$, $u = U/V$, $q = Q/S_i^2$. The local defining equation for W is $uq(s) - t^2 = 0$. Thinking of W' as $\mathrm{Proj}(\bigoplus_{p \geq 0} I^p \mathcal{O}_W)$ with $I = (u, t)$, we have open sets $\mathcal{U}_1 : \tilde{t} \neq 0$ and $\mathcal{U}_2 : \tilde{u} \neq 0$. (The tilde indicates we view these as projective coordinates on the Proj.) We have the following coordinates and equations for W' and Y':

$$
\begin{aligned}
\mathcal{U}_1; \quad & u't = u; \quad W' : u'q(s) - t = 0; \quad Y' : t = 0, \\
\mathcal{U}_2; \quad & t'u = t; \quad W' : q(s) - ut'^2 = 0; \quad Y' : u = 0.
\end{aligned}
\tag{5.21}
$$

The strict transform Z' of Z lies in the locus $\tilde{t} = 0$ and so does not meet $\mathcal{U}_1$. Both defining equations for W' are smooth, and Y' is smooth on $\mathcal{U}_2$. Finally, $Z' \cap Y' : q = t' = u = 0$ is also smooth. $\quad\square$

Write
$$W^0 := W' - Z'; \quad Y^0 = Y' - Y' \cap Z'.$$
We want to show that the chains over which we integrate can be understood as chains on the topological pair $(W^0 - U, Y^0)$ for some open U (cf. Lemma 5.6 below). In z-coordinates, we deal with chains γ_k, (5.6), which are parametrized $\alpha_k = r_k e^{i\theta_k}, 0 \leq r_k < \infty$ for fixed θ_k. Note that for $r \gg 1$ the real part of f on γ_k will $\to -\infty$. By abuse of notation, we write γ_j also for the closure of this chain on $\mathbb{P}^1$, i.e. including the point $r = \infty$.

Write $F(z) = F(b) + Q(S_1, \ldots, S_{m-2})$. It is easy to check by looking at weights that $|S_k| = O(|r|^k)$ as $|r| \to \infty$. On the other hand, because the paths are chosen so the real parts of $a_{m-1}\alpha_k^{m-1}$ are all negative we find there exist positive constants C, C' such that $C|r|^{m-1} \leq |F(z)| = |Q(S(z))| \leq C'|r|^{m-1}$. In homogeneous coordinates
$$(S_k, T), (U, V),$$

the point associated to a point with coordinates z on our chain is

$$S_k = S_k(z) = O(|r|^k), \quad 1 \le k \le m - 2; \ T = 1;$$
$$U = 1; \ V = Q(S(z)) \ge C|r|^{m-1}. \tag{5.22}$$

With reference to the coordinates in (5.21) we see that

$$|u| = |U/V| \le C^{-1}|r|^{1-m}, \quad |t| = |T/S_i| \ge C_1|r|^{-i},$$
$$|t'| = |t/u| \ge C_2|r|^{m-1-i}. \tag{5.23}$$

In particular, the limit as $|r| \to \infty$ does not lie on $\mathcal{U}_2$. Since, near ∞ $Z \subset \mathcal{U}_2$, we conclude our chains stay away from Z at infinity.

We fix $\epsilon \ll 1$ and $N \gg 1$ and define a connected, simply connected domain $D \subset \mathbb{A}^1 \subset \mathbb{P}^1$ by

$$D = \left\{ re^{i\theta} \mid r > N, \ -\tfrac{\pi}{2} - \epsilon < \theta < \tfrac{\pi}{2} + \epsilon \right\}$$

thus, D is an open sector at infinity, and $\exp(z)$ is rapidly decreasing as $|z| \to \infty$ in the complement of D. In what follows, let $g : W \to \mathbb{P}^1$ be the projection.

LEMMA 5.6. *The assignment*

$$\gamma \mapsto \int_\gamma \exp\left(Q(S_1/T, \dots, S_{m-2}/T)d(S_1/T) \wedge \dots \wedge d(S_{m-2}/T)\right)$$

defines a functional $H_{m-2}(W^0 - g^{-1}(D), Y^0; \mathbb{Q}) \to \mathbb{C}$.

Proof. Write τ for the above integrand. Let M be some neighborhood of Z. Then τ is rapidly decreasing on $W^0 - g^{-1}(D)$ near $Y^0 - M \cap Y^0$, where the size is defined by some metric on the holomorphic $m - 2$ forms on W^0. Since the chains are compact, a chain γ on W^0 will be supported on $W - M$ for a sufficiently small neighborhood M of Z. Thus, integration defines a functional

$$C_{m-2}(W^0 - g^{-1}(D), Y^0) \to \mathbb{C}.$$

It remains to show $\int_{\partial\Gamma} \tau = 0$ for an $m - 1$ chain Γ. Let M be an open neighborhood of Z not meeting Γ. Let R be an open neighborhood of Y^0 in $W - M$. Write $\Gamma = \Gamma_1 + \Gamma_2$ where $\Gamma_1 \subset \bar{R}$ and $\Gamma_2 \cap R = \emptyset$. Since τ is closed, $\int_{\partial\Gamma_2} \tau = 0$. On the other hand, the volume of $\partial\Gamma_1$ can be taken to be bounded independent of R. It follows that $\int_{\partial\Gamma} \tau = 0$. $\square$

LEMMA 5.7. $H_{m-2}(W^0 - g^{-1}(D), Y^0; \mathbb{Q}) \cong \mathbb{Q}$.

Proof. Let $p \in D$ be a point. We first show

$$H_{m-2}(W^0 - g^{-1}(p), Y^0; \mathbb{Q}) \cong \mathbb{Q}. \tag{5.24}$$

We calculate $H^*(W^0 - g^{-1}(p))$ using the Leray spectral sequence. For $x \neq 0, \infty$, $g^{-1}(x)$ is a smooth, affine quadric of dimension $m - 3$. So $R^p g_* \mathbb{Q} = (0)$ away from $0, \infty$ for $p \neq 0, m - 3$, and $R^{m-3} g_* \mathbb{Q}|_{\mathbb{P}^1 - \{0,\infty\}}$ is a rank 1 local system. The monodromy about 0 and ∞ is induced by $(S, T) \mapsto (S, -T)$ and $(S, T) \mapsto (-S, T)$, respectively. Both actions give -1 on the fibres. It follows that, writing $j : \mathbb{P}^1 - \{0, \infty\} \hookrightarrow \mathbb{P}^1$, we have

$$j_! R^{m-3} g_* \mathbb{Q}|_{\mathbb{P}^1 - \{0,\infty\}} \cong j_* R^{m-3} g_* \mathbb{Q}|_{\mathbb{P}^1 - \{0,\infty\}} \cong Rj_* R^{m-3} g_* \mathbb{Q}|_{\mathbb{P}^1 - \{0,\infty\}} . \tag{5.25}$$

It follows that the natural map $R^{n-1} g_* \mathbb{Q} \to j_* R^{n-1} g_* \mathbb{Q}|_{\mathbb{P}^1 - \{0,\infty\}}$ is surjective and we get a distinguished triangle in the derived category

$$\mathcal{P} \to R^{n-1} g_* \mathbb{Q} \to Rj_* R^{n-1} g_* \mathbb{Q}|_{\mathbb{P}^1 - \{0,\infty\}} \tag{5.26}$$

where $\mathcal{P}$ is a sheaf supported over $0, \infty$. In particular,

$$H^1(\mathbb{P}^1, R^{m-3} g_* \mathbb{Q}) \cong H^1(\mathbb{P}^1 - \{0, \infty\}, R^{n-1} g_* \mathbb{Q}) = (0) , \tag{5.27}$$

where the vanishing comes by identifying with group cohomology of $\mathbb{Z}$ acting on $\mathbb{Q}$ with the generator acting by -1. An easy Gysin argument yields

$$H^1(\mathbb{P}^1 - \{p\}, R^{m-3} g_* \mathbb{Q}) \cong H^1(\mathbb{P}^1 - \{0, \infty, p\}, R^{m-3} g_* \mathbb{Q}) = \mathbb{Q} . \tag{5.28}$$

It also follows from (5.26) that $H^2(\mathbb{P}^1, R^{m-3} g_* \mathbb{Q}) = (0)$. The spectral sequence thus gives

$$H^{m-2}(W^0, \mathbb{Q}) \cong (R^{m-2} g_* \mathbb{Q})_{\{0,\infty\}} . \tag{5.29}$$

To compute these stalks, note the fibre of $W' \to \mathbb{P}^1$ over 0 is a singular quadric with singular point $S_1 = \ldots = S_{m-2} = 0$, $T = 1$ away from Z. Thus the fibre of $g : W^0 \to \mathbb{P}^1$ over 0 is the homogeneous affine quadric $Q(S) = 0$ which is contractible. Further, because Z meets the fibre of W' smoothly, one has basechange for the non-proper map g, so $(R^{m-2} g_* \mathbb{Q})_0 = (0)$. At infinity, we have seen again that Z meets the fibre smoothly, so again one has basechange for g. Let $h : W^0 - Y^0 \hookrightarrow W^0$. It follows that

$$\left(R^{m-2} g_*(h_! \mathbb{Q}) \right)_{\{0,\infty\}} = (0) . \tag{5.30}$$

Combining (5.28),(5.29),(5.30) yields (5.24).

To finish the proof of the lemma, we must show the inclusion

$$\left(W^0 - g^{-1}(D), Y^0 \right) \hookrightarrow \left(W^0 - g^{-1}(p), Y^0 \right)$$

is a homotopy equivalence. We can define a homotopy from $D - \{p\}$ to ∂D by flowing along an outward vector field v. E.g. if $p = 0$ and one has cartesian coordinates x, y, one can take $v = x \frac{d}{dx} + y \frac{d}{dy}$. Since $W'/\mathbb{P}^1$ is smooth over D, one can lift v to a vector field w on $g'^{-1}(D)$. Since Z' meets the fibres of g' smoothly over some larger $D_1 \supset D$, we can arrange for w to be tangent to Z' along Z'. Let h be a smooth function on $\mathbb{P}^1$ which is positive on D and vanishes on $\mathbb{P}^1 - D$. We view $g'^*(h)w$ as a vector field on $W' - g'^{-1}(p)$. Flowing along $g'^*(h)w$ lifts the flow along hv, carries $g'^{-1}(\bar{D})$ into $g'^{-1}(\partial D)$ and stabilizes $W' - Z'$ over $\bar{D}$. This is the desired homotopy equivalence. □

REMARK 5.8. *It follows from Theorem 2.3.3 in [T] that*

$$\int_{\Theta} \exp(Q) dt \in \left(\frac{2\pi}{(m-1)a_{m-1}} \right)^{m-2/2} \cdot \mathbb{Q}.$$

References

[BE1]   S. BLOCH, H. ESNAULT, A Riemann-Roch theorem for flat bundles, with values in the algebraic Chern-Simons theory, Annals of Mathematics, to appear.

[BE2]   S. BLOCH, H. ESNAULT, Gauß–Manin determinants for rank 1 irregular connections on curves, preprint (1999).

[D]   P. DELIGNE, letter to J.-P. Serre, 8 février 1974.

[E]   H. ESNAULT, Algebraic differential characters, Proceedings of the Conference on Motives at the Landau Institute in Jerusalem, March 1996, Birkhäuser Verlag, in print.

[K]   N. KATZ, Gauß Sums, Kloosterman Sums, and Monodromy Groups, Annals of Math. Studies 116, Princeton University Press, 1988.

[Ko]   M. KONTSEVICH, Periods, preprint (1999).

[L]   G. LAUMON, Transformation de Fourier, constantes d'équations fonctionnelles, et conjecture de Weil, Publ. Math. IHES 65 (1987), 131–210.

[M]   B. MALGRANGE, Équations différentielles à coefficients polynomiaux, Progress in Math. 96, Birkhäuser, (1991).

[ST]   T. SAITO, T. TERASOMA, Determinant of period integrals, Journ. AMS 10:4 (1997), 865–937.

[T]   T. TERASOMA, Confluent hypergeometric functions and wild ramification, Journ. of Alg. 185 (1996), 1–18.

SPENCER BLOCH, Dept. of Math., University of Chicago, Chicago, IL 60637, USA
bloch@math.uchicago.edu

HÉLÈNE ESNAULT, Mathematik, FB6, Universität Essen, 45117 Essen, Germany
esnault@uni-essen.de

GAFA, Geom. funct. anal.
Special Volume – GAFA2000, 32 – 56
1016-443X/00/S1032-25 $ 1.50+0.20/0

ΓGAFA Geometric And Functional Analysis

PROBLEMS IN HAMILTONIAN PDE'S

J. Bourgain

1 Introduction

The purpose of this exposé is to describe a line of research and problems, which I believe, will not be by any means completed in the near future. As such, we certainly hope to encourage further investigations. The list of topics in this field is fairly extensive and only a few will be commented on here. Their choice was mainly dictated by personal research involvement. It should also be mentioned that the different groups of researchers may have very different styles and aims. As a science, claims and results range from pure experimentation to rigorous mathematical proofs. Although my primary interest is this last aspect, I have no doubt that numerics or heuristic argumentation may be equally interesting and important. The history of the Korteweg-de-Vries equation for instance is a striking example of how a problem may evolve through these different interacting stages to eventually create a beautiful theory. As a mathematician, I feel however that it is essential one remains fully aware of what is rigorous mathematical argumentation and what is not. Failure to do so would result in general confusion about the nature of the statements and a great loss of challenging mathematics.

Some of my coworkers believe today's availability of powerful computational means is partly responsible for a declining interest in the often difficult rigorous work. An amazing comment here is that theoretical computer science has been to the contrary mathematically invigorating with no consensus problem about rigor. It is also true that evidence of certain phenomena gathered from extensive computation is often received by the pure mathematician with certain scepticism or dismissed as unreliable. At this point, there does not seem to be such a thing as a truly certified numerical PDE experiment.

The discussion below is purely mathematically oriented. We mainly aim to highlight a set of problems that are both physically important and offer an analytic challenge. These problems relate to the time evolution in Hamiltonian PDE's and some of the themes are

(i) Existence and break-down of solutions to the Cauchy problem for data in various thresholds, in particular, the existence of global classical solutions.

(ii) In case of singularity formations such as blowup of local solutions, one may investigate blowup speed, profile, etc.

(ii) Assuming global solutions exist for a class of data, one may ask for further details on the dynamics. A distinction needs to be made here between problems on unbounded spatial domains (such as $\mathbb{R}^d$) and problems on bounded domains (for instance periodic or Dirichlet boundary conditions). In the first case, one often encounters a scattering phenomenon with an asymptotically linear flow. In the second case, no dispersion is possible and it is natural to look for certain recurrence in the dynamics, such as invariant measures on KAM-type behaviors.

We will center the discussion around two types of model-equations, namely the nonlinear wave equations (NLW) and nonlinear Schrödinger equations (NLS). The NLW-case is a model with finite speed propagation, the NLS-case is not. In certain issues, such as the Cauchy problem, infinite speed propagation makes the problem harder. In other issues, such as the stability of quasi-periodic solutions on bounded domains for instance, the stronger separation properties of normal modes in the NLS-case plays a positive role. Although the NLW and NLS models described below have been extensively studied over previous years, they still offer a large collection of unsolved basic problems. Past research has also shown that arguments and methods discovered in this context usually apply to broader classes of equations. We do however exclude from our discussion the few isolated instances of integrable equations that do present lots of atypical structure and features.

A comprehensive treatment of the topics brought up below is certainly beyond the scope of the paper. The reader interested in the subject may consult for instance the recent books [SuS], [Bo1] and references cited. The reference list in the present paper is very incomplete and only complements this short exposé.

2 Two Hamiltonian Models

We consider NLW of the form

$$y_{tt} - \Delta y + \rho y \pm y^{p-1} = 0 \tag{2.1}$$

where $\rho \geq 0$. The Hamiltonian formulation of (2.1) is

$$\begin{cases} y_t = -v = -\frac{\partial H}{\partial v} \\ y_t = -\Delta y + \rho y \pm y^{p-1} = \frac{\partial H}{\partial y} \end{cases} \tag{2.2}$$

with Hamiltonian

$$H = \int \left[\tfrac{1}{2}|\nabla y|^2 + \tfrac{1}{2}v^2 + \tfrac{1}{2}\rho y^2 \pm \tfrac{1}{p}y^p \right] \tag{2.3}$$

and canonical coordinates y (= position) and v (= speed).

Notice the 2 possible signs $+$ (= defocusing case) and $-$ (= focusing case).

In the (classical) NLS type equation

$$iu_t + \Delta u \mp u|u|^{p-2} = 0 \tag{2.4}$$

the function u is complex valued. The Hamiltonian format of (2.4) is thus

$$iu_t = \frac{\partial H}{\partial \bar{u}} \tag{2.5}$$

with Hamiltonian

$$H = \int \left[|\nabla u|^2 \pm \tfrac{2}{p}|u|^p \right] \tag{2.6}$$

and canonical coordinates $(\mathrm{Re}\, u, \mathrm{Im}\, u)$. Again there is the defocusing (resp. focusing) case corresponding to $+$ (resp. $-$) sign in (2.6).

Thus in the defocusing case for NLW (resp. NLS), conservation of the Hamiltonian implies in particular an a priori bound on $\int |\nabla y|^2 + v^2$ (resp. $\int |\nabla u|^2$), assuming initial data $(y(0), v(0)) = (y(0), \dot{y}(0))$ in $H^1 \times H^0$ (resp. $u(0) \in H^1$). We denote here by H^s the usual order s Sobolev space, i.e.

$$\|\phi\|_{H^s} = \|\partial^s \phi\|_2. \tag{2.7}$$

In the focusing case, the Hamiltonian may not be bounded from below for sufficiently large p (depending on dimension d) in the nonlinearity and smooth solutions may develop singularities in finite time.

Besides (2.1), (2.4) there are other important examples of NLW and NLS type equations involving nonlocal nonlinearities or nonlinearities containing derivatives. We do not intend to discuss them here.

3   The Cauchy Problem

We will mainly center the discussion around the NLS

$$iu_t + \Delta u + \lambda u|u|^{p-2} = 0. \tag{3.1}$$

There is a parallel theory for NLW. The reader may consult the exposé of S. Klainerman in this volume [K] for a review of recent progress in the field of hyperbolic equations.

Our spatial domain will be $\mathbb{R}^d$ in this section. Observe that the equation (3.1) is invariant under the scaling

$$u \to u_a(x,t) = a^{\frac{2}{p-2}} u(ax, a^2 t) \tag{3.2}$$

which also leaves the (homogeneous) Sobolev space H^{s_0} with

$$p - 2 = \frac{4}{d - 2s_0} \tag{3.3}$$

invariant. Notice the particular cases

$$p = 2 + \tfrac{4}{d} \quad \text{(the conformally invariant NLS) where } s = 0 \tag{3.4}$$

and

$$p = 2 + \frac{4}{d-2} \quad (d \geq 3) \text{ where } s = 1. \tag{3.5}$$

Thus in (3.4), the L^2-norm (which is a conserved quantity for NLS equations of the form $iu_t + \Delta u + \frac{\partial}{\partial \bar{u}} f(|u|^2) = 0$) is the scale invariant space. In case (3.5), we get the space H^1 and the H^1-norm may be a priori controlled from the Hamiltonian in the defocusing case.

We will say that the IVP (Initial value problem)

$$\begin{cases} iu_t + \Delta u + \lambda u |u|^{p-2} = 0 \\ u(0) = \phi \in H^s \end{cases} \tag{3.6}$$

is locally wellposed, provided (3.6) has a unique solution $u \in C_{H^s}([0,T[)$ on a nontrivial time interval $(T > 0)$ and $u = u_\phi$ depends continuously on the initial data ϕ (in most positive results, the dependence of u on ϕ will be at least Lipschitz).

If $[0,T[= \mathbb{R}_+$ (the time T may be taken arbitrarily large) we call the problem globally wellposed.

Essentially speaking, one should not expect a wellposedness theory below the scale invariant threshold. In this respect, the situation for NLS is well understood local in time.

Theorem 3.7. *Assume in (3.6) that $s \geq 0$ and $s \geq s_0$ defined by (3.3).*

Assume also $p - 2 > [s]$ if p is not an even integer (a smoothness compatibility condition between s and the nonlinearity). Then there is $T^* > 0$ s.t. (3.6) is wellposed on $[0, T^*[$ in the sense described above. Moreover $T^* = T^*(\|\phi\|_{H^s})$ if $s > s_0$.

If $\|\phi\|_{H^s}$ is sufficiently small, then $T^* = \infty$.

See [C] or [SuS] for details.

REMARKS. (i) The result is dependent on whether the equation is focusing ($\lambda > 0$) or defocusing ($\lambda < 0$).

(ii) The key ingredients in the proof are estimates on the linear group $e^{it\Delta}$. More precisely

– The decay estimate:

$$\|e^{it\Delta}\phi\|_{L^\infty} \leq C|t|^{-d/2}\|\phi\|_{L^1} \tag{3.8}$$

– Strichartz inequality (which is global in space-time):

$$\|e^{it\Delta}\phi\|_{L^{2\frac{(d+2)}{d}}}(dx\,dt) \lesssim C\|\phi\|_2. \tag{3.9}$$

From the L^2-conservation

$$\|u(t)\|_2 = \|\phi\|_2 \tag{3.10}$$

there is the following immediate consequence of Theorem 3.7.

COROLLARY 3.11. If $p < 2 + \frac{4}{d}$, then (3.6) is globally wellposed for $\phi \in L^2$ and also for $\phi \in L^2 \cap H^s$, $s > 0$. In this last case, we get a solution $u \in C_{H^s}(\mathbb{R}_+)$.

It is well known that for $p \geq 2 + \frac{4}{d}$, in the focusing case, sufficiently large (smooth) initial data may lead to solutions blowing up in finite time (see next section). One conjectures however that in the defocusing case, the local solution given by Theorem 3.7 extends to a global one. Thus

PROBLEM. Assume $\lambda < 0$ in (3.1). Can one take $T^* = \infty$ in Theorem 3.7?

Assuming the equation defocusing ($\lambda < 0$), let us summarize some positive results and more specific problems. We will call the problem H^1-subcritical (resp. critical, supercritical) if $s_0 < 1$ (resp. $s_0 = 1$, $s_0 > 1$). Thus in dimension $d = 1, 2$, the problem is always subcritical. The next statement is a consequence of Theorem 3.7 and the energy concentrations.

COROLLARY 3.12. In the defocusing H^1-subcritical case, the IVP is globally wellposed for data in $H^1 \cap H^s$, $s \geq 1$. In particular, classical solutions exist for all time.

In the critical case, there is the following partial result (see [Bo2], [Gr2]).

Theorem 3.13. Let $d \geq 3$ and $p = 2 + \frac{4}{d-2}$ (defocusing case). Then there is global wellposedness of the IVP for radial $\phi \in H^1 \cap H^s$, $s \geq 1$.

REMARKS.   (i) Observe that the local result Theorem 3.7 provides in the critical case $s = s_0$ an existence time $T^* = T^*(\phi)$ (not only dependent on $\|\phi\|_{H^{s_0}}$). Thus the difficulty is to deal with possible energy conservations on small balls. One of the additional ingredients involved here is Morawetz dispersive inequality $(d \geq 3)$

$$\iint \frac{|u(x,t)|^p}{|x|} dx\, dt < \infty \tag{3.14}$$

which is also the basis of global scattering.

The analogue for defocusing NLW

$$y_{tt} - \Delta y + y^5 = 0 \qquad (d = 3) \tag{3.15}$$

$$y_{tt} - \Delta y + y^3 = 0 \qquad (d = 4) \tag{3.16}$$

$$\text{etc.}$$

is unrestricted to radial data. Thus there is wellposedness for arbitrary data $\phi \in H^1 \cap H^s$, $s \geq 1$. The particular case of radial data is due to M. Struwe [St] and the general case was solved by M. Grillakis [Gr1]. This is an instance where the infinite speed propagation in NLS makes the problem (of controlling energy concentrations) harder.

PROBLEM. *Does Theorem 3.13 remain valid for general data?*

(ii) Observe that, since we do not have at our disposal any conserved quantity stronger than the energy, the statement in Theorem 3.13 remains nontrivial for classical (i.e. smooth) solutions. Both proofs [Bo2], [Gr2] contain in fact an important component which is purely H^1. The same consideration of lack of conserved quantities explains also why, both for supercritical NLW and NLS, no global results seem available (except for small data cf. Theorem 3.7). Thus the following issue is widely open.

PROBLEM. *Global existence of classical solutions for defocusing supercritical NLW, NLS.*

(iii) Some comments about scattering in the energy space.

Consider the defocusing NLS

$$iu_t + \Delta u - u|u|^{p-2} = 0 \qquad u(0) = \phi \in H^1 \cap H^s \quad (s \geq 1). \tag{3.17}$$

Assume

$$p > 2 + \frac{4}{d} \quad \text{and} \quad p < 2 + \frac{4}{d-2} \quad \text{if} \quad d \geq 3. \tag{3.18}$$

Denoting the wave map operator

$$\Omega_+ \varphi = \varphi + i \int_0^\infty e^{i\tau\Delta}(u|u|^{p-2})(\tau)d\tau \tag{3.19}$$

one has that

$$\Omega_+\varphi \in H^1 \cap H^s \tag{3.20}$$

$$\left\|e^{it\Delta}(\Omega_+\varphi) - u(t)\right\|_{H^1\cap H^s} \overset{t\to\infty}{\longrightarrow} 0. \tag{3.21}$$

For $d \geq 3$ these are classical results going back to Ginibre-Velo [GV] and Lin-Strauss [LiS]. In the radial case, the result remains valid for $p = 2 + \frac{4}{d-2}$ ($d = 3, 4$), see [Bo2]. Recently, scattering in the energy space was also proven by Nakaniski [N] in dimension $d = 1, 2$ ($p > 2 + \frac{4}{d}$).

PROBLEM. *Is there global scattering in the energy space for $p = 2 + \frac{4}{d}$?*

(See also [C1,2] for other results on scattering).

(iv) We like to sketch the theoretical possibility for computer assisted proofs of global existence and scattering, for a given data ϕ. Consider for instance the 3D supercritical problem

$$\begin{cases} iu_t + \Delta u - u|u|^6 = 0 \\ u(0) = \phi \end{cases} \tag{3.22}$$

where ϕ is a given smooth function. We do expect a global smooth solution + scattering. For this to hold, it is sufficient to show that for some time, $0 < T < \infty$,

(a) (3.22) has a smooth solution on [0, T]. Equivalently, $T^* > T$, where T^* refers to Theorem 3.7

(b) The norm $\left\|e^{i(t-T)\Delta}u(T)\right\|_{L^{15}_{t\geq T}L^{15}_x} < \delta$

where $\delta > 0$ is some numerical constant (we do not explain the role of the L^{15}-norm here). About step (a). If we fix a time T, one may establish the result numerically. To do this, one first gathers sufficiently many discrete data and interpolates them with a (smooth) function $v = v(x,t)$, $t < T$. Assuming (3.22) has indeed a smooth solution, the function v will eventually necessarily satisfy the equation approximately, i.e.

$$\begin{cases} iv_t + \Delta v - v|v|^6 = \varepsilon(x,t) \\ v(0) = \phi. \end{cases} \tag{3.23}$$

Here ε may be made arbitrarily small in any chosen space-time norm ($t \leq T$), by pushing the numerics far enough. Denoting u the "true" solution, the difference $w = u - v$ satisfies an equation

$$\begin{cases} iw_t + \Delta w - w|w|^6 + P(w,\overline{w}) = \varepsilon(x,t) \\ w(0) = 0 \end{cases} \tag{3.24}$$

where P is a polynomial in $w, \overline{w}$ of degree ≤ 6, no constant terms and coefficients depending on v. The wellposedness of (3.24) on a given interval $[0, T]$ results then form the same analysis as the small data theory, provided $\varepsilon(x, t)$ is sufficiently small (depending on the size of T).

Next, consider step (b). It may be derived from Strichartz' inequalities that

$$\left\| e^{i\tau\Delta}(u(T) - \psi) \right\|_{L^{15}_{x,\tau}} \leq C \| u(T) - \psi \|_{H^{7/6}} \qquad (3.25)$$

so that the issue is stable under appropriate approximation. For given (smooth) ψ, there is the stationary phase evaluation for $\tau \to \infty$

$$(e^{i\tau\Delta}\psi)(x) = \int e^{i(x \cdot \xi + \tau|\xi|^2)} \hat{\psi}(\xi) \sim |\tau|^{-\frac{3}{2}} e^{-i\frac{|x|^2}{4\tau}} \hat{\psi}\left(-\frac{x}{2\tau}\right) \qquad (3.26)$$

hence

$$\left| (e^{i\tau\Delta}\psi)(x) \right| \lesssim |\tau|^{-3/2} \left| \hat{\psi}\left(-\frac{x}{2\tau}\right) \right|. \qquad (3.27)$$

It suffices therefore clearly to evaluate the oscillatory integral in (3.26) on a large enough finite space-time region. Observe that in this second step, the function $\psi = v(T)$ is an explicitly given object, gotten in particular from the step (a) construction (practically, the functions v and ψ may be given by finite wavelet expansions).

4 Singularity Formation for NLS

The only known mechanisms in the NLS case for singularity formation relate to focusing NLS

$$iu_t + \Delta u + u|u|^{p-2} = 0, \quad p \geq 2 + \frac{4}{d}. \qquad (4.1)$$

In this case, classical solutions of large L^2-norm may blowup in finite time T. A general criterion result from Glassey's viriel inequality [Gl]

$$\frac{d^2}{dt^2} \left[\int |u(t)|^2 |x|^2 dx \right] \leq cH(\phi) \qquad \text{(for some } c > 0) \qquad (4.2)$$

for data ϕ with $H(\phi) < 0$. One deduces indeed easily from (4.2) that there has to be a singularity formation in finite time T^*.

Assuming $p < 2 + \frac{4}{d-2}$, it follows in particular from the local existence theory that

$$\lim_{t \overset{\to}{<} T^*} \| u(t) \|_{H^1} = \infty. \qquad (4.3)$$

PROBLEM. *What may be said about the blowup speed, i.e. the growth of $\| \nabla u(t) \|_2$ for $t \overset{\to}{<} T^*$?*

Let us consider the critical nonlinearity $p = 2 + \frac{4}{d}$. Returning to the local existence theorem Theorem 3.7, following fact may be shown ([Bo3]).

FACT. *If the solution u for an L^2-data ϕ given by Theorem 3.7 has maximal existence time $[0, T^*[$, $T^* < \infty$, there is the following L^2-concentration phenomenon*

$$\limsup_{t \vec{<} T^*} \sup_{\substack{I \subset \mathbb{R}^d \\ |I| < (T^*-t)^{1/2}}} \int_I |u(t)|^2 dx > c \tag{4.4}$$

where $c > 0$ is a constant. We denote here by I some d-dimensional cube and $|I|$ its size.

In the case of a classical data, this concentration phenomenon had been observed earlier, in particular in the works of Cazenave-Weissler, Merle and Tsutsumi. In this case, (4.4) implies that

$$\lim_{t \to T^*} (T^* - t)^{1/2} \|u(t)\|_{H^1} > c > 0. \tag{4.5}$$

In fact, it is known that in this case, there has to be an L^2-concentration of at least $\|Q\|_2$ on arbitrary small balls for $t \vec{<} T^*$. Here Q refers to the ground-state, i.e. the positive radial solution of

$$\Delta Q - Q + Q^{1+4/d} = 0. \tag{4.6}$$

In the 2D-case for instance, there is quite a bit of numerical and meta-mathematical work related to the cubic NLS

$$iu_t + \Delta u + u|u|^2 = 0. \tag{4.7}$$

Self-similar blowup solutions are suggested with various blowup speeds: $|\ell n(T^* - t)|^{1/2} \cdot (T^* - t)^{-1/2}$, $|\ell n(T^* - t)|^\gamma \cdot (T^* - t)^{-1/2}$ for certain $\gamma > 0$, $\left[\frac{\ell n|\ell n(T^*-t)|}{T^*-t}\right]^{1/2}$ among others. The reader may consult [SuS] for a more detailed discussion and related literature.

The following conjecture seems consistent with all heuristic predictions.

PROBLEM. *Prove that if u blows up at time T^*, then necessarily*

$$\varlimsup_{t \vec{<} T^*} (T^* - t)^{1/2} \|u(t)\|_{H^1} = \infty. \tag{4.8}$$

This would be a strengthening of inequality (4.5).

In the conformal case $p = 2 + \frac{4}{d}$, the NLS has an additional symmetry given by the so-called pseudo-conformal transformation

$$u \to Cu(x,t) = t^{-d/2} e^{\frac{|x|^2}{4it}} u\left(\frac{x}{t}, -\frac{1}{t}\right) \qquad (t > 0) \tag{4.9}$$

mapping a solution u to a solution Cu. In the focusing case, (4.9) permits us to construct explicit blowup solutions for the NLS

$$iu_t + \Delta u + u|u|^{4/d} = 0.\tag{4.10}$$

Let $u_0(x,t) = e^{it}Q(x)$ be the ground state solution, Q given by (4.6). Then, by (4.9)

$$u(x,t) = t^{-d/2}\, e^{\frac{|x|^2-4}{4it}} Q\left(\tfrac{x}{t}\right)\tag{4.11}$$

yields an explicit solution of (4.10), blowing up at time $t = 0$.

Thus in (4.11) we have $\|u(t)\|_2 = \|Q\|_2$.

FACT. *If u is a classical solution of* (4.10) *blowing up in finite time, then*

$$\|u(t)\|_2 \geq \|Q\|_2.\tag{4.12}$$

Blowup solutions for which $\|u(t)\|_2 = \|Q\|_2$ have been extensively studied by F. Merle [Me1] who proved in particular the following (roughly stated) uniqueness property.

Theorem 4.13. *If $\|u(t)\|_2 = \|Q\|_2$ and u blows up in finite time, then u coincides with* (4.11), *up to the symmetries of the equation.*

Statement (4.12) is a consequence of the earlier stated fact that the amount of the L^2-norm absorbed in the blowup is at least $\|Q\|_2$. There is the following question about a possible quantization.

PROBLEM. *Assume u a solution of* (4.10) *blowing up at time T. Is the amount of L^2-norm absorbed in the blowup necessarily of the form $k.\|Q\|_2$ with $k \in \mathbb{Z}_+$?*

REMARKS. (i) Merle [Me2] has constructed examples of blowup solutions with an arbitrary number of blowup sites (at a fixed time). These examples are essential obtained by superposition of translates of (4.11) and satisfy thus the quantization property cited above.

(ii) In [BoW] examples are constructed of solutions u of (4.10) blowing up at some time T with only part of $\|u(t)\|_2$ absorbed in the blowup. Essentially the solutions obtained in [BoW] are of the form

$$u(x,t) = t^{-\frac{d}{2}} e^{\frac{|x|^2-4}{4it}} Q\left(\tfrac{x}{t}\right)\chi_{[t<0]} + v(x,t)\tag{4.14}$$

where $v \neq 0$ is smooth for all time. Also solutions are constructed with several distinct blowup times.

(iii) It is clear that in (4.11)

$$\|u(t)\|_{H^1} = 0\left(\tfrac{1}{t}\right) \text{ for } t \to 0.\tag{4.15}$$

Clearly these minimum L^2-norm blowup solutions are unstable. However one may conjecture a stability property in the following restricted sense (we formulate it as a problem on the ground state solution and may be converted to a stability statement of (4.11) by applying the pseudo-conformal transformation).

PROBLEM. *There is a finite codimensional manifold $\mathcal{M}$ in a neighborhood of Q such that if $\phi \in \mathcal{M}$, then the solution u of the IVP*

$$iu_t + \Delta u + u|u|^{4/d} = 0, \quad u(0) = \phi \tag{4.16}$$

is of the form

$$u(x,t) = e^{it}Q(x) + (\text{dispersive}). \tag{4.17}$$

An important element here is the work of M. Weinstein [W] on the stability analysis of the linearization of equation (4.10) at the groundstate solution, i.e. the non-selfadjoint linear problem

$$\begin{cases} iv_t - Lv = 0 \\ v(0) = \phi \end{cases} \tag{4.18}$$

where

$$Lv = -\Delta v + v - \left(\tfrac{2}{d} + 1\right)Q^{4/d}v - \tfrac{2}{d}Q^{4/d}\bar{v}. \tag{4.19}$$

It is also the main ingredient in [BoW] mentioned above.

5 Periodic Boundary Conditions

Let us replace the spatial domain $\mathbb{R}^d$ by a bounded domain such as the d-torus $\mathbb{T}^d$, $\mathbb{T} = \mathbb{R}/\mathbb{Z}$. As mentioned before, the absence of dispersion here gives an entirely different outlook on the time-dynamics. In this and the next 2 sections, we plan to review some explorations in this context. Starting with the Cauchy problem, especially for NLS, there is the issue of understanding the periodic analogues of Strichartz' inequality

$$\|e^{it\Delta}\phi\|_{L^{\frac{2(d+2)}{d}}(dx\,dt)} \le C\|\phi\|_2 \tag{5.1}$$

for the linear group $e^{it\Delta}$ (which is the basic ingredient). Recall that (5.1) is a global inequality in space-time. In the periodic setting, the oscillatory integral in (3.26) is replaced by an exponential sum

$$(e^{it\Delta}\phi)(x) = \sum_{n \in \mathbb{Z}^d} \hat{\phi}(n)e^{i(n.x+|n|^2t)} \tag{5.2}$$

which is natural to study as a function on $\mathbb{T}^{d+1}$.

PROBLEM (Conjecture). *The following inequalities hold*

(i) $\|e^{it\Delta}\phi\|_{L^q(\mathbb{T}^{d+1})} \le C\|\phi\|_{L^2(\mathbb{T}^d)}$ *for* $q < \frac{2(d+2)}{d}$ (5.3)

(ii) $\forall \varepsilon > 0,\ \|e^{it\Delta}\phi\|_q \ll N^\varepsilon\|\phi\|_2$ *for* $q = \frac{2(d+2)}{d}$, supp $\hat{\phi} \subset B(0, N)$.   (5.4)

REMARKS. (i) Inequality (5.3) fails at the endpoint $q = \frac{2(d+2)}{d}$ and (5.4) seems the proper modification.

(ii) Statement (5.3) is correct for $d = 1$, $q \le 4$ and statement (5.4) for $d = 1, 2$.

These facts result from elementary arithmetics.

Combining exponential sum estimates and techniques from Fourier Analysis, one may obtain the following distributional inequality related to (5.4)

$$\text{mes } \left[(x, t) \in \mathbb{T}^{d+1} \mid |e^{it\Delta}\phi| > \lambda\right] \ll N^\varepsilon \lambda^{-\frac{2(d+2)}{d}} \text{ for } \lambda > N^{d/4}. \quad (5.5)$$

The restriction on λ in (5.5) is reminiscent (and related) to the results and problems on large values estimates for Dirichlet sums in number theory see [Mo]).

As a consequence of the preceding, the local and global Cauchy problem for periodic NLS

$$\begin{cases} iu_t + \Delta u + \lambda u|u|^{p-2} = 0 \\ u(0) = \phi \in H^s(\mathbb{T}^d) \\ u \text{ periodic in } x \end{cases} \quad (5.6)$$

is not as well understood as the $\mathbb{R}^d$ case, especially when $d \ge 3$. However, what is known, for instance, is the global wellposedness for the periodic cubic NLS

$$iu_t + \Delta u - u|u|^2 = 0 \quad (5.7)$$

in dimension $d = 1, 2, 3$, for data $\phi \in H^s$, $s \ge 1$ (the $d = 1$ case corresponds to the integrable Zakharov-Shabat equation and should be omitted from the discussion below). Thus there is a solution $u \in C_{H^s}(\mathbb{R}_+)$ (which depends real analytically on the data) and satisfies moreover an estimate

$$\|u(t)\|_{H^s} \le |t|^{C(s-1)}\|\phi\|_{H^s} \text{ for } t \to \infty. \quad (5.8)$$

Observe that for $s > 1$, the $\|u(t)\|_{H^s}$ is not subject to an a priori bound (except for $d = 1$).

PROBLEM. *Is there a possible growth of* $\|u(t)\|_{H^s}$ *for* $t \to \infty$. *If so, how fast may this growth be?*

Thus the issue here is how fast energy stored in low Fourier modes may escape to higher modes when $t \to \infty$.

REMARKS. (i) This problem may also be addressed in certain $\mathbb{R}^d$-models. However, if scattering in the H^s-norm occurs, it also implies that $\sup_{t>0} \|u(t)\|_{H^s} < \infty$.

(ii) Estimates of the form (5.8) may be derived for a large class of H^1-subcritical equations including NLW. One should observe that power-like growth of higher Sobolev norms may actually occur in close relatives of the 1D cubic NLW, i.e. examples of the form

$$y_{tt} + B^2 y + y^3 = 0 \qquad (5.9)$$

where $B = \sqrt{-\Delta} + 0(1)$ is given by a selfadjoint Fourier multiplier. We conjecture however that in NLS equations, say of the form (5.7) an estimate

$$\|u(t)\|_{H^s} \ll t^\varepsilon \|\phi\|_{H^s}, \quad t \to \infty \text{ for all } \varepsilon > 0 \qquad (5.10)$$

holds, improving on (5.8). The reason for this is the stronger arithmetic separation properties of normal frequencies $\mu_n = |n|^2$ for NLS, compared with NLW ($\mu_n = |n|$) that do force a certain localization of the energy over long time scales. In this respect, inequality (5.10) was proven for linear time dependent Schrödinger equations of the form

$$iu_t + \Delta u + V(x, t)u = 0 \qquad (5.11)$$

with V an arbitrary real, bounded, smooth, x-periodic potential (amazingly no specified behavior in t seems needed, besides smoothness).

For more details and relevant literature on the material presented in this section, the reader may consult [Bo1,4].

6   Invariant Measures

In the understanding of long time dynamics in bounded spatial domains, existence of invariant measures on various phase spaces is a natural thing to look for. From a classical point of view, the most desirable situation should be to construct such a measure on smooth functions. Except for the integrable model of the 1D cubic NLS $iu_t + u_{xx} \pm u|u|^2 = 0$, whether this may be done or not is unknown.

Essentially speaking, in the other cases, the only known examples of invariant measures are either living on spaces of rough data or fields (those produced from the Gibbs-measure) or live on finite dimensional or infinite dimensional tori with very strong compactness properties (obtained from KAM tori). Results along this line will be described in this and next section.

In finite dimensional phase space, it follows from Liouville's theorem that the Lebesgue measure on $\mathbb{R}^{2n}$ is invariant under the Hamiltonian flow

$$\dot{p}_i = -\frac{\partial H}{\partial q_i} \qquad \dot{q}_i = \frac{\partial H}{\partial p_i} \qquad (i = 1, \ldots, n) \tag{6.1}$$

with Hamiltonian $H(p_1, \ldots, p_n, q_1, \ldots, q_n)$. Hence so is the Gibbs measure

$$d\nu = e^{-\beta H} \, \Pi_{i=1}^n dp_i dq_i \,. \tag{6.2}$$

Consider the case of NLS with periodic boundary conditions in dimension d

$$iu_t = \frac{\partial H}{\partial \bar{u}} \,. \tag{6.3}$$

Writing a Fourier expansion

$$u(t) = \phi = \sum_{n \in \mathbb{Z}^d} \widehat{\phi}(n) e^{in.x} \tag{6.4}$$

we choose

$$p_n = \operatorname{Re} \widehat{\phi}(n) \,, \qquad q_n = \operatorname{Im} \widehat{\phi}(n) \tag{6.5}$$

as canonical coordinates for our infinite dimensional phase space. This choice (rather than $(\operatorname{Re} u, \operatorname{Im} u)$) has the advantage that it is easy to pass to finite dimensional models by projecting on finitely many Fourier modes. The formula (6.2) may be written as

$$d\nu = e^{\mp \frac{1}{p} \int |\phi|^p} \left(e^{-\frac{1}{2} \int |\nabla \phi|^2} \Pi d^2 \phi \right) = e^{\mp \frac{1}{p} \int |\phi|^p} d\mu \tag{6.6}$$

which has the tentative interpretation of a weighted Wiener measure. This measure is formally invariant under the flow of (6.3). In order to develop this idea rigorously, the two problems to resolve are the normalization of (6.6) and next the construction of a well defined dynamics on the support of that measure. This program was initiated in the papers [LRS]. The problem is strongly dimensional dependent and also reflects certain aspects of the Cauchy problem, especially in the focusing case (such as blowup behavior).[1]

In dimension $d = 1$, Wiener measure lives on functions of class $H^{\frac{1}{2}-}(\Pi)$ and hence $\int |\phi|^p dx$ is almost surely finite, for any p. Thus in the defocusing case, with weight $e^{-\frac{1}{p} \int |\phi|^p}$ in (6.6), the formula produces trivially a well defined measure that is absolutely continuous w.r.t. Wiener measure $d\mu$. One of the main results in [LRS] is that in the focusing case, (6.6) may be normalized for $p \leq 6$ by restricting the L^2-norm. Thus $e^{\frac{1}{p} \int |\phi|^p}$ is replaced by

$$e^{\frac{1}{p} \int |\phi|^p} \chi_{[\|\phi\|_2 < B]} \tag{6.6'}$$

---

[1]Observe also that these measures are significantly different from those constructed for diffusive PDE's.

where B is an arbitrary cutoff for $p < 6$ and taken sufficiently small for $p = 6$. As mentioned in section 4, $p = 6 = 2 + \frac{4}{d}$ is critical in the sense of possible blowup behavior for large data in the classical theory. It turns out indeed that the restrictions in the previous statement are necessary.

In [Bo5], the author established a unique dynamics. This amounts to proving global wellposedness of the Cauchy problem

$$\begin{cases} iu_t + u_{xx} \pm u|u|^{p-2} = 0 \\ u(0) = \phi \end{cases} \tag{6.7}$$

for almost all data ϕ in the support of the Wiener measure. This task in fact reduces to verifying the result local in time, since the measure invariance may be exploited similarly to a conservation law.

For $d = 2$ and $d = 3$, the expression $\int |\phi|^p$ diverges. The problem is resolved by the well-known process of Wick ordering, assuming $|\phi|^p$ a polynomial in $\phi, \bar\phi$, hence p an even integer. Thus $\int |\phi|^p$ is replaced by $\int : |\phi|^p:$ which is almost surely finite. Moreover, in the 2D defocusing case, the formula

$$e^{-\int :|\phi|^p:} d\mu \tag{6.8}$$

still defines a weighted Wiener measure with density in $\cap_{p<\infty} L^p(\mu)$.

Construction of the dynamics for $p = 4$ was performed in [Bo6].

Considering the 2D-focusing case equation

$$iu_t + \Delta u + u|u|^2 = 0 \tag{6.9}$$

a natural construction would consist of considering measures

$$\chi_{[\int :|\phi|^2: < B]} \left(e^{\int :|\phi|^4:} \right) d\mu \tag{6.10}$$

obtained by restriction of the Wick ordered L^2-norm,

$$\int : |\phi_\omega|^2 := \sum_n \frac{|g_n(\omega)|^2 - 1}{|n|^2 + \rho} \tag{6.11}$$

and generating Wiener measure by the random Fourier series

$$\phi_\omega = \sum_{n \in \mathbb{Z}^2} \frac{g_n(\omega)}{\sqrt{|n|^2 + \rho}} e^{inx}. \tag{6.12}$$

Here $\{g_n(\omega)\}$ refers to a system of standard, independent complex Gaussian random variables. Observe indeed that $\int |\phi|^2 = \infty$ a.s. if $d > 1$ while $\int :|\phi|^2: < \infty$ a.s. for $d = 2, 3$.

This idea was explored by A. Jaffe [J] for cubic nonlinearities. Unfortunately, the construction barely misses the quartic nonlinearity. It does apply however in the context of Hartree-equations involving a nonlocal nonlinearity obtained by convolving with a suitable potential. Consider the equation

$$iu_t + \Delta u + (|u|^2 * V)u = 0. \tag{6.13}$$

In 2D, with potential V satisfying any condition of the form

$$|\widehat{V}(n)| < |n|^{-\varepsilon} \text{ for } |n| \to \infty \quad (\varepsilon > 0). \tag{6.14}$$

Wick ordering of the nonlinearity and restriction of the Wick ordered L^2-norm as in (6.10) produces then an invariant normalized measure for (6.13).

Similar constructions were performed in [Bo7] for the 3D case.

Coming back to the 2D normalization problem for focusing NLS, the preceding discussion leads naturally to the question how to Wick order (or develop a substitute for Wick ordering) in the case of powers $|\phi|^p$ where p is not an even integer (or ϕ^p, $p \notin \mathbb{Z}_+$, in the real case). Those attempts seem only to produce the free measure on phase space however [S], which is physically not very interesting.

PROBLEM. *What are the ergodicity properties of the Gibbs measures in the typical (i.e. non-integrable) case?*

PROBLEM. *Existence of invariant measures on smooth phase space, for instance for defocusing NLS*

$$iu_t + \Delta u - u|u|^2 = 0$$

in $d = 2$, $d = 3$.

PROBLEM. *Existence of a limiting invariant Gibbs-measure dynamics when the period tends to infinity.*

About the last problem in the case of the 1D wave equation, results were obtained by McKean, Vaninski [MV]. The problem for NLS in this context is again the lack of finite speed propagation. In [Bo8], results are obtained in 1D for the defocusing equation $iu_t + u_{xx} - u|u|^{p-2} = 0$ with $p \leq 4$.

7   KAM Tori and Nekhoroshev Stability

Over recent years, there has been a certain amount of research related to the existence and persistency of invariant tori for Hamiltonian perturbations of linear or integrable PDE's. The basic model would be that of a perturbed linear equation of the form

$$iu_t + Au + \varepsilon \frac{\partial H}{\partial \bar{u}} = 0 \tag{7.1}$$

where A is a selfadjoint operator containing parameters. Examples related to the Schrödinger equation would be:

$$A = \text{ Fourier multiplier with } A_n \sim |n|^2 \text{ for } n \to \infty \qquad (7.2)$$

$$A = -\Delta + V \text{ with } V \text{ a potential} \qquad (7.3)$$

Given a nonlinear equation, parameters may often be extracted from the nonlinearity using partial Birkhoff normal forms and amplitude-frequency modulation. This has been pursued for instance in [KuP] for 1D NLS of the form

$$iu_t + \Delta u + mu + u|u|^2 + \text{(higher order)} = 0 \qquad (7.4)$$

and in [Bo9] for the 2D-equation. We shall give some more details at the end of this section.

A basic difference with the corresponding topic in the theory of classical smooth dynamical systems are new resonance problems due to an infinite dimensional phase space. These difficulties are particularly severe in space dimension $D \geq 2$,

Several methods have been developed. In [Ku], the classical KAM and Melnikov scheme based on eliminating the perturbation by consecutive canonical transformations has been reworked in the context of finite dimensional tori in infinite dimensional phase space, provided there are no multiplicities or near-multiplicities in the normal frequencies. In the context of (7.1) this restriction limits the application to 1D-models with Dirichlet of Neumann bc's. Recall that the periodic spectrum (λ_n) of $-\frac{d^2}{dx^2} + V$ appears typically in pairs $\lambda_{2n-1}, \lambda_{2n}$, where roughly

$$|\lambda_{2n} - \lambda_{2n-1}| \sim |\widehat{V}(n)| \to 0 \qquad (7.5)$$

rapidly for $n \to \infty$.

In higher dimension, unbounded multiplicities result from a possible large number of lattice points on a circle (or sphere) of radius R when $R \to \infty$.

One of the most interesting developments in this research is the emerging of new techniques to overcome this problem. The pioneering paper here is that of W. Craig and E. Wayne [CrW] in which the simplest case (presenting the multiplicity issue) of time periodic solutions of the 1D wave equations with periodic bc of the form

$$y_{tt} - y_{xx} + \rho y + y^3 + \text{higher order}(x, y) = 0 \qquad (7.6)$$

is investigated. The method used is a variant of the Nash-Moser scheme and is a priori independent of Hamiltonian structure. It is based on a

Liapounov-Schmidt decomposition and a treatment of small divisor problems by multiscale analysis very reminiscent of the arguments used in localization theory for lattice Schrödinger operators. In particular in the context of a quasi-periodic potential, such as the almost Mathieu operator

$$\varepsilon\Delta + \cos(n\lambda + \sigma) \tag{7.7}$$

(cf. [FSWi]).

This technique has lead to considerable progress. Time periodic solutions (i.e. 1 dim tori) for (7.1) may be produced in any dimension and the full theory of quasi-periodic solutions is available for $D = 1,2$ NLS with periodic bc (cf. [Bo9]).

See also [Cr] for an updated survey.

PROBLEM. *Construction of quasi-periodic solutions for 3D NLS with periodic bc.*

REMARK. The difference between the 2D and 3D case is the larger multiplicity of normal frequencies. In 2D, these basically correspond to the integer solutions of an equation

$$n_1^2 + n_2^2 = R^2 \tag{7.8}$$

with fixed $R \to \infty$. The set of those lattice points is at most of size $\exp \frac{\log R}{\log \log R} \ll R^\varepsilon$ (the divisor function) and may be partitioned in $R^{1/5}$-distant sets say containing at most 2 sites. In 3D, (7.8) is replaced by the equation

$$n_1^2 + n_2^2 + n_3^2 = R^2 \tag{7.9}$$

which may have more than R solutions.

PROBLEM. *Construction of quasi-periodic solutions for 2D NLW with periodic bc.*

REMARK. The normal frequencies here are essentially $\mu_n = \sqrt{n_1^2 + n_2^2}$ and $\{\mu_n - \mu_{n'}\}$ is clearly a dense set in $\mathbb{R}$. This fact seems to require substantial improvement of the known technologies.

PROBLEM. *How abundant are the quasi-periodic solutions in phase space? The question may be posed in the context of any nonintegrable Hamiltonian PDE.*

REMARK. The methods of construction are perturbative and typically require us to proceed in a neighborhood of the origin. The following particular issue for instance is unknown.

PROBLEM. *Consider an NLS*

$$iu_t + u_{xx} + u|u|^2 + (\text{higher order terms}) = 0 \qquad (7.10)$$

and denote QP the union of all smooth quasi-periodic trajectories in the phase space. Is QP dense for the weak topology?

Let us also mention that in fact, the first new application of the methods developed for PDE's relates to the classical Melnikov theorem in finite dimensional phase space, which remains valid in the presence of multiple normal frequencies.

REMARK. Let us point out an important difference between the standard KAM situation where 2(dim invariant tori) = (dim phase space) and the Melnikov problem where 2(dim invariant tori) < dim phase space. In the first situation, the admissible frequencies for the perturbed tori contains in particular sets of frequencies characterized by diophantine conditions (independent of the perturbation) and one obtains also families depending smoothly on the perturbation parameter ε. This is not the case in the Melnikov setting and one finds sets of admissible frequencies depending on the perturbation. This problem was observed in the work of Moser [Mos], Eliasson [E] and in [Bo10].

PDE's also provide a natural setting for problems of existence of infinite-dimensional invariant tori. There has been research on this topic by various authors. Some of the models considered in mathematical physics only present finite range interactions, cf. [FSW]. In other (more recent) investigations, frequencies are chosen sufficiently lacunary to avoid diophantine problems. My particular interest here goes to KAM tori constructed on the full set of frequencies (only of quadratic growth) for an NLS of the form

$$iu_t + u_{xx} - V(x)u + \varepsilon\frac{\partial H}{\partial \bar{u}} = 0. \qquad (7.11)$$

Here the nonlinearity is fixed and V is taken to be a "typical" potential. Construction of such KAM tori was performed independently by J. Pöschel and the author using different methods (see [P], [Bo11]). In both constructions, there is a very rapidly (hardly explicit) decay of the amplitudes, hence leading to strong compactness properties. Imposing such fast decay appears here as trade-off for the severe small divisor issues. Refining the techniques to reach a more realistic decay condition on the action variables is certainly a most interesting project.

While proving almost periodic behaviors for large classes of data seems difficult (and probably is untrue), one may in several interesting instances expect such approximative behavior over very long time scales. In the classical finite dimensional theory, this issue is addressed by the Nekhoroshev stability theory and our aim here is to discuss some investigations in this spirit for Hamiltonian PDE's.

Consider a Hamiltonian of the form

$$H(q, \bar{q}) = \sum \lambda_n I_n + F(I_n) + \varepsilon H'(q, \bar{q}) \tag{7.12}$$

where $I_n = |q_n|^2$, thus a perturbation of the integrable case.

Remark that we have not introduced action-angle variables in H' to avoid the 0-singularity when $|q_n| \to 0$ for $|n| \to \infty$. Assume $|\lambda_n| \geq 1$, F bounded from below and $H'(q, \bar{q})$ real analytic in the topology induced by the unperturbed Hamiltonian. Obviously

$$\left| I_n(T) - I_n(0) \right| = \left| \int_0^T \frac{dI_n}{dt} \right|$$

$$= \left| \int_0^T \operatorname{Re} \bar{q}_n \dot{q}_n \right|$$

$$= \left| \int_0^T \operatorname{Im} \bar{q}_n \frac{\partial H'}{\partial q_n} \right|$$

$$\leq \varepsilon \int_0^T |H(q, \bar{q})| = 0(\varepsilon T). \tag{7.13}$$

So that stability for time scales $0(1/\varepsilon)$ is clear.

Let us recall two mechanisms leading to stability for longer time scales $(T \ll (1/\varepsilon)^A, A$ large or even $T < e^{(1/\varepsilon)^\gamma}, \gamma > 0)$. The first is that of a perturbation of a nonresonnant linear system, thus with Hamiltonian of the form

$$H = \sum \lambda_n I_n + \varepsilon H'(q, \bar{q}) \tag{7.14}$$

where the frequency vector $\bar{\lambda} = (\lambda_n)$ had "good" diophantine properties. In this case, one aims to remove by the standard mechanism of symplectic transformations the perturbation in (7.14) up to high order in ε. These symplectic transformations Γ are obtained as time-1 shift for the Hamiltonian flow with Hamiltonian F of the form

$$F(q, \bar{q}) = \sum \frac{a_{\bar{n}}}{\nu_1 \lambda_{n_1} + \nu_2 \lambda_{n_2} + \cdots} q_{n_1}^{\nu_1} q_{n_2}^{\nu_2} q_{n_3}^{\nu_3} q_{n_4}^{\nu_4} \cdots \tag{7.15}$$

$(\nu = \pm 1, q^{+1} \equiv q, q^{-1} \equiv \bar{q})$

$$H_0 \Gamma = H + \varepsilon \{H, F\} + \frac{\varepsilon^2}{2!} \{\{H, F\}, F\} + \cdots \tag{7.16}$$

where $\{H_1, H_2\} = \sum_n \left(\frac{\partial H_1}{\partial q_n} \frac{\partial H_2}{\partial \bar{q}_n} - \frac{\partial H_1}{\partial \bar{q}_n} \frac{\partial H_2}{\partial q_n} \right)$ is the Poisson bracket.

In high or infinite dimensional phase space, the resonance problems may require us to allow in the remainder term also certain expressions involving monomials $\prod_j q_{n_j}^{\nu_j}$ of large degree and certain monomials involving q_n's with n large, besides $0(\varepsilon^A)$-terms. Assuming the q_n small with rapid decay at infinity these additional terms turn out also to have an $0(\varepsilon^A)$-effect, A large. Thus in the new coordinates, the Hamiltonian has the form

$$H = \sum \tilde{\lambda}_n I_n + 0(\varepsilon^A) \,. \tag{7.17}$$

(with $\tilde{\lambda}_n = \lambda_n + 0(\varepsilon)$ the modulated frequencies) and we get stability for times $T = 0(\varepsilon^{-A})$. In this setup, the stability result may relate to topologies different from the Hamiltonian one. For instance, in [Bo11], results along this line were obtained for Hamiltonian perturbations of nonresonant linear Schrödinger and wave equations (1D with Dirichlet bc) of the form

$$iu_t + u_{xx} + V(x)u + \varepsilon\frac{\partial H}{\partial \bar{u}} = 0 \tag{7.18}$$

$$y_{tt} - y_{xx} + \rho y + \varepsilon\partial_y f(y) = 0 \tag{7.19}$$

Let us recall here that in PDE issues, the Hamiltonian $H'(q, \bar{q})$ in (7.12) does not involve only short range interactions.

As mentioned before one may sometimes reduce the Hamiltonian to the form (7.19) with a diophantine frequency vector $\bar{\lambda}$ by exploiting nonlinear parts of the Hamiltonian and amplitude frequency modulation. The typical example is that of a Birkhoff normal form

$$H = \sum \lambda_n I_n + \sum I_n^2 + \varepsilon H'(q, \bar{q}) \,. \tag{7.20}$$

Define

$$J_n = I_n - I_n(0) \tag{7.21}$$

so that

$$H = \text{ const } + \sum \tilde{\lambda}_n J_n + \sum J_n^2 + \varepsilon H'(q, \bar{q}) \tag{7.22}$$

with modulated frequencies

$$\tilde{\lambda}_n = \lambda_n + 2I_n(0) \,. \tag{7.23}$$

For typical data $q(0)$, $I_n(0) = |q_n(0)|^2$ one may thus expect certain diophantine properties from the new frequencies, which reduces the question to case (7.14).

There are PDE-models particularly suitable to this approach. Consider for instance the 1D NLS

$$iu_t + u_{xx} \pm u|u|^2 + \text{ (higher order) } = 0 \qquad u(x,t) = \sum q_n(t) e^{inx} \,. \tag{7.24}$$

The normal form transformation observed by Kuksin-Pöschel [KuP] permits us to express the Hamiltonian of this model in the form

$$H(q, \bar{q}) = \sum (n^2 + \rho)|q_n|^2 + \nu \sum |q_n|^4 + \text{ (higher order)} \qquad (7.25)$$

Assuming $\|q\| < \varepsilon$, appropriate rescaling leads thus to a normal form (7.20).

Longtime stability results along these lines were obtained by Bambusi [B1] considering typical data q that are almost fully localized on a finite set of Fourier modes, thus

$$q = \sum_{n \in \mathcal{F}} q_n(0) e^{inx} + 0(\varepsilon) \qquad (7.26)$$

where $\#\mathcal{F} < B(\varepsilon) < \log 1/\varepsilon$.

In [Bo12], one considers typical data q in H^s, s large, of size ε and obtain stability times $(1/\varepsilon)^A$ with $A = A(s) \to \infty$ for $s \to \infty$.

There is a second mechanism for stability (sometimes referred to as the "geometric" Nekhoroshev theorem). The Hamiltonian considered is for instance of the form (7.20)

$$H = \sum \lambda_n I_n + \sum I_n^2 + \varepsilon H'(q, \bar{q}) \qquad (7.27)$$

(no diophantine assumption on λ). Here the nonlinearity (convexity in case (7.27) but weaker assumptions may be exploited too, [Ne]) is used to derive stability of the action variables (in the Hamiltonian topology). There is an elegant argument due to Lochak [Lo], based on approximation by periodic orbits. It was also reworked in PDE context (related to NLS (7.24)) to obtain exponentially long stability times considering again data of the form (7.26) (see [B2]).

PROBLEM. *To what extent does the geometric Nekhoroshev theorem apply to the PDE context?*

What seems clear is that in both stability mechanisms described above, strong restrictions need to be imposed on the metrical entropy of the unperturbed torus compared with the size ε of the perturbation. In this respect, it is possible to construct examples in $2(N + 1)$-dim phase space of Hamiltonians

$$H = \sum_{n=0}^{N} \lambda_n |q_n|^2 + \sum |q_n|^4 + \varepsilon H'(q, \bar{q}) \qquad (7.28)$$

$(\lambda_n \sim 1)$ with H' a polynomial, bounded in $\ell_{N+1}^2(\mathbb{C})$-norm, such that for $T \sim 1/\varepsilon$

$$\left| |q_0(T)|^2 - |q_0(0)|^2 \right| > 0(1) \qquad (7.29)$$

(with frequencies λ_n chosen randomly). What happens here is that the unperturbed torus is too large (in the sense of ℓ^2-entropy) compared with ε. The construction may probably be improved in this respect, but it certainly does not come close to explain fully the strong limitations of the positive results available so far.

In the classical theory, a drift of action-variables over exponentially long times in $1/\varepsilon$ is explained by the mechanism of Arnold diffusion, involving a motion on trajectories essentially contained in alternating stable and unstable manifolds of KAM tori. Of course, one may expect that this mechanism will also be present in infinite Hamiltonian models. Observe here that this method requires a certain abundance of invariant KAM tori, where our understanding is certainly more limited in the PDE-setting at this point. It is however more natural to look here for mechanisms of diffusion that are due to large space dimension and interactions with many modes. In the example mentioned above for instance, q_0 interacts with all the modes $q_1, \ldots, q_N$ and the instability time is much shorter $0(1/\varepsilon)$. It is likely the precise nature of the nonlinearity (for instance short range interactions) plays a role here.

References

[B1]   D. BAMBUSI, Long time stability of some small amplitude solutions in nonlinear Schrödinger equations, CMP 189 (1997), 205–226.

[B2]   D. BAMBUSI, Nekhoroshev theorem for small amplitude solutions in nonlinear Schrödinger equations, Math. Z. 230 (1999), 345–387.

[BeFG]   G. BENETTIN, J. FRÖHLICH, A. GIORGILLI, A Nekhoroshev-type theorem for Hamiltonian systems with infinitely many degrees of freedom, CMP 119 (1989), 95–108.

[Bo1]   J. BOURGAIN, Global solutions of nonlinear Schrödinger equations, AMS Colloquium Publications 46 (1999).

[Bo2]   J. BOURGAIN, Global wellposedness of defocusing 3D critical NLS in the radial case, J. AMS, 12:1 (1999), 145-171.

[Bo3]   J. BOURGAIN, Refinements of Strichartz' inequality and applications to 2D-NLS with critical nonlinearity, IMRN 1998, N5, 253–283.

[Bo4]   J. BOURGAIN, Nonlinear Schrödinger equations, in Hyperbolic equations and frequency interactions (Park City, UT, 1995), 3-157, IAS/Park City Math. Ser. 5, AMS (1999).

[Bo5]   J. BOURGAIN, Periodic nonlinear Scrödinger equations and invariant measure, Comm. Math. Phys. 160 (1994), 1–26.

[Bo6]   J. BOURGAIN, Invariant measures for the 2D-defocusing nonlinear

Schrödinger equation, CMP 176 (1996), 421–445.

[Bo7]   J. BOURGAIN, Invariant measures for the Gross-Piatevskii equation, J. Math. Pures et Appliquées, T76, Fase 8 (1997), 649–702.

[Bo8]   J. BOURGAIN, Invariant measures for NLS in infinite volume, preprint 98, to appear in CMP.

[Bo9]   J. BOURGAIN, Quasi-periodic solutions of Hamiltonian perturbations of 2D linear Schrödinger equations, Annals of Math. 148 (1998), 363–439.

[Bo10] J. BOURGAIN, On Melnikov's persistency problem, Math. Research Letters 4 (1997), 445–458.

[Bo11] J. BOURGAIN, Construction of approximative and almost-periodic solutions of perturbed linear Schrödinger and wave equations, GAFA 6:2 (1996), 201–230.

[Bo12] J. BOURGAIN, On diffusion in high dimensional Hamiltonian systems and PDE, to appear in J. Analyse Jerusalem.

[BoW]   J. BOURGAIN, W. WANG, Construction of blowup solutions for the nonlinear Schrödinger equation with critical nonlinearity, Annali Scuola Norm. Sup. Pisa (Vol. dedicated to De Giorgio), 25:1-2 (1997), 197–215.

[C]   T. CAZENAVE, An introduction to nonlinear Schrödinger equation, Textos de Métodos Mathemáticos 26, Instituto de Matematica, UFRJ, Rio de Janeiro.

[Cr]   W. CRAIG, Problémes de petits diviseurs dans les equations aux dérivies partielles, preprint.

[CrW]   W. CRAIG, C. WAYNE, Newton's method and periodic solutions of nonlinear wave equations, Comm. Pure Appl. Math. 46 (1993), 1409–1501.

[E]   L.H. ELIASSON, Perturbations of stable invariant tori, Ann. Scuola Norm. Sup. Pisa, Cl. Sci. (4) 15 (1988), 115–147.

[FSW]   J. FRÖHLICH, T. SPENCER, E. WAYNE, Localization in disordered, nonlinear dynamical systems, J. Stat. Ph. 42 (1986), 247–274.

[FSWi] J. FRÖHLICH, T. SPENCER, P. WITTWER, Localization for a class of one-dimensional quasi-periodic Schrödinger operators, CMP 132 (1990), 5–25.

[GV]   J. GINIBRE, G. VELO, Scattering theory in the energy space for a class of nonlinear Schrödinger equations, J. Math Pure Appl. 64 (1985), 363–401.

[Gl]   R. GLASSEY, On the blowing up of solutions to the Cauchy problem for nonlinear Schrödinger operators, J. Math. Phys. 8 (1977), 1794–1797.

[Gr1]   M. GRILLAKIS, Regularity and asymptotic behaviour of the wave equation with a critical nonlinearity, Annals of Math. 132 (1990), 485–509.

[Gr2]   M. GRILLAKIS, Classical solutions for the nonlinear Schrödinger equation, preprint.

[J]   A. JAFFE, Private communication.

[K]   S. KLAINERMAN, PDE as a unified subject, in these Proceedings.

[Ku]   S. KUKSIN, Nearly integrable infinite-dimensional Hamiltonian systems,

Springer Lecture Notes in Math. 1556 (1993).

[KuP]   S. KUKSIN, J. PÖSCHEL, Invariant Cantor manifolds of quasi-periodic oscillations for a nonlinear Schrödinger equation, Annals of Math. 143:1 (1996), 149–179.

[LRS]   J. LEBOWITZ, R. ROSE, E. SPEER, Statistical mechanics of the nonlinear Schrödinger equation, J. Stat. Phys. 50 (1998), 657–687.

[LiS]   J. LIN, W. STRAUSS, Decay and scattering of a solution of a nonlinear Schrödinger equation, J.FA 30, 245-263.

[Lo]   P. LOCHAK, Canonical perturbation theory via simultaneous approximation, Uspekhi Math. Nauk 47:6 (1992), 59–140.

[MV]   H. MCKEAN, K. VANINSKI, Statistical mechanics of nonlinear wave equations, in 'Stochastic Analysis', Proc. Symp. Pure Math. 57 (1996), 457–463.

[Me1]   F. MERLE, Determination of blow-up solutions with minimal mass for nonlinear Schrödinger equation with critical power, Duke Math. J. 69 (1993), 427–453.

[Me2]   F. MERLE, Construction of solutions with exact k blow-up points for the Schrödinger equation with critical power nonlinearity, Comm. Math. Phys. 149 (1992), 205–214.

[Mo]   H. MONTGOMERY, Ten Lectures on the Interface Between Analytic Number Theory and Harmonic Analysis, CBMS, Regional Conf. Ser. in Math, 84 (1994).

[Mos]   J. MOSER, On the theory of quasi-periodic motions, Siam Review 8 (1966), 145–172.

[N]   N. NAKANISHI, Energy scattering for nonlinear Klein–Gordon and Schrödinger equations in spatial dimensions 1 and 2, preprint UTMS 98-50 (1998).

[Ne]   N. NEKHOROSHEV, An exponential estimate of the time of stability of nearly integrable Hamiltonian systems, Uspekhi Math. Nauk 32:1 (1977), 5–66.

[P]   J. PÖSCHEL, On the construction of almost periodic solutions for a nonlinear Schrödinger equation, preprint (1995).

[S]   T. SPENCER, Private communication.

[St]   M. STRUWE, Globally regular solutions to the u^5-Klein-Gordon equation, Ann. Scuola Norm Sup. Pisa, Ser. 4, 15 (1988), 495–513.

[SuS]   C. SULEM, P-L. SULEM, The nonlinear Schrödinger equation, Self-Focusing and Wave Collapse, Springer Applied Math. Sciences 139 (1999).

[W]   M. WEINSTEIN, Modulational stability of ground states of nonlinear Schrödinger equations, Siam J. Math. Anal. 16 (1985), 25–40.

Jean Bourgain, Institute for Advanced Study, Princeton, NJ 08540, USA
and University of Illinois, Urbana, IL 61901, USA
bourgain@ias.edu

GAFA, Geom. funct. anal.
Special Volume – GAFA2000, 57 – 78
1016-443X/00/S1057-22 $ 1.50+0.20/0

© Birkhäuser Verlag, Basel 2000

∎GAFA Geometric And Functional Analysis

ON A CLASSICAL LIMIT OF QUANTUM THEORY
AND THE NON-LINEAR HARTREE EQUATION

JÜRG FRÖHLICH, TAI-PENG TSAI AND HORNG-TZER YAU

1   General Perspective, Description of the Problem

Historically, the description of Nature in theoretical physics began with
an encoding of rather special, simple natural phenomena exhibiting high
approximate symmetry in special, unstable mathematical structures. For
example, the shape of the orbits of planets around the sun was conceived
to be circular. As the ambition to describe more general classes of natural
phenomena grew those special mathematical structures had to be deformed
and enlarged into more general, more stable mathematical structures. Kepler discovered that the idea of elliptic orbits, with the sun in one of the
foci, provides a better model for the motion of planets.

With the advent of the idea of *dynamical* Laws of Nature, one discovered
that it is these laws that may exhibit high symmetry, rather than special
consequences of these laws. Newtonian mechanics led to a first model of
space and time. The orbits of planets and moons could be understood as
trajectories in Newtonian space-time which are solutions of the Newtonian
equations of motion. These equations of motion exhibit high symmetry:
In all inertial frames of reference they take the same form, the transition
from one to another inertial frame being given by a Galilei transformation.
These symmetries are accompanied by conservation laws.

During the twentieth century, our understanding of Nature underwent
revolutionary developments and changes which led to at least three fundamental *deformations* of the mathematical structure underlying physical
theory. The first one originated from the study of the laws of black-body
radiation and resulted in the discovery of *quantum theory*. As early as 1900,
it uncovered new fundamental constants of Nature, the elementary electric
charge e, the Boltzmann constant k_B, Planck's constant $\hbar$ – all related to

---

Notes accompanying a lecture presented by J. F. at the conference "Visions in Mathematics", Tel Aviv, August 1999.

the atomic constitution of matter – and the Planck length ℓ_P, which is connected with gravity. Planck's law of black-body radiation, combined with previously discovered laws of thermodynamics and statistical physics, enabled him to determine the approximate values of e, k_B and $\hbar$, and from the speed of light, c, and Newton's law of gravitational attraction he could then infer the approximate value of ℓ_P.

The second revolution arose from attempts to find a unified description of mechanical and electromagnetic phenomena and resulted in Einstein's special theory of relativity and in the deformation of Newtonian space-time to Minkowski space-time and of the Galilei group to the Poincaré group. The underlying deformation parameter is the speed of light, c. The third one arose from trying to find a relativistic theory of gravitation and an invariant (coordinate-independent) formulation of fundamental dynamical laws and resulted in Einstein's general theory of relativity. In this theory, Minkowski space-time is deformed to a dynamical space-time, described as a Lorentzian manifold, whose topology and metric shape are determined, in part, by the distribution of matter. *Global* space-time symmetries are promoted to *local* symmetries (gauge symmetries of the second kind). The underlying deformation parameter is ℓ_P.

The deformation of the mathematical structure underlying Hamiltonian mechanics to the one underlying non-relativistic quantum mechanics was accomplished last and has remained, perhaps, the most mysterious one. The underlying deformation parameter is Planck's constant $\hbar$. It represents the first example of a deformation of classical geometry to noncommutative geometry, namely the deformation of symplectic geometry to the noncommutative geometry of quantized phase spaces.

The three deformations mentioned above and their diverse, fundamental physical consequences have set much of the agenda for the theoretical physics of the twentieth century and are likely to occupy an important place in our agenda for many years to come.

From the experience gained from the deformations of Newtonian to relativistic mechanics, of special to general relativity and of Hamiltonian to quantum mechanics, one may infer that *dimensionful, fundamental constants of Nature*, c, $\hbar$, ℓ_P, play the role of *deformation parameters* (to be distinguished from scale hierarchy parameters, such as the Rydberg, the QCD scale or the mass of the Z boson, and from dimensionless coupling constants, such as the feinstructure constant $\alpha \simeq 1/137$, or Yukawa couplings). To "turn on" c^{-1} or $\hbar$ or ℓ_P necessitates a fundamental change in the mathematical structure of physical theory, the passage from one family to a new family of theories (while ratios of energy scales and dimensionless coupling constants can be viewed as *moduli* parametrizing a family of theories). Dimensionful, fundamental constants of nature permit us to introduce a natural system of units in which they take the value 1, i.e. $c = 1$, $\hbar = 1$ and $\ell_P = 1$, in natural units (while ratios of energy scales and dimensionless coupling constants are pure numbers independent of our choice of units).

Deforming one family of theories to a new family of theories is the subject of *deformation theory*, while the passage from an effective description of natural phenomena at a high energy scale to an effective description of phenomena at a lower energy scale (within one family of effective theories) is the subject of *renormalization group theory*. Deformation theory is algebra, renormalization group theory is analysis. Progress in theoretical physics appears to rely on progress in deformation – and in renormalization group theory!

The sketch in Fig. 1 may convey a very rough impression of where we stand, at present.

A   Newtonian mechanics, Euclidian geometry, Hamiltonian mechanics, symplectic geometry

B   Special relativistic classical field theory, Minkowski geometry, infinite-dimensional symplectic geometry

C   Non-relativistic quantum mechanics, noncommutative geometry of quantized phase spaces

D   Special relativistic quantum field theory, noncommutative geometry of infinite-dimensional quantized phase spaces, functional integration, BRST and BV cohomology

E   Newtonian gravity and its reformulation as a geometrical theory, making use of Riemannian geometry

F   General Relativity, general Lorentzian geometry, ...

G   "Non-relativistic quantum gravity" (?), noncommutative Riemannian geometry

M   "Quantum gravity" or "M-theory" (?), infinite-dimensional noncommutative differential geometry (?), ...

The question marks indicate that the corresponding terms do not have a mathematically precise definition, yet. The "Escherian" aspect of the cube in Fig. 1 is intended to indicate that we are still facing obstacles or,

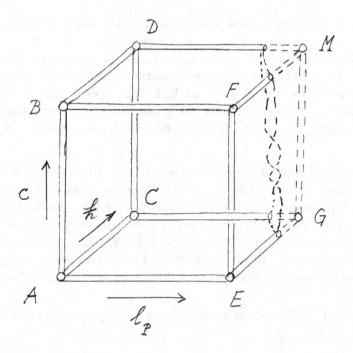

Figure 1

perhaps, obstructions in attempting to add the missing vertices G and M, and a link between them, to the cube.

Our discussion makes it clear that the following have been and will remain *important tasks* for theoretical physicists:

(1) Find suitable stable mathematical structures as foundations for physical theory by *deforming* more special, less stable mathematical structures known to underly physical theory in various asymptotic regimes.

For the future (ℓ_P): Find the deformations leading to vertex G and, most importantly, to vertex M, ("M-theory").

(2) Once a physical theory founded on a general, stable mathematical structure (e.g., non-relativistic quantum mechanics) has been found, one must understand in which asymptotic regime it reduces to one of its ancestor theories (e.g., Hamiltonian mechanics) founded on a less stable structure from which the former has arisen by deformation.

For the future (length scales $\gg \ell_P$):

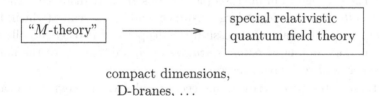

compact dimensions,
D-branes, ...

We don't really know precisely what "M-theory" is, and we don't understand how as yet unobserved symmetries (in particular, space-time supersymmetry) are broken and how a hierarchy of low-energy scales is generated.

For now ($c \gg$ typical velocities):

weak interaction strengths between
compounds, saturation of forces
$\alpha \simeq 1/137$ *small* $\Rightarrow v \ll c$.

For some results, see [D].

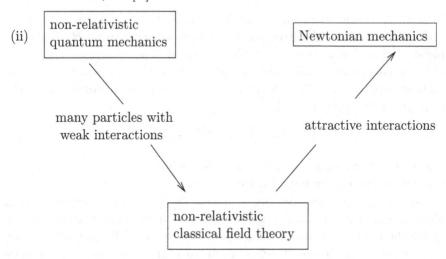

A simple example of a solution of problem (ii) is the *main subject* of these notes.

Warning: Since, in natural units, $c = 1$, $\hbar = 1$, $\ell_P = 1$, the problems formulated above *cannot* be solved by letting c^{-1} or $\hbar$ or ℓ_P tend to 0 !

(3) The passage from microscopic theories at short distance – and time scales to *effective theories* at large distance – and time scales (which, hopefully, are easier to analyze) necessitates a deeper and mathematically more precise understanding of *renormalization group methods*. (See the notes of T. Spencer, and refs. given there.)

In these notes, we describe some recent results which provide an example of a solution of problem (2), (ii). We consider a system of very many non-relativistic bosons with very weak two-body interactions and exhibit a class of states with the property that the time evolution of expectation values of certain operators in these states is approximately described by classical Hamiltonian equations of motion, more specifically by the *non-linear Hartree equations*. We then sketch how certain solutions of the non-linear Hartree equations approach solutions of *Newtonian equations of motion* for point particles when the two-body forces are attractive at short distances. Details may be found in [FrTY]. (The reader is invited to consult the notes of J. Bourgain for mathematical results on related Hamiltonian partial differential equations.)

2   Large Systems of Non-Relativistic Bosons

In non-relativistic quantum mechanics, pure states are unit rays in a separable Hilbert space, $\mathcal{H}$, mixed states are density matrices on $\mathcal{H}$. A specific physical situation is encoded in the choice of some concrete *algebra $\mathcal{A}$ of bounded operators contained in (or equal to) the algebra, $B(\mathcal{H})$, of all bounded operators on $\mathcal{H}$. The time-evolution of an operator $a \in \mathcal{A}$ in the Heisenberg picture is given by

$$\mathcal{A} \ni a \mapsto a(t) := e^{i\widehat{H}t/\hbar} a\, e^{-i\widehat{H}t/\hbar}, \tag{1}$$

where $\widehat{H}$ is a selfadjoint operator densely defined on $\mathcal{H}$ (whose spectrum is usually assumed to be bounded from below).

As an example, we consider a system of N spinless, non-relativistic, identical quantum mechanical (point) particles in physical space $\mathbb{E}^3$ moving under the influence of an external force with potential W and of two-body interactions with potential $-\Phi$. As an algebra $\mathcal{A}$ we may, e.g., choose the algebra of smooth, bounded functions $a(x_1, \ldots, x_N)$ on $\mathbb{R}^{3N}$ which are totally symmetric in their N arguments. Its spectrum corresponds to the space of configurations of N identical point particles.

The Hilbert space of state vectors, $\mathcal{H} = \mathcal{H}_\sigma^{(N)}$, is chosen to be a subspace of the space, $L^2(\mathbb{R}^{3N}, \Pi_{j=1}^N d^3 x_j)$, of square-integrable functions on $\mathbb{R}^{3N}$ of a specific symmetry type, σ, under permutations of their N arguments ("quantum statistics"); i.e., a function $\psi \in \mathcal{H}_\sigma^{(N)}$ transforms under an irreducible representation σ of the group of permutations of $\{1, \ldots, N\}$ when its arguments are permuted. When σ is the trivial representation ψ is totally symmetric in its N arguments, and we speak of *Bose–Einstein statistics*; when σ is the alternating representation ψ is totally anti-symmetric in its N arguments, and we speak of *Fermi–Dirac statistics*. (Higher-dimensional representations σ are sometimes called parastatistics.)

The Hamiltonian of the system is given by

$$\widehat{H} \equiv H^{(N)} := \sum_{j=1}^N \left(-\frac{\hbar^2}{2} \Delta_j + W(x_j) \right) - \sum_{1 \le i < j \le N} \Phi\left(x_i - x_j\right), \qquad (2)$$

and we have chosen units so that the mass of a particle is $= 1$.

In eq. (2), $\Delta_j = \sum_{a=1}^3 \partial^2/(\partial x_j^a)^2$, and $x_j = (x_j^1, x_j^2, x_j^3)$ denotes the position of the jth particle in $\mathbb{E}^3$ (in a fixed system of Cartesian coordinates). If W is, e.g., a smooth positive function on $\mathbb{E}^3$ and Φ is "small" relative to $-\Delta$, in the sense of Kato, then $H^{(N)}$ is essentially selfadjoint on a natural domain dense in $L^2(\mathbb{R}^{3N}, \Pi d^3 x_j)$. Since $H^{(N)}$ is symmetric under permutations of the particle labels $1, \ldots, N$, the time evolution, $\exp(-iH^{(N)}t/\hbar)$, leaves the subspaces $\mathcal{H}_\sigma^{(N)}$ of $L^2(\mathbb{R}^{3N}, \Pi d^3 x_j)$ invariant, for arbitrary σ.

There are straightforward generalizations of this class of systems to systems of several species of identical particles some of which may have non-zero quantum mechanical spin, e.g., to Coulomb system; see [L].

Two important problems in the analysis of such systems are (see [L]):

(A) *Stability* (i.e., the energy per particle is bounded from below by a constant $-\varepsilon$, uniformly in $N \Leftrightarrow H^{(N)} \ge -\varepsilon N, \forall N$); and

(B) *Existence of thermodynamic limit* (choose $W = W_\Lambda$ so as to confine the particles to a bounded region $\Lambda \subset \mathbb{E}^3$; keep the particle density $\rho := N/|\Lambda|$ fixed, where $|\Lambda|$ is the volume of Λ, and try to pass to the limit $\Lambda \nearrow \mathbb{E}^3$. – A necessary condition for the existence of a thermodynamic limit is stability.)

Much of *theoretical condensed matter physics* is the study of large systems of identical, non-relativistic, quantum mechanical point particles for which (A) and (B) hold. Such systems exhibit fascinating physical phenomena such as *Bose–Einstein condensation* (σ = trivial representation, two-body interactions repulsive); *superconductivity, Landau-* and *marginal*

Fermi liquids, Hall liquids, different forms of magnetism, ... (σ = alternating representation). Precise mathematical understanding of some of these phenomena is only beginning to emerge; see, e.g., [FST] (and [CFS] for a review).

The topic of these notes is a study of dilute systems of a large number of neutral atoms with Bose–Einstein statistics (σ = trivial rep.) and with weak two-body interactions. If, in the initial state, the mean distance between atoms is sufficiently large and the average kinetic energy per atom is large then one may argue that, on sufficiently short time scales, short-range repulsion between the atoms can be neglected. Then the system can be described by a Hamiltonian $H^{(N)}$ of the form (2), where $-\Phi$ is, e.g., the potential of a long-range attractive force. In this idealization, stability (A) is, however, usually violated.

We propose to identify a class of initial states for such systems with the property that the time evolution of the expectation values of certain operators in such states approaches one described by some classical, non-relativistic Hamiltonian PDE, in the limit where the strength, κ, of the two-body interaction tends to 0, with $N \propto \kappa^{-1}$.

In order to analyze such systems, it is convenient (though dispensable) to make use of the formalism of "second quantization". We imagine that the number, N, of atoms may fluctuate (e.g., because the system is coupled to a reservoir of atoms not included in our description). It is then convenient to introduce the Hilbert space

$$\mathcal{F} = \bigoplus_{N=0}^{\infty} \mathcal{H}^{(N)}, \quad \mathcal{H}^{(0)} = \mathbb{C}, \tag{3}$$

where $\mathcal{H}^{(N)} = \mathcal{H}^{(N)}_{\sigma=\text{trival rep.}}$, which is commonly called *Fock space*. A vector $\underline{\psi} \in \mathcal{F}$ is a sequence

$$\underline{\psi} = \{\psi^{(N)} \mid \psi^{(N)} \in \mathcal{H}^{(N)}\}_{N=0}^{\infty} \tag{4}$$

with

$$\|\underline{\psi}\|^2 := \sum_{N=0}^{\infty} \|\psi^{(N)}\|_2^2 < \infty.$$

The scalar product of two vectors $\underline{\psi}$ and $\underline{\varphi}$ is given by

$$\langle \underline{\psi}, \underline{\varphi} \rangle := \sum_{N=0}^{\infty} (\psi^{(N)}, \varphi^{(N)}), \tag{5}$$

where $(\cdot, \cdot)$ is the usual L^2 scalar product. Let f be a test function on $\mathbb{E}^3$. We define an *"annihilation operator"* $a(f)$ on $\mathcal{F}$ by setting

$$a(f)\underline{\psi} = \underline{\psi}_f,$$

where

$$\psi_f^{(N)}(x_1, \ldots, x_N) := \sqrt{N+1} \int d^3x\, \overline{f(x)}\, \psi^{(N+1)}(x, x_1, \ldots, x_N). \quad (6)$$

The operator $a(f)$ is unbounded, but closable. The *"creation operator"* $a^*(f)$ is defined to be the adjoint of $a(f)$ on $\mathcal{F}$. It is an unbounded closable operator. It is easy to verify that

$$[a(f), a(g)] = [a^*(f), a^*(g)] = 0,$$
$$[a(f), a^*(g)] = (f, g)\mathbb{1}, \quad (7)$$

for arbitrary test functions f and g. Eqs. (7) are called *canonical commutation relations*. The Hamiltonian, $\widehat{H}$, on Fock space is given by

$$\widehat{H} = \bigoplus_{N=0}^{\infty} H^{(N)}, \text{ with } H^{(0)} = 0.$$

In terms of annihilation- and creation operators, $\widehat{H}$ can be expressed as follows:

$$\widehat{H} = \frac{1}{2} \int d^3x \big[(\nabla a^*)(x)(\nabla a)(x) + W(x)\, a^*(x)\, a(x) \big]$$
$$- \frac{1}{2} \int d^3x \int d^3y\, a^*(x) a^*(y) \Phi(x - y) a(y) a(x), \quad (8)$$

and we now use units such that $\hbar = 1$. The two-body potential $-\Phi$ is chosen as follows:

$$\Phi = \kappa \Phi_c, \quad \kappa \ll 1, \quad (9)$$

where $\Phi_c \in L^{3/2} + L^{\infty}$ (with $L^p = L^p(\mathbb{R}^3, d^3x)$) (depending on the analytical methods employed, one may need to also impose some regularity properties on first and second derivatives of Φ; see, e.g., [H]). An example for Φ_c is

$$\Phi_c(x) \simeq |x|^{-6} + g|x|^{-1}, \quad g \ll 1, \quad (10)$$

for $|x|$ large compared to the diameter of an atom, and $\Phi_c(x)$ is regular near $x = 0$. The first term in Φ_c describes van der Waals attraction, the second term gravitational attraction between atoms. The external potential W is assumed to be a smooth, positive function.

It is easy to check that

$$H^{(N)} \geq -C\kappa N^2,\tag{11}$$

for some finite constant C. For $N \propto \kappa^{-1}$, kinetic and potential energy terms in $H^{(N)}$ both scale like κ^{-1}.

Next, we introduce rescaled annihilation and creation operators,

$$A_\kappa(f,t) := \kappa^{1/2} e^{i\hat{H}t} a(f) e^{-i\hat{H}t}\tag{12}$$

($\hbar = 1$!), and

$$A_\kappa^*(f,t) := (A_\kappa(f,t))^*.\tag{13}$$

Let $\underline{\varphi}$ be a vector in Fock space $\mathcal{F}$ with a finite number of particles and smooth k-particle wave functions of compact support, for all k. We define a vector $\underline{\theta}_\kappa$ by setting

$$\underline{\theta}_\kappa := C_\kappa (a^*(\psi_0))^{N_\kappa} \underline{\varphi},\tag{14}$$

where ψ_0 is a smooth function of compact support which, for convenience, we choose to have $\|\psi_0\|_2 = 1$,

$$N_\kappa = [\kappa^{-1}],\tag{15}$$

where $[a]$ denotes the integer part of a positive number a, and the constant C_κ is chosen such that $\langle \underline{\theta}_\kappa, \underline{\theta}_\kappa \rangle = 1$.

Let $\psi(x,t)$ be a solution of the *non-linear Hartree equation*

$$i\partial_t \psi = -\tfrac{1}{2}\Delta\psi + W\psi - (\Phi_c * |\psi|^2)\psi\tag{16}$$

with initial condition

$$\psi(x,0) = \psi_0(x).\tag{17}$$

We define

$$\psi(f,t) := \int d^3x \overline{f(x)}\psi(x,t),$$

and

$$\psi^*(f,t) := \overline{\psi(f,t)},$$

where f is a test function.

The following theorem is proven in [H].

Theorem 2.1. *Under the hypotheses described above,*

$$\lim_{\kappa\to 0}\left\langle \underline{\theta}_\kappa, \prod_{j=1}^m A_\kappa^*(f_j,t_j)A_\kappa(g_j,s_j)\underline{\theta}_\kappa \right\rangle = \prod_{j=1}^m \psi^*(f_j,t_j)\psi(g_j,s_j),\tag{18}$$

for arbitrary $m = 1, 2, \ldots$, *arbitrary test functions* $f_1, \ldots, f_m$, $g_1, \ldots, g_m$ *and finite times* $t_1, \ldots, t_m$, $s_1, \ldots, s_m$.

Thus, a classical limit described by the non-linear Hartree equation (16) *is approached in systems of* $N \sim \kappa^{-1}$ *bosons interacting through two-body forces of strength* κ, *as* κ *tends to 0, for an appropriate choice of the initial state of the system.*

In the following, we show that the non-linear Hartree equation is an example of a Hamiltonian PDE and then explore a Newtonian point-particle limit of the non-linear Hartree equation, for two-body potentials which are *attractive* at short distances.

The theorem quoted above is but one example of a result on the approach of quantum dynamics towards some classical Hamiltonian dynamics in a suitable limit. Another example would be a quantum spin system whose dynamics is expected to approach the one described by the Landau–Lifshitz equation in the limit where the local magnetization becomes large.

An unsatisfactory aspect of results of this type is that one must choose rather special sequences of states, $\{\underline{\theta}_\kappa\}$, in order to approach a classical limit. States leading to a classical limit do *not* form a linear space (which may explain why there are no puzzles related to the superposition principle). We do not understand much, yet, about a general class of states leading to classical limits, and in which physical situations such states appear naturally.

The material in the following sections is motivated by the above theorem, eq. (18): We study the NL Hartree equation (16) and the question under what circumstances it reduces to the Newtonian mechanics of point particles. We begin with a brief review of the *Hamiltonian nature* of the *non-linear Hartree equations.*

3   NL Hartree Equations as a Hamiltonian System

The phase space for the non-linear Hartree equation (16) is the energy (Sobolev) space

$$\Gamma = H^1(\mathbb{R}^3) \tag{19}$$

with complex coordinates $\psi(x), \bar{\psi}(x)$. The symplectic two-form is given by $\frac{i}{2} d\psi \wedge d\bar{\psi}$. It leads to the following Poisson brackets

$$\{\psi(x), \psi(y)\} = \{\bar{\psi}(x), \bar{\psi}(y)\} = 0$$
$$\{\psi(x), \bar{\psi}(y)\} = 2i\delta(x - y). \tag{20}$$

The Hamilton functional $\mathcal{H}$ leading to the NL Hartree equations (16) is given by

$$\mathcal{H}(\bar{\psi}, \psi) = \frac{1}{4} \int \left[|\nabla \psi|^2 + 2W|\psi|^2 - (\Phi_c * |\psi|^2)|\psi|^2 \right], \quad (21)$$

with $\psi = \psi(x)$, $\int(\cdot) = \int_{\mathbb{R}^3} d^3 x(\cdot)$, and $*$ denotes convolution. For $\Phi_c \in L^{3/2} + L^\infty$, $\mathcal{H}$ is well defined on Γ (Sobolev and Hölder inequalities), and it is bounded below on the spheres

$$\mathcal{S}_N := \{\psi \mid \psi \in \Gamma, \, \mathcal{N}(\bar{\psi}, \psi) = N < \infty\} \qquad (22)$$

in phase space Γ, where

$$\mathcal{N}(\bar{\psi}, \psi) := \int |\psi|^2; \qquad (23)$$

see, e.g., [GV]. Hamilton's equations of motion for ψ are

$$\dot{\psi}(x, t) = \{\mathcal{H}(\bar{\psi}, \psi), \psi\}(x, t) \qquad (24)$$

and are easily seen to be equivalent to the NL Hartree equation (16). From the form of $\mathcal{H}$ we infer the following *symmetries* and corresponding *conservation laws*.

(1) Gauge invariance of the first kind,

$$\psi(x) \mapsto e^{i\theta}\psi(x), \quad \bar{\psi}(x) \mapsto e^{-i\theta}\bar{\psi}(x), \qquad (25)$$

leaves $\mathcal{H}$ invariant; (25) is a symplectic flow generated by the Hamiltonian vector field associated with the functional $\frac{1}{2}\mathcal{N}(\bar{\psi}, \psi)$. Thus $\mathcal{N}(\bar{\psi}, \psi)$ must be conserved by the time evolution generated by (the Hamiltonian vector field determined by) $\mathcal{H}$, which means that

$$\{\mathcal{H}, \mathcal{N}\} = 0. \qquad (26)$$

It follows that the spheres $\mathcal{S}_N$ defined in (22) are *invariant* under the time evolution.

(2) If $W = 0$ and if Φ_c is rotation-invariant then arbitrary Galilei transformations are symmetries of $\mathcal{H}$. Space translations are generated by the momentum functional

$$\mathcal{P}(\bar{\psi}, \psi) := \frac{i}{2} \int \bar{\psi} \nabla \psi, \qquad (27)$$

and rotations by the angular momentum functional

$$\mathcal{L}(\bar{\psi}, \psi) := \frac{i}{2} \int \bar{\psi}(x \wedge \nabla)\psi. \qquad (28)$$

All components of $\mathcal{P}$ and $\mathcal{L}$ Poisson commute with $\mathcal{H}$. Velocity transformations (boosts), i.e. $x \mapsto x - vt$, are represented on time-dependent trajectories, $\psi(x,t)$, in Γ by

$$\psi(x,t) \mapsto \psi(v;x,t) := \psi(x - vt, t)e^{i(v \cdot x - (v^2/2)t)} \,. \qquad (29)$$

These transformations map solutions of the equations of motion (24) to other solutions of (24). It follows that the "center of mass motion" of a solution $\psi(x,t)$ of (24) is inertial.

Instead of the Hamiltonian formalism, we may make use of the *Lagrangian formalism* to study the Hartree equations (16). These equations are then obtained as the Euler–Lagrange equations corresponding to the *action functional*

$$S(\bar{\psi}, \psi) = \int_{t_1}^{t_2} dt \left[\frac{i}{2} \int \bar{\psi}(\cdot, t) \frac{\partial}{\partial t} \psi(\cdot, t) - \mathcal{H}\big(\bar{\psi}(\cdot, t), \psi(\cdot, t)\big) \right], \qquad (30)$$

by variation with respect to $\bar{\psi}$ (with $\delta\bar{\psi}(\cdot, t_i) = 0$, for $i = 1, 2$).

Existence and regularity of global solutions of the Cauchy problem associated with the non-linear Hartree equation, under the conditions on Φ_c described above, is known. If $-\Phi_c$ is radially symmetric, decreasing (*repulsive* two-body forces) and of sufficiently short range then the scattering map associated with the non-linear Hartree equations exists and satisfies asymptotic completeness. These results are reviewed in [GV]. In the situation of interest in these notes ($-\Phi_c$ radially symmetric, but *increasing*, i.e., for *attractive* two-body forces) existence and regularity are known from the results reviewed in [GV], but the scattering theory is much more subtle, and only very preliminary results are available. Our main interest is in the Newtonian point-particle limit of the NL Hartree equations, for two-body forces which, at small distances, are *attractive*.

4   The Point-Particle Limit of the NL Hartree Equations with Attractive Two-Body Forces

For reasons that will become apparent shortly, we start by trying to find (local and absolute) minima of the Hamilton functional $\mathcal{H}$ restricted to a sphere $\mathcal{S}_N$, as defined in (22), for a vanishing external potential, $W = 0$. In this situation, the momentum and angular momentum functionals, $\mathcal{P}$ and $\mathcal{L}$, are conserved. We should therefore study the (generalized) energy functional

$$\mathcal{E}(\bar{\psi}, \psi; \nu, \pi, \lambda) := \mathcal{H}(\bar{\psi}, \psi) + \tfrac{\nu}{2}\big(N - \mathcal{N}(\bar{\psi}, \psi)\big)$$
$$+ \pi \cdot \big(P - \mathcal{P}(\bar{\psi}, \psi)\big) + \lambda \cdot \big(L - \mathcal{L}(\bar{\psi}, \psi)\big), \quad (31)$$

where ν, π and λ are Lagrange multipliers. Variation with respect to $\psi, \bar{\psi}, \nu, \pi$ and λ yields the following variational equations:

$$-\tfrac{1}{2}\triangle\psi - \big(\Phi_c * |\psi|^2\big)\psi - \nu\psi - i\pi \cdot \nabla\psi - i\lambda \cdot (x \wedge \nabla)\psi = 0 \quad (32)$$

as well as the complex conjugate equation, and

$$\mathcal{N}(\bar{\psi}, \psi) =, N, \quad \mathcal{P}(\bar{\psi}, \psi) = P, \quad \mathcal{L}(\bar{\psi}, \psi) = L. \quad (33)$$

We do not know much about the general solutions to these equations. But, for the purposes of these notes, the following special case is particularly important: We search for an absolute minimum, Q_N, of $\mathcal{H}(\bar{\psi}, \psi)|_{S_N}$ which is rotation-invariant and has zero momentum. The function Q_N is the solution of the equations

$$-\tfrac{1}{2}\triangle\psi - \big(\Phi_c * |\psi|^2\big)\psi = \nu\psi,$$
$$\mathcal{N}(\bar{\psi}, \psi) = N, \quad (34)$$

and we henceforth set $\nu =: E$. It follows that

$$E = \frac{1}{2N} \int \big[|\nabla Q_N|^2 - (\Phi_c * |Q_N|^2)|Q_N|^2\big], \quad (35)$$

which is strictly *negative*, for a non-trivial minimum Q_N.

Fairly standard methods of variational calculus show that there exists some finite $N_* = N_*(\Phi_c)$ such that, for $N > N_*$, eqs. (34) have a non-trivial, smooth, positive solution $Q_N(x)$ which decays exponentially in $|x|$, with decay rate $\sqrt{-E}$. From the theory of binding in Schrödinger quantum mechanics one infers that $N_*(\Phi_c) = 0$, if $\Phi_c(x)$ is positive and decreases in $|x|$ more slowly than $|x|^{-2}$; but $N_*(\Phi_c) > 0$ if Φ_c is of short range. Covariance under translations and velocity transformations then implies that

$$\psi_N(v, a; x, t) := Q_N(x - a - vt)e^{i(v \cdot x - [\frac{v^2}{2} + E]t)} \quad (36)$$

solves the non-linear Hartree equations (16), for an arbitrary velocity $v \in \mathbb{R}^3$. It describes a *"soliton"* traveling with velocity v.

Let $\mathcal{H}''$ denote the Hessian of $\mathcal{H}|_{S_N}$ at Q_N. Then one finds that

$$(Q_N, \mathcal{H}''Q_N) = \varepsilon_0(Q_N, Q_N), \quad \varepsilon_0 < 0. \quad (37)$$

In fact, $\mathcal{H}''$ has precisely one eigenvector, ϕ_0, corresponding to a negative eigenvalue. Furthermore,

$$\mathcal{H}'' \frac{\partial}{\partial x^a} Q_N = 0, \quad \text{for} \quad a = 1, 2, 3, \tag{38}$$

which follows from the translation invariance of $\mathcal{H}$, and the continuous spectrum of $\mathcal{H}''$ covers the interval $[-E, \infty)$, with $-E > 0$.

Assumption*. It will be assumed henceforth that Φ_c can be chosen such that the multiplicity of the eigenvalue $\varepsilon_1 = 0$ of $\mathcal{H}''$ is equal to 3.

The eigenspace of $\mathcal{H}''$ corresponding to the eigenvalue 0 is then spanned by the vectors $\frac{\partial}{\partial x^1} Q_N$, $\frac{\partial}{\partial x^2} Q_N$, $\frac{\partial}{\partial x^3} Q_N$, and it follows that

$$(\phi, \mathcal{H}''\phi) \geq \varepsilon_2(\phi, \phi), \quad 0 < \varepsilon_2 \leq -E, \tag{39}$$

for arbitrary $\phi \in L^2(\mathbb{R}^3)$, with $\phi \perp Q_N$, $\frac{\partial}{\partial x^a} Q_N$, $a = 1, 2, 3$.

We are now prepared to discuss the Newtonian point-particle limit of the non-linear Hartree equation. Our discussion is heuristic; for mathematical results see [FrTY].

We choose an external potential W of the form

$$W(x) \equiv W^{(\varepsilon)}(x) := V(\varepsilon x),$$

where V is a positive, smooth function on $\mathbb{R}^3$, and $\varepsilon \ll 1$ is a parameter. Furthermore, the two-body potential $-\Phi_c$ is chosen to be

$$\Phi_c(x) = \Phi_s(x) + \Phi_\ell(\varepsilon x), \tag{40}$$

where $\Phi_s(x)$ is a rotation-invariant, smooth function decaying rapidly in $r := |x|$, as $r \to \infty$, and with the properties that

$$\frac{\partial \Phi_s(r)}{\partial r} < 0, \quad \text{for } r > 0, \tag{41}$$

and that Assumption*, above, holds for $\Phi_c = \Phi_s$ The potential Φ_ℓ is rotation-invariant and may be of long range, e.g.,

$$|\Phi_\ell(x)| \sim r^{-1}, \quad \text{as } r \to \infty.$$

For technical convenience, we also assume that Φ_ℓ is once continuously differentiable, with $|\partial \Phi_\ell / \partial r|$ uniformly bounded in r.

Next, we pick a positive integer k and k positive numbers $N_1, \ldots, N_k$, with $N_j > N_*(\Phi_s)$, for all j. For $\varepsilon = 0$ and $N > N_*(\Phi_s)$, we define

$$\delta_N := \sqrt{N^{-1} \int d^3 x Q_N(x)^2 x^2}. \tag{42}$$

We consider a configuration of k solitons, $Q_{N_j}(x-q_j)$, $q_j \in \mathbb{R}^3$, $j = 1, \ldots, k$, with the following properties: Each $Q_{N_j}(x)$ is a solution of equations (34), with $N = N_j$, for $\Phi_c = \Phi_s + \Phi_\ell(0)$, i.e., $\varepsilon = 0$; and

$$\left(\max_{j=1,\ldots,k} \delta_{N_j} \right) \Big/ \left(\min_{1 \leq i < j \leq k} |q_i - q_j| \right) \leq \varepsilon. \tag{43}$$

Our goal is to construct a solution, ψ_ε, of the non-linear Hartree equation (16) of the form

$$\psi_\varepsilon(x,t) = \sum_{j=1}^{k} Q_{N_j}(x - q_j(t)) e^{i\theta_j(x,t)} + h_\varepsilon(x,t), \tag{44}$$

where $q_j(0) = q_j$, $\dot{q}_j(0) = v_j$, $j = 1, \ldots, k$, with the following properties: There is a positive constant T such that, for all times t with $|t| < T/\varepsilon$,

(i) $\|h_\varepsilon(\cdot, t)\|_{H^1} = o(\varepsilon)$,

(ii) $\theta_j(x,t) = \dot{q}_j(t) \cdot [x - q_j(t)] + \vartheta_j(t)$,

where $\vartheta_j(t)$ is independent of x; the trajectories $q_1(t), \ldots, q_k(t)$ and the phases $\vartheta_1(t), \ldots, \vartheta_k(t)$ will turn out to be determined by the non-linear Hartree equations (16) and properties (i) and (ii).

In order to convince ourselves that the ansatz (44) with properties (i) and (ii) is plausible, we evaluate the action functional $S(\bar{\psi}, \psi)$ introduced in eq. (30) on the ansatz functions $(\bar{\psi}, \psi) = (\bar{\psi}_\varepsilon, \psi_\varepsilon)$, with ψ_ε as in (44). We shall regard the soliton trajectories $q_1(t), \ldots, q_k(t)$, the phases $\vartheta_1(t), \ldots, \vartheta_k(t)$ and the small-amplitude wave h_ε as variational parameters of the ansatz $(\bar{\psi}_\varepsilon, \psi_\varepsilon)$. After a moderately lengthy calculation one finds that, for a certain choice of $\vartheta_j(t)$,

$$S(\bar{\psi}_\varepsilon, \psi_\varepsilon) = \int_{-\tau}^{\tau} dt \sum_{j=1}^{k} \left[\frac{N_j}{2} \dot{q}_j(t)^2 - N_j V(\varepsilon q_j(t)) \right.$$
$$\left. + \sum_{i:i<j} N_i N_j \Phi_\ell(\varepsilon(q_i(t) - q_j(t))) + o(\varepsilon) \right] \tag{45}$$

for $\tau < T/\varepsilon$. Up to error terms $\sim o(\varepsilon)$, which are the source of a variety of mathematical subtleties, this is the action for a system of k Newtonian point particles with masses $N_1, \ldots, N_k$ moving under the influence of an external acceleration field with potential $W^{(\varepsilon)}$ and of two-body forces with potential $-N_i N_j \Phi_\ell(\varepsilon(q_i - q_j))$, $1 \leq i < j \leq k$; $q_j(t)$ is the position of the jth particle at time t.

Expression (45) for $S(\bar{\psi}_\varepsilon, \psi_\varepsilon)$ suggests that, for $S(\bar{\psi}, \psi)$ to have a critical point at $(\bar{\psi}, \psi) = (\bar{\psi}_\varepsilon, \psi_\varepsilon)$, with ψ_ε given by (41) and satisfying properties (i) and (ii), the particle trajectories $\{q_j(t)\}_{j=1}^k$ must be solutions of Newton's equations of motion

$$\ddot{q}_j(t) = -\varepsilon\left[\nabla V(\varepsilon q_j(t)) - \sum_{i \neq j} N_i \nabla \Phi_\ell\big(\varepsilon(q_i(t) - q_j(t))\big)\right]$$
$$+ a_j(t), \quad j = 1, \ldots, k, \quad (46)$$

where the *"friction force"* $a_j(t)$ satisfies a bound $|a_j(t)| = o(\varepsilon)$, for $|t| < T/\varepsilon$.

These ideas are made precise (in the form of mathematical theorems) in [FrTY], using methods partly based on the work of Weinstein [W], for just one soliton in an external potential $W^{(\varepsilon)}$. But the methods in [FrTY] can be extended to general k.

Apparently, we have gained insight into how Newtonian point-particle mechanics with small friction forces can emerge from the non-linear Hartree equation with *attractive* two-body forces of short range (Φ_s) and weak attractive *or* repulsive two-body forces of very long range ($\Phi_\ell(\varepsilon \cdot)$) by considering the evolution of configurations of far-separated solitons over large, finite intervals of time. The friction forces ($a_j(t)$) appearing in Newton's equations of motion originate in the emission of small-amplitude, dispersive radiation from the solitons. This is an example of a general and quite fundamental phenomenon: *friction* (or dissipation) through *emission of radiation* in *conservative* (classical and quantum mechanical) *systems* of infinitely many degrees of freedom.

For similar considerations in the context of vortex dynamics in a Ginzburg–Landau superfluid see [OS].

The results discussed above appear to be a special case of more general results on the Newtonian limit of the non-linear Hartree equation. One might imagine that it is possible to construct *"rotating-body"* solutions of the non-linear Hartree equations with non-zero angular momentum relative to their center of mass. Such solutions are, in general, unstable. Nevertheless, one could attempt to construct solutions of the non-linear Hartree equation which, over a large, finite interval of time, describe a configuration of finitely many rotating bodies, whose motion is described by a system of Euler-type equations with small friction terms, emitting small-amplitude dispersive radiation.

The results reviewed in eqs. (39) through (46) and the speculations just described raise the issue of *asymptotic properties* of the dynamics determined by the non-linear Hartree equation in the limit, as time tends to $\pm\infty$. Such problems belong to the realm of non-linear *scattering theory*. In [FrTY], a result on the scattering of small-amplitude waves off a single soliton has been established. It requires $W(x)$ and $\Phi_c(x)$ to be smooth and of short range. We consider an "asymptotic profile" given by

$$\psi_{as}(x,t) = Q_N(x-vt)e^{i(x\cdot v-[(v^2/2)+E]t)} + h(x,t), \qquad (47)$$

where h is a solution of the free-particle Schrödinger equation

$$i\partial_t h(x,t) = -\tfrac{1}{2}\triangle h(x,t),$$

with $h(x,0) \in H^3(\mathbb{R}^3) \cap W^{3,1}(\mathbb{R}^3, (1+|x|^2)d^3x)$, and $\hat{h}(v,0) = 0$. The claim is that there are solutions, $\psi_\pm(x,t)$, of the non-linear Hartree equation (16) such that

$$\psi_\pm(x,t) \underset{t\to+\infty}{\longrightarrow} \psi_{as}(x,t) \qquad (48)$$

in $H^1(\mathbb{R}^3)$.

Thus, (non-linear) *Møller wave maps*, $\Omega_\pm$, exist (as symplectic maps) on asymptotic profiles of the form (47); (see also [GV] for background material on non-linear scattering theory). Of course, the results described here do not bring us anywhere close to an understanding of asymptotic completeness of the scattering map.

5   More About Open Problems, Conclusions

In this last section, we draw the reader's attention to some interesting open problems.

(1) Clearly, a deeper understanding of the non-linear scattering theory of solitons and waves for the non-linear Hartree equation would be highly desirable. In particular, it would be important to gain more detailed insight into the domain of definition and the *image* of the Møller wave maps $\Omega_\pm$.

A comparatively simple situation has been studied by Soffer and Weinstein [SW]. They consider a smooth external potential W of short range with exactly one quantum mechanical bound state, i.e., the spectrum of the operator $-\tfrac{1}{2}\triangle + W$ consists of one simple eigenvalue $\varepsilon < 0$ and of purely absolutely continuous spectrum covering $[0,\infty)$. One must also assume that there is no zero-energy resonance. They consider a local non-linearity (NL Schrödinger equation); but their techniques can be extended to the NL Hartree equation with Φ_c of short range. They construct small-amplitude non-linear bound states (stationary solutions of (16)) by standard bifurcation theory techniques. They then exhibit a class of initial conditions with the property that the asymptotic behavior of solutions of (16) with these initial conditions is that of a non-linear bound state plus a purely dispersive wave which tends to zero (pointwise) at the free dispersion rate, as $t \to \pm\infty$. They also show that initial conditions close to a (small-amplitude) non-linear bound state have this property.

Obviously, one would like to extend these results to external potentials W with more than one bound state and understand the radiative decay of initial conditions close to an *excited* boundstate of $-\frac{1}{2}\triangle + W$ to a non-linear groundstate plus dispersive radiation. One would also like to go *beyond* results for small-amplitude initial conditions. In this case, it can happen that the restriction of the Hamilton functional $\mathcal{H}(\bar{\psi}, \psi)$, see eq. (21), of the non-linear Hartree equation (16) to a sphere $\mathcal{S}_N$ in phase space Γ (see (22)) has several distinct local quadratic minima, provided N is large enough; (choose W to have several minima separated by large barriers and $-\Phi_c$ attractive). One would then like to understand the shape of the *"basins of attraction"* in phase space of these local minima. [The forward (backward) basin of attraction of a one-parameter family of local minima of $\mathcal{H}\big|_{\mathcal{S}_N}$ parametrized by N consists of all initial conditions in phase space Γ which approach an element of this family plus dispersive radiation decaying to zero, as $t \to +\infty$ ($t \to -\infty$, respectively). More generally, one may introduce a notion of a *center manifold* of (asymptotically) attracting configurations of solitons (freely moving, for $W = 0$) to which solutions of the non-linear Hartree equation with initial conditions sufficiently close to the center manifold converge, as $t \to +\infty$, or $t \to -\infty$. In this connection, see, e.g., [PW].]

(2) Consider a configuration, ψ_ε, of, say, just two far-separated solitons with initial positions q_1, q_2 and initial velocities $v_1 = \dot{q}_1(0)$, $v_2 = \dot{q}_2(0)$, as described in eq. (44), with $h_\varepsilon(x, 0) = 0$. Let us assume that the two-body force with potential $-\Phi_\ell$ is purely attractive and of short range. We choose the initial conditions, (N_1, q_1, v_1) and (N_2, q_2, v_2) in such a way that the two solitons form a bound system, in the sense that

$$\frac{N_1}{2}v_1^2 + \frac{N_2}{2}v_2^2 - N_1 N_2 \Phi_\ell\big(\varepsilon(q_1 - q_2)\big) < 0 \,. \tag{49}$$

We would like to understand the fate of this configuration, as $t \to \infty$. To begin with, we would like to estimate the power, $P_R(t)$, of dispersive radiation through a sphere of radius $R \gg \max(|q_1|, |q_2|)$; see [OS] for some results on a related problem. Furthermore, we would like to show that a configuration of two bound solitons collapses to a single soliton moving through space with a constant velocity, for $W = 0$ (!), plus a dispersive, outgoing wave tending to O (pointwise) at the free dispersion rate; ("radiative collapse of a binary system").

(3) It appears reasonable to imagine that if the potential Φ_ℓ is chosen to be given by

$$\Phi_\ell(x) \simeq |x|^{-1},$$

for $|x|$ large (Φ_s as in sect. 4 and $W = 0$), the non-linear Hartree equation (16) can be used to describe Newtonian gravity of a giant cloud of bosonic atoms, with built-in dispersive friction. It would be interesting to understand how, on intermediate time scales, inhomogeneities in the initial conditions grow to form a structure consisting of rotating bodies plus dispersive radiation ("structure formation"), before the system collapses into a number of far separated solitons escaping from each other. Perhaps, one may gain some insight into such problems with the help of numerical simulations; see, e.g., [MKNZ].

(4) It is of interest to study transport phenomena described by the non-linear Hartree equation with a *random* external potential W whose distribution is homogeneous and has correlations of short range. For some results in this direction see [AF].

Perhaps most importantly, we would like to understand much more about how non-linear equations of classical physics emerge as descriptions of quantum systems in certain asymptotic regimes.

Progress in theoretical physics is achieved in many different ways. If one is very lucky (and ingenious) one may discover new Laws of Nature or *new* physical theories or a new mathematical structure useful to formulate such theories. More modestly, one may attempt to improve the mathematical understanding of *known* physical theories or derive new consequences of such theories. It appears safe to predict that we will be sailing well into the 21st century with plenty of interesting questions about physical theories discovered during the past three hundred years to worry about, and that "analysis and physics" will be a worthy partner of the currently more popular "geometry and physics".

Acknowledgments. We thank J. Bourgain and A. Soffer and, especially, I.M. Sigal and T. Spencer for very helpful discussions. The work of H.T. Yau has been partially supported by the US NSF under grant DMS-9703752.

References

[AF]   C. ALBANESE, J. FRÖHLICH, Periodic solutions of some infinite-dimensional Hamiltonian systems associated with non-linear partial difference equations, I, Commun. Math. Phys. 116 (1988), 475–502; C. ALBANESE, J. FRÖHLICH, T. SPENCER, Periodic solutions of some infinite-dimensional Hamiltonian systems associated with non-linear partial difference equations, II, Commun. Math. Phys. 119 (1988), 677–699.

[CFS]   T. CHEN, J. FRÖHLICH, M. SEIFERT, Renormalization group methods: Landau-Fermi liquid and BCS superconductor, in Proc. of Session LXII of the Les Houches summer schools – "Fluctuating Geometries in Statistical Mechanics and Field Theory" (F. David, P. Ginsparg, J. Zinn-Justin, eds.), Elsevier, Amsterdam, New York, 1996.

[D]   J. DIMOCK, The non-relativistic limit of $P(\phi)_2$ quantum field theories: two particle phenomena, Commun. Math. Phys. 57 (1977), 51–66; QED$_2$ in the Coulomb gauge, Ann. Inst. H. Poincaré 43 (1985), 167–179.

[FST]   J. FELDMAN, M. SALMHOFER, E. TRUBOWITZ, Renormalization of the Fermi surface, in Proc. of the XIIth Intl. Congress of Math. Physics, 24–34, International Press, Cambridge, MA, 1999; and references to previous work given there.

[FrTY] J. FRÖHLICH, T.-P. TSAI, H.-T. YAU, The point-particle limit of the non-linear Hartree equation, to appear.

[GV]   J. GINIBRE, G. VELO, On a class of nonlinear Schrödinger equations with nonlocal interaction, Math. Z. 170 (1980), 109–136; see also J. GINIBRE, G. VELO, Scattering theory in the energy space for a class of nonlinear Schrödinger equations, J. Math. Pure Appl. 64 (1985), 363–401; Scattering theory in the energy space for a class of Hartree equations, preprint 1998.

[H]   K. HEPP, The classical limit for quantum mechanical correlation functions, Commun. Math. Phys. 35 (1974), 265–277; some of the ideas in this paper can be traced back to: E. SCHRÖDINGER, Der stetige Übergang von der Mikro- zur Makromechanik; Die Naturwissenschaften 28 (1926), 664–669.

[L]   E.H. LIEB, The Stability of Matter: From Atoms to Stars (Selecta of Elliott H. Lieb), Springer-Verlag, Berlin, Heidelberg, New York, 1991.

[MKNZ] A.V. MIKHAILOV, E.A. KUZNETSOV, A.C. NEWELL, V.E. ZA-KHAROV, eds., "The Nonlinear Schrödinger equation", Proc. of conference in Chernogolovka, Physics D 87:1-4, North Holland, Amsterdam, New York, 1995.

[OS]   YU.N. OVCHINNIKOV, I.M. SIGAL, Dynamics of localized structures, Physica A 261 (1998), 143–158; and references given there; e.g., YU.N. OVCHINNIKOV, I.M. SIGAL, Ginzburg-Landau equation, I, General discussion, in "PDE's and their Applications" (L. Seco, et al., eds.), CRM Proceedings and Lecture Notes 12 (1997), 199–220.

[PW]   C.-A. PILLET, C.E. WAYNE, Invariant manifolds for a class of dispersive, Hamiltonian partial differential equations, J. Diff. Equations 141 (1997), 310–326.

[SW]   A. SOFFER, M.I. WEINSTEIN, Multichannel nonlinear scattering for nonintegrable equations, Commun. Math. Phys. 133 (1990), 119–146; Multichannel nonlinear scattering for nonintegrable equations, II, the case of anisotropic potentials and data, J. Diff. Equations 98:2 (1992), 376–390.

[W]   M.I. WEINSTEIN, Modulation stability of groundstates of nonlinear Schrödinger equations, SIAM J. Math. Annal. 16 (1985), 472–490.

JÜRG FRÖHLICH, Theoretical Physics, ETH-Hönggerberg, CH–8093 Zürich, Switzerland

TAI-PENG TSAI AND HORNG-TZER YAU, Courant Institute of Mathematical Sciences, 251 Mercer Street, New York University, New York, NY 10471, USA

GAFA, Geom. funct. anal.
Special Volume – GAFA2000, 79 – 117
1016-443X/00/S1079-39 $ 1.50+0.20/0

© Birkhäuser Verlag, Basel 2000

I GAFA Geometric And Functional Analysis

ROUGH STRUCTURE AND CLASSIFICATION

W.T. GOWERS

1 Introduction

When I was first asked to speak at the "Visions in Mathematics" conference, I had what I believe was a typical reaction. I wanted to try to emulate Hilbert a century ago, but since I knew that I could not possibly match his breadth of vision, I was forced to make some sort of compromise. In this paper I shall discuss several open problems, not always in areas I know much about, but they are not intended as a list of the most important questions in mathematics, or even the most important questions in the areas of mathematics that I have worked in. Rather, they are a personal selection of problems that, for one reason or another, have captured my attention over the years.

The title of this paper refers to a general theme that links several unsolved problems in combinatorics. I believe that what makes them difficult is, in many cases, the lack of adequate "rough classification" theorems. I shall explain what I mean by this in §3 and give examples of problems for which classification might very well be useful. In §4 I give a more miscellaneous list of questions, with brief accounts of why I find them interesting. Most, but not all, of the questions on this list are very well known.

As the conference drew nearer, I decided that I could not resist at least some attempt at "the vision thing" [Bu]. I am not in favour of predicting which areas of mathematics are likely to be fashionable in fifteen (let alone fifty) years' time, but in §2 I have made a more general prediction about mathematics as a whole. Briefly, my contention is that during the next century computers will become sufficiently good at proving theorems that the practice of pure mathematical research will be completely revolutionized, quite possibly to the point where we would be reluctant to recognize

The research for this article was partially supported by the Clay Mathematics Institute.

it as mathematical research at all. My reasons for this opinion are outlined in the next section.

I do not claim any originality for my views on this topic. Indeed, by the time of my lecture at the conference, some of the ideas had been touched on by other speakers and during discussion sessions. Moreover, interest in automatic theorem proving has existed for many years, and since the conference I have been referred to similar views expressed by people who work in the area. (One reference is given at the end of the next section.)

2   Will Mathematics Exist in 2099?

For at least a century, mathematicians have thought about the possibility of automating mathematics. Hilbert famously asked in his tenth problem whether there was an algorithm for solving Diophantine equations, and later extended this to the question of whether there was an algorithm which would find a proof of any mathematical statement that had one. In 1936, Turing, equally famously, formalized the notion of algorithm and soon afterwards demonstrated the insolubility of the halting problem, thus showing that no such algorithm existed. While this result, and later demonstrations that several natural and well known problems in mathematics were also impossible to solve systematically, may have initially seemed somewhat negative, they also had a positive side. The idea that all our creativity and insight might be reduced to something mechanical was, after all, not very appealing. Turing's result therefore came as a relief, since it left mathematicians with something to do.

More recently, with the rise of the theory of computational complexity and NP-completeness, it has been widely understood that this relief was somewhat misplaced, because, as Gödel in fact appreciated, an algorithm for finding proofs is not much use if it takes an unfeasibly long time. Far more threatening to mathematicians would be an algorithm which could determine, for any given mathematical statement, whether or not it had a proof of length at most n, and do this in polynomial time (ideally with a polynomial of small degree). The existence of an algorithm of this kind is one of the major open problems of mathematics, since it is equivalent to whether P=NP, but many mathematicians take comfort from the fact that nobody has found a short algorithm for solving any NP-complete problem and most experts believe that no such algorithm exists.

However, this comfort, even if you are convinced that P$\neq$NP, is still misplaced, because searching for shortish proofs of arbitrary mathematical statements is a very bad model of what mathematicians actually do. Indeed, there is absolutely no evidence that where mathematicians score over computers is in their ability to solve NP-complete questions quickly. If somebody were to unearth an *idiot savant* who could determine in his or her head whether collections of Boolean formulae were satisfiable, or even factorize large integers, then one would be forced to ask some profound questions, but that will not happen (a prediction I make with great confidence). Therefore, the only possible explanation for the great success of mathematicians over the centuries is that our search for proofs of theorems is very restricted. Indeed, this is obviously true – we tend to prefer questions that are interesting, comprehensible, and seemingly not completely out of reach, and we like our proofs to provide *explanations* rather than just formal guarantees of truth.

This line of thought leads to the following question: can the notion of a "good proof" be usefully formalized? If so, then the impact on mathematics could well be comparable to the impact of the formalizing of the notions of proof and algorithm. It is perhaps unlikely that a definition will emerge which is as indisputably correct as that of a recursive function or Turing machine, but it might be realistic to try to isolate certain properties of what we commonly regard as good proofs and formalize them. Here is an incomplete list of features often associated with good proofs.

(i) Proofs are usually clearer if they have a hierarchical structure. For example, if the main theorem follows from three lemmas that all have clear statements, then one can isolate parts of the proof, think about them separately and understand the argument without having to hold too much in one's head at once.

(ii) Many good proofs are variants of existing better known arguments. This makes them easier to understand, because all one has to do is concentrate on the parts that are new.

(iii) It is easier to understand a proof that can be reduced to a few basic ideas. (Needless to say, it is not easy to formalize what one means by a "basic idea".)

(iv) Sometimes the form of a proof is dictated to some extent by the existence of certain examples and counterexamples. These clarify the proof by suggesting approaches that might work and demonstrating that other approaches cannot work.

(v) A good definition can greatly improve a proof. In extreme cases, choosing the right definition can reduce a difficult problem to a simple exercise in checking.

(vi) Many of the best proofs, or at least the best-written proofs, come with some indication of how they were discovered. To put that the other way round, proofs that depend on magic tricks, while they may be impressive and elegant, are not completely satisfactory as explanations.

The above features are interrelated, and they all have in common that a proof with one of the features is, other things being equal, easier to understand than one without. I would like to concentrate on (vi), and adopt it as a working definition of "explanation" as opposed to "guarantee of truth". That is, I shall think of an explanation as a proof together with an account of how the proof was discovered. This account should not be something like "I was sitting in the bath and suddenly had the following brilliant idea", but rather a demystification of the proof: that is, a convincing demonstration of how a mere (mathematically trained) human being could search for a proof and find this one, without relying on excessive amounts of luck.

I am aware that this definition (which will be made more precise in a moment) does not capture all our associations with the word "explanation". In particular, it might be easier to show how one came up with a brute force argument than to show how one thought of a neat definition which, after one had seen it, made the result obvious. Nevertheless, if we wish to teach computers to find proofs, it is likely to be a good idea to reflect on how we do so ourselves.

One way of explaining a proof, in the above sense, is to show that it can be generated using what one might think of as standard mathematical reflexes. Some of these might be very general methods of approach, and others might be trying to apply well known techniques from the area of mathematics in question. Here are a few examples of what I mean by general methods of approach. For my lecture I drew flow charts, but my TeX skills are very basic, so I shall represent them here as pseudo-algorithms. A much more detailed and sophisticated discussion of mathematical reflexes can be found in Polya's classic books, such as [P, vols. I and II].

A. Generalization.

    1. Is p a special case of q?
       If it seems not to be, then goto 2, else goto 3.

    2. Can q be appropriately modified?
       If it can then goto 1, else goto 5.

3. Is q true?

   If it might be, then goto 4, else goto 2.
4. Try to prove q instead of p.
5. Try a different approach.

B. Modification of Known Argument.

1. Find a known statement q that resembles p.
2. Can the proof of q be easily modified to give p?

   If yes then END, else goto 3.
3. Identify a statement r, forming part of the proof of q, such that the analogous statement s, which one would like as part of the proof of p, is not easy to prove.
4. If s is obviously false then goto 6, else goto 5.
5. Try to prove s.
6. Try a different approach.

C. Spotting Patterns in Special Cases.

1. Find a special case r of p.
2. If r is trivial, then goto 3, else goto 4.
3. Can I identify the next simplest special case?

   If so, call it r and goto 2, else goto 8.
4. Can I see how to prove r?

   If so, then goto 5, else goto 7.
5. Does the proof of r generalize to a proof of p?

   If so, then END, else goto 6.
6. Can I find a different, more instructive proof of r?

   If so, then goto 5, else goto 3.
7. Try to prove r instead of p.
8. Try a different approach.

D. Habituation, or Getting Bored.

1. Have I managed to think of any interesting ideas connected with p in the last couple of hours?

   If so, then goto 2, else goto 3.
2. Think about one of the interesting ideas.
3. Make a nice cup of coffee and think about something else for a while.

As for reflexes of a more specific kind, I mean things like trying to apply well known theorems or techniques. For this, one needs to have some ability to classify problems, so that one has an idea of which theorems or techniques may be helpful. Rather than giving several examples of the use of standard methods to solve problems, let me return to the question of automating mathematics and present an imagined dialogue between a mathematician and a computer in two or three decades' time. The idea of the dialogue is that the computer is very helpful to the mathematician, while not doing anything particularly clever. This represents an unthreatening intermediate stage between what we have now, computers that act as slaves doing unbelievably boring calculations for us, and full automation of mathematics. I have written the dialogue in English, but this is supposed to be a translation of a more formal language which has not yet been invented. (I shall discuss this point a little more later.)

Mathematician. Is the following true? Let $\delta > 0$. Then for N sufficiently large, every set $A \subset \{1, 2, \ldots, N\}$ of size at least δN contains a subset of the form $\{a, a + d, a + 2d\}$?
Computer. Yes. If A is non-empty, choose $a \in A$ and set $d = 0$.
M. All right all right, but what if d is not allowed to be zero?
C. Have you tried induction on N, with some $\delta = \delta(N)$ tending to zero?
M. That idea is no help at all. Give me some examples please.
C. The obvious greedy algorithm gives the set

$$\{1, 2, 4, 5, 10, 11, 13, 14, 28, 29, 31, 32, 37, 38, 40, 41, \ldots\}.$$

I notice that large parts of the set are translations of other parts. In fact, this set is very like the Cantor set, so this gives a bound of $\delta \geqslant N^{(\log 2/ \log 3)-1}$.

As for random methods, let us choose each element of $\{1, 2, \ldots, N\}$ to be in A with probability δ, the choices being independent. The expected number of subsets of A of the form $\{a, a + d, a + 2d\}$ is $\delta^3 N^2$, to within an absolute constant. Hence, if $\delta < N^{-2/3}$ it is possible for there to be none. The expected size of A is δN, and standard estimates tell us that A is very likely indeed to be of about this size.

Applying one further standard idea, we note that if the number of sets in A of the form $\{a, a + d, a + 2d\}$ is at most half the number of points in A, then we can delete one element from each of them and still be left with a substantial proportion of A. This tells us that we can choose any δ that satisfies $C\delta^3 N^2 \leqslant \delta N$, so in particular we can find some δ of the form $cN^{-1/2}$.
M. Well, random methods often give the best answer for problems like this. Let's try to prove that $cN^{-1/2}$ is best possible.

C. [*Pauses for 0.001 seconds*]   Actually it isn't. Behrend found a much better bound in 1946. [*Downloads paper*]

M.   Oh dear, I'm out of ideas then. Could you give me a suggestion by any chance?

C.   We have a set A. We want to prove that a subset of a certain form exists. The best way of proving existence is often to count.

M. [*Intrigued*]   Yes, but what would that mean for a problem like this?

C.   Here we wish to count the number of solutions (x, y, z) of the single linear equation $2y = x + z$. A standard tool in such situations is Fourier analysis, exponential sums, the circle method, whatever you like to call it.

M.   Where is the group structure?

C.   Identify $\{1, 2, \ldots, N\}$ with $\mathbb{Z}/N\mathbb{Z}$. It isn't pretty but it sometimes works. I write $\mathbb{Z}_N$ for $\mathbb{Z}/N\mathbb{Z}$ and $\hat{A}(r)$ for $\sum_{s \in A} \exp(2\pi i r s / N)$. Then the number of triples $(a, a + d, a + 2d) \in A^3$ is $N^{-1} \sum_{r \in \mathbb{Z}_N} \hat{A}(r)^2 \overline{\hat{A}(-2r)}$.

M.   We are trying to show that that expression cannot be zero unless δ is very small.

C. [*Trained to humour mathematicians*]   That is indeed almost equivalent to our original problem. Thank you. Notice that $N^{-1}\hat{A}(0)^3 = \delta^3 N^2$, exactly the number we had before when A was chosen randomly. This number is large and positive – a good sign.

M.   So we would like the rest of the sum to be small enough not to cancel out the zero term.

C.   Yes. I am trying a few obvious ideas such as the Cauchy-Schwarz inequality. Unfortunately, there does not seem to be any reason for the rest of the sum to be small.

M.   Please show me your calculations.

C.   Here they are. [*Displays them.*] They do enable me, or rather us, to prove a partial result. It states that if $\max_{r \neq 0} |\hat{A}(r)| \leqslant c\delta^2 N$, then the image of A in $\mathbb{Z}_N$ contains many sets of the desired form.

This dialogue could be continued to show what to do when $\hat{A}(r)$ is large for some r. However, the computer has already guided the mathematician to an important insight which is the basis for Roth's proof of his theorem on arithmetic progressions [Ro1]. Finishing the proof is not all that difficult, though perhaps it might be for the mathematician above, who seems a bit too reliant on his computer.

Notice that the computer is at every stage trying standard ideas: induction, a greedy algorithm, random methods, counting, taking the Fourier transform and applying Cauchy-Schwarz to a sum of products. What makes it think of *these* standard ideas, rather than some other completely inappropriate ones? Part of the answer lies in how the problem is initially put to the computer. I would envisage not the formal statement given at the beginning of the dialogue, but something more interactive. For example, this stage of the process could be menu-driven. Initially one would choose Combinatorics, or perhaps Combinatorial Number Theory. This would bring up another menu from which one would choose Extremal Problems. For the next menu one would choose the word Maximize. The next choice would be Over Sets A of Integers. Then one would choose Cardinality of A. Then a menu would come up with a choice of constraints, or perhaps one would have to type them in: $A \subset \{1, 2, \ldots, N\}$ and A contains no arithmetic progression of length three.

At the end of a process like this, the computer would have many ideas about how the problem was conventionally classified. For example, it would be natural for the computer to think of the circle method because many problems in additive number theory would involve concepts such as the number of arithmetic progressions of length three in a set A. The computer could then look for a list of standard tricks, and find the slogan in the above dialogue to do with counting solutions of a linear equation.

Notice that it is important for the computer that it has access to a mathematical database which is much more sophisticated than anything we have at present. One might wonder how the computer could have discovered Behrend's paper so quickly, since understanding ordinary text is notoriously difficult for computers. I was imagining that a great deal of work had gone into classifying papers so that Behrend's paper appeared on the database not as Proc. Nat. Acad. Sci. 23 (1946), 331–332, but rather as something like Maximize – Size of A – Subject To – (1) $A \subset \{1, 2, \ldots, N\}$ & (2) A contains no 3-AP – Lower Bound. As well as the paper, there would be a summary of the main result – Lower Bound At Least $n \exp(-c\sqrt{\log n})$.

I do not see any obstacle to creating a useful database of the above kind, other than the time and effort that would be involved. The structure of the database would be very well suited to the internet, which would also make it easier to distribute the work amongst many people, so perhaps even this obstacle is not all that great. There are of course parts of mathematics that do not lend themselves particularly easily to classification of the above kind, but this is no reason to dismiss the whole idea. Even a database covering only certain areas of mathematics would be useful.

Encoding and classifying general methods, tricks, rules of thumb and so on would obviously be harder. One of the main reasons for this is that they are often not precisely formulated, and dealing with vague statements is a well known hard problem in computer science. However, hard does not mean impossible and there has already been important progress on it. Moreover, at least some of the vague principles of mathematics do not appear to present serious difficulties. Consider for example the following simple but extremely useful consequence of the Cauchy-Schwarz inequality:

$$\left(\sum_{i=1}^{n}|a_i|\right)^2 \leqslant n\sum_{i=1}^{n}|a_i|^2.$$

This is a useful inequality because the ℓ_2-norm is often easier to handle than the ℓ_1-norm. However, in many contexts the inequality is far too weak, so one tends to use it only when there is some reason to believe that the a_i are of roughly equal size. This sounds like a vague idea, and many mathematicians, when thinking about a problem, will not need to clarify it further, but further clarification is perfectly possible. For example, another very simple but surprisingly useful inequality is

$$\sum_{i=1}^{n}|a_i|^2 \leqslant \max_{1\leqslant i\leqslant n}|a_i|\sum_{i=1}^{n}|a_i|,$$

which tells us that the first inequality will be sharp to within a constant C if $\max_{1\leqslant i\leqslant n}|a_i| \leqslant Cn^{-1}\sum_{i=1}^{n}|a_i|$. This is not a necessary condition, but it could form one of a number of tests that a computer could apply (or rather, attempt to apply) whenever it used the first inequality, to see whether it was likely to be expensive.

If there were to be a serious attempt to codify the many principles and working methods of mathematicians, then it would be an immensely useful exercise to return to the mathematics we already know and try to rewrite the proofs as explanations (in the restricted sense that I defined earlier). Many textbooks begin with definitions that justify themselves later by leading to interesting results: this would be forbidden in the kind of textbook I have in mind. Instead, every definition should be presented as arising naturally in the search for the solution of some problem. Of course, the problem itself should also be justified; or alternatively one could take as one's starting point that the problem is worth studying, leaving the justification to somebody else. (The question of justifying open problems was discussed by Gromov at the conference. See his contribution to this volume, especially the ends of §1 and §2. For the sake of brevity I have ignored many aspects of mathematical practice, such as making conjectures and proposing definitions. These must also be taken seriously if mathematics is to be automated.)

Even if the automation of mathematics turns out to be much more difficult than I have supposed, this exercise of rewriting could have enormous benefits, because, unsurprisingly, the question of how we might teach mathematics to computers is closely related to the question of how we should teach it to human beings. I have no time for those who believe that talent for mathematical research is a mysterious unteachable quality which either you have or you don't. Of course there is such a thing as natural ability, but most mathematicians develop their ability as they mature. At the moment, they tend to do this with very little external guidance and very little help from the literature. A series of books aimed at making transparent the thought-processes that led to certain results could show to young mathematicians why research is not as impossible as it looks and convey much more efficiently the tricks used by those with more experience.

Two examples of what I mean about the justification of definitions are provided by simplicial homology and the Riemann zeta function. When I was taught simplicial homology, I was presented with a complicated (as it seemed at the time) definition which then, miraculously, led to interesting and far more concrete results such as fixed-point theorems. Similarly, a typical introduction to the zeta function begins with Euler's product formula, which leads rapidly to interesting theorems about the primes, but is not presented as the inevitable result of thinking about the primes. For these two concepts, I can give plausible (but completely ahistorical and by no means unique) accounts of how they might have been invented. I intend to put them on my personal web page (http://www.dpmms.cam.ac.uk/~wtg10/), but do not expect to get round to this before about January 2001. I also hope to give some examples of how apparent bright ideas can be explained as the natural result of familiar methods of searching for proofs.

I am fully aware that what I have described so far is a long way from the full automation of mathematics. In particular, I have ignored some notorious problems in artificial intelligence. I have briefly discussed vagueness, but the example I gave was a simple one, and does not obviously generalize. I have not mentioned the difficulty of pattern-recognition, although recognising similarity is a very important part of mathematical research. What I have tried to show is that these difficulties are not a good reason for doing nothing. Reflecting on one's thought processes does not require any expertise in computing and is potentially very beneficial to others; and a computer with a great deal of information but only very modest intelligence could still be extremely useful.

I have also not discussed the many efforts that have already been made in the general area of automatic theorem proving, an area which (as is probably obvious) I know little about. My impression is that most of the progress so far has been from the bottom up – trying to get computers to discover simple but interesting theorems starting almost from scratch. What I have talked about is more of a top-down project to which any mathematician could contribute. It might be that eventually such a top-down approach could reduce the general problem of automating mathematics to a large number of smaller, more manageable problems, so that perhaps the mathematical community could meet the automatic theorem provers half way. (One way to learn about the state of the art in automatic theorem proving is to visit the website http://www.math.temple.edu/~zeilberg/OPINIONS.html and follow links, especially from the rather extreme opinion 36.)

Finally, then, here is my guess about the impact of computers on mathematics over the next century. Although some of the difficulties involved in automatic theorem proving are formidable, a century is a very long time in computing, so I expect computers to be better than humans at proving theorems in 2099. I do not feel particularly happy about this, but I expect the impact on my own mathematical life to be entirely positive, because a semi-intelligent database of the kind I have described would be a wonderful resource and would take a great deal of the drudgery out of research. There might even be a golden age when computers were good at exercises but not yet good at having deep insights. Then, instead of wasting a week not noticing that a hoped-for lemma had a simple counterexample, one could get the computer to check it. In other words, computers would still do the boring bits for us, but these would not be quite as boring as they are now. However, such a golden age, if it occurs, is unlikely to last for long. The next stage might be one where only a very few outstanding mathematicians could discover proofs that were inaccessible to computers. Perhaps others would concentrate on teaching, but computers would probably become better than us at this as well, at least if they themselves had been taught in anything like the way I have outlined. In the end, the work of the mathematician would be simply to learn how to use theorem-proving machines effectively and to find interesting applications for them. This would be a valuable skill, but it would hardly be pure mathematics as we know it today.

3 Rough Structure and Classification, and Related Problems

In his talk, Gromov drew attention to the fact that not many ways were known of defining manifolds. (See the subsection "Randomization" in §5 of his contribution to this volume.) For a different reason, something similar is true for graphs. This might at first seem a ridiculous assertion – all you have to do to define a graph is take the set of all pairs of elements of the set $\{1, 2, \ldots, n\}$ and pass to an arbitrary subset. However, while it is true that a mathematical object does not have to satisfy delicate constraints in order to be a graph, one still faces the purely logical problem that the number of comprehensible descriptions is limited. Moreover, while graph theorists often talk as though they are considering individual graphs, most problems in graph theory are, strictly speaking, about *families* $(G_n)_{n=1}^{\infty}$ of graphs, where the number of vertices of G_n is usually n, or perhaps some simple function of n that tends to infinity.

Here is a list of a few common ways of defining families of graphs.

 (i) *Repeating patterns.* Some examples of this are

    (a) K_n, the complete graph of order n;
    (b) P_n, the path with n vertices;
    (c) C_n, the cycle with n vertices;
    (d) $K_{1,n}$, the "star" with one vertex joined to n others, and no further edges;
    (e) $\{0, 1\}^n$, the discrete n-dimensional cube;
    (f) a dyadic tree with a root and n further layers.

 (ii) *Simple constructions from known graphs.* These include products, quotients, disjoint unions, removal of a few edges and so on.

(iii) *Use of algebra and number theory.* Two notable examples are the Paley graph and the Ramanujan graphs of Lubotzky, Phillips and Sarnak [LPS]. In general, Cayley graphs often have interesting properties.

(iv) *Random graphs.* It is not quite clear in what sense this phrase defines a family of graphs: see the discussion below.

This list could undoubtedly be continued, but I would be surprised if somebody could give me a further ten genuinely distinct and interesting classes of definitions.

Let us consider for a moment what it means to say that a graph is "random". We would like to make statements such as "A random graph on

n vertices with edge-probability $\frac{1}{3}$ has approximately $\frac{1}{3}\binom{n}{2}$ edges." Sometimes, one can make sense of such a statement by inserting the words "with high probability" in the appropriate place. However, there are contexts in which this is not adequate. For example, there is a class of graphs, known as expanders, which have direct practical applications in the design of certain algorithms. It is known that graphs chosen randomly are, with high probability, good expanders. On the other hand, faced with a particular randomly chosen graph, how can one be sure that it is not one of the exceptions? This is perhaps not a genuine practical problem, since a graph that works with probability $1 - 10^{-8}$ is just as useful as one that works with complete certainty (though I have heard that people in industry who buy computer systems are very suspicious of random elements in the design). Even so, a very important development in graph theory was the realization that several properties shared by almost all graphs are equivalent, and serve as a useful definition of "randomness".

These properties apply to a graph G with n vertices such that almost every vertex has degree approximately $n/2$. They show that G resembles a graph where each edge is chosen independently with probability $1/2$. (Similar results hold for other probabilities as well.) Here are three of them.

(i) For any two large sets A and B of vertices, the number $e_G(A, B)$ of pairs $(x, y) \in A \times B$ such that xy is an edge of G is $|A||B|/2 + o(n^2)$ (and thus roughly the number one would expect if G was a random graph).

(ii) For any fixed (small) graph H, the number of (induced) subgraphs of G isomorphic to H is roughly what the expected number would have been if G had been chosen randomly.

(iii) Define the adjacency matrix A of G by setting $A(x, y)$ to be 1 if xy is an edge and 0 otherwise. Then the second largest eigenvalue (in modulus) of A is $o(n)$.

In connection with property (i) above, define the *density* $\delta_G(A, B)$ to be $e_G(A, B)/|A||B|$. Another way of stating the property is to say that $\delta_G(A, B)$ is approximately $1/2$ for all large sets A and B. Interestingly, one can replace (ii) by the weak-seeming assumption that G contains roughly the expected number of cycles of length four.

One of the first papers in which it was shown that the smallness of the second largest eigenvalue implied randomness properties was by Alon and Milman [AM]. The first paper dealing explicitly with the idea of pseudo-random graphs was by Thomason [T1]. The paper of Chung, Graham and Wilson [CGW] made the whole area much more systematic, giving a long list of equivalent randomness properties. They called a graph satisfying any one (and hence all) of the properties *quasirandom*, perhaps to make clear the distinction between this idea and the rather different concept of pseudorandomness in computer science.

There are two morals I would like to draw from the above results. The first is that in many contexts they allow one to give a clear meaning to the concept of a random graph, even when the graph has been explicitly chosen. (For example, the Paley graph is known to be quasirandom in this sense.) The second is that they support our intuition that two random graphs are basically the same. Of course, they will not be isomorphic, but they share all the properties that interest us, so the difference between them is not important.

I can now explain what I mean by "rough classification". It is clear that there are too many graphs for it to be sensible to try to classify them all up to isomorphism. However, as just commented, we often like to regard non-isomorphic graphs as basically the same. Implicit in that discussion was a metric which can be defined as follows. Given two graphs G and H, both with n vertices, and a bijection ϕ between $V(G)$ and $V(H)$, define

$$d_\phi(G, H) = n^{-2} \max_{A, B \subset V(G)} \left| e_G(A, B) - e_H(\phi A, \phi B) \right|$$

and let $d(G, H)$ be the minimum of all the $d_\phi(G, H)$. If G and H are close in this metric, it means that there is a one-to-one correspondence between their vertices such that the densities $\delta_G(A, B)$ in G are approximately the same as the corresponding densities in H, at least when A and B are large. A graph is quasirandom if and only if it is close to a typical random graph in this metric.

By a rough classification of graphs, I mean a choice of a small number of representative graphs, such that every graph is close in the above metric to one of the representatives. Such a classification will be useful if the representative graphs can be easily described and analysed. It is not at all obvious that a useful classification should exist, but an important result, Szemerédi's regularity lemma [S1], shows that it does.

To state the result, let us introduce a further definition. If G is a graph and A and B are sets of vertices in G, let us say that the pair (A, B) is *ε-regular* if, whenever A' and B' are subsets of A and B with $|A'| \geqslant \epsilon|A|$ and $|B'| \geqslant \epsilon|B|$, we have $|d_G(A', B') - d_G(A, B)| \leqslant \epsilon$. If A and B are disjoint, this condition, which is very similar to the first property of quasirandom graphs listed above, states that the edges between A and B behave very like those of a random bipartite graph on A and B with edge probability $d_G(A, B)$. Szemerédi's regularity lemma is as follows.

LEMMA. *Let $\epsilon > 0$. Then there exists a positive integer $k = k(\epsilon)$ such that, given any finite graph G, the vertices of G can be partitioned into k classes $V_1, \ldots, V_k$, all of size $\lfloor n/k \rfloor$ or $\lceil n/k \rceil$, such that all but at most $\epsilon \binom{k}{2}$ of the pairs (V_i, V_j) (with $i < j$) are ϵ-regular.*

A common way of defining graphs is to choose a few sets $V_1, \ldots, V_k$ and a symmetric $[0, 1]$-valued matrix $(p_{ij})_{i,j=1}^{k}$, and to join $x \in V_i$ to $y \in V_j$ with probability p_{ij}, all choices being independent. Szemerédi's regularity lemma states that *every* graph is like that, at least up to closeness in the metric defined earlier. Since random graphs are quite well understood, this gives us a useful classification.

One consequence of this result is that if G is a graph, and H is an induced subgraph of G formed by choosing half the vertices of G at random, then H looks very similar to G. The reason is that, with high probability, the vertex set of H will be made up of roughly half of each of the sets V_i in the partition given by Szemerédi's lemma, and the regularity of a pair (V_i, V_j) is inherited by large subsets (V_i', V_j'), at the cost of a slight worsening of the parameter ϵ. This general principle about random subgraphs is hard to demonstrate without the use of some result along the lines of the regularity lemma (though there are weaker versions that suffice).

A major drawback with the regularity lemma is that the dependence of k on ϵ is very bad – a tower of twos of height proportional to ϵ^{-5}. Moreover, it has been shown that this unfortunate situation is necessary. In [G1] a graph is constructed which gives a lower bound for k of a tower of twos of height proportional to $\epsilon^{-1/16}$. This limits the use of the lemma to situations where the dependence on ϵ is not important.

Before I move on to talk about open problems, let me briefly discuss the other half of my title: rough structure. Mathematicians vary considerably in the amount of structure they like, and this allows them to be roughly classified in a useful way. At one end of the spectrum are those who start with so many assumptions that even finding a new object that satisfies the assumptions is a great achievement. Those who manage to do so are often rewarded with the discovery of beautiful and detailed patterns. At the other end are mathematicians who like more elbow room. They operate under fewer constraints, so the objects they study exist in great profusion and the problems that arise are of a very different character.

As my discussion of classification already indicates, this distinction, though it undoubtedly exists, is not quite as great as it might seem. I would like to blur it a little further by showing that patterns can emerge unexpectedly even in very combinatorial problems.

There is a theorem of Freiman [Fr1,2] that illustrates this extremely well. Let A be a set of integers, and write $A + A$ for the set $\{x + y : x, y \in A\}$. It is an easy exercise to show that if $|A| = n$ then $2n - 1 \leqslant |A + A| \leqslant \frac{1}{2}n(n + 1)$ and that both bounds are tight. Freiman's theorem gives a complete description of A when $A + A$ is small, or to be precise, when $|A + A| \leqslant Cn$ for some fixed constant C. An obvious way to construct such a set is to let P be an arithmetic progression of size at most $Cn/2$ and to let A be an arbitrary subset of P of size n. However, this is not the only source of examples: one can take a "two-dimensional" arithmetic progression, which is a set of the form

$$\{a + r_1 d_1 + r_2 d_2 : 0 \leqslant r_1 < m_1, 0 \leqslant r_2 < m_2\}.$$

It can be checked that if A is a two-dimensional progression, then $|A+A| \leqslant 4|A|$. More generally, $A + A$ will be small if A is a large subset of a d-dimensional arithmetic progression for some small d. (The definition should be obvious.)

A further easy exercise is to show that if $|A + A| = 2n - 1$, then A must be an arithmetic progression of length n. Harder than this, but still not too hard, is to show that if $|A + A|$ is just a little larger than $2n$, at most $2.1n$ say, then there is an arithmetic progression of length slightly greater than n that contains A. The strongest result of this kind is due to Freiman.

Theorem. *Let A be a set of integers of size n. If $0 \leqslant b \leqslant n - 3$ and $|A + A| \leqslant 2n - 1 + b$, then A is contained in an arithmetic progression of length $n + b$.*

This result shows that A is forced to be one-dimensional if $|A + A| \leqslant 3n - 4$. This bound is tight, because if m is any integer larger than $2n - 3$ and A is the set $\{1, 2, \ldots, n - 1, m\}$, then $|A + A| = 3n - 3$.

Freiman's main theorem is much deeper, and tells us what happens when we gradually relax the condition on the size of the sumset. It states that the *only* sets with small sumsets are large subsets of low-dimensional arithmetic progressions.

Theorem. *For every C there exist constants $K = K(C)$ and $d = d(C)$ such that if A is any set of n integers and $|A + A| \leqslant Cn$, then A is a* *subset of a generalized arithmetic progression of dimension at most d and cardinality at most Kn.*

Notice that the assumption made in Freiman's theorem is combinatorial, but the conclusion is very definitely structural, in an appropriate rough sense. This is not particularly surprising when C is close to the extremal value of 2, when A is, just as one might expect, a small perturbation of some easily described extremal set, but for larger values of C the result is much less to be expected, and is not at all easy to prove. A beautiful and completely different proof of Freiman's theorem was obtained by Ruzsa [Ru], and a considerably cleaned up version of Freiman's original argument has recently been given by Bilu [Bi].

I have presented Szemerédi's regularity lemma and Freiman's theorem in a very positive way, and they both have important applications. However, they do not solve everything. For the rest of this section I shall discuss problems which appear to be hard to solve without some sort of rough classification theorem. In each case, the classification theorem would have to be better than anything we have at present. I begin with three problems in elementary Ramsey theory.

1. Bounds for diagonal Ramsey numbers. Ramsey's theorem asserts that for every positive integer k there exists a positive integer N such that if the edges of the complete graph on N vertices are coloured red and blue, then there must be k of the vertices with all edges joining them of the same colour. (The graph spanned by these vertices is called *monochromatic.*) The smallest N with this property is denoted by $R(k,k)$. It is known that $R(3,3) = 6$ and $R(4,4) = 18$, but $R(k,k)$ is not known for any $k \geqslant 5$. It is unlikely that a reasonable formula for $R(k,k)$ exists, so a much more interesting question is how it behaves asymptotically. Simple arguments show that $2^{k/2} \leqslant R(k,k) \leqslant 2^{2k}$, but, despite strenuous efforts on the part of many mathematicians, no significant improvement on either bound has been obtained. (By significant I mean a lower bound of $2^{\alpha k}$ with $\alpha > 1/2$ or an upper bound of $2^{\beta k}$ with $\beta < 2$.)

There is an equivalent definition of $R(k,k)$, which I shall sometimes use without further comment. It is the smallest N such that any graph with N vertices contains a clique or independent set with k vertices (that is, k vertices either completely joined or not joined at all).

2. Covering by triangle-free graphs. The Ramsey number $R(3,3,...,3)$ (where 3 occurs r times) is the smallest integer N such that if the complete graph on N vertices is coloured with r colours, then there must be a triangle with all its edges of the same colour. Relatively straightforward arguments show that this number lies between e^r and r^r. It is an open question whether there is an upper bound of the form C^r for some absolute constant C.

An equivalent formulation of this problem is as follows: if the complete graph on n vertices is to be written as a union of k triangle-free graphs, how large must k be? One can do it if $k = \lceil \log_2 n \rceil$. Is it necessary to use at least $c \log n$ graphs?

There is a further equivalent formulation of the problem in terms of Shannon capacity – see the contribution of Alon in this volume for a discussion of this concept.

3. Ramsey numbers under additional constraints. The following is a beautiful problem of Erdős and Hajnal. Let H be a fixed finite graph. If G is assumed not to contain any induced copy of H, then by how much does the bound for $R(k,k)$ improve? In particular, does there exist a constant C depending on H only such that every graph with at least k^C vertices containing no induced H has a complete or empty subgraph with k vertices?

What is it about these three problems that makes them hard? The most important reason seems to be that it is not obvious what the extremal graphs should be. For the first problem, the lower bound $R(k,k) \geqslant 2^{k/2}$ was proved by Erdős [E] in 1947. His proof can be summarized as follows: colour each edge red or blue randomly and independently with probability one half, and check that the expected number of monochromatic complete subgraphs with k vertices is less than one. To put that even more loosely: a random colouring works.

Once one has seen this proof, it comes to seem so natural that it is tempting to conjecture that the best possible bound, or something close to it, is attained by a random graph, and indeed many who have thought about the problem believe this to be so. However, an unsettling result of Thomason [T2] suggests that the proof of such a fact will not be easy. A natural approach to estimating $R(k,k)$ is to consider the following more general problem. Let k and n be positive integers, and suppose that the edges of a complete graph on n vertices are coloured red and blue. What is the smallest possible number of monochromatic complete subgraphs of order k? Erdős conjectured that a random colouring should be best possible, which would suggest a bound of approximately $2.2^{-\binom{k}{2}}\binom{n}{k}$. This result is true when $k = 3$, but Thomason showed that it is false when $k = 4$.

It is important to emphasize the sense in which random graphs are not best. Some graphs, such as the remarkable Ramanujan graphs, improve on random constructions by being "more random than random". For example, Ramanujan graphs are d-regular and have a second largest eigenvalue of $2\sqrt{d-1}$, which is asymptotically as small as possible. Although random graphs also tend to have a second largest eigenvalue of size $C\sqrt{d-1}$, the constant C is larger than 2. This is not the nature of the non-randomness of Thomason's example, and indeed cannot be, because any graph which is *better* than random will be quasirandom in the sense defined earlier, and therefore it and its complement will contain about the same number of copies of K_4 as a random graph.

Thomason's example is constructed as follows. He first chooses a fixed graph H with k vertices. (One example that works is a certain product of seven triangles.) He then partitions the vertices of a complete graph into equal-sized sets $V_1, \ldots, V_k$ and colours an edge between $x \in V_i$ and $y \in V_j$ blue if ij is an edge of H and red otherwise. Details can be found in [T2,3]. Notice that Szemerédi's regularity lemma and the facts about quasirandom graphs imply that any example is *necessarily* similar to Thomason's in that it will involve partitioning the vertices into a few sets and changing the edge probabilities between the sets from $1/2$ to something else, although it is quite surprising that Thomason manages with probabilities of zero and one.

As for the second problem, one obvious way to cover n edges with $\lceil \log_2 n \rceil$ triangle-free graphs is to use bipartite ones. One can let the vertices of G be all 01-sequences $(\epsilon_1, \ldots, \epsilon_k)$ of length k, and for each i define a bipartite graph H_i by joining $(\epsilon_1, \ldots, \epsilon_k)$ to $(\eta_1, \ldots, \eta_k)$ if and only if $\epsilon_i \neq \eta_i$. Once again, we have a very natural example which looks as though it might be best possible. However, it is not.

In order to demonstrate this, we make a simple observation. Suppose that G is a complete graph on m vertices which can be covered by r triangle-free graphs $H_1, \ldots, H_r$. Now let $k > 1$ and consider the complete graph with vertex set G^k. For $1 \leqslant i \leqslant r$ and $1 \leqslant j \leqslant k$ define a graph H_{ij} by joining $(x_1, \ldots, x_k)$ to $(y_1, \ldots, y_k)$ if and only if $x_j y_j$ is an edge of H_i. It is easy to check that the graphs H_{ij} are triangle-free and cover all edges of G^k. The number of graphs H_{ij} is rk and G^k has m^k vertices. If $n = m^k$ then $rk = \log n / ((\log m)/r)$, so to beat the first bound, all we need to do is find an example where $(\log m)/r$ is greater than $\log 2$. One such example exists when $m = 5$. The complete graph on five vertices can be covered with two five-cycles, and $(\log 5)/2 > \log 2$. A better example still is obtained as follows. Let G be the complete graph K_{41}, and identify the vertices of G with the integers modulo 41. The quartic residues form a subgroup Q of the multiplicative group of $\mathbb{F}_{41}$. This subgroup has index four, and it can be checked that no two of its elements differ by one. We now colour the edges xy of G with four colours according to the coset of $x - y$. (Since -1 is a quartic residue, the coset of $x - y$ equals the coset of $y - x$.) It is easy to see that a monochromatic triangle would imply the existence of four non-zero elements a, u, v, w of $\mathbb{F}_{41}$ such that $au^4 + av^4 = aw^4$. This in turn would imply the existence of non-zero p and q such that $1 + p^4 = q^4$, which were checked not to exist. Therefore, each colour class represents a triangle free graph, and in this case $(\log m)/r$ is $(\log 41)/4$, which is larger than $(\log 5)/2$.

Even this is not the best known example, which (as I learned from Noga Alon at the conference) is related to a certain Cayley graph with over three hundred vertices and was discovered by a computer.

For the third problem, if H is a moderately complicated graph with seven vertices, say, then there is no hope of guessing what the extremal graph G will be. The result is plausible, because the condition that G contains no induced copy of H rules out almost any use of randomness, which seems to be very important if a graph is not to contain large cliques or independent sets.

It is interesting to think about why for some extremal problems it is easy to identify the best possible examples, while for others it is much harder. The three problems above are all hard because there are two completely different techniques for producing examples, and one can produce a whole spectrum of further examples by mixing them. For example, the following graph G shows that $R(k, k) \geqslant k^2$. Let the vertices of G be $V_1 \cup \cdots \cup V_k$, where the V_i are disjoint sets of size k. Join $x \in V_i$ to $y \in V_j$ if and only if $i \neq j$. At one time Turán believed that the bound of k^2 was sharp, and although we now know that he was completely wrong, Thomason's example shows that it is difficult to dismiss examples that mix this sort of explicit approach with a probabilistic one. This also accounts for the difficulty with the third problem. A similar phenomenon occurs with the second, in that there is a spectrum of methods of constructing triangle-free graphs. At one end is a complete bipartite graph and at the other is a random graph with appropriate edge-probabilities (and a few deletions). In between, one can construct a triangle-free graph H on r vertices using random methods, and then blow it up to a graph G on n vertices by partitioning the vertices of

G into sets $V_1, \ldots, V_r$ and joining $x \in V_i$ to $y \in V_j$ if and only if ij is an edge of H.

It is because the extremal graphs for these problems are not obvious that there seems to be a need for some sort of classification. Returning to the problem of estimating $R(k, k)$, although the random graph may well give the right answer, it appears to be necessary to take account of graphs such as the one constructed by Thomason, even if only to rule them out. One might imagine that Szemerédi's regularity lemma was ideally suited to the purpose, but the poor bounds in the lemma make it far too crude. In fact, even the definition of quasirandom graphs is too crude, because it is perfectly possible for a graph G on n vertices to be $n/2$-regular and have a second largest eigenvalue proportional to $\sqrt{n}$, but to contain a complete subgraph with $\sqrt{n}$ vertices. In other words, the distinction between two quasirandom graphs as previously defined can be very important in the context of Ramsey's theorem. This suggests the need for a stronger definition: as far as I know, nothing suitable has been proposed.

I have concentrated so far on graph theory, but similar difficulties arise in many other contexts. Here are two problems in combinatorial number theory.

4. Roth's theorem on arithmetic progressions. In 1974, Szemerédi [S2] proved the following theorem, which had been conjectured by Erdős and Turán in 1936 [ET].

Theorem. *Let $k \geqslant 3$ be an integer and let $\delta > 0$. Then if N is sufficiently large, every subset A of $\{1, 2, \ldots, N\}$ of cardinality at least δN contains an arithmetic progression of length k.*

The correct asymptotic dependence of N on k and δ is still a major open problem, but the problem I wish to concentrate on here is more specific. It is to determine the dependence of N on δ when $k = 3$.

The first progress towards the Erdős-Turán conjecture was due to Roth [Ro1], who showed that the conjecture was true when $k = 3$, and that N could be taken to be $\geqslant \exp\exp(C/\delta)$. This was later improved to $\exp(1/\delta^C)$ by Szemerédi [S3] and Heath-Brown [He]. Very recently, it has been shown by Bourgain [Bo3] that C can be $2 + \epsilon$. More precisely, if $\delta \geqslant c \log\log N(\log N)^{-1/2}$, then A must contain an arithmetic progression of length three.

The best known bound in the other direction is given by an ingenious example discovered by Behrend in 1946 [B]. It shows that the inequality $N \geqslant \exp(C \log(\delta^{-1})^2)$, or equivalently $\delta \geqslant \exp(-c\sqrt{\log N})$, is not enough to guarantee an arithmetic progression of length three.

This situation is interesting for a number of reasons. First, the bounds leave open the question of whether a density $\delta = (\log N)^{-1}$ is enough. This is an important question because $(\log N)^{-1}$ is the density of the primes in the interval $\{1, 2, \ldots, N\}$. A result of van der Corput, based on techniques of Vinogradov, shows that the primes contain infinitely many arithmetic progressions of length three. It is still possible, however, that there is a proof of this fact which depends only on the density of the primes and not on more detailed facts about their distribution.

A second reason is that although Behrend's example gives a very weak bound compared with what is known in the other direction, it does at least show that pure random methods do not give an extremal set. (This point was discussed by the mathematician and the computer in the dialogue in the previous section.)

5. Bounds for Freiman's theorem.

In the statement of Freiman's theorem given earlier, no mention was made of the dependence of K and d on C. This is still far from being completely resolved. See the article of Bilu [Bi] for details about what is known. Improving the known bounds would have important applications. To give some idea of what is *not* known, here are two specific questions. If A is a set of n integers and $|A + A| \leqslant C|A|$, can anything be said about the structure of A if $C = n^\alpha$ for some $\alpha > 0$, or even if $C = \log n$? The existing proofs give no information at all in this situation. For the second question, let A satisfy the same conditions. Does A have a subset B of size cn such that B is contained in a generalized arithmetic progression P of cardinality K and dimension d, where c and K have some reasonable dependence on C and $d = A \log C$ for some absolute constant A? For applications, it would still be very interesting if d could be bounded above by $(\log C)^A$.

These two problems raise difficulties that are similar in spirit to the difficulties in the graph-theoretic problems. Behrend's example shows that there is not an obvious extremal set for Problem 4, and once again there is a conflict between random and explicit techniques which partially accounts for this situation. Similarly, the sets that arise in any consideration of Freiman's theorem are very varied. So for both problems a classification would almost certainly be useful.

More specifically, for both problems one finds oneself wanting to understand subsets $A \subset \mathbb{Z}_N$ of size αN, where α is a constant, or perhaps a number that tends to zero very slowly as N tends to infinity. (Here $\mathbb{Z}_N$

is the group of integers modulo N.) A good way to do this is to look at the Fourier transform of the characteristic function of A, defined by $\hat{A}(r) = \sum_{s \in A} e(rs/N)$, where $e(x)$ is notation for $\exp(2\pi i x)$. The coefficient $\hat{A}(0)$ is just the cardinality of A. If $\hat{A}(r)$ is significantly smaller than $\alpha^2 N$ for every non-zero r, then A behaves in many ways like a random subset of $\mathbb{Z}_N$, where every element is chosen with probability α. We therefore have a good definition of quasirandomness for subsets of $\mathbb{Z}_N$. See [CG] for more details about this.

Notice that the information conveyed by Fourier coefficients about subsets of $\mathbb{Z}_N$ is very similar to the information conveyed by eigenvalues about graphs. This is not a coincidence: if A is a subset of $\mathbb{Z}_N$, one can define a directed graph G_A with vertex set $\mathbb{Z}_N$ by having an edge from x to y if and only if $y - x \in A$, and it is easy to check that the eigenvalues of G_A are the Fourier coefficients of A. The analogies go further. Just as one can try to classify graphs, one can also try to classify subsets of $\mathbb{Z}_N$. For this one needs a notion of "not interestingly different": it turns out that the difference between two sets A and B is for many purposes unimportant if $\hat{A}(r) - \hat{B}(r)$ is small for every r.

Fourier analysis has been an essential part of the best arguments known for analysing subsets A of $\mathbb{Z}_N$ in connection with Problems 4 and 5. The reason is that, because perturbing the Fourier coefficients of a set A does not make an important difference, we only have to worry about the large coefficients. There are very few of these, so we have the following efficient way of encoding the essential information about A: choose a constant β, let $K = \{r : |\hat{A}(r)| \geqslant \beta N\}$ and for $r \in K$ let $f_A(r) = \hat{A}(r)$. It might seem as though mapping sets A to their functions f_A is a ready-made classification scheme, but I would regard "semi-classification" as a better description, because it is not easy to use all the information contained in f_A. The known arguments tend to throw away all information about the set K except for its cardinality, but the structure of K is very important as well. In particular, the behaviour of A depends very much on whether there are simple arithmetical relationships between the elements of K, such as two of them adding to a third. Unfortunately, it is not easy to see how to incorporate this information into arguments, and this seems to be a barrier to further progress on the two problems. Once again, what appears to be missing is a better classification scheme. (Similar remarks apply to eigenvalues of graphs. Useful though they are, for many problems it is important to know how the corresponding eigenvectors fit together, and this information is hard to use.)

The next two problems are designed to show that difficulties similar to the ones discussed so far can arise in many contexts.

6.   A problem about subsets of the n-sphere.   Denote by S_{n-1} the Euclidean unit sphere of $\mathbb{R}_n$. Let $m < n$, let $\epsilon > 0$ and let A be a subset of S_{n-1} that intersects every m-dimensional subspace of $\mathbb{R}_n$. Write A_ϵ for the set $\{x : d(x, A) \leqslant \epsilon\}$. There are interesting questions one can ask about the set A_ϵ. For example, how small can its measure be? For given values of m and ϵ, what is the largest r such that A_ϵ is forced to contain the unit sphere of an r-dimensional subspace?

These questions are hard, because there are two natural ways of constructing a set A with the given property. The first is to take the unit sphere of an $(n - m)$-dimensional subspace, and the second is to take a union of randomly placed spherical caps of some given radius. Again we are in a situation where many other constructions can be devised that mix these two ideas (for example, instead of caps, which are expansions of zero-dimensional sets, one could take expansions of k-dimensional sets for some small k). It seems likely that if A is required to be centrally symmetric, then the measure of A_ϵ is minimized when A is the sphere of an $(n - m)$-dimensional subspace. On the other hand, it does not seem necessary for A_ϵ to contain the sphere of such a large subspace. Perhaps one can do better by allowing A_ϵ to be larger, but less regular.

For some values of m and ϵ, a rough classification of subsets of S_{n-1} might help to determine a good bound for r. One would like to exploit the fact that sets A for which A_ϵ is small have some structure, and deduce from this that r is larger than is implied by measure-theoretic considerations alone.

One possible method of providing *upper* bounds for r is to let B be the unit sphere of an $(n - m)$-dimensional subspace, let ϕ be a continuous antipodal map from S_{n-1} to itself and let $A = \phi(B)$. This makes A a sort of topological sphere of dimension $n - m$. Unfortunately, it is hard to think of a choice for ϕ which will provide a good bound and be easy to analyse.

7.   A problem about layers of the discrete cube.   Let $r < s < t < n$ be positive integers, write $[n]^{(s)}$ for the set of all subsets of $[n] = \{1, 2, \ldots, n\}$ of size s and let $\mathcal{A}$ be a subset of $[n]^{(s)}$. That is, $\mathcal{A}$ is a collection of sets of size s. Write $\partial_r \mathcal{A}$ for the collection of all sets $B \in [n]^{(r)}$ such that $B \subset A$ for some $A \in \mathcal{A}$, and $\partial^t \mathcal{A}$ for the collection of all sets $C \in [n]^{(t)}$ such that $A \subset C$ for some $A \in \mathcal{A}$. Suppose that every set in

$[n]^{(t)}$ contains some set in $\mathcal{A}$, so $\partial^t \mathcal{A} = [n]^{(t)}$. How small can $\partial_r \mathcal{A}$ be?

The nature of this problem depends a great deal on the values of r, s, t and n. Notice that, once again, there are two completely different ways of constructing a set $\mathcal{A}$ that satisfies the conditions of the problem. If one wishes to minimize the cardinality of $\mathcal{A}$, then it is best to choose it randomly. However, amongst all sets $\mathcal{A}$ of a given cardinality, random ones are those for which $\partial_r \mathcal{A}$ is largest. An alternative construction exploits the pigeonhole principle in a simple way. For convenience let us suppose that s divides t and that $k = t/s$ divides n. Partition the set $[n]$ into k equal-sized subsets $V_1 \cup \cdots \cup V_k$ and let $\mathcal{A}$ be the collection of all sets A of size r such that $A \subset V_i$ for some i. Every set of size t intersects at least one of the V_i in at least $n/k = s$ points, so $\mathcal{A}$ satisfies the required condition.

Let us suppose for a moment that s is much smaller than t. If r is only very slightly smaller than s, then it looks as though a random choice of $\mathcal{A}$ is best, since each set of $\mathcal{A}$ will contain only a few sets in $[n]^{(r)}$. If on the other hand r is much smaller than s, then the explicit construction turns out to be better. Between these two extremes, there is a range of values of r where it is not obvious which construction is better, or whether something "in between" is more suitable. In fact, I do not even know how to prove what appears to be the case in the extreme situations – that the examples I have sketched are best possible.

This problem is perhaps slightly artificial, but it is a simple example of a type of difficulty I have met several times when thinking about other questions, and I doubt whether I am alone in this. Whatever one thinks about the interest or otherwise of the problem, it is hard to imagine a solution being dull. Once again, it looks as though a rough classification would be very useful, this time of subsets of $[n]^{(s)}$. (There do exist versions of Szemerédi's regularity lemma when s is fixed, but they would not help if one took $s = \sqrt{n}$, for example.)

8. Finding rough classifications. Arising from the previous problems is the general problem of finding new and useful rough classifications. Amongst the objects that one might like to classify are the following.

(i) Graphs. The problem here is to find a useful replacement of Szemerédi's regularity lemma. Because the bad bound in the lemma is (in a very weak sense!) tight, replacing the lemma would mean replacing the description it gives of a general graph by something more complicated, in return for much better bounds. Another interesting project is to find alternative proofs for results that use the lemma

(just as it is interesting to replace a proof that relies on the classification of finite simple groups by one that does not). See [KoS], [GrRR]
and [G2, Prop. 12] for some examples of how it can be avoided. For
more general information on the regularity lemma and its variants,
see [KoS].

(ii) Triangle-free graphs. Two ways of constructing these are to use random methods, or to take a given triangle-free graph H with vertices
$x_1, \ldots, x_k$ and to define a graph G with vertex set partitioned into
sets $V_1, \ldots, V_k$ by joining $x \in V_i$ to $y \in V_j$ if and only if $x_i x_j$ is an edge
of H. Combining these methods, and other simple operations such
as subgraphs and disjoint unions, yields a large class of triangle-free
graphs. Do techniques such as these get anywhere near to generating
all triangle-free graphs?

(iii) Sparse graphs. It has long been recognised that a major drawback
with Szemerédi's regularity lemma is that the only graphs for which it
gives any information are those with a number of edges proportional
to $\binom{n}{2}$, while many problems of great interest concern graphs with,
say, n^α edges for some $\alpha \in (1, 2)$. Some variants of the regularity
lemma do exist for such sparse graphs, but they fall well short of
anything that one might wish to call a classification.

(iv) Subsets of Abelian groups, especially $\mathbb{Z}_N$ and $\mathbb{Z}$, with two sets regarded as similar if their Fourier transforms are close.

(v) Subsets of $\mathbb{Z}_N$, with two sets regarded as similar if their characteristic
functions are close in the following norm:

$$\|f\| = N^{-4} \sum_{x,a,b,c \in \mathbb{Z}_N} \prod_{\epsilon \in \{0,1\}^3} f_\epsilon(x - \epsilon_1 a - \epsilon_2 b - \epsilon_3 c).$$

Here f_ϵ denotes f if $\epsilon_1 + \epsilon_2 + \epsilon_3$ is even and the complex conjugate
of f otherwise. It turns out that, while the large Fourier coefficients
of a set A contain all the information about arithmetic progressions
of length three, they do not do so for longer progressions. If A and
B are close in the above norm, then they will contain approximately
the same number of arithmetic progressions of length four – hence its
importance. For more details see [G2]. A classification with respect
to this norm would necessarily be much more sophisticated than a
classification with respect to closeness of Fourier coefficients.

(vi) Subsets of metric spaces such as $[0,1]^2$ and S_{n-1} with two sets regarded as similar if their metric properties are roughly the same.
There are many ways of making this question precise, some of which

have been studied extensively. One possibility is to say that subsets
A and B of a metric space X are similar if there is a small $\epsilon > 0$ and
there are 1-Lipschitz maps $\phi : A \to X$ and $\psi : B \to X$ such that
every point of B is within ϵ of some point of ϕA, every point of A
is within ϵ of some point of ψB, and $d(x, \psi\phi x)$ and $d(y, \phi\psi y)$ are at
most ϵ for every $x \in A$ and $y \in B$.

(vii) Subsets of combinatorial structures such as the discrete cube $\{0, 1\}^n$
and $[n]^{(r)}$.

4   A Further Selection of Problems

Every problem in this section had to satisfy two criteria. First, I myself
should have spent at least an hour trying to solve it, and second, I should
have at least some intention of spending more time on it in the future. These
were enough to rule out most of the interesting problems in mathematics,
leaving me with a manageable list.

9.   The Riemann hypothesis.   This is so well known that I shall not
even bother to state it here. I include it because I do not wish to make
some kind of statement by not doing so. However, I have nothing much
to say about it. Instead, I refer the reader to the vast literature on the
subject, including parts of the articles of Connes, Iwaniec and Sarnak in
this volume.

10.   The Kakeya problem.   This is closely related to the Riemann
hypothesis. It has many formulations, of which here are three.

(i) Let A be a subset of $\mathbb{R}^d$ such that for every line L in $\mathbb{R}^d$, A contains a
line L' parallel to L. Such a set is often called a Kakeya set, or a Besicovitch
set. Besicovitch showed that the measure of A can be zero, but all known
examples have Hausdorff dimension d. Bourgain [Bo4] has shown that the
Hausdorff dimension of A must be at least $\frac{13d}{25} + \frac{12}{25}$, which is a significant
improvement on a relatively easy bound of $\frac{d}{2} + 1$. (Very recently this bound
was improved further by Katz and Tao to $\frac{6d}{11} + \frac{5}{11}$ [KaT].) It is known that
the dimension must be d when $d = 2$, but for $d > 2$ the problem is open.
The first non-trivial lower bound when $d = 3$ was discovered by Bourgain
[Bo2]. His argument showed that the dimension of a three-dimensional
Kakeya set was at least $7/3$. This was improved by Wolff to $5/2$ [W].
This bound was recently improved very slightly by Katz, Laba and Tao
[KaLT], an interesting result as $5/2$ was a natural limit in Wolff's proof
and surpassing it was very difficult.

(ii) Let S be the square $\{(x, y, z) \in \mathbb{R}^3 : z = 0, 0 \leqslant x, y \leqslant 1\}$ and let P be the point $(1/2, 1/2, 1)$. Now divide S into n^2 subsquares S_{ij} of sidelength n^{-1} in the obvious way. For each of these subsquares, form a cone C_{ij} by taking the convex hull of S_{ij} and P. Translate each cone C_{ij} in an xy-direction (or in other words, slide its base to somewhere else in the xy-plane) and call the translated cone C'_{ij}. Then how small can the volume of $\bigcup_{i,j=1}^n C'_{ij}$ be? If the smallest possible volume behaves like $n^{-\alpha}$ for some $\alpha > 0$, then there is a Kakeya set in $\mathbb{R}^3$ with Hausdorff dimension less than three. The converse is also true. The obvious d-dimensional analogue of this equivalence holds as well.

(iii) Let m and r be integers, and for every d between 1 and r, let $P_d \subset \mathbb{Z}$ be an arithmetic progression of length m and common difference d. How small can the union $\bigcup_{d=1}^r P_d$ be? In particular, if $r = m^C$ for some fixed integer $C \geqslant 2$, then can one improve on the trivial bound of mr by a factor of m^α for some $\alpha > 0$? Once again, a construction that achieved this could be translated into an example of a Kakeya set with less than full dimension.

The third version of the problem gives some clue that it has a number-theoretical content. A close connection with a conjecture of Montgomery (see [Mo, Chapter 7]), and hence with the zeros of the zeta function, was uncovered by Bourgain in his modestly titled paper [Bo1]. Ever since a famous paper of Fefferman [F], it has been realized that Kakeya sets are also of great importance in harmonic analysis. A solution to the Kakeya problem would therefore be a major advance for several different reasons.

11. Does P equal NP? I have already mentioned this problem in §2. Since not all mathematicians feel comfortable stating it, let me give what I find the cleanest definition of the class of NP problems. First, it is convenient to think not about problems but about Boolean functions, that is, functions $\phi : \{0, 1\}^n \to \{0, 1\}$. A problem such as the travelling salesman problem can easily be regarded as such a function: encode your graph G with n vertices as a 01-sequence η_G of length $\binom{n}{2}$ in an obvious way and define $\phi(\eta_G)$ to be 1 if G contains a Hamilton cycle and 0 otherwise. A Boolean function ϕ belongs to P if there is a polynomial-time algorithm for computing ϕ. (Strictly speaking, one should talk about a sequence ϕ_n of functions, where ϕ_n has $f(n)$ variables, and the time taken to compute ϕ_n is bounded above by a polynomial function of $f(n)$.) It belongs to NP if it is a *projection* of a function in P, in the following sense: there is an integer m bounded above by a polynomial in n and a function $\psi : \{0, 1\}^m \times \{0, 1\}^n$ which belongs to P such that $\phi(\eta) = 1$ if and only if there exists $\xi \in \{0, 1\}^m$

such that $\psi(\xi, \eta) = 1$.

For the travelling salesman problem, ξ will represent an ordering of the vertices, η a graph, and $\psi(\xi, \eta) = 1$ if and only if the ordering of the vertices given by ξ defines a Hamilton cycle in the graph given by η. This is very easy to check, so ψ belongs to P, and the function ϕ defined earlier, which is 1 if and only if $\psi(\xi, \eta) = 1$ for some ξ, is in NP.

One further definition, which I shall not give precisely, is that of an NP-complete function. An important discovery of Cook [Co] and Karp [K], which explains the interest in the P=NP problem, is that many NP problems (or corresponding Boolean functions) are *universal*, in the sense that a polynomial-time algorithm for solving one of them can be converted into a polynomial-time algorithm for solving an arbitrary problem in NP. Such a problem is called NP-complete. In fact, hundreds of such problems are now known, and many of them are of direct practical importance.

The P=NP problem may look off-putting because the set of all functions Turing-computable in polynomial time is somehow not very mathematical. I have sometimes seen it written that the problem is difficult because it is hard to rule out *all* possible polynomial-time algorithms. However, there is a very clean and completely mathematical reformulation of the problem in terms of so-called circuit complexity which shows that this is not really the difficulty at all. I shall give the definition in an equivalent form.

Let $\phi : \{0, 1\}^n \to \{0, 1\}$ be a Boolean function and let $A = \{\eta : \phi(\eta) = 1\}$. For each $1 \leqslant i \leqslant n$, let $E_i = \{\eta : \eta_i = 1\}$, and call the sets E_i and their complements *elementary* sets. The *circuit complexity* of ϕ, or A, is the length N of the shortest sequence $A_1, A_2, \ldots, A_N$ such that $A_N = A$ and each A_i is either an elementary set, the complement of some earlier A_j or the union or intersection of two earlier A_js.

A problem which is almost, but not quite, equivalent to whether P=NP is the following. Let ϕ be a Boolean function in NP. Must the circuit complexity of ϕ be bounded above by a polynomial in the number of variables? A polynomial-time algorithm for computing ϕ can be converted into a sequence $A_1, \ldots, A_N$ of polynomial length, so a superpolynomial bound for the circuit complexity would imply that P and NP are not equal. It is theoretically possible that the converse to this is false, but it is hard to envisage a construction of a sequence $A_1, \ldots, A_N$ with so little regularity that it, or some adaptation of it, could not be efficiently generated algorithmically. I am therefore convinced that circuit complexity is the right approach to the problem. (This is the standard view of theoretical computer scientists, but not of all mathematicians.)

Unfortunately, it does not make the problem easy, and the best known lower bound for the circuit complexity of an NP-function is still only linear. A beautiful and important recent paper of Razborov and Rudich [RR] goes a long way towards explaining why the problem is difficult, and is required reading for anybody who would like to prove that P≠NP. A natural approach to proving lower bounds for circuit complexity is to define some measure of simplicity, prove that sets A with small circuit complexity must be simple, and prove that the set A corresponding to some NP-function is not simple. Razborov and Rudich define a notion of "natural proof", which basically means a proof resulting from a "natural" measure of simplicity. They show that all the known bounds in circuit complexity have been achieved using natural proofs in their sense, and they then show that any natural proof that P≠NP could be converted into a polynomial-time algorithm for factorizing large integers. This is convincing evidence that a proof that P≠NP would have to be very strange indeed. For more details, see the contributions of Razborov and Wigderson in this volume, or the paper of Razborov and Rudich itself, which is well written and not hard to read.

The following problem (mentioned by Wigderson at the conference) has a similar flavour to the P=NP problem, but has nothing to do with computers and appears at first glance to be much simpler. Let M be an $n \times n$ invertible matrix over $\mathbb{F}_2$. Gaussian elimination tells us that M can be converted into the identity matrix by a sequence of at most n^2 elementary row operations. Moreover, a simple counting argument (there are fewer than $2n^2$ elementary row operations and almost 2^{n^2} non-singular matrices) shows that almost all matrices need at least $cn^2/\log n$ operations. The problem is to give an explicit example of a matrix that needs a superlinear number. ("Explicit" does not mean "unambiguously defined", or one could just enumerate all the matrices and take the first one that worked. One possible interpretation is that it should be possible to generate an explicit matrix, or rather sequence of matrices, with a polynomial-time algorithm.) This problem remains stubbornly open, and a solution might well shed light on the P=NP problem itself.

12. An effective version of Roth's theorem.

By Roth's theorem I now mean not his theorem on arithmetic progressions but his much more famous theorem on rational approximation of algebraic numbers [Ro2]. This states that for every algebraic number α and every $\epsilon > 0$ there exists

$c > 0$ such that for any pair p, q of positive integers we have the inequality $|\alpha - p/q| \geqslant c/q^{2+\epsilon}$. This is a very considerable strengthening of Liouville's theorem. The problem is that the proof gives no information at all about the dependence of c on α and ϵ. The same is true of earlier, weaker results of Thue and Siegel.

The known results in this direction are very modest. For example, Baker [B] has proved an effective version of Roth's theorem in the case $\alpha = 2^{1/3}$, obtaining the inequality $|2^{1/3} - p/q| \geqslant 10^{-6}/q^{2.995}$, which should be compared with the lower bound from Liouville's theorem of c/q^3, where c can be taken to be $1/16$.

A closely related problem, mentioned by Furstenberg in his talk at this conference, concerns the continued-fraction expansion of algebraic numbers. If α is a quadratic irrational, then everything is known, and the partial quotients are eventually periodic. Roth's theorem implies that the partial quotients of an algebraic number cannot grow too quickly, but, except in the quadratic case, there is no example for which it is known whether they are bounded.

13. Littlewood's conjecture.
This problem was discussed by Margulis at the conference. Littlewood conjectured that if α and β are any two real numbers and if $\|x\|$ denotes the distance from x to the nearest integer, then $\liminf_{n \in \mathbb{N}} n \, \|n\alpha\| \, \|n\beta\| = 0$.

To understand the problem, consider a quadratic irrational α. Liouville's theorem is equivalent to the statement that $\|n\alpha\|$ is always bounded below by cn^{-1}. On the other hand, usually $\|n\alpha\|$ is much bigger than this. One can therefore ask whether it is possible to find two numbers α and β, neither of which can be well approximated by rationals, with the further property that $\|n\beta\|$ is large whenever $\|n\alpha\|$ is small and vice-versa. Littlewood's conjecture is a precise formulation of this question.

The existence of such a pair of numbers α, β would have the more combinatorial consequence that for every n one could find a set X of n points in $[0, 1]^3$ and an absolute constant $c > 0$ such that the inequality $\prod_{i=1}^{3} |x_i - y_i| \geqslant cn^{-1}$ held for any two distinct points $x, y \in X$. Indeed, writing $\langle t \rangle$ for the fractional part of t, a suitable set would be $\{(mn^{-1}, \langle \alpha m \rangle, \langle \beta m \rangle) : 1 \leqslant m \leqslant n\}$. As far as I know, the existence of such a set X is an open question, so it is possible that Littlewood's conjecture, if it is true, has a combinatorial proof. A simple argument shows that one can find $n/\log n$ points with the given property. Any improvement would be interesting.

If Littlewood's conjecture is false, then a good candidate for a counterexample would be the pair $(\sqrt{2}, \sqrt{3})$. It seems to be very hard to analyse concrete examples such as this, and perhaps this is the main difficulty with the problem.

14.   Self-avoiding random walks.   In common with many mathematicians, I am fascinated by the long list of results in statistical physics that have been obtained by physicists using non-rigorous arguments. One of the most appealing source of problems is the area of self-avoiding random walks. A self-avoiding walk of length n is a path $x_0, x_1, \ldots, x_n$ of vertices in the lattice $\mathbb{Z}^d$, where $x_0 = 0$, x_{i-1} and x_i are adjacent for $1 \leqslant i \leqslant n$ and there are no repeats. If one places the uniform distribution on the set of all self-avoiding walks of length n in $\mathbb{Z}^2$ or $\mathbb{Z}^3$, then one obtains an important probabilistic model about which a great deal is known to physicists and almost nothing to mathematicians. I will mention just a few problems.

Nearly all the most interesting questions are special cases of the following vague one: what does a typical self-avoiding walk of length n look like? Perhaps the most obvious question is the expected distance from the origin to x_n. In two dimensions, the arguments of physicists show that the answer ought to be $n^{3/4}$, but no upper bound of the form $n^{1-\epsilon}$ has been proved rigorously. The situation for lower bounds is even worse. One might expect a trivial argument to show a lower bound of $cn^{1/2}$, since a self-avoiding walk surely travels further than a pure random walk. However, it is not known how to prove rigorously that a typical self-avoiding walk does not tend to fold back on itself. Of course, most points of a self-avoiding walk must have distance $cn^{1/2}$ from the origin by a trivial counting argument, so this last question is specifically to do with the end point.

Another question concerns the number $f(n)$ of self-avoiding walks of length n. It is easy to see that f is submultiplicative, so a standard argument shows that $f(n) = C^n g(n)$ for some constant C, known as the connective constant, and some function g that grows subexponentially. The exact value of C is not known, but this is not a particularly interesting problem for two reasons. First, C can in principle be determined to any accuracy by an exact counting of all walks of some suitable length n_0 (although this rapidly becomes impossible in practice). More importantly, C does not have a property known as *universality*.

One can change the model of self-avoiding walks in many ways, such as changing the lattice in which it lives. It is known by physicists that certain parameters are unaffected by such changes. For example, the expected end-to-end distance is always proportional to $n^{3/4}$. These unchanging parameters are called universal, and are regarded as having physical significance. The connective constant is not universal in this sense, as it certainly does depend on the underlying lattice.

On the other hand, the function g does seem to be universal. Once again, much more is known to physicists than to mathematicians: they tell us that $g(n)$ grows like $n^{11/32}$. One might at first wonder how one could hope to study the function g when the connective constant C is not known. However, it is a simple exercise to show that g has the following interpretation: $g(n)$ grows like n^{α} if and only if the probability that two self-avoiding random walks of length n do not intersect (except at the origin) behaves like $n^{-\alpha}$. This shows that g does have some sort of independent significance. From a pure mathematical point of view, it is a very interesting open problem to prove that g is bounded above by a polynomial. The best known bound in two dimensions is due to Hammersley and Welsh [HaW], who showed by an ingenious elementary argument that $g(n) \leqslant \exp(cn^{1/2})$.

The reason the hypothesis of universality is plausible is that one would expect the macroscopic behaviour of a very long self-avoiding walk not to depend too critically on the precise behaviour at very small magnitude scales. Connected with this belief is the view that there should be some sort of appropriately scaled limit of self-avoiding random walks, playing the role that Brownian motion and Wiener measure play for ordinary random walks. This is another important open problem.

These problems are all at their most interesting in two and three dimensions. Since the Hausdorff dimension of a Brownian path is two, one does not expect the self-avoidance constraint to be all that significant in four dimensions or more. This intuition has in fact been justified rigorously when the dimension is at least five, by Hara and Slade [HarS1,2]. Even though their results are not very surprising, they are a major achievement and the proofs are not at all easy. For much more detail on all these problems and results, see the excellent book of Madras and Slade [MS].

15.   Untying the unknot.   There has been a great deal of excitement about knot theory in the last twenty years, including the discovery of important new knot invariants. However, the following basic problem is open: does there exist a polynomial-time algorithm for recognising a diagram of the unknot, and for producing a sequence of Reidemeister moves that unties it? (The time is a function of the number of crossings in the diagram.) If such an algorithm could be found, it would open up a completely new approach to recognising and classifying knots. I know at least one expert in knot theory who strongly believes in the existence of such an algorithm. It is by no means trivial to find *any* algorithm for recognising the unknot, but this has been done by Haken [H]. I myself have performed the following scientific experiment. I take a long piece of string, tie the two ends together with a small knot, and then entangle the resulting loop, trying to make it hard to disentangle (but without pulling anything tight). I have never managed to cause myself the slightest difficulty.

16.   The Knaster hypothesis. This is the following question. Let $f : S_n \to \mathbb{R}$ be continuous and let $x_1, \ldots, x_{n+1}$ be any $n + 1$ points in S_n. Must there exist an orthogonal map α such that the values $f(\alpha x_i)$ are all equal? When $n = 1$ a positive answer follows trivially from the intermediate value theorem. I think the conjecture is known to be true when $n = 2$ but open for $n \geqslant 3$. It is also known for certain very special sets of points $x_1, \ldots, x_{n+1}$.

One motivation for the Knaster hypothesis, aside from the intrinsic appeal of the question, is that, as observed by Milman [Mi], a positive answer would provide a completely new proof of one of the central results of Banach space theory – Dvoretzky's theorem [D]. This states that for any k and $\epsilon > 0$ there exists n such that every n-dimensional normed space has a k-dimensional subspace whose Banach-Mazur distance from a Hilbert space is at most $1 + \epsilon$. (Geometrically, it says that an n-dimensional centrally symmetric convex body has an almost ellipsoidal k-dimensional cross section.) Dvoretzky's theorem follows from the Knaster hypothesis as follows. Consider S_{n-1} as the unit sphere of ℓ_2^n, let $x_1, \ldots, x_n$ be a δ-net of the unit sphere of a k-dimensional subspace V, and let f be the restriction of an arbitrary norm $\|.\|$ on $\mathbb{R}^n$ to S_{n-1}. By the Knaster hypothesis, there is a rotation α such that all the values $\|\alpha x_i\|$ are equal. Standard arguments then show that all points x in the subspace αV have roughly the same ratio $\|x\| / \|x\|_2$. This argument, when made precise, gives the best possible estimate for the asymptotic dependence of n on ϵ for fixed k, which is still an open problem.

Certain natural approaches to the Knaster hypothesis turn out to fail for good reasons, and many people have the feeling that it may well be false. However, even a considerable weakening, such as proving it for $\sqrt{n}$ points in S_n, would be useful for Dvoretzky's theorem.

17.   The density Hales-Jewett theorem. The Hales-Jewett theorem concerns colourings of the set $\{1, 2, \ldots, k\}^N$. If x belongs to this set, A is a subset of $\{1, 2, \ldots, N\}$ and $1 \leqslant j \leqslant k$, then let us write $x \oplus jA$ for the point y obtained from x by changing x_i to j if $i \in A$ and leaving it unaltered otherwise. A *combinatorial line* is a set of the form $\{x \oplus jA : 1 \leqslant j \leqslant k\}$. For example, when $k = 3$ and $N = 5$, then the point $x = (3, 3, 1, 1, 2)$ and the set $\{1, 3, 5\}$ give rise to the combinatorial line

$$\{(1, 3, 1, 1, 1), (2, 3, 2, 1, 2), (3, 3, 3, 1, 3)\} .$$

Notice that if the set $\{1, 2, \ldots, k\}^N$ is imagined as an N-dimensional grid, then combinatorial lines are special cases of geometrical lines, that is, sets of k collinear points.

The Hales-Jewett theorem states that for any k and r there exists N such that if the set $\{1, 2, \ldots, k\}^N$ is coloured with r colours, then there must be a combinatorial line consisting of points of the same colour. This theorem is easily seen to be a strengthening of van der Waerden's theorem, and like van der Waerden's theorem it has a density version: for every k and every $\delta > 0$ there exists N such that every subset $A \subset \{1, 2, \ldots, k\}^N$ with at least δk^N elements contains a combinatorial line.

This density theorem is due to Furstenberg and Katznelson [FuK] and makes heavy use of ergodic theory. Moreover, no proof is known that does not use these techniques. It would be very interesting to find a non-ergodic proof even in the case $k = 3$. From the theorem in this case it is a simple exercise to deduce the following result. For every $\delta > 0$ there exists N such that every subset of the two-dimensional grid $\{1, 2, \ldots, N\}^2$ of size at least δN^2 contains a triple of the form $\{(x, y), (x, y + d), (x + d, y)\}$, that is, an aligned isosceles rightangled triangle. Even this statement lacks any proof that gives a reasonable upper bound for N in terms of δ.

18. Danzer's problem. Is it possible to find a set X of points in $\mathbb{R}^2$ such that every convex body of area 1 contains at least one point of X and not more than C points, where C is an absolute constant? If K is a convex body of area 1, then there are homothetic rectangles R_1 and R_2 such that $R_1 \subset K \subset R_2$ and such that the area of R_2 is at most four times that of R_1. It follows that it is enough to consider rectangles in the statement of the problem.

There is a small connection between this problem and some of the questions in the previous section, because the obvious random and explicit methods for constructing such a set X both fail, but for very different reasons.

Danzer's actual question is weaker than the above. He asks for a set X which intersects every convex body of area 1 and also satisfies the inequality

$|X \cap RD| \leqslant CR^2$ for every R, where D is the unit disc about the origin. A beautiful paper of Bambah and Woods [BW] shows that X cannot be a finite union of translates of lattices, thus ruling out what at first appears to be a promising source of potential examples.

19. Billiards in polygons. There is a famous example, known as Kafka's study, which shows that one can design a connected (but non-convex) room such that, even if the walls are entirely lined with mirrors, no single light source can illuminate the whole of the room. The example exploits the fact that a beam of light which passes between the two foci of an ellipse will return between the two foci after it has been reflected off the boundary of the ellipse. It is not known whether there is a polygonal example.

A closely related question is the following. Suppose an infinitesimally small billiard ball is set moving on a frictionless (and pocketless) polygonal table. Will it return arbitrarily close to its starting point and direction? The answer is known to be positive if the angles between all the sides of the polygon are rational multiples of π, as then there are only finitely many directions to consider, and the problem can easily be reduced to a known and easy result about so-called interval exchange maps.

Let $[0, 1)$ be partitioned into m half-open intervals in two different ways, $I_1 \cup \cdots \cup I_m$ and $J_1 \cup \cdots \cup J_m$ in such a way that I_r and J_r have the same length for every $r \leqslant m$. Let ϕ_r be the translation of I_r to J_r. Let ϕ be the map which takes $x \in I_r$ to $\phi_r(x)$. An *interval exchange map* from $[0, 1)$ to $[0, 1)$ is any map of this kind. It is quite straightforward to prove that such maps are recurrent, in the sense that for every x and every $\epsilon > 0$ there exists $N > 0$ such that the distance from $\phi^N x$ to x is at most ϵ.

I have asked around and found nobody who can tell me whether the obvious two-dimensional generalization of the above result is true or false. That is, suppose one chops a square into rectangles and rearranges them (by translation) so that they form the original square again. Is the resulting rectangle exchange map recurrent? I believe that it is, but some experts in dynamical systems to whom I have asked the problem have the opposite instinct.

References

[AM]   N. ALON, V.D. MILMAN, Eigenvalues, expanders and superconcentrators, Proc. 25th Annual FOCS, Singer Island, FL, IEEE, New York (1984), 320–322.

[BW]   R.P. BAMBAH, A.C. WOODS, On a problem of Danzer, Pacific J. Math. 37 (1971), 295–301.

[B]   F.A. BEHREND, On sets of integers which contain no three in arithmetic progression, Proc. Nat. Acad. Sci. 23 (1946), 331–332.

[Bi]   Y. BILU, Structure of sets with small sumset, Astérisque 258 (1999), 77–108.

[Bo1] J. BOURGAIN, Remarks on Montgomery's conjectures on Dirichlet series, Geometric Aspects of Functional Analysis (1989-1990), Springer Lecture Notes in Mathematics 1469 (1991), 153–165.

[Bo2] J. BOURGAIN, Besicovitch type maximal operators and applications to Fourier analysis, GAFA 1 (1991), 147–187.

[Bo3] J. BOURGAIN, On triples in arithmetic progression, GAFA 9:5 (1999), 968–984.

[Bo4] J. BOURGAIN, On the dimension of Kakeya sets and related maximal inequalities, GAFA 9:2 (1999), 256–282.

[Bu]   George Bush, Interview, Time, 26th Jan. 1987.

[CG]   F.R.K. CHUNG, R.L. GRAHAM, Quasi-random subsets of $\mathbb{Z}_n$, J. Comb. Th. A 61 (1992), 64–86.

[CGW] F.R.K. CHUNG, R.L. GRAHAM, R.M. WILSON, Quasi-random graphs, Combinatorica 9 (1989), 345–362.

[Co]   S.A. COOK, The complexity of theorem proving procedures, Proc. 3rd Annual ACM Symposium on the Theory of Computing (1971), 151–158.

[D]   A. DVORETZKY, Some results on convex bodies and Banach spaces, Proc. Symp. on Linear Spaces, Jerusalem (1961), 123–160.

[E]   P. ERDŐS, Some remarks on the theory of graphs, Bull. Amer. Math. Soc. 53 (1947), 292–294.

[ET]   P. ERDŐS, P. TURÁN, On some sequences of integers, J. London Math. Soc. 11 (1936), 261–264.

[F]   C. FEFFERMAN, The multiplier problem for the ball, Annals of Math. 94 (1971), 330–336.

[FuK] H. FURSTENBERG, Y. KATZNELSON, A density version of the Hales-Jewett theorem, J. D'Analyse Math. 57 (1991), 64–119.

[Fr1] G.A. FREIMAN, Foundations of a Structural Theory of Set Addition (in Russian), Kazan Gos. Ped. Inst., Kazan 1966.

[Fr2] G.A. FREIMAN, Foundations of a Structural Theory of Set Addition, Translations of Mathematical Monographs 37, Amer. Math. Soc., Providence, R.I., USA, 1973.

[G1]   W.T. GOWERS, Lower bounds of tower type for Szemerédi's uniformity lemma, GAFA 7 (1997), 322–337.

[G2]   W.T. GOWERS, A new proof of Szemerédi's theorem for arithmetic progressions of length four, GAFA 8 (1998), 529–551.

[GrRR] R.L. GRAHAM, V. RÖDL, A. RUCIŃSKI, On graphs with linear Ramsey numbers, preprint.

[H]   W. HAKEN, Theorie der Normalflächen, ein Isotopiekriterium für den Kreisnoten, Acta Math. 105, 245–375.

[HaW]   J.M. HAMMERSLEY, D.J.A. WELSH, Further results on the rate of convergence to the connective constant of the hypercubical lattice, Quart. J. Math. Oxford Ser. (2) 13 (1962), 108–110.

[HarS1] T. HARA, G. SLADE, Self-avoiding walk in five or more dimensions, Comm. Math. Phys. 147 (1992), 101–136.

[HarS2] T. HARA, G. SLADE, The lace expansion for self-avoiding walk in five or more dimensions, Rev. Math. Phys. 4 (1992), 235–327.

[He]   D.R. HEATH-BROWN, Integer sets containing no arithmetic progressions, J. London Math. Soc. (2) 35 (1987), 385–394.

[K]   R.M. KARP, Reducibility among combinatorial problems, Complexity of Computer Computations, Proc. Sympos., IBM Thomas J. Watson Res. Centr, Yorktown Heights, N.Y., 1972, (R.E. Miller, J.W. Thatcher, eds.) Plenum Press, New York 1972, 85–103.

[KaLT] N.H. KATZ, I. LABA, T. TAO, An improved bound on the Minkowski dimension of Besicovitch sets in $\mathbb{R}^3$, Annals of Math., to appear.

[KaT]   N.H. KATZ, T. TAO, A new bound on partial sum-sets and difference-sets, and applications to the Kakeya conjecture, submitted.

[KoS]   J. KOMLÓS, M. SIMONOVITS, Szemerédi's Regularity Lemma and its applications in Graph Theory, in "Combinatorics, Paul Erdős is 80 (Vol 2)", Bolyai Society Math. Studies 2, 295–352, Kesthely (Hungary) 1993, Budapest 1996.

[LPS]   A. LUBOTZKY, R. PHILLIPS, P. SARNAK, Explicit expanders and the Ramanujan conjectures, Proceedings of the 18th ACM Symposium on the Theory of Computing 1986, 240–246; also Combinatorica 8 (1988), 261–277.

[MS]   N. MADRAS, G. SLADE, The Self-Avoiding Random Walk, Birkhäuser, Boston, 1992.

[Mi]   V.D. MILMAN, A few observations on the connections between local theory and some other fields, in Geometric Aspects of Functional Analysis, Israel seminar (GAFA) 1986-1987 (J. Lindenstrauss, V.D. Milman, eds.), Springer LNM 1317, (1988), 283–289.

[Mo]   H.L. MONTGOMERY, Ten Lectures on the Interface Between Analytic Number Theory and Harmonic Analysis, CBMS Regional Conference Series in Math. 84, AMS 1994.

[P]   G. POLYA, Mathematics and Plausible Reasoning, Vols. I and II, Princeton University Press, 1954.

[RR]   A.A. RAZBOROV, S. RUDICH, Natural proofs, in 26th Annual ACM Symposium on the Theory of Computing (STOC '94, Montreal, PQ, 1994); also J. Comput. System Sci. 55 (1997), 24–35.

[Ro1]   K. ROTH, On certain sets of integers, J. London Math. Soc. 28 (1953), 245–252.

[Ro2]   K. ROTH, Rational approximations to algebraic numbers, Mathematika 2 (1955), 1–20 (with corrigendum p. 168).

[Ru]   I.Z. RUZSA, Generalized arithmetic progressions and sumsets, Acta Math. Hungar. 65 (1995), 379–388.

[S1]   E. SZEMERÉDI, Regular partitions of graphs, Colloques Internationaux C.N.R.S. 260 – Problèmes Combinatoires et Théorie des Graphes, Orsay 1976, 399–401.

[S2]   E. SZEMERÉDI, On sets of integers containing no k elements in arithmetic progression, Acta Arith. 27 (1975), 299–345.

[S3]   E. SZEMERÉDI, Integer sets containing no arithmetic progressions, Acta Math. Hungar. 56 (1990), 155–158.

[T1]   A.G. THOMASON, Pseudo-random graphs, Proceedings of Random Graphs, Poznán 1985 (M. Karonski, ed.), Annals of Discrete Mathematics 33, 307–331.

[T2]   A.G. THOMASON, A disproof of a conjecture of Erdős in Ramsey theory, J. London Math. Soc. 39 (1989), 246–255.

[T3]   A.G. THOMASON, Graph products and monochromatic multiplicities, Combinatorica 17 (1997), 125–134.

[W]   T.H. WOLFF, An improved bound for Kakeya type maximal functions, Revista Mat. Iberoamericana 11 (1995), 651–674.

W.T. GOWERS, Centre for Mathematical Sciences, Wilberforce Road, Cambridge CB3 0WA, UK

GAFA, Geom. funct. anal.
Special Volume – GAFA2000, 118 – 161
1016-443X/00/S10118-44 $ 1.50+0.20/0

GAFA Geometric And Functional Analysis

SPACES AND QUESTIONS

Misha Gromov

1 Dawn of Space

Our Euclidean intuition, probably inherited from ancient primates, might have grown out of the first seeds of geometry in the motor control systems of early animals who were brought up to sea and then to land by the Cambrian explosion half a billion years ago. The primates' brain had been idling for 30-40 million years. Suddenly, in a flash of one million years, it exploded into growth under the relentless pressure of sexual-social competition and sprouted a massive neocortex (70% neurons in humans) with an inexplicable capability for language, sequential reasoning and generation of mathematical ideas. Then Man came and laid down space on papyrus in a string of axioms, lemmas and theorems around 300 B.C. in Alexandria.

Projected to words, the brain's model of space began to evolve by dropping, modifying and generalizing its axioms. The Parallel Postulate fell first: Gauss, Schweikart, Lobachevski,[1] Bolyai (who else?) came to the conclusion that *there is a unique non-trivial one-parameter deformation of the metric on $\mathbb{R}^3$ keeping the space fully homogeneous.*[2]

It is believed, that Gauss, who convinced himself of the validity of hyperbolic geometry somewhere between 1808 and 1818, was disconcerted by the absence of a Euclidean realization of the hyperbolic plane H^2. By that time, he must have had a clear picture of the geometry of surfaces in $\mathbb{R}^3$ (exposed in his "Disquisitones circa superficies curvas" in 1827), where the (intrinsic) distance between two points on a surface is defined as the length of the shortest (better to say "infimal") curve *in* the surface between these points. (This idea must have been imprinted by Nature in the brain, as most animals routinely choose shortest cuts on rugged terrains.) Gauss discovered the following powerful and efficient criterion for isometry between surfaces, distinguishing, for example, a piece of a round sphere $S^2 \subset \mathbb{R}^3$

---

[1] Incidentally, the first mathematics teacher of Gauss ($\approx$ 1790), Johann Martin Bartels, later on became the teacher of Lobachevski ($\approx$ 1810) in Kazan.

[2] A metric space X is *fully homogeneous* if every partial isometry $X \supset \Delta \leftrightarrow \Delta' \subset X$ extends to a full isometry of X (as for Euclid's triangles with equal sides in $\mathbb{R}^2$).

from an arbitrarily bent sheet of paper (retaining its intrinsic Euclideanness under bendings).

Map a surface $S \subset \mathbb{R}^3$ to the unit sphere S^2 by taking the vectors $\overline{\nu}(s) \in S^2$ parallel to the unit normal vectors $\nu(s)$, $s \in S$. If S is C^2-smooth,

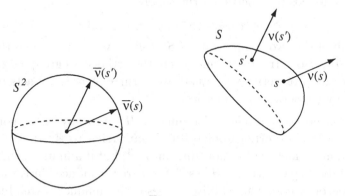

Figure 1

the Gauss map $G : S \to S^2$ for $s \mapsto \overline{\nu}(s)$ is C^1 and its Jacobian, i.e. the infinitesimal area distortion, comes with a non-ambiguous sign (since the directions of ν's give coherent orientations to S and S^2) and so S appears with a real function, called *Gauss curvature* $K(s) \underset{def}{=} \operatorname{Jac} G(s)$.

Theorema Egregium. *Every isometry between surfaces, say $f : S \to S_1$, preserves Gauss curvature, $K(f(s_1)) = K(s_1)$ for all $s_1 \in S_1$.*

For example, the plane has $K \equiv 0$ (as the Gauss map is constant) and so it is not (even locally) isometric to the unit sphere where $K \equiv 1$ (for the Gauss map is identity on S^2). More generally, no strictly convex surface is locally isometric to a saddle surface, such as the graph of the function $z = xy$ for instance, since strict convexity makes $K > 0$ while saddle points have $K \leq 0$.

Gauss was well aware of the fact that the hyperbolic plane H^2 would have constant negative curvature if it were realized by a surface in $\mathbb{R}^3$. But he could not find such a surface! In fact, there are (relatively) small pieces of surfaces with $K = -1$ in $\mathbb{R}^3$ investigated by Beltrami in 1868 and it is hard to believe Gauss missed them; but he definitely could not realize the whole H^2 by a C^2-surface in $\mathbb{R}^3$ (as is precluded by a theorem of Hilbert (1901). This could be why (besides his timidity in the face of the

Kantian guards of Trilobite's intuition)[3] Gauss refrained from publishing his discovery.

Probably, Gauss would have been delighted to learn (maybe he knew it?) that the flat Lorenz-Minkowski "metric" $dx^2 + dy^2 - dz^2$ on $\mathbb{R}^{2,1} = \mathbb{R}^3$ induces a true positive metric on the sphere

$$S^2_- = \{x, y, z \mid x^2 + y^2 - z^2 = -1\},$$

where each of the two components of S^2_- (one is where $z > 0$ and the other with $z < 0$) is isometric to H^2 and where the orthogonal group $O(2,1)$ (i.e. the linear group preserving the quadratic form $x^2 + y^2 - z^2$) acts on these two H^2's by (hyperbolic) isometrics.

There is no comparable embedding of H^2 into any $\mathbb{R}^N$ (though H^2 admits a rather contorted isometric C^∞-immersion to $\mathbb{R}^5$ (to $\mathbb{R}^4$?) and, incredibly, an isometric C^1-embedding into $\mathbb{R}^3$) but it admits an embedding into the Hilbert space, say $f : H^2 \to \mathbb{R}^\infty$, where the induced intrinsic metric is the hyperbolic one, where all isometries of H^2 uniquely extend to those of $\mathbb{R}^\infty$ and such that

$$\mathrm{dist}_{\mathbb{R}^\infty}(f(x), f(y)) = \sqrt{\mathrm{dist}_{H^2}(x, y)} + \delta(\mathrm{dist}(x, y))$$

with *bounded* function $\delta(d)$ (where one can find f with $\delta(d) = 0$, but this will be *not* isometric in our sense as it blows the lengths of all curves in H^2 to infinity). Similar embeddings exist for metric trees as well as for real and complex hyperbolic spaces of all dimensions but not for other irreducible symmetric spaces of non-compact type. (This is easy for trees: arrange a given tree in $\mathbb{R}^\infty$, such that its edges become *all* mutually orthogonal and have prescribed lengths.)

Summary. Surfaces in $\mathbb{R}^3$ provide us with a large easily accessible *pool of metric spaces*: take a domain in $\mathbb{R}^2$, smoothly map it into $\mathbb{R}^3$ and, voilà, you have the induced Riemannian metric in your lap. Then study the isometry problem for surfaces by looking at *metric invariants* (curvature in the above discussion), relate them to *standard spaces* ($\mathbb{R}^N, \mathbb{R}^\infty, \mathbb{R}^{2,1}$), and consider *interesting* (to whom?) *classes* of surfaces, e.g. those with $K > 0$ and with $K < 0$.

REMARKS AND REFERENCES. (a) It seems to me that the reverence for human intuition and introspective soul searching stand in the way of any attempt to understand how the brain does mathematics. Hopefully, the

---

[3]Of all people, had he been alive, Kant himself could have been able to assimilate, if not accept, the non-Euclidean idea.

experience of natural scientists may lead us to a meaningful model (a provisional one at this stage, say in the spirit of Kanerva's idea of distributed memory, see [K]).

(b) Our allusions to the history of mathematics are borrowed from [Kl], [N] and [V].

(c) Little is known of what kind of maps $S \to S^2$ can serve as Gauss maps G of *complete* surfaces in $\mathbb{R}^3$. For example, given a domain $U \subset S^2$, one may ask whether there exists an *oriented closed immersed* (i.e. with possible self-intersections) surface $S \subset \mathbb{R}^3$ with $G(S) \subset U$ (where $U \neq S^2$ forces S to be topologically the 2-torus). This now appears to me a typical misguided "natural" question; yet I have not lost hope it may have a revealing solution (compare 2.4.4. in [G7]).

(d) It is unknown if every surface with a C^∞-smooth Riemannian metric can be isometrically C^∞-immersed into $\mathbb{R}^4$. (Another "natural" question?) But isometric immersions into high dimensional spaces are pretty well understood (see 2.4.9 – 2.4.11 and Part 3 in [G7]).

(e) The above equivariant embedding $H^2 \to \mathbb{R}^\infty$ tells us that the isometry group $\mathrm{Iso}H^2 = PSL_2\mathbb{R}$ is *a-T-menable*, opposite to *Kazhdan's property T* (defined in (A) of §V) satisfied by the majority of groups. (*A-T*-menability generalizes amenability. This property was first recognized by Haagerup, I presume, who used different terminology.)

2   Spirit of Riemann

The *triangle inequality* is not always easy to verify for a given function in two variables $d : X \times X \to \mathbb{R}_+$ as it is a non-local property of d on X; thus one cannot create metrics at will. Yet, an arbitrary metric d on a connected space can be made "local" by replacing it by the supremal metric d^+ agreeing with d on an "arbitrary fine" covering of X (as we pass, for example, from the restriction of the Euclidean metric on a submanifold, e.g. a surface, in $\mathbb{R}^N$, to the induced Riemannian, or *intrinsic*, metric). More generally, following Riemann (Habilitationsschrift, June 10, 1854) one starts with an arbitrary field g of Euclidean metrics on a domain $U \subset \mathbb{R}^n$, i.e. a continuous map $u \mapsto g_u$ from U to the $\frac{n(n+1)}{2}$-dimensional space G of positive definite quadratic forms on $\mathbb{R}^n$. One measures distances in small neighbourhoods $U_\varepsilon(u) \subset U$, $u \in U$, by setting $d_{u,\varepsilon}(u', u'') = \|u' - u''\|_u = (g_u(u' - u'', u' - u''))^{1/2}$ for $u', u' \in U_\varepsilon$ and then defines the Riemannian (geodesic) distance dist_g on U as the supremal metric for these $d_{u,\varepsilon}$ as

$\varepsilon \to 0$. Finally, Riemannian manifolds V appear as metric spaces locally isometric to the above U's. (The latter step is slick but hard to implement. Try, for example, to show that the induced (intrinsic) metric on a smooth submanifold $V \subset \mathbb{R}^N$ is Riemannian in our sense.)

The magic power of this definition is due to the infinitesimal kinship of "Riemannian" to "Euclidean". If V is smooth (i.e. all g's on U's are at least C^2) then locally near each point v, one can represent V by a neighbourhood U of the origin $0 \in \mathbb{R}^n$ with $v \mapsto 0$, so that the corresponding g on U agrees with the Euclidean (i.e. constant in v) metric $g_0 = g_0(x,y) = \langle x,y \rangle = \sum_{i=1}^n x_i y_i$ up to the *first* order,

$$g_u = g_0 + \frac{1}{2} \sum_{i,j=1}^n \left(\frac{\partial^2 g(0)}{\partial u_i \partial u_j} \right) u_i u_j + \dots ,$$

i.e. with the first order Taylor terms missing and where, moreover, only $\frac{n^2(n^2-1)}{12}$ terms among $\left(\frac{n(n+1)}{2} \right)^2$ second derivatives $\frac{\partial^2 g_{\mu\nu}(0)}{\partial u_i \partial u_j}$ do not vanish. The resulting $\frac{n^2(n^2-1)}{12}$ functions on U, when properly organized, make the *Riemann curvature tensor* of V (which reduces to the Gauss curvature for $n = 2$) measuring the deviation of (V, g) from flatness (i.e. Euclideanness).

The (polylinear) algebraic structure built into g allows a fully fledged analysis on (V, g), such as the Laplace-Hodge operator, potential theory etc. This turned out to be useful for particular classes of manifolds distinguished by additional (global, local or infinitesimal) symmetry, where the major achievements coming to one's mind are:

- Hodge decomposition on the cohomology of Kähler manifolds V and a similar (non-linear) structure on the spaces of representations of $\pi_1(V)$.

- Existence of Einstein metrics on Kähler manifolds with algebra-geometric consequences.

- Spectral analysis on locally symmetric (Bruhat-Tits and adelic as well as Riemannian) spaces leading, for instance, to various cohomology vanishing theorems, T-property (with applications to expanders) and (after delinearization) to super-rigidity of lattices in semi-simple Lie groups.

The linear analysis on *general* Riemannian manifolds pivots around the Atiyah-Singer-Dirac operator and the index (Riemann-Roch) theorem(s). These originated from Gelfand's question (raised in the late fifties) aiming at an explicit topological formula for the index of an elliptic operator (which is easily seen to be deformation invariant) and became a central theme in mathematics starting from the 1963 paper by Atiyah and Singer.

The *non-linear* Riemannian analysis on general V's followed for the most part the classical tradition concentrating around elliptic variational problems with major advances in the existence and regularity of solutions: minimal subvarieties, harmonic maps, etc. The most visible "external" application, in my view, concerns manifolds with *positive scalar curvature* – the subject motivated by problems (and ideas) coming from general relativity – resolved by Schoen and Yau with a use of minimal hypersurfaces.

Manifolds of each dimension *two, three* and *four* make worlds of their own, richer in structure than all we know so far about $n \geq 5$.

In dimension two we possess the Cauchy-Riemann equations and are guided by the beacon of the Riemann mapping theorem, the crown jewel of differential geometry.

The four-dimensional peculiarity starts with algebra: the orthogonal group $O(4)$ locally decomposes into two $O(3)$. This allows one to split (or rather to square-root) certain natural (for the $O(4)$-symmetry) non-linear second order operators in a way similar to how we extract the Cauchy-Riemann $\overline{\partial}$ from the boringly natural selfadjoint Laplacian Δ on $\mathbb{R}^2$. The resulting first order operators (may) have non-zero indices and satisfy a kind of non-linear index theorem discovered by Donaldson in 1983 for the Yang-Mills and then extended to the Seiberg-Witten equation. (Both equations were first written down by physicists, according to the 20th century lore.)

Manifolds of dimension three borrow from their two- and four-dimensional neighbours: Thurston's construction of hyperbolic metrics on basic 3-manifolds relies on geometry of surfaces while Floer homology descends from Yang-Mills.

Shall we ever reach spaces beyond Riemann's imagination?

REMARK AND REFERENCES. It will need hundreds of pages to account for the above forty lines. Here we limit ourselves to a few points.

(a) The Riemannian metric g naturally (i.e. functorially) defines *parallel transport* of vectors along smooth curves in V which is due to the absence of first derivatives in an appropriate Taylor expansion of g. This can be seen clearly for V realized in some $\mathbb{R}^N$ (which is not a hindrance according to the Cartan-Janet-Burstin-Nash isometric embedding theorem) where a family $X(t)$ of tangent vectors is parallel *in* V along our curve γ parametrized by $t \in \mathbb{R}$ iff the ordinary (Euclidean) derivative $\frac{dX(t)}{dt} \in \mathbb{R}^N$ is normal to V at $\gamma(t) \in V$ for all t. (This is independent of the isometric

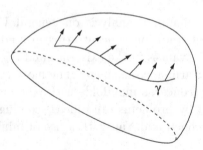

Figure 2

embedding $V \to \mathbb{R}^N$.) If a curve $\gamma : [0,1] \to V$ comes back making a loop (i.e. $\gamma(0) = \gamma(1)$), every tangent vector $X = X(0) \in T_v(V)$, $v = \gamma(0) \in V$, transforms to $\gamma_*(X) \overset{def}{=} X(1) \in T_v(V)$ and we obtain a homomorphism from the "group" of loops at v to the linear group of isometric automorphisms of the tangent space $T_v(V) = \mathbb{R}^N$, i.e. to the orthogonal group $O(n)$; its image $H \subset O(n)$ is called the *holonomy group* of V (which is independent of v for connected V). Generically, $H = O(n)$ ($SO(n)$ for orientable V), but sometimes H has positive codimension in $O(n)$. For example, $\dim H = 0$ iff V is locally Euclidean (Parallel Postulate is equivalent to $H = \{\mathrm{id}\}$) and if $V = V_1 \times V_2$, then $H = H_1 \times H_2 \subset O(n_1) \times O(n_2) \underset{\neq}{\subset} O(n)$ for $n_1, n_2 \neq 0$. Then there are several *discrete* series of *symmetric spaces* – monumental landmarks towering in the vastness of all Riemannian metrics $\mathbb{R}^n, S^n, H^n, \mathbb{C}P^n$, $SL(n)/SO(n)$... It is natural to think that these are essentially all V's with small holonomy, since $\mathrm{codim}\, H > 0$ implies a rather over-determined system of P.D.E. for g (for example, $\dim H = 0 \Leftrightarrow$ curvature $(g) = 0$, i.e. $\frac{n^2(n^2-1)}{12}$ equations against mere $\frac{n(n+1)}{2}$ components of g. Yet flat metrics exists!). But lo and behold: lots of even dimensional manifolds carry *Kähler metrics* where $H \subset U(n) \underset{\neq}{\subset} SO(2n)$. Just take a *complex analytic* submanifold V in $\mathbb{C}^N$ (or in $\mathbb{C}P^N$) and observe (which is obvious once being said) that the parallel transport in the induced metric preserves the complex structure in the tangent spaces. (This may be not so striking, perhaps, for those who have absorbed the impact of holomorphic functions, over-determined by Cauchy-Riemann in many variables, but there are less expected beautiful exotic holonomy beasts predicted by Berger's classification and brought to life by Bryant, see [Br].)

The Kähler world is tightly knit (unlike the full Riemannian universe) with deep functorial links between geometry and topology. For example, the first cohomology of a compact Kähler V comes by the way of a holomorphic (!) map to some complex torus, $V \to \mathbb{C}^d/\text{lattice}$ for $d = \frac{1}{2}\,\text{rank}H^1(V)$. This extends to (non-Abelian) representations $\pi_1(V) \to GL(n)$ for $n \geq 2$ (Siu, Corlette, Simpson ..., see [ABCKT]) furnishing something like an "*unramified* non-Abelian Kählerian class field theory" (in the spirit of Langlands' program) but we have no (not even conjectural) picture of the "transcendental part" of $\pi_1(V)$ (killed by the profinite completion of π_1). Can, for example, π_1 (Kähler) have an unsolvable word problem? Is there an internal structure in the category of Kähler fundamental groups functorially reflecting the geometry of the Kähler category? (All *known* compact Kähler manifolds can be deformed to complex projective ones and it may be preferable to stay within the complex algebraic category, with no fear of ramifications, singularities and non-projectiveness.)

(b) There exists a unique (up to normalization) second order differential operator $\mathbf{S}$ from the space of positive definite quadratic differential forms (Riemann metrics) g on V to the space of functions $V \to \mathbb{R}$ with the following two properties.

$\mathbf{S}$ is Diff-equivariant for the natural action of diffeomorphisms of V on both spaces.

$\mathbf{S}$ is linear in the second derivatives of g (being a linear combination of components of the full curvature tensor).

Then $\mathbf{S}(g)$ (or $\mathbf{S}(V)$) is named the *scalar curvature* of (V, g) with the customary normalization $\mathbf{S}\,(S^n) = n(n-1)$.

If $n = 2$, $\mathbf{S}$ coincides with Gauss curvature, it is additive for products, $\mathbf{S}(V_1 \times V_2) = \mathbf{S}(V_1) + \mathbf{S}(V_2)$ and it scales as g^{-1}, i.e. $\mathbf{S}(\lambda g) = \lambda^{-1}\,\mathbf{S}(g)$ for $\lambda > 0$.

The following question proved to be more to the point than one could expect.

What is the geometric and topological structure of manifolds with $\mathbf{S} > 0$? (This comes from general relativity as $\mathbf{S} > 0$ on world sheets reflects positivity of energy.)

The condition $\mathbf{S} > 0$ appears quite plastic for $n \geq 3$, where one can rather freely manipulate metrics g keeping $\mathbf{S}(g) > 0$, e.g. performing geometric surgery; besides, every compact V_0 turns into V with $\mathbf{S}(V) > 0$ when multiplied by a small round sphere of dimensions ≥ 2. Yet this plasticity has its limits: Lichnerowicz found in 1963 a rather subtle topological obstruction $(\widehat{A}(V) = 0$ if V is spin) with the use of the index theorem. Then in 1979 Schoen and Yau approached the problem from another angle (linked to ideas in general relativity) and proved, among other things, that n-tori (at least for $n \leq 7$) carry no metrics with $\mathbf{S} > 0$ thus answering a question by Geroch. Inspired by this, we revived with Blaine Lawson in 1980 Lichnerowicz' idea, combined it with the Lusztig-Mistchenko approach to the *Novikov conjecture* on homotopy invariance of Pontryagin classes of non-simply connected manifolds and found out that the bulk of the topological obstructions for $\mathbf{S} > 0$ comes from a "limit on geometric size" of V induced by the inequality $\mathbf{S} > 0$ (similar to but more delicate than that for $K > 0$, see III).

Yet, the above question remains open with an extra mystery to settle: what do minimal hypersurfaces and the Dirac operator have in common? (Seemingly, nothing at all but they lead to almost identical structure results for $\mathbf{S} > 0$, see [G6] for an introduction to these issues.)

It appears that an essential part of the difficulty in understanding $\mathbf{S} > 0$ (and the Novikov conjecture) is linked to the following simple minded question: *what is the minimal $\lambda > 0$, such that the unit sphere $S^N_\infty(1)$ in the Banach space $\ell^N_\infty = (\mathbb{R}^N, \|x\| = \sup_i |x_i|)$ admits a λ-Lipschitz map into the ordinary n-sphere $S^N(1)$ in $\mathbb{R}^N = \ell^N_2$ with non-zero degree?* Probably, $\lambda \to \infty$ for $N \to \infty$ (even if we stabilize to maps $S^N_\infty(1) \times S^M(R) \to S^{N \times M}(1)$ with arbitrarily large M and R) and this might indicate new ways of measuring "size of V" in the context of $\mathbf{S} > 0$ and the Novikov conjecture.

"Soft and hard". Geometric (and some non-geometric) spaces and categories (of maps, tensors, metrics, (sub)varieties...) can be ranked, albeit ambiguously, according to plasticity or flexibility of (the totality of) their members. (1) Topology could appear flabby and structureless to Poincaré's contemporaries but when factored by homotopies (the very source of flexibility) it crystallizes to a rigid algebraic category as hard and symmetric as a diamond.

(2) Riemannian manifolds, as a whole, are shapeless and flexible, yet they abide "conservation laws" imposed by the Gauss-Bonnet-Chern identities.

Deeper rigidity appears in the presence of elliptic operators extracting finite dimensional structure out of infinite dimensional depth of functional spaces. Also we start seeing structural rigidity (e.g. Cheeger compactness) by filtering metrics through the glasses of (say, sectional) curvature.

(3) Kähler metrics and algebraic varieties seem straight and rigid in the Riemannian landscape (never mind a dense set of Riemannian spaces appearing as real loci of complex algebraic ones) but they look softish to the eye of an algebraic geometer. He/she reinforces rigidity with the Calabi-Yau-Aubin theorem turning Kähler to Einstein-Kähler. (Nothing of the kind seems plausible in the full Riemannian category for $n > 4$.)

(4) Homogeneous, especially symmetric, spaces stand on the top of the geometric rigidity hierarchy (tempting one to q-deform them) and (sometimes hidden) symmetries govern integrable (regarded rigid) systems. (Softness in dynamics is associated with hyperbolicity.)

(5) Lattices Γ in semi-simple Lie groups grow in rigidity with dimension, passing the critical point at $\Gamma \subset SL_2(\mathbb{C})$, where they flourish in Thurston's hyperbolic land. A geometer unhappy with Mostow (over)rigidity for $n > 3$ is tempted to switch from lattices to (less condensed) subgroups with infinite covolumes and more balanced presentations (to the dismay of a number theorist thriving on the full arithmetic symmetry of Γ). Most flexible among all groups are the (generalized) small cancellation ones, followed by higher dimensional hyperbolic groups while lattices and finite simple groups are most rigid. Vaguely similarly, the rigidity of Lie algebras increases with decrease of their growth culminating in Kac-Moody and finite dimensional algebras.

(6) Holomorphic functions on Stein manifolds V are relatively soft (Cartan's theory) as well as holomorphic maps $f : V \to W$ for homogeneous and elliptic (i.e. with a kind of exponential spray) W by the (generalized) Grauert theorem allowing a homotopy of every continuous map $f_0 : V \to W$ to a holomorphic one. Holomorphic maps moderated by bounds on growth become more rigid (e.g. functions of finite order have essentially unique Weierstrass product decomposition). Algebraic maps, ordinarily rigid, sometimes turn soft, e.g. for high degree maps of curves to $\mathbb{P}^1$ by the Segal theorem. And Voevodski theory (if I interpret correctly what little I understood from his lecture) softens the category of algebraic varieties by injecting some kind of homotopies into these.

(7) **Big three.** There are three outstanding instances where striking structural patterns emerge from large and flexible geometric spaces: symplectic/contact, dimension 4 and $\mathbf{S} > 0$, conducted in all three cases by "Riemannian" and "elliptic". We met $\mathbf{S} > 0$ earlier (which seems least conceptually understood among the big three), symplectic and contact belong to Eliashberg and Hofer at this meeting (with "soft" versus "hard"

discussed (in [G10]) and nobody, alas, gave us a panorama of $n = 4$).

(8) **h-Principle.** Geometers believed from 1813 till 1954, since Cauchy (almost) proved rigidity of closed convex polyhedral surfaces in $\mathbb{R}^3$, that isometric immersions are essentially rigid. Then Nash defied everybody's intuition by showing that every smooth immersion of a Riemannian manifold $f_0 : V \to \mathbb{R}^N$ can be deformed, for $N - 2 \geq n = \dim V$, to a C^1-smooth (not C^2!) isometric $f : V \to \mathbb{R}^N$ with little limitation for this deformation, allowing one in particular, to freely C^1-deform all $V \subset \mathbb{R}^N$ keeping the induced (intrinsic) metric intact. (This is sheer madness from a hard-minded analyst's point of view as the N components of f satisfy $\frac{n(n+1)}{2}$ partial differential equations comprising an over-determined system for $N < \frac{n(n+1)}{2}$, where one expects no solutions at all!) The following year (1955) Kuiper adjusted Nash's construction to $N = n + 1$ thus disproving C^1-rigidity of convex surfaces in $\mathbb{R}^3$.

Next, in 1958, Smale stunned the world by turning the sphere $S^2 \subset \mathbb{R}^3$ inside out. He did it not by exhibiting a particular (regular) homotopy (this was done later and only chosen few are able to follow it through) but by developing the homotopy theoretic approach used by Whitney for immersions of curves into $\mathbb{R}^2$. Then Hirsch incorporated Smale into the obstruction theory and showed that a continuous map $f_0 : V \to W$ can be homotoped into an immersion *if* the obvious necessary condition is satisfied: f_0 lifts to a fiberwise injective homomorphism of tangent bundles, $T(V) \to T(W)$, with the exception of the case of closed equidimensional manifolds V and W where the problem is by far more subtle.

It turned out that many spaces X of solutions of partial differential equations and inequalities abide the homotopy principle similar to that of Nash, Smale-Hirsch and Grauert: every such X is canonically homotopy equivalent to a space of continuous sections of some (jet) bundle naturally associated to X. (For example, the space of immersions $V \to W$ is homotopy equivalent to the space of fiberwise injective morphisms $T(V) \to T(W)$ by the Hirsch theorem.)

The geometry underlying the proof of the h-principle is shamefully simple in most cases: one creates little (essentially 1-dimensional, à la Whitney) wrinkles in maps $x \in X$ which are spread all over by homotopy and render X soft and flexible. But the outcome is often surprising as seen in Milnor's (intuitively inconceivable) two *different* immersed disks in the plane with common boundary which come up with logical inevitability in Eliashberg's folding theorem.

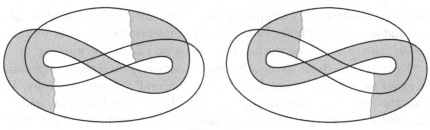

Figure 3

Despite the growing array of spaces subjugated by the h-principle (see [Gro]$_{PDR}$, [S]) we do not know how far this principle (and softness in general) extends (e.g. for Gauss maps with a preassigned range, see I). Are there sources of softness not issuing from dimension one? Some encouraging signs come from Thurston's work on foliations, Gao-Lohkamp h-principle for metrics with Ricci < 0, and especially from Donaldson's construction of symplectic hypersurfaces (where "softness" is derived from a kind of "ampleness" not dissimilar in spirit to Segal's theorem, see (6)).

A tantalizing wish is to find new instances, besides the big three, where softness reaches its limits with something great happening at the boundary. Is there yet undiscovered life at the edge of chaos? Are we for ever bound to elliptic equations? If so, what are they? (There are few globally elliptic non-linear equations and no general classification. But even those we know, coming from Harvey-Lawson calibrations, remain mainly unexplored.) And if this wish does not come true we still can make a living in soft spaces exploring their geometry (their topology is completely accounted for by the h-principle) as we do in anisotropic spaces (see III).

(9) Our "soft and hard" are not meant to reveal something profound about the nature of mathematics, but rather to predispose us to acceptance of geometric phenomena of various kinds. Besides, it is often more fruitful to regard "numbers", "symmetric spaces", "Gal$\overline{\mathbb{Q}}/\mathbb{Q}$", "$SL_n(\mathbb{A})/SL_n(\mathbb{Q})$"... as "true mathematical entities" rather than descendants of our general "spaces", "groups", "algebras" etc. But one cannot help wondering how these perfect entities could originate and survive in the softly structured brain hastily assembled by blind evolution. Some basic point (scientific, not philosophical) seems to completely elude us.

Nature and naturality of questions.   Here are (brief, incomplete, personal and ambiguous) remarks intended to make clearer, at least terminologically, the issues raised during discussions we had at the meeting.

"Natural" may refer to the structure or *nature of mathematics* (granted this exists for the sake of argument), or to "natural" for *human nature*. We divide the former into (pure) mathematics, logic and philosophy, and the latter, according to (internal or external) reward stimuli, into intellectual, emotional, and social. **E**(motional) plays the upper hand in human decision (and opinion) (except for a single man you might have the privilege to talk mathematics to) and in some people (Fermat, Riemann, Weil, Grothendieck) **i-e** naturally converges to **m-l-p**. But for most of us it is not easy to probe the future by conjecturally extrapolating mathematical structures beyond the present point in time. How can we trust our mind overwhelmed by **i-e-s** ideas to come up with true **m-l-p** questions?(An **e-s**-minded sociologist would suggest looking at trends in fund distribution, comparable weights of authorities of schools and individuals and could be able to predict the influential role of Hilbert's problems and Bourbaki, for example, without bothering to read a single line in there.) And "**i-e-s-**natural" does not make "a stupid question": the 4-color problem, by its sheer difficulty (and expectation for a structurally rewarding proof) has focused attention on graphs while the solution has clarified the perspective on the role of computers in mathematics. But this being unpredictable, and unrepeatable, cannot help us in **m-l-p**-evaluation of current problems which may look **i-e** deceptively 4-colored. (As for myself, I love unnatural, crazily unnatural problems but you stumble upon them so rarely!)

3 $K \gtrless 0$ and Other Metric Stories

What are the "most Euclidean" Riemannian manifolds?

We have already been acquainted with the fully homogeneous spaces also called, for a good reason, (complete simply connected) *of constant curvature K*: the round S^n with $K = +1$, the *flat* $\mathbb{R}^n$ with $K = 0$, and the hyperbolic H^n with $K = -1$.* (Observe that λS^n and λH^n converge to $\mathbb{R}^n$ for $\lambda \to \infty$ in a natural sense, where $\lambda(X, \text{dist}) \overset{def}{=} X(\lambda \text{ dist})$ and $K(\lambda X) = \lambda^{-\frac{1}{2}} K(X)$, as is clearly seen, for example, for λ-scaled surfaces $X \subset \mathbb{R}^3$.) Now, somewhat perversely, we bring in topology and ask for

---

*The fourth and the last fully homogeneous Riemannian space is $P^n = S^n/\{\pm 1\}$.

compact manifolds with constant curvature, i.e. *locally* isometric to one of the above S^n, $\mathbb{R}^n$, or H^n. Letting S^n go, we start with the flat (i.e. $K = 0$) case and confirm that compact locally Euclidean manifolds exist: just take a lattice Λ in $\mathbb{R}^n$ (e.g. $\Lambda = \mathbb{Z}^n$) and look at the *torus* $T = \mathbb{R}^n/\Lambda^n$. Essentially, there is little else to see:

F-theorem. *Every compact flat manifold X is covered by a torus with the number of sheets bounded by a universal constant $k(n)$.*

This sounds dry but it hides a little arithmetic germ on the bottom: *there is no regular k-gon with vertices in a lattice Λ (e.g. $\Lambda = \mathbb{Z}^2$) in $\mathbb{R}^2$ for $k \geq 7$ (or for $k = 5$).* Indeed, transported edges of such k-gon R would make a smaller regular $R' \subset \Lambda$ and the contradiction follows by iteration $R'', R''', \ldots$.

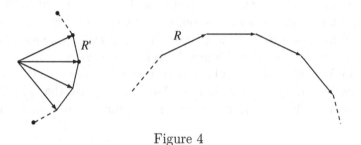

Figure 4

The tori T themselves stop looking flat as they all together make the marvellous moduli space $SO(n)\backslash SL_n(\mathbb{R})/SL_n(\mathbb{Z})$ (of isometry classes of T's with $\mathrm{Vol}\,T = 1$) locally isometric to $SL_n(\mathbb{R})/SO(n)$ apart from mild (orbifold) singularities due to elements of finite order (< 7) in $SL_n(\mathbb{Z})$.

Turning to $K = -1$, we may start wondering if such spaces exist in a compact form at all. Then, for $n = 2$, we observe that the angles of *small* regular k-gons $R \subset H^2$ are almost the same as in $\mathbb{R}^2$ while *large* $R \subset H^2$ have almost zero angles: thus, by continuity, for every $k \geq 5$, there exists $R_\square \subset H^2$ with $90°$ angles. We reflect H^2 in (the lines extending the) sides of $R_\square$ and take the subgroup $\Gamma \subset \mathrm{Isom}(H^2)$ generated by these k reflections. This Γ is *discrete* on H^2 with $R_\square$ serving as a fundamental domain similarly to the case of the square $R_\square \subset \mathbb{R}^2$ and the quotient space H^2/Γ (equal R) becomes an honest manifold (rather than orbifold) if we take instead of Γ a subgroup $\Gamma' \subset \Gamma$ without torsion (which is not hard to find).

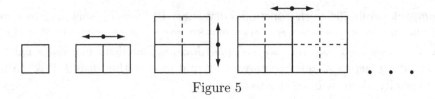

Figure 5

The same idea works for dodecahedra in H^3 and some other convex polyhedra in H^n for small n, but there are no compact hyperbolic reflection groups for large n (by a difficult theorem of Vinberg). The only (known) source of high dimensional Γ comes from arithmetics, essentially by intersecting $SO(n,1)$ somehow embedded into $SL_N(\mathbb{R})$ with $SL_N(\mathbb{Z})$ (where the orthogonal group $SO(n,1)$ doubly covers the isometry group $PSO(n,1)$ of $H^n \subset \mathbb{R}^{n,1}$). Non-arithmetic Γ are especially plentiful for $n = 3$ by Thurston's theory and often have unexpected features, e.g. some $V = H^3/\Gamma$ fiber over S^1 (which is hard to imagine ever happening for large n). Moreover, the topological 3-manifolds fibered over S^2 *generically*, (i.e. for pseudo-Anosov monodromy) admit metrics with $K = -1$. Unbelievable – but true by Thurston (who himself does not exclude that finite covers of most atorical V fiber over S^1; yet this remains open even for V with $K = -1$).

Alexandrov's spaces. *What are the most general (classes of) spaces similar to those with $K = \pm 1$?*

Alexandrov suggested an answer in 1955 by introducing spaces with $K \geq 0$, where the geodesic triangles have the sum of angles $\geq 2\pi$, and those with $K \leq 0$, where (at least small) triangles have it $\leq 2\pi$. But we take another, more functorial route departing from the following

Euclidean K-theorem. *Every 1-Lipschitz (i.e. distance non-increasing) map f_0 from a subset $\Delta \subset \mathbb{R}^n$ to some $\mathbb{R}^m$ admits a 1-Lipschitz extension $f : \mathbb{R}^n \to \mathbb{R}^m$ for all $m, n \leq \infty$.*

This is shown by constructing f point by point and looking at the worst case at each stage where extendability follows from an obvious generalization of the pretty little lemma:

LEMMA. *Let Δ and Δ' in $S^{n-1} \subset \mathbb{R}^n$ be the sets of vertices of two simplices (inscribed into the sphere S^{n-1}) where the edges of Δ are correspondingly $\leq$ than those of Δ'. Then Δ is congruent to Δ' (i.e. $\leq \Rightarrow =$), provided Δ is not contained in a hemisphere.*

Now we say that a metric space X has $K \geq 0$, if every partial 1-Lipschitz map $X \supset \Delta \to \mathbb{R}^m$ extends to a 1-Lipschitz $f : X \to \mathbb{R}^m$, for all m, while $K \leq 0$ is defined with such extensions for $\mathbb{R}^n \supset \Delta \to X$, where for $K \leq 0$ we additionally require *uniqueness* of middle points in X (see below).

To make this worthwhile, one adds the metric completeness of X and the *locality property*: $\text{dist}(x, x')$ should equal the infimal length of curves in X between x and x'. (Equivalently, there is a *middle point* $y \in X$ where $\text{dist}(x, y) + \text{dist}(y, x') = \text{dist}(x, x')$.) Then, one arrives at the following elegant proposition justifying the definitions:

K-Theorem. *If $K(X) \geq 0$ and $K(Y) \leq 0$, then every partial 1-Lipschitz map $X \supset \Delta \to Y$ admits a 1-Lipschitz extension $X \to Y$.*

To apply this one needs examples of spaces with $K \gtrless 0$ and these are easier to observe with Alexandrov's definition. Fortunately, both definitions are equivalent and we have:

Complete (e.g. closed) convex hypersurfaces in $\mathbb{R}^n$ have $K \geq 0$, while saddle surfaces in $\mathbb{R}^2$ have (at least locally) $K \leq 0$. Symmetric spaces of compact type have $K \geq 0$ while those of non-compact type have $K \leq 0$.

Take a 2-dimensional polyhedron V assembled of convex Euclidean k-gons and observe that the link L of each vertex in V is a graph (i.e. 1-complex) with a length assigned to every edge e equal the k-gonal angle corresponding to e. Then $K(V) \geq 0 \Leftrightarrow$ each L is isometric to a segment $\leq \pi$ or to a circle no longer than 2π.

$K(V) \leq 0 \Leftrightarrow$ all cycles in all L are longer than 2π and V is simply connected.

Finally, if X has locally $K \geq 0$, then the same is true globally while globally $K \leq 0 \Leftrightarrow$ (locally $K \leq 0$) + ($\pi_1 = 0$).

The $\pi_1 = 0$ condition breaks the harmony (I guess it was upsetting Alexandrov) and brings confusion to the notion of $K \leq 0$ as the "local" and "global" meanings diverge. But at the end of the day the π_1-ripple makes the geometry of $K \leq 0$ much richer (and softer) than all we know of $K \geq 0$, since there are lots of spaces V with $K_{\text{loc}} \leq 0$ (as already seen in the 2-dimensional polyhedra) where the group theoretic study of $\pi_1(V)$ may rely on geometry (for example, in the Novikov conjecture). Besides, the global definition of $K \leq 0$ can be relaxed by, roughly, allowing λ-Lipschitz extension of partial 1-Lipschitz maps with $1 \leq \lambda \leq \text{const} < \infty$ bringing along larger classes of (hyperbolic-like) spaces and groups where geometry and algebra are engaged in a meaningful conversation.

3.1   Anisotropic spaces.   There is a class of metrics which can be analytically generated with the same ease as the Riemannian ones; besides, we
find among them spaces X in some way closer to $\mathbb{R}^n$ than S^n and H^n: these
X are metrically homogeneous as well as *self-similar*, i.e. λX is isometric
to X for all $\lambda > 0$. (In the Riemannian category there is nothing but $\mathbb{R}^n$
like that.)

A *polarization* on a smooth manifold V (e.g. $\mathbb{R}^n$ or a domain in $\mathbb{R}^n$)
is a subbundle H of the tangent bundle $T(V)$, i.e. a field of m-planes on
V, where for $1 \leq m \leq n-1$, $n = \dim V$ in the case at hand. Besides H
we need an auxiliary Riemannian metric g on V but what matters is the
restriction of g on H. We define $\mathrm{dist}(v,v') = \mathrm{dist}_{H,g}(v,v')$ by taking the
infimum of g-lengths of piecewise smooth curves between v and v' which are
chosen among H-*horizontal* curves, i.e. those which are *everywhere tangent*
to H. It may happen that this distance is infinite (even for connected V)
if some points admit no horizontal connecting paths between them, as it
happens for *integrable* H, where $\mathrm{dist}(v,v') < \infty \Leftrightarrow v$, and v' lie in the
same leaf of the foliation integrating H. This is not so bad as it seems
but we want $\mathrm{dist} < \infty$ at the moment and so we insist on the existence of
a horizontal path between every two points in V. It is not hard to show
that *generic* C^∞-smooth polarizations H do have this property for $m \geq 2$,
where "generic" implies, in particular, that the space of "good" H's is open
and dense in the space of all C^∞-polarizations on V.

A practical way for checking this is to take $m_+ \geq m$ vector fields tangent
to H, and spanning it (these always exist), say $X_1, \ldots, X_{m_+}$. Then the
sufficient criterion for our H-connectivity (also called *controllability*) is as
follows: The successive commutators $X_i, [X_i, X_j], [[X_i, X_j], X_k] \ldots$ of the
fields span the tangent bundle $T(V)$.

The simplest instance of the above is the pair of the fields $X_i = \frac{\partial}{\partial x_1}$ and
$X_2 = \frac{\partial}{\partial x_2} + x_1 \frac{\partial}{\partial x_3}$ in $\mathbb{R}^3$, where $[X_1, X_2] = -\frac{\partial}{\partial x_3}$. Here H can be presented
as the kernel of the *standard contact form* $dx_3 + x_1 dx_2$ and, in fact, contact
fields H are H-connected for all contact manifolds of dimension ≥ 3.

Next, look at a left invariant polarization H on a Lie group G defined
by the left translates of a linear subspace h in the Lie algebra $L = L(G)$.
It is not hard to see that the commutators condition holds $\Leftrightarrow$ there is
no Lie subalgebra in L containing h besides L itself. Take, for instance,
the 3-dimensional *Heisenberg* group G (homeomorphic to $\mathbb{R}^3$) with $L = $
$L(G)$ generated by x_1, x_2, x_3, where $[x_1, x_2] = x_3$ and x_3 commutes with

x_1 and x_2. Here one takes h spanned by x_2, x_3 and observe that this h is invariant under *automorphisms* of L defined by $x_1 \mapsto \lambda^2 x_1^2$, $x_2 \mapsto \lambda x_2$, $x_3 \mapsto \lambda x_3$ for all $\lambda > 0$. Then, for each left invariant metric g on G, the corresponding *automorphisms* $A_\lambda : G \to G$ preserves H and scale g on H by λ. Thus G is selfsimilar but quite different from $\mathbb{R}^3$. For example, the Hausdorff dimension of $(G, \mathrm{dist}_{H,g})$ is 4 rather than 3.

What are natural maps between the above non-isotropic spaces? Lipschitz maps do not serve as well here as in the Riemannian category since our new spaces are usually *not* mutually bi-Lipschitz equivalent. On the other hand typical spaces X and Y of the same topological dimension are locally Hölder homeomorphic with some positive exponent $\alpha < 1$ with a bound $\alpha > \alpha_n > 0$. But the optimal value of α for given spaces (or classes of those) remains unknown. A similar problem is that of finding a sharp lower bound for the Hausdorff dimension of a subset $Y \subset X = (X, \mathrm{dist}_{H,g})$ by the topological dimension in terms of commutation properties of fields spanning H. (This is quite easy if Y is a *smooth* submanifold in X.)

The large scale geometry of $(X, \mathrm{dist}_{H,g})$ is rather close to the Riemannian geometry of (X, dist_g) and so H does not matter much. Conversely, the *local* geometry essentially depends on H (and very little on g) and the main open problems are basically local.

Concentrated mm spaces. Let X be a metric space which is also given a Borel measure μ, often assumed to be a probability measure, i.e. $\mu(X) = 1$. This μ may come out of the metric (e.g. the Riemannian measure, sometimes normalized to have $\mu(X) = 1$), but often μ has little to do with the original distance as, for example, the Gaussian measure on $\mathbb{R}^n$. We want to study (X, dist, μ) in a probabilistic fashion by thinking of functions f on X as random variables, being concerned with their distributions, i.e. push-forward measures $f_*(\mu)$. Here is our

Basic problem. Given a map $f : X \to Y$, relate *metric* properties of f to the structure of the *measure* $f_*(\mu)$ on Y.

Here "metric" refers to how f distorts distance (expressed, for instance, by the Lipschitz constant $\dot\lambda(f)$), where we distinguish the case of functions, i.e. of $Y = \mathbb{R}$. On the probabilistic side we speak of "structure of $f^*(\mu)$" expressed entirely in terms of Y and where a typical question is how *concentrated* $f_*(\mu)$ is, i.e. how close it is to a point measure in Y.

Gaussian example. Take $\mathbb{R}^N$ for X with the measure $(\exp -\|x\|^2)dx$. *Then, every 1-Lipschitz function $f : \mathbb{R}^N \to \mathbb{R}$ is at least as much concen-*

trated as the orthogonal projection $f' : \mathbb{R}^N \to \mathbb{R}$. (There is a *1-Lipschitz* self-mapping $\mathbb{R} \to \mathbb{R}$ pushing forward $f'_*(\mu)$ to $f_*(\mu)$.)

A parallel example, where the geometry is seen more clearly, is $X = S^N$ with the normalized Riemannian measure. Here again every 1-Lipschitz function $f : S^N \to \mathbb{R}$ is concentrated as much as the linear one; consequently, $f^*(\mu)$ converges, for $N \to \infty$, to a δ-measure on $\mathbb{R}$ (with the rate $\approx \sqrt{N}$).

The essence of the above concentration is the sharp contrast between the spread of the original measure on X (e.g. the distance between μ-random points in S^N is $\approx \pi/2$ for large N) and strong localization of $f_*(\mu)$ on Y (e.g. the characteristic distance on $\mathbb{R}$ with respect to $f_*(\mu)$ is $\approx 1/\sqrt{N}$ in the spherical case). The following definition is aimed to capture this phenomenon in the limit for $N \to \infty$ by interbreeding metric geometry with the ergodic theory (not quite as in the ergodic theorem where $f = f_N$ appears as the average of N transforms of a given f_0).

Let X be a (probability) measure space and $d : X \times X \to \mathbb{R}_+ \cup \infty$ a function satisfying the standard metric axioms except that we allow $d(x, x') = \infty$. In fact, we are keen for the (apparently absurd) situation of $d = \infty$ almost everywhere on X, moderated by the

Ergodicity axiom. For every $Y \subset X$ with $\mu(X) > 0$ the distance to Y,
$$d_Y(x) \underset{def}{=} \inf_{y \in Y} d(x, y)$$
is measurable and a.e. finite on X.

EXAMPLE. Let X be a foliated measure space where each leaf is (measurably in $x \in X$) assigned a metric. Then we define
$$d(x, y) = \begin{cases} \infty \text{ if } x, y \text{ are not in the same leaf} \\ d(x, y) = \text{dist}_L(x, y) \text{ if } x \text{ and } y \text{ lie in some leaf } L, \end{cases}$$
and observe that our ergodicity axiom amounts here to the ordinary ergodicity.

Next we distinguish *concentrated spaces* by insisting on the universal bound on the distances between subsets in X in terms of the measures of these subsets. Here
$$\text{dist}(Y, Y') \overset{def}{=} \inf d(y, y') = \inf_{x \in Y'} d_Y(x)$$
and the bound is given by a function $C(a, a')$, $a, a' > 0$, (where C may go to infinity for $a, a' \to 0$), so that $\text{dist}(Y, Y') \leq C(\mu(Y), \mu(Y'))$.

EXAMPLES. If X in the previous example is foliated (i.e. partitioned) into the orbits of an *amenable* group G acting on X, then the resulting d on

X is, essentially, never concentrated. But if G has property T, then it is concentrated.

Let X_1, X_2, ... be a sequence of Riemannian manifolds and X be the infinite Cartesian product, $X_1 \times X_2 \times \ldots$, where the "metric" between infinite sequences $x = (x_1, x_2 \ldots)$ and $y = (y_1, y_2, \ldots)$ is the Pythagorean one,

$$\mathrm{dist}(x, y) \overset{def}{=} \left(\mathrm{dist}^2(x_1, y_1) + \mathrm{dist}^2(x_2, y_2) + \ldots \right)^{1/2}.$$

What goes wrong here is that $\mathrm{dist}(x, y) = \infty$ for most x and y but we can tolerate this in the presence of the product measure μ on X coming from normalized Riemannian measures on X_i. If the first (sometimes called second) eigenvalues of (the Laplace operator Δ on) X_i are separated from zero, $\lambda_1(X_i) \geq \varepsilon > 0$, then the product is concentrated and, moreover, one can make a meaningful analysis (on functions and often on forms) on X. Furthermore, if $\lambda_1(X_i) \to \infty$ for $i \to \infty$, then Δ has discrete spectrum on X with finite multiplicity (computed by the usual formula for products). For instance, if X_i are n_i-spheres of radii R_i, then $\lambda_1(X_i) = n_i R_i^{-2}$ and their product is concentrated for $R_i \leq \varepsilon \sqrt{n_i}$ with extra benefits for $\sqrt{n_i}/R_i \to \infty$.

One can deform and modify products, retaining concentration, e.g. for projective limits of some towers of smooth fibration, such as the infinitely iterated unit tangent bundles of Riemannian manifolds.

Similarly to products, the spaces X of maps between Riemannian manifolds, $A \to B$, carry (many different) "foliated Hilbert manifold" structures which in the presence of measures (e.g. Wiener measure for 1-dimensional A) allow analysis on X.

Now comes a painful question: are these X good for anything? Do they possess a structural integrity or just encompass (many, but so what?) examples? A convincing theorem is to be proved yet.

REMARKS AND REFERENCES. (a) The F-theorem makes the core of Bieberbach's solution to a problem on Hilbert's list (N. 18, where the n-dimensional hyperbolic case is also mentioned and dismissed as adding little new to the results and methods of Fricke and Klein).

(b) Our definition of $K \gtrless 0$ is motivated by [LS], where the authors prove the K-theorem for maps from spaces with $K \geq \lambda$ to those with $K \leq \lambda$ (in the sense of Alexandrov) under a mild restriction ruling out, for example, maps $S^n \to S^m$ for $m < n$, where K-property obviously fails to be true but, unfortunately, missing maps $S^n \to S^m$ for $m \geq n$ where it is known to be true (a conclusive version seems not hard but no published proof is available). Also, there is a Lipschitz extension result from arbitrary metric spaces to those with $K \leq 0$ (due to Lang, Pavlovic and Schroeder).

(c) The theory of spaces with $K \geq 0$ (as well as with $K \leq \lambda < \infty$) is by now well developed (see [P]) and intrinsically attractive. Yet it suffers from the lack of a systematic process of generating such spaces apart from convex hypersurfaces (and despite several general constructions: products, quotients, spherical suspension ...). Also, there is no serious link with other branches of geometry, not even with the local theory of Banach spaces, and there are few theorems (K-theorem is a happy exception) where the conclusion is harder to verify (in the available examples) than the assumptions (as is unfortunately frequent in global Riemannian geometry). A possible way of enriching (and softening) $K \geq 0$ is letting $n \rightarrow \infty$ (and taking $n = \infty$ seriously) as is done for Banach spaces. (See [Ber] for a broader perspective, [Pe] for a most recent account and [G9] for a pedestrian guide to curvature.)

(d) **Ricci**. Analytically speaking, the most natural of curvatures is the Ricci tensor that is the quadratic differential form on V associated to g via a (essentially unique) Diff-invariant second order quasi-linear differential operator (like **S**) denoted $\mathrm{Ri} = \mathrm{Ri}(g)$. Manifolds with positive (definite) Ri generalize those with $K \geq 0$ but they admit no simple metric description as their essential features involve the Riemannian measure on V, e.g. R-balls in V with $\mathrm{Ri} \geq 0$ have smaller volumes than the Euclidean ones. (If $K \geq 0$ is motivated by convexity, then $\mathrm{Ri} \geq 0$ can be traced to positive mean curvature of hypersurfaces.) Thus it remains unclear how far the idea of $\mathrm{Ri} \geq 0$ extends beyond the Riemannian category: what are admissible singularities and what happens for $n = \infty$? (See [G9] and [G1] for an introduction and [CC] for the present state of the art.)

Encouraged by $\mathrm{Ri} \geq 0$ one turns to $\mathrm{Ri} \leq 0$ formally generalizing $K \leq 0$ but the naive logic does not work: *every metric can be approximated by those with $\mathrm{Ri} < 0$ by the Lohkamp h-principle* and no hard structural geometry exist.

(e) **Einstein and the forlorn quest for the best metric.** It is the geometers dream (first articulated by Heinz Hopf, I believe) to find a canonical metric g_{best} on a given smooth manifold V so that all topology of V will be captured by geometry. This happened to come true for surfaces as all of them carry (almost unique) metrics of constant curvature and is predicted for $n = 3$ by Thurston's geometrization conjecture. Also, there is a glimpse of hope for $n = 4$ (Einstein, self-dual) but no trace of g_{best} has ever been found for $n \geq 5$. What is the reason for this? Let us take some (energy) function E on the space $\mathcal{G}$ of metrics, say built of the full curvature tensor, something like $\int (\mathrm{Curv}(g))^{\frac{n}{2}} dv$ (where the exponent $n/2$ makes the integral scale invariant). Imagine, the gradient flow of E brings all of $\mathcal{G}$ to a "nice" subspace $\mathcal{G}_{\mathrm{best}} \subset \mathcal{G}$ (ideally, a single point or something not very large anyway). Then the group Diff V would act on $\mathcal{G}_{\mathrm{best}}$ (as all we do should be Diff-invariant) with compact isotropy subgroups (we assume V is compact at the moment), e.g. if $\mathcal{G}_{\mathrm{best}}$ consisted of a single point, then Diff V would isometrically act on $(V, \mathcal{G}_{\mathrm{best}})$. But the high dimensional topology (unknown to Hopf) tells us that the space Diff V is too vast, soft and unruly to be contained in something nice and cosy like the desired $\mathcal{G}_{\mathrm{best}}$. (Diff is governed by Waldhausen K-theory bringing lots of homotopy to Diff which hardly can be accounted for by a rigid geometry. And prior to Waldhausen our dream was shattered by Milnor's spheres ruling out smooth canonical deformations of general metrics on the spheres S^n, $n \geq 6$, to the standard ones.)

Besides topology, there is a geometric reason why we cannot freely navigate in the rugged landscapes of spaces like Diff. To see the idea, let us look at the simpler problem of finding the "best" closed curve in a given free homotopy class of loops in a Riemannian manifold V. If $K(V) \leq 0$, for instance, the gradient flow (of the energy function on curves) happily terminates at a unique geodesic: this is the best we could hope for. But suppose $\pi_1(V)$ is computationally complicated, i.e. the word problem cannot be solved by a fast algorithm, say, unsolvable by any algorithm at all. Our flow (discretized in an obvious way) is a particular algorithm, we know it must *badly* fail, and the only way for it to fail is to get lost and confused in deep local minima of the energy. Thus our V must harbor lots and lots of locally minimal geodesics in each homotopy class, in particular, infinity of contractible closed geodesics, disrupting the route from topology to simple geometry.

Well, one may say, let us *assume* $\pi_1(V) = 0$. But what the hell does "assume" mean? Given a V, presented in any conceivable *geometric* form (remember, we are geometers here, not shape-blind topologists), there is no way to check if $\pi_1(V) = 0$ since this property is not algorithmically verifiable. Consequently, there are innocuously looking metrics on such manifolds as S^n for $n \geq 5$, teeming with short closed curves which no human being can contract in a given stretch of time. (In fact, a predominant majority of metrics are like this for $n \geq 5$.) A similar picture arises for higher dimensional (e.g. minimal) subvarieties (with extra complication for large dimension and codimension, even for simplest V such as flat tori, where the trickery of minimal subvarieties was disclosed by Blaine Lawson), and by the work of Alex Nabutovski the spaces like $\mathcal{G}(V)/\operatorname{Diff}(V)$ harbor the same kind of complexity rooted in the Gödel-Turing theorem.

Following Alex we (I speak for myself) are lead to the pessimistic conclusion that there is no chance for a distinguished g_{best} (or even $\mathcal{G}_{\text{best}} \subset \mathcal{G}$) for $n \geq 5$ and that "natural" metrics, e.g. Einstein $\mathcal{G}$ with $\operatorname{Ri}(g) = \lambda g$ for $\lambda < 0$, must be chaotically scattered in the vastness of $\mathcal{G}$ with no meaningful link between geometry and topology. (This does not preclude, but rather predicts, the existence of such metrics, e.g. Einstein, on all V of dimension ≥ 5: the problem is there may be too many of them.)

On the optimistic side, we continue searching for g_{best} in special domains in $\mathcal{G}$, e.g. following Hamilton's Ricci flow (say, for Ricci ≥ 0 or $K \leq 0$) or stick to dimensions three (where Michael Anderson makes his theory) or four (where Taubs discovered certain softness in selfduality). Alternatively, we can enlarge (rather than limit) the category and look for extremal (possibly) singular varieties with only partially specified topology, i.e. with a prescribed value of a certain topological invariant, such as a characteristic number or the simplicial volume. For example, each (decent) topological space X is, rather canonically, accompanied by metric spaces homotopy equivalent to it, such as a suitably subdivided semisimplicial model of X which is an infinite dimensional simplicial complex, call it $X_\Delta \overset{\text{hom}}{\sim} X$, where the metric on X_Δ comes from a choice of a standard metric on each k-simplex, $k = 1, 2, \ldots$. This X_Δ is too large to please our eye, but it may contain some distinguished subvarieties, e.g. minimizing homology classes in $H_*(X_\Delta) = H_*(X)$, at least for certain spaces X (see 5 H_+ in [G1] and references therein).

(f) Anisotropic metrics appear under a variety of names: "non-holonomic", "control", "sub-elliptic", "sub-Riemannian", "Carnot-Caratheodory", bearing the traces of their origin. They were extensively studied by analysts since Hörmander's work on hypo-elliptic operators but I do not know where and when they were promoted from technical devices to full membership in the metric community. Now, besides P.D.E., they help group theorists in the study of discrete nilpotent groups Γ since the *local* geometry of (especially self-similar) metrics on nilpotent Lie groups $G \supset \Gamma$ adequately reflects the *asymptotic* geometry of Γ. The local geometry of dist_H reduces, in turn, to that of the polarization $H \subset T(G)$ which is rather soft as far as low dimensional H-horizontal (i.e. tangent to H) subvarieties in G are concerned as follows from the Nash implicit function theorem. This provides some information on dist$_H$-minimal subvarieties in G (alas, only the case of surfaces is understood) and allows us to evaluate the *Dehn* function of Γ (see [G3], [G2] and references therein). Also observe that (however meager) results and problems in the anisotropic geometry may serve as model for other soft spaces (of solutions of P.D.E. as H-horizontality for H).

(g) It is helpful to think of a (metric) measure space (X, μ) as a high dimensional configuration space of a physical system (which is, indeed, often the case). Here $f : X \to Y$ is an *observable* projecting X to a low dimensional "screen", our Y such as the space-time $\mathbb{R}^4$, for instance, where $f_*(\mu)$ is what we see on the screen. The concentration of $f_*(\mu)$, ubiquitous in the probability theory and statistical mechanics, was brought to geometry (starting from Banach spaces) by Vitali Milman following the earlier work by Paul Levy. The Levy-Milman concentration phenomenon has been observed for a wide class of examples, where, besides the mere concentration, one is concerned with large deviations and fluctuations. (See [Mi], [G1] and references herein.) Unfortunately, our definition of concentrated spaces does not capture large deviations (which are more fundamental than fluctuations). Possibly, this can be helped by somehow enriching the structure. Besides, one can proceed by allowing variable measures (states) μ as in the cube $\{0, 1\}^N$ with the product μ_p of N-copies of the $(p, 1 - p)$-measure on $\{0, 1\}$ and in Gibbs measures parametrized by temperature. Then, inspired by physics, one wonders what should be general objects responsible for concentration in quantum statistical mechanics (where the concentration is limited by the Heisenberg principle) and, finally, one may turn to non-metric structures which may come in probability and geometry along with concentration.

4 Life Without Metric

It is (too) easy to concoct invariants of a metric space V, e.g. by looking at the ranges of Cartesian powers of the distance function mapping V^k to $\mathbb{R}^{\frac{k(k-1)}{2}}$ for

$$d^k : (v_1, \dots, v_i, \dots, v_k) \mapsto \{\mathrm{dist}(v_i, v_j)\}_{1 \le i < j \le k}.$$

(The *diameter* of V appears, for example, as the maximal segment in the image of d^k in $\mathbb{R}^1_+$ for $k = 2$.) Then, there are various (often positive) "energy functions" E on spaces of subvarieties W in X and maps $f : V \to V_0$ such as $W \mapsto \mathrm{Vol}_m W$ for $m = \dim W$, and $\int \|df\|^2 dv$ for Riemannian

V and V_0. Each E generates invariants of V, for instance, infima of E on given classes of W's or f's, (e.g. $\inf \mathrm{Vol}_m W$ for $[W] = h \in H_m(V)$) or, more generally, the full *Morse landscape* of E including the spectrum of the critical values of E (e.g. the spectrum of the Laplace operator on functions and form on a Riemannian V). These invariants I provide us with the raw material for asking questions and making conjectures: what are possible values of $I = I(E)$ and relations between I's for different E's? What are spaces V with a given behaviour of an I?, etc. But spaces without metrics become rather slippery, hard to grasp and assess. Just look at a foliation or dynamical system F on a manifold V. The essential invariants (entropy, asymptotics of periodic orbits) often change discontinuously under deformations (perturbations) of the structure and are hard to evaluate, even approximately, for a given F. And it is not easy at all to come up with new numerical invariants making sense for all objects in the category. This is due to the fact that the (local) group of (approximate) automorphisms of such a structure (at a point $v \in V$) is potentially *non-compact*. Consequently, the action of $\mathrm{Diff}(V)$ on the space $\mathcal{F}$ of our structures F on V may have non-trivial dynamics (e.g. non-compact isotropy groups $\mathrm{Is}_F = \mathrm{Aut}(V, F)$, $F \in \mathcal{F}$) making the quotient space $\mathcal{F}/\mathrm{Diff}\,V$, where our invariants are supposed to live, non-separable. (Intuitively, invariants should be independent of observers attached to different coordinate systems in V; if there are non-compactly many equivalent observers it becomes difficult to reconcile their views, as is in the special and general relativity, for example.)

Now we glance at a couple of H-*structures* for interesting non-compact subgroups $H \subset GL_n(\mathbb{R})$ (where the compact case of $H = O(n)$ corresponds to the Riemannian geometry).

$\mathbb{C}$-structures. These, customarily called *almost complex* structures J on V, are fields of $\mathbb{C}$-linear structures J_v in $T_v(V)$, $v \in V$. Such a J may be expressed by an anti-involution, also called $J : T(V) \to T(V)$ (corresponding to $\sqrt{-1}$), where the pertinent H is (non-compact!) $GL_m(\mathbb{C}) \subset GL_n(\mathbb{R})$ for $n = 2m$. Morphisms, called $\mathbb{C}$-*maps* $f : V_1 \to V_2$, where the differential $Df : T(V_1) \to T(V_2)$ is $\mathbb{C}$-linear (i.e. commute with J), are rare for (non-integrable) V_1 and V since the corresponding (elliptic) P.D.E.-system is over-determined for $\dim_{\mathbb{R}} V_1 > 2$. So we stick to $\mathbb{C}$-*curves*, maps of Riemann surfaces $S \to V = (V, J)$, also called J-*curves* if this matters. We mark each S with a point $s \in S$ and then the totality of $\mathbb{C}$-curves makes a huge space $\mathcal{S} = \mathcal{S}(V, J)$ foliated into surfaces, where each leaf in $\mathcal{S}$ is represented by a fixed $S \to V$ with variable marking $s \in S$.

What is the (possible) global geometry (e.g. dynamics) of S and how can it be read from J?

Start with the subfoliation $\mathcal{C} \subset \mathcal{S}$ of *closed* leaves in $\mathcal{S}$ corresponding to closed (Riemann) surfaces $S \to V$, and try to mimic the geometry of curves in complex algebraic (first of all, projective) varieties V. This $\mathcal{C}$ is filtered by the *degrees* d of $S \subset V$ (playing the role of periods of closed orbits, say for actions of $\mathbb{Z}$ on some space) where curves of degree d may degenerate to (several) curves of lower degree thus compactifying each (moduli) space $\mathcal{C}_d$ by low dimensional strata built of $\mathcal{C}_i$ with $i < d$. Conversely, one can often *fuse* lower degree curves to higher degree d by deforming their (reducible) unions to irreducible $S \in \mathcal{C}_d$.

What happens to this web of algebraic curves of degree d when we slightly perturb the underlying J_0?

The answer depends on the *virtual* dimension of the space $\mathcal{C}_d$, i.e. the Fredholm index of the elliptic system defining $S \in \mathcal{C}_d$. For example, the curves in abelian varieties $\mathbb{C}^m/\Lambda$ are unstable under (even integrable) deformations of J_0, but the curves in certain V's (with sufficiently ample anti-canonical bundles) such as $\mathbb{C}P^m$, remain essentially intact under small (and large as we shall see below) deformations of J_0 (yet the shape and position of $\mathcal{C}_d$ in $\mathcal{S}$ may be, a priori, greatly distorted by ε-deformations of J_0 for d large compared to ε^{-1}). As we follow a deformation J_t moving it further away from the original J_0, a curve $S_0 \subset V$, persistent for small t, may eventually perish by blowing up to something non-compact as t reaches some critical value t_c. What is needed to keep S alive (as a *closed* J_t-curve) is an a priori bound on the area of $S \to (V, J = J_t)$ (measured with some background Riemannian metric g in V, where a specific choice of g is not important as we deal here with compact V). Such a bound is guaranteed by the following *tameness* assumption of J which limits the area of (even approximately) J-holomorphic curves S by their topology, namely by $[S] \in H_2(V)$.

Call J *tamed* by a *closed* 2-form ω on V, if ω is positive on all J-curves in V (i.e. $\omega(\tau, J\tau) > 0$ for all non-zero vectors $\tau \in T(V)$). If so, J is *tame with respect to the cohomology class* $h = [\omega] \in H^2(V)$: the area of each closed oriented "approximately J-holomorphic" curve $S \subset V$ is bounded in terms of the value $h[S] \in \mathbb{R}$. To make it precise, denote by $S_\varepsilon \subset S$ the set of points $s \in S \subset V$, where the plane $T_s(S) \subset T_S(V)$ is ε-close to a $\mathbb{C}$-line (for a fixed background metric). Then "tame" means the existence of ε, δ, $C > 0$, such that the inequality area $S_\varepsilon \geq (1-\delta)$ area S implies $h(S) \geq C^{-1}$ area S for all closed oriented surfaces $S \subset V$. Clearly, "ω-tame" $\Rightarrow$ "h-tame" but the converse ($\forall h \exists \omega \ldots$) remains questionable. (One may try the Hahn-Banach theorem, especially for dim $V = 4$.)

If (V, J_0) is a Kähler (e.g. algebraic) manifold, then J_0 is tamed by the (symplectic) Kähler form ω and as far as J_t remains $[w]$-tame we have decent moduli spaces of J_t-holomorphic curves in (V, J_ε) (where $J = J_t$ may be quite far from J_0, e.g. $J = A J_0$, for an arbitrary symplectic automorphism $T(V) \rightarrow T(V)$). For example, if we start with the standard $(\mathbb{C}P^m, J_0)$ and $J = J_1$ is joined with J_0 by a homotopy of $[\omega_0]$-tame structure J_t (for the standard symplectic 2-form ω_0 on $\mathbb{C}P^m$), then $(\mathbb{C}P^m, J_1)$ admits a rational J_1-curve S of degree 1 through each pair of points, where, moreover, S is unique for $m = 2$. (This remains true for all ω-tame structures on $\mathbb{C}P^2$, with *no* a priori assumption $\omega = \omega_0$, by the work of Taubs and Donaldson.) But what happens to closed $\mathbb{C}$-curves at the first moment t_c when J_t loses tameness? What kind of subfoliation $S_d \subset S$ is formed by the limits of $S \in \mathcal{C}_d$ for $t \rightarrow t_c$? It seems, at least for dim $V = 4$ (e.g. for $\mathbb{C}P^2$ and $S^2 \times S^2$), that most of the closed $\mathbb{C}$-curves blow up simultaneously forming a regular (foliated-like) structure in V. (This is reminiscent of how Kleinian groups degenerate remaining discrete and beautiful at the verge of extinction.)

Are there non-tame (V, J) with rich moduli spaces of closed (especially rational) curves, say having such a curve passing through each pair of points in V? (If a 4-dimensional (V, J) has many J-curves, it is tame by an easy argument. On the other hand the majority of higher dimensional (V, J) contain isolated pockets of J-curves with rather shapeless and useless $\mathcal{C}_d$ like closed geodesics in (most) Riemannian manifolds lost in accidental wells of energy.)

Turning to non-closed $\mathbb{C}$-curves we find a prerequisite for the Nevanlinna theory as they share (the principle symbol of) $\bar{\partial}$ with ordinary holomorphic functions and maps. (This is also crucial for the study of closed $\mathbb{C}$-curves.) For example, we can define *hyperbolic* (V, J) which receive *no* non-constant J-maps $\mathbb{C} \rightarrow V$ and these V (we assume compactness) carry a non-degenerate *Kobayashi metric*, i.e. the supremal metric for which the $\mathbb{C}$-maps $H^2 \rightarrow V$ are 1-Lipschitz. This hyperbolicity has a point in common with tameness: the space of $\mathbb{C}$-maps $f : S^2 \rightarrow V$ with $\bar{\partial}f$ ranging in a compact set is compact for hyperbolic V. (This is also implied by the "tame" bound area $S \leq \text{const}([S])$ for approximately J-holomorphic spheres in V, provided there is no J-holomorphic spheres in V.) Consequently, for each $\mathbb{C}$-structure on $W = V \times S^2$ compatible with J on the fibers $V \times s$, $s \in S^2$, there is a rational (i.e. spherical) $\mathbb{C}$-curve in W passing through a given point $w \in W$ that *contractibly* projects to W. (This remains valid for irrational curves S if the Teichmüller space of S is incorporated into W.) Another link between "tame" and "hyperbolic" is expressed by the following (easy to prove) topological criterion for hyperbolicity. Let $\tilde{V}$ be a Galois covering of V and $\tilde{\ell}$ be a 1-form with sublinear growth ($\|\tilde{\ell}(\tilde{v})\|/\operatorname{dist}(\tilde{v}, \tilde{v}_0) \to 0$ for $\tilde{v} \to \infty$), and with invariant (under the deck transformation group) differential $\tilde{w} = d\tilde{\ell}$. *If J is tamed by the corresponding class $[w] \in H_2(V)$, then (V, J) is hyperbolic.* (For example, if J is tame, $\pi_2(V) = 0$ and $\pi_2(V)$ is hyperbolic, then (V, J) is hyperbolic.)

Are there further topological criteria for (non)-hyperbolicity (where $\pi_1(V)$ is not so large)? For example, are there ω-tame hyperbolic structures on the 4-torus? (Parabolic curves, i.e. $\mathbb{C}$-maps into such torus (T^4, w, J) could help us to study ω, where the ultimate goal is to show that every symplectic structure ω on T^4 is isomorphic to the standard one as is known for $\mathbb{C}P^2$, for instance by the work of Taubs and Donaldson.)

What are essential metric properties of $(V, \operatorname{dist}_{\mathrm{Kob}})$ for hyperbolic V? Nothing is known about it. *How much can one deform a hyperbolic J keeping hyperbolicity?* It is clear (by Brody's argument) that small perturbations do not hurt. Take, for instance, a (necessarily singular) J-curve S_0 in an ω-tame $(\mathbb{C}P^2, J)$, where the fundamental group of the complement is hyperbolic (e.g. free non-cyclic which is the case for S_0 consisting of $d + 1$ rational curves of degree 1 where exactly d of them meet at a single point and $d \geq 3$). Then every non-constant J-map $\mathbb{C} \to (\mathbb{C}P^2, J)$ meets S_0 and if we slightly deform S_0, the complement of the resulting complement $\mathbb{C}P^2 \backslash S_\varepsilon$ remains hyperbolic, provided each irreducible component of S_0 minus the remaining components were hyperbolic (as happens if the above $d + 1$ lines are augmented by another one meeting each at a single point). Thus, for every $d \geq 5$, there is a non-*empty* open subset $\mathcal{H}_d \subset \mathcal{C}_d$ of smooth J-curves S in $(\mathbb{C}P^2, J)$ of degree d where every non-constant J-map $\mathbb{C} \to \mathbb{C}P^2$ meets S. Probably, this $\mathcal{H}_d$ is dense as well as open for large d possibly, depending on J, (where the case of curves with many, depending on J, irreducible components seems within reach). A similar observation can be made for other compact V and differences of these, e.g. for those associated to tori, but nowhere does one come close to what is known in the classical algebraic case. Here is another kind of (test) question with no classical counterpart:

Given a hyperbolic (V, J_0), when and how can one modify J_0 inside a (arbitrarily) small neighbourhood $U \subset V$, such that the resulting (V, J_1) admits a parabolic (or even rational) curve through each point $v \in V$?

If (V, J) is hyperbolic then the space $\mathcal{H}$ of J-maps $H^2 \to V$ is *compact* and projects *onto* $\mathcal{S}$ with circle fibers. The group $G = PSL_2\mathbb{R} = \mathrm{Iso}\, H^2$ naturally acts on $\mathcal{H}$ and periodic (i.e. compact) orbits correspond to closed J-curves $S \subset V$, where genus $(S) \geq \mathrm{const}(\mathrm{Area}\, S)$. If (V, J) is algebraic, then periodic orbits are dense in $\mathcal{H}$; besides, there are many finite dimensional invariant subsets in $\mathcal{H}$ corresponding to (solutions of) algebraic O.D.E. over V. *Are finite dimensional invariant subsets dense in $\mathcal{H}$ for all (tame) V? When does V contain a metrically complete J-curve $S \subset V$, preferably with locally bounded geometry? What are (if any) invariant measures in $\mathcal{H}$? ω out of $\mathcal{H}$ for ω-tame J in the absence of closed J-curves?* Take, for example an algebraic (V, J_0) with a nice foliation, e.g. a flat connection over a Riemann surface or the standard foliation on a Hilbert modular surface. *How does such a foliation fare under tame homotopies J_t?*

If (V, J) is non-hyperbolic then the main issue is to understand the space $\mathcal{P} \subset \mathcal{S}$ of parabolic leaves, or equivalently, the space $\mathcal{P}'$ of non-constant J-maps $\mathbb{C} \to V$ with the action of $G = \mathrm{Aff}\,\mathbb{C}$. One knows, for the standard $\mathbb{C}P^n$, that rational maps are dense in $\mathcal{P}'$ (and hence in $\mathcal{H}$) by the Runge theorem. This, probably, is true for all tame (V, J) with "sufficiently many" rational curves (but this seems unknown even for rationally connected *algebraic V*). Let, for instance, $V = (\mathbb{C}P^n, J)$ where J is tamed by the standard ω. Then, by fusing (a sequence of) rational curves, one can obtain a parabolic one containing a given countable subset in V. Most likely, one can prove the Runge theorem(s) for V in this way following Donaldson's approach to Runge for Yang-Mills.

Denote by $\mathcal{B} \subset \mathcal{P}'$ the space of *non-constant* 1-Lipschitz $\mathbb{C}$-maps $\mathbb{C} \to V$ and observe that the closure of every G-orbit in $\mathcal{P}'$ meets $\mathcal{B}$ by the Bloch-Brody principle. If V contains (many) rational curves, then $\mathcal{B}$ is quite large, e.g. if $V = (\mathbb{C}P^n, J_{\mathrm{stand}})$, then there is a map $f \in \mathcal{B}$ interpolating from an arbitrary δ-separated subset $\Delta \subset \mathbb{C}$; probably, this remains true for all J tamely homotopic to the standard one as well as for more general "rationally connected" V. Conversely, (a strengthened version of) this interpolation property is likely to imply the existence of (many) rational curves in V.

Apart from rational curves, one can sometimes produce $\mathbb{C}$-maps $\mathbb{C} \to V$ by prescribing some asymptotic boundary conditions, e.g. by proving non-compactness of the space of J-disks with boundaries on a non-compact family of Lagrangian subvarieties. This works for many J on the *standard* symplectic $\mathbb{R}^{2m}$ and yields, for instance, parabolic curves for J on tori tamed by the standard ω. *How large is $\mathcal{B}$ for these tori?* Hopefully, the mean (ordinary?) dimension of $\mathcal{B}$ (for the action of $\mathbb{C} \subset G$ on $\mathcal{B}$) is finite and the natural map from $\mathcal{B}$ to the projectivized $H_2(V)$ is non-ambiguous and somehow represents the (homology) class of $\mathbb{C}P^{m-1}$ corresponding to $\mathbb{C}$-lines in $H_1(V) = \mathbb{C}^m = \mathbb{R}^{2m}$.

Pseudo-Riemannian manifolds. Given an H-structure g on a manifold V for a non-compact group $H \subset GL_n(\mathbb{R})$, $n = \dim V$, one may rigidify (V, g) by reducing H to a maximal compact subgroup $K \subset H$, i.e. by considering a Riemannian metric g_K on V compatible with g. For example, if $g(= J)$ is a $\mathbb{C}$-structure, then $K = U(m) \subset GL_m(\mathbb{C})$, $m = \frac{n}{2}$, and g_K is an Hermitian metric; if g is pseudo-Riemannian, i.e. a quadratic differential form on V of type (p, q) with $p + q = n$, then $K = O(p) \times O(q) \subset O(p, q)$ and g_K is a Riemannian metric such that there is a frame $\tau_1^+, \ldots, \tau_p^+, \tau_1^-, \ldots, \tau_q^-$ at each point, where both forms g and g_K become diagonal with $g_K(\tau_i^+, \tau_i^+) = g(\tau_i^+, \tau_i^+)$ and $g_K(\tau_j^-, \tau_j^-) = -g(\tau_j^-, \tau_j^-)$. *What properties (invariants) of g can be seen in an individual g_K and/or in the totality $\mathcal{G}_K = \mathcal{G}_K(g)$ of all g_K?* It may happen that two structures g and g' are virtually indistinguishable in these terms, namely when for each $\varepsilon > 0$, there exist g_K and g_K' which are $(1+\varepsilon)$-bi-Lipschitz equivalent. (The Diff-orbit of $\mathcal{G}_K$ might be C^0-dense in the space of all Riemannian metrics under the worst scenario.) There are few known cases where $\mathcal{G}_K$ tells you something useful about g. An exceptionally pleasant example is given by *conformal* structures g with $\mathcal{G}_K(g)$ telling you everything about g (i.e. the C^0-closures of the Diff-orbits of $\mathcal{G}_K$ and $\mathcal{G}_K'$ are essentially disjoint unless g and g' are isomorphic; furthermore the action of Diff V on the space of conformal structures is proper apart from the standard conformal S^n). Next, if H preserves an exterior r-form on $\mathbb{R}^n$, and the corresponding form $\Omega = \Omega(g)$ on V is closed, then g_K is bounded from below by volumes of r-cycles $C \subset V$ with $\Omega(C) \neq 0$ and, at least, Diff $\mathcal{G}_K$ is not dense. Finally, for certain symplectic structures $g = \omega$, the minimal g_K-areas of some 2-cycles C (realizable by $\mathbb{C}$-curves) equal $\omega(C)$ for all g_K, thus limiting geometry of (V, g_K) in a more significant way.

Having failed with (robust) Lipschitz geometry of g_K one resorts to curvature and looks at the subsets $\mathcal{G}_K(\kappa) \subset \mathcal{G}_K$ of adapted metrics $g_K = g_K(g)$ with some (norm of) curvature bounded by $\kappa \in \mathbb{R}$. Here the mere fact of $\mathcal{G}_K(\kappa)$ being empty for a given κ, gives one a non-trivial complexity bound on g. For example, one may study the infimal $\kappa \in \mathbb{R}_+$, such that g admits g_K with the sectional curvatures between $-\kappa$ and κ (with normalized volume if $H \not\subset SL_n\mathbb{R}$). And again one is tempted to search for the "best" g_K adapted to g with non-zero expectation for "mildly non-compact" H inspired by conformal metrics of constant scalar curvature delivered by the Schoen-Aubin solution to the Yamabe problem.

Now we turn to the case at hand, pseudo-Riemannian g of type (p, q) on a (typically compact) manifold V having closer kinship to Riemannian geometry than general g. *What are (most general) morphisms in the pseudo-Riemannian category comparable to "Lipschitz" for Riemannian manifolds?*

Since we can compare metrics on a fixed manifold by $g \leq g'$ for $g - g'$ being positive semidefinite, we may speak of $(+)$-*long* maps $f : V \to V'$ where $f^*(g') \geq g$. However, unlike the Riemannian case where short (rather than long) maps are useful for all V and V', this makes sense only for (p, q) not being too small compared to (p', q'): if $p' > p$ then every isometric immersion $f_0 : V \to V'$ can be a little C^0-perturbed to some f with $f^*(g')$ being as large as you want (actually equal to a given $g_1 > f_0^*(g')$ homotopic to g by the Nash-Kuiper argument), and if also $q' > q$, everything becomes soft and one gets all metrics on V (homotopic to g) by C^1-immersions arbitrarily C^0-close to f_0. Thus we stick to $p = p'$ and start with *positive slices* in V, i.e. immersed p-dimensional $W \subset V$ with $g|W > 0$. If W is connected with *non-empty* boundary, we set $R(W, w) = \text{dist}(w, \partial W)$ for the induced Riemannian metric ($\sup_w R(w, W)$ is called "in-radius" of W), and define $R_+(V, v)$ as the supremum of $R(v, W)$ for all positive slices through v. (Connected *Riemannian* manifolds with *non-empty* boundaries have $R_+ = R < \infty$ at all their points while closed manifolds have $R_+ < \infty$ if and only if $\pi_1 < \infty$. But even for $\pi_1 = \{e\}$ there is no *effective* bound on R_+ due to non-decidability of $\pi_1 \overset{?}{=} \{e\}$. Moreover, there are rather small metrics on S^3 with almost negative curvature and thus with arbitrarily large R_+). Next, we dualize by taking $(+)$-long maps f from (V, v) to Riemannian (W, w) of dimension p and denote by $\underline{R}_+(V, v)$ the infimum of $R_+(W, w)$ over all possible (W, w) and f, where, clearly, $\underline{R}_+ \geq R_+$.

A $(+)$-long map $V \to W$ is necessarily a *negative* submersion, i.e. with g-negative fibers. Conversely, given a negative submersion f (proper, if V is non-compact) of V to a smooth manifold W, one can find a Riemannian metric on W making f long. (There is a unique supremal Finsler metric on

W making f long, which can be minorized by a Riemannian one.) Thus, for example, if a compact (possibly with a boundary) V admits a negative submersion into a connected manifold with finite π_1 (or with non-empty boundary), then $R_+(V) < \infty$.

Manifolds with $R_+(V,v) \leq \text{const} < \infty$ for all $v \in V$ are kind of hyperbolic in the $(+)$-directions (e.g. this condition is C^0-stable). If $p = 1$ and V is closed, then it always has a circular positive slice and $R_+(V,v) = \infty$ for all $v \in V$, but if $p \geq 2$, then every $g = g_0$ admits a deformation g_t with $R_+ \to 0$ for $t \to \infty$: take a generic (and thus non-integrable) p-plane field $S_+ \subset T(V)$ with $g|S > 0$ and make $g_t = g + tg_-$ where g_- is the negative part of g for the normal splitting $T(V) = S_+ \oplus S_+^\perp$. (This works even locally and shows that the majority of g have R_+ small in some regions of V and infinite in other regions.)

It seems by far more restrictive to require that both R_+ and R_- (i.e. R_+ for $-V$) are bounded on V, where a sufficient condition for compact V is the existence of a negative submersion $V \to W_+$ as well as a positive submersion $V \to W_-$, where W_+ and W_- have $\pi_1 < \infty$ or non-empty boundaries. This can be slightly generalized by allowing somewhat more general pairs of $\pm$-foliations with uniformly compact fibers (e.g. coming from submersion to simply connected or to bounded *orbifolds*) and sometimes one submersion suffices. For example, start with a negative submersion $f : V \to W_+$ where W_+ is simply connected (or bounded) and the (negative) fibers of f are also simply connected (or bounded). Deform the original g on V to g_1 agreeing with g on the fibers while being very positive normally to the fibers. Then all negative slices in (V, g_1) keep C^1-close to the fibers of f and so $R_-(V, g_1)$ is bounded as well as R_+. (Probably, there are more sophisticated, say closed manifolds V, where the bound on $R_\pm(V, v)$ comes from different sources at different $v \in V$.)

Despite some hyperbolic features, $\pm$ bounded pseudo-Riemannian metrics on closed manifolds are reminiscent of positive curvature, e.g. they are accompanied in known examples by *closed* positive or negative slices with $\pi_1 < \infty$ (are these inevitable?) and seem hard to make on aspherical V.

Besides taming g by $\pm$ foliations, one may try (pairs of) differential forms on V, where a closed p-form ω is said to be (strictly) $(+)$-*tame* g if it is (strictly) non-singular on the positive slices. For example, if V is metrically split, $V = V_+ \times V_-$, the pull-back ω_+ of the volume form of V_+ strictly $(+)$-tames g; similarly ω_- strictly $(-)$-tames g, while $\omega_+ + \omega_-$ is

$\pm$ taming, albeit non-strictly. *How much do closed positive slices persist under (strictly) (+)-tame homotopies of g? In particular, what happens to area maximizing closed positive 2-slices in $(2, q)$-manifolds under strictly (+)-tame homotopies?*

The tangent bundle to the space $\mathcal{S}_-$ of negative slices in V carries a natural (positive!) L_r-norm (we use r as p is occupied) since g is positive normally to $S \subset V$ for all $S \in \mathcal{S}_-$ (as well as for all $S_{p',q} \subset V$ with $0 \leq p' < p$). But the associated (path) metric in $\mathcal{S}_-$ may, a priori, degenerate and even become everywhere zero. Yet, there are some positive signs.

Let V metrically split $V = [0, 1] \times V_-$ for V_- closed. Then the L_1-distance between $0 \times V_-$ and $1 \times V_-$ equals the volume of V and hence > 0. Consequently all L_r-distances are > 0.

Let $V = V_+ \times V_-$ with closed V_- and arbitrary Riemannian V_+. Then $\operatorname{dist}_{L_\infty}(v_+ \times V_-, v'_+ \times V_-) = d = \operatorname{dist}_{V_+}(v_+, v'_+)$, since the projection $V \to V_-$ together with a 1-Lipschitz retraction

$$V_+ \to [0, d] = [v_+, v'_+]$$

give us a (+)-short (i.e. (−)-long) map $V \to [0, d] \times V_-$ and the above applies (only to L_∞ not necessarily to other L_r).

Let V fiber over a compact p-manifold with closed negative fibers. Then the L_∞-distance is > 0 between every two distinct fibers, essentially by the same argument (also yielding positivity of L_∞-distances between general closed negative slices isotopic to the fibers).

How much does the metric geometry of $(\mathcal{S}_-, \operatorname{dist}_{L_\infty})$ (and of $\mathcal{S}_+$) capture the structure of g? How degenerate can $\operatorname{dist}_{L_\infty}$ be for general (V, g)? (This $\operatorname{dist}_{L_\infty}$ is vaguely similar to Hofer's metric in the space of Lagrangian slices in symplectic V, which also suggests vanishing of our $\operatorname{dist}_{L_r}$, $r < \infty$, for most V.)

As we mentioned earlier, a general complication in the study of H-structures g with non-compact $H \subset GL_n\mathbb{R}$ is a possible *non-stability* (or recurrence) of g due to certain unboundedness of the set of diffeomorphisms of V moving g (or a small perturbation of g) close to g. The simplest manifestation of that is non-compactness of the automorphisms (isometry) group of (V, g) which may have different nature for different structures g. For example, non-compactness of the conformal transformations f of S^n is seen in the graphs $\Gamma_f \subset S^n \times S^n$ as degeneration of these to unions of two fibers $(s_1 \times S^n) \cup (S^n \times s_2)$ with an uniform bound on $\operatorname{Vol}\Gamma_f$ for

$f \to \infty$. On the other hand, graphs of isometrics f of (p, q)-manifolds V are represented by (totally geodesic) *isotropic* (where the metric vanishes) n-manifolds $\Gamma_f \subset V \times -V$ and their volumes (as well as in-radii, both measured with respect to some background Riemannian metric in $V \times V$) go to infinity for $f \to \infty$, while their local geometry remains bounded (unlike the conformal case). With this in mind, we call g *stable*, if it admits a C^0-neighbourhood $\mathcal{U}$ in the space $\mathcal{G}$ of all g's, such that the graphs of isometrics $f : (V, g') \to (V, g'')$ with $g', g'' \in \mathcal{U}$ have $\mathrm{Vol}_n(\Gamma_f) \leq \mathrm{const} < \infty$ (where the background metric is not essential as we assume V compact).

EXAMPLE. Start with a metrically split $V = V_+ \times V_-$. The isometrics here are essentially the same as those of the Riemannian manifold $V^+ = V_+ \times (-V_-)$, since $V \times (-V) = V_+ \times V_- \times (-V_+) \times (-V_-) = V_+ \times (-V_-) \times (-V_+) \times V_- = V^+ \times (-V^+)$, and so $V \times (-V)$ and $V^+ \times (-V^+)$ share the same isotropic submanifolds. If V is closed simply connected (or f does not mix up $\pi_1(V_+)$ and $\pi_1(V_-)$ too much) one sees, by looking at (local) isometries of V^+, that all isotropic submanifolds in $V \times (-V)$ have bounded in-radii (as well as volumes, if they are closed) and so $\mathrm{Iso}(V, g)$ is compact. One sees equally well that $g = g_+ \oplus g_-$ is stable, and, moreover, one gets a good control over the stability domain $\mathcal{U}$ of g. Namely, take $\lambda > 0$ and let $\mathcal{U}_\lambda = \mathcal{U}_\lambda(g)$ consist of those g', where the fibers $V_+ \times v_-$ and $v_+ \times V_-$ are g'-positive and g'-negative correspondingly and the projections of these fibers to (V_+, g_+) and (V_-, g_-) are λ-bi-Lipschitz for g' restricted to these fibers. Then (at least for $\pi_1(V) = 0$) the graphs of diffeomorphisms $f : V \to V$ with $\mathcal{U}_\lambda \cap f(\mathcal{U}_\lambda) \neq \emptyset$ are uniformly bi-Lipschitz to V^+, and so all $g' \in \bigcup_{\lambda < \infty} \mathcal{U}_\lambda$ are stable.

What are the most general stable g? Are simply connected manifolds of type $(1, q)$ always stable? (Of course, *generic* g are stable, but we are concerned with exceptional (V, g), e.g. with non-compact group $\mathrm{Iso}(V, g)$.)

The above motivates the idea of *iso-stability* for V of type (p, q) with $p = q$ (e.g. $V = V_0 \oplus -V_0$) limiting the size of ϵ-isotropic submanifolds in V. This is enhanced in the presence of 0-*taming* p-forms ω on V, which do not vanish on the isotropic p-planes in $T(V)$. For example, if $V \times (-V)$ is 0-tame (with $\deg \omega = n$ on this occasion), then V is stable with respect to the diffeomorphisms with the graphs homologous to the diagonal in $V \times V$, as is the case for the above $g \in \mathcal{U}_\lambda$ and for (V, g) tamed by $\pm$-foliations with p and q-volume preserving (or just uniformly bounded) holonomies.

REMARKS AND REFERENCES. (a) Closed $\mathbb{C}$-curves in tame manifolds exhibit a well organized structure with intricate interaction between moduli spaces C_d for different d and regular asymptotics for $d \to \infty$: quantum multiplication, mirror symmetry etc (see [MS]). This also applies to non-closed curves with prescribed Fredholm boundary (or asymptotic) condition, e.g. J-maps of Riemann surfaces with boundaries, $(S, \partial S) \to (V, W)$ for a given *totally real* $W \subset V$, where everything goes as in the boundary free case (including Kobayashi metric, Bloch-Brody etc). Less obvious conditions come up in the study of fixed points of Hamiltonian transformations and related problems: Floer homology, A-categories, contact homology of Eliashberg and Hofer. But it is unclear (only to me?) what is the most general Fredholm condition in the $\mathbb{C}$-geometry.

(b) The questions concerning unbounded $\mathbb{C}$-curves, which parallel (Nevanlinna kind) complex analysis rather than algebraic geometry, remain as widely open as when I collected them for (then expected) continuation of [G6]. *Do the spaces $S, \mathcal{H}, \mathcal{P}, \mathcal{B}$ possess geometric structure comparable to (and compatible with) what we see in $\mathcal{C} = \cup_d C_d$? What is the right language to describe such a structure (if it exists at all)?*

Even in the classical case of algebraic V boasting of lots and lots of deep difficult theorems, there is no hint of the global picture in sight, not even a conjectural one (see [Mc] for the latest in the field).

(c) Hyperbolicity of (V, J) can sometimes be derived from negativity of a suitable curvature of a (Riemann or Finsler) metric adapted to J, either on V itself or on some jet space of $\mathbb{C}$-curves in V. A most general semi-local hyperbolicity criterion is expressed by the *linear isoperimetric inequality* in $\mathbb{C}$-curves $S \subset V$. If such an inequality holds true on relatively small J-surfaces $S \subset V$, then it propagates to all S in the same way as the real hyperbolicity does (see [G5]) which makes it, in principle, verifiable for compact V. (The linear inequality seems to *follow* from hyperbolicity by a rather routine argument, but I failed to carry it through. Possibly, one should limit oneself to closed V and integrability of J may be also helpful.)

It would be amusing to find a sufficiently general (positive) curvature condition for the existence of many rational curves in (V, J), encompassing complex hypersurfaces V with $\deg V$ (much) smaller than $\dim V$, for example, and allowing singular spaces in the spirit of Alexandrov's $K \geq 0$. Conversely, in the presence of many closed (especially rational) curves, one expects extra local structures on V, e.g. taming forms ω (which easily come from closed curves for $\dim V = 4$, see [G6]).

(d) One can sometimes make foliations (or at least, laminations) out of parabolic curves in V as is done in [B] for J on tori tamed by the *standard* symplectic ω (whereas the original question aimed at eventually *proving* that ω is standard).

(e) *How much do we gain in global understanding of a compact (V, J) by assuming that the structure J is integrable (i.e. complex)?* It seems nothing at all: there is no single result concerning *all* compact complex manifolds. (If $\dim V = 4$, then the Kodaira classification tells us quite a bit, say for *even* $b_1 \geq 2$, especially if there are 4 elements in $H^1(V)$ with non-zero product yielding a *finite* morphism of V onto $\mathbb{C}^2/\Lambda$.) This suggests the presence of (unreachable?) pockets of (moduli spaces of) integrable J's with weird properties (like those produced by Taubs on 6-manifolds); but there is no general existence theorem for complex structures either (not even for open V's, compare p. 103 in [Gro]$_{PDR}$) and even worse, no systematic way to produce them. So far, COMPACT COMPLEX MANIFOLDS have not stood up to their fame.*

(f) $\mathbb{C}$-curves, defined by restricting their tangent planes to the subvariety $E_0 = \mathbb{C}P^m \subset Gr_2\mathbb{R}^{2m}$, owe their beauty and power to the *ellipticity* of E_0: there is a single plane $e \in E_0$ through each line in $\mathbb{R}^{2m}$. One can deform E_0 by keeping this condition thus arriving at generalized $\mathbb{C}$-structures where the resulting E-*curves* are similar to $\mathbb{C}$-curves, and where the picture is the clearest for $m = 2$ (see [G6] and references therein). In general, a field E of subsets $E_v \subset Gr_kT_v(V) = Gr_k\mathbb{R}^n$, $n = \dim V$, defines a class of k-dimensional E-*subvarieties* in V, said "directed by E" (e.g. $W \subset \mathbb{R}^n$ with Gauss image in E_0), which seem most intriguing under *ellipticity* assumptions on E. To formulate these, let $F_0 = F_0(E_0)$, $E_0 \subset Gr_k\mathbb{R}^n$, denote the space of pairs (e, h) for $e \in E_0$ and all hyperplanes $h \subset e$, and look at the tautological map $\pi_0 : F_0 \to Gr_{k-1}\mathbb{R}^n$. Ideally, F_0 is a smooth closed manifold and π is *stably* one-to-one and onto (i.e. remains such under small perturbations of E_0) as in the $\mathbb{C}$-case. Such E_0 do not come cheaply: all known instances of them appear as (deformations of) Lie group orbits in $Gr_k\mathbb{R}^n$, e.g. in Harvey-Lawson calibrated geometries (where a most tantalizing $E_0 \subset Gr_3\mathbb{R}^7$ is associated to rational J-curves in S^6 for the standard G_2-invariant $\mathbb{C}$-structure J on S^6). One gains more examples by dropping "onto" thus arriving at over-determined elliptic systems ("isotropic in pseudo-Riemannian", for instance) which need integrability in order to have solutions (as "special Lagrangian" of Harvey-Lawson). Also, one may

---

*Fedia Bogomolov suggested to look at manifolds appearing as spaces of leaves of foliations in pseudoconvex bounded domains in $\mathbb{C}^N$ with algebraic tangent bundles.

allow E_0 and π_0 to have some singularities (similar to those present in Yang-Mills in a different setting), but in all cases one is stuck with two problems: *what are possible elliptic E_0, and what are (global and local) analytic properties (especially singularities) of the corresponding E-subvarieties W in* $(V, E = \{E_v\})$? (If E is over-determined, one looks at W's directed by a small neighbourhood $E_\varepsilon \supset E$.)

(g) The radii $\underline{R}_\pm$ are less useful than $R_\pm$ as they make sense only for rather special pseudo-Riemannian manifolds $V = (V, g)$. Yet $\underline{R}_\pm$ can be used for characterization of such V, e.g. the equality $\underline{R}_\pm = R_\pm < \infty$ seems to (almost?) distinguish metrically split manifolds. Also one can generalize $\underline{R}_\pm$ by considering submersions $V \to W_\pm$ being $\pm$ long *normally* to the fibers, where $1 \le \dim W_+ \le p$ and $1 \le \dim W_- \le q$ and where g is $\pm$ definite on the bundle of vectors normal to the fibers. Then the resulting radii $\underline{R}_\pm^\perp(V)$ satisfy $\underline{R}_\pm \ge \underline{R}_\pm^\perp \ge R_\pm$.

(h) The finiteness of $R_\pm(V)$ does not ensure stability of V for $\dim V \ge 4$ as simple (e.g. split) examples show but this seems to "limit instability to codimension two". *Can one go further with stronger radii-type invariants?*

(i) Besides the in-radius, there are other Riemannian invariants to gauge pseudo-Riemannian metrics such as the macroscopic dimension (see [G6]) of (complete) positive slices in V, or the maximal radius of a *Euclidean p-ball* in V. (The Euclidean metric on $\mathbb{R}^p$ dominates other g: there is a long map $\mathbb{R}^p \to (\mathbb{R}^p, g)$ for every Riemannian metric g on $\mathbb{R}^p$. But one can go beyond $\mathbb{R}^p$ by admitting slices with non-trivial topology. Some V may contain lots of these, e.g. some V of large dimension support (p, q)-metrics g so that *every* Riemannian p-manifold admits an isometric immersion into (V, g).) Furthermore, one may look at homotopies and extendability of slices with controlled size thus getting extra invariants of V. Actually, the mere topology of the space of, say, closed positive slices can be immensely complicated encouraging us to seek conditions limiting this complexity (e.g. in the spirit of diagram groups, see [GuS]).

(j) If V is compact, one may distinguish *complete* slices for the metric induced from some Riemannian background h in V and then compare these induced Riemannian metrics with those induced from our g of type (p, q). (Besides completeness, h brings forth other classes of slices, e.g. those with some bounds on curvature.)

(k) The isometrics of (non-stable!) (V, g) have attractive geometry and dynamics (see [DG] for an introduction and references) with many elementary questions remaining open, e.g. *does every isometry of the interior of V extend to the boundary?* (This comes from relativity, I guess.)

(1) Most of the current pseudo-Riemannian research is linked to general relativity focused on the Einstein equation (see[BeEE]).

5   Symbolization and Randomization

A common way to generate questions (not only) in geometry is to confront properties of objects specific to different categories: what is a possible *topology* (e.g. *homology*) of a manifold with a given type of *curvature?* How is the dynamics of the *geodesic* flow correlated with *topology* and/or *geometry?* How fast can a *harmonic function* decay on a *complete* manifold with a certain asymptotic *geometry*, e.g. curvature? How many *critical points* may *geometrically* defined energy have on a given space of maps or subvarieties? What are possible *singularities* of the *exponential map* or of the *cut locus?* Does every (almost complex) *manifold* (of dim ≥ 6) support a *complex structure?* etc. These seduce us by simplicity and apparent naturality, sometimes leading to new ideas and structures (tangentially related to the original questions, as in the Morse-Lusternik-Schnirelmann theory motivated by closed geodesics), but often the mirage of naturality lures us into a featureless desert with no clear perspective where the solution, even if found, does not quench our thirst for structural mathematics. (Examples are left to the reader.)

Another approach consists in interbreeding (rather than intersecting) categories and ideas. This has a better chance for a successful outcome with questions following (rather than preceding) construction of new objects. Just look at how it works: symbolic dynamics, algebraic arithmetic and non-commutative geometry, quantum computers, differential topology, random graphs, p-adic analysis ... Now we want to continue with *symbolic geometry* and *random groups*.

Given a category of "spaces" X with *finite* Cartesian products, we consider formal *infinite* products $\mathcal{X} = \underset{i \in I}{\times} X_i$, where the index set has an additional (discrete) structure, e.g. being a graph or a discrete group Γ. In the latter case we assume that all X_i are the same, $\mathcal{X} = X^\Gamma$ consists of functions $\chi : \Gamma \to X$, and Γ acts naturally on $\mathcal{X}$. Nothing happens unless we start looking at morphisms $\Phi : \mathcal{X} \to \mathcal{Y}$ over a fixed Γ. Such a Φ is given by a finite subset $\Delta \subset \Gamma$ of cardinality d and a map $\varphi : X^\Delta = X^d \to Y$, i.e. a function $y = \varphi(x_1, \dots, x_d)$, where $\Phi(\chi)(\gamma)$ is defined as the value of φ on the restriction of χ to the γ-translate $\gamma\Delta \subset \Gamma$ for all $\gamma \in \Gamma$. Thus we enrich the original category by making single variable (Γ-equivariant) functions $\Phi(\chi)$ out of functions $\varphi(x_1, \dots, x_d)$ in several variables.

Take a particular category of X's, e.g. algebraic varieties, smooth symplectic or Riemannian manifolds, (smooth) dynamical systems, whatever you like, and start translating basic constructions, notions and questions into the "symbolic" language of $\mathcal{X}$'s. This is pursued in [G4] and [G11] with an eye on continuous counterparts to $\mathcal{X}$, e.g. spaces of holomorphic maps $\mathbb{C} \to X$ for algebraic varieties X with a hope to make "algebraic" somehow reflected in such spaces. I have not gone far: a symbolic version of the Ax mapping theorem for amenable Γ (similar to the Garden of Eden in cellular automata) and a notion of *mean dimension* defined for all compact Γ-spaces $\mathcal{X}$ with amenable Γ (in the spirit of topological entropy) recapturing $\dim_{\text{top}} X$ for $\mathcal{X} = X^\Gamma$ (applicable to spaces like $\mathcal{B}$ of J-maps $\mathbb{C} \to (X, J)$ for instance). That's about it. (The reader is most welcome to these $\mathcal{X}$; if anything, there is no lack of open questions; yet no guarantee they would lead to a new grand theory either.)

Randomization. Random lies at the very source of manifolds, at least in the smooth and the algebraic categories: general smooth manifolds V appear as pull-back of special submanifolds under *generic* (or random) smooth maps f between standard manifolds, e.g. zeros of generic functions $f : \mathbb{R}^N \to \mathbb{R}^{N-n}$ or (proper) generic maps from $\mathbb{R}^N$ to the canonical vector bundles W over Grassmann manifolds $\text{Gr}_{N-n} \mathbb{R}^M$ (Thom construction), where V come as $f^{-1}(0)$ for the zero section $0 = \text{Gr}_{N-n}\mathbb{R}^M \subset W$. (Other constructions in differential topology amount to a little tinkering with V's created by genericity. Similarly, the bulk of algebraic manifolds comes from intersecting ample generic hypersurfaces in standard manifolds, e.g. in $\mathbb{C}P^N$, and the full list of known constructions of, say non-singular, algebraic varieties is dismally short.)

One may object by pointing out that every (combinatorial) manifold can be assembled out of simplices. Indeed, it is easy to make polyhedra, but no way to recognize *manifolds* among them (as eventually follows from undecidability of triviality for finitely presented groups). Here is another basic problem linked to "non-locality of topology". *How many triangulations may a given space X (e.g. a smooth manifold, say the sphere S^n) have?* Namely, let $t(X, N)$ denote the number of mutually combinatorially non-isomorphic triangulations of X into N simplices. Does this t grow *at most* exponentially in N?, i.e. whether $t(X, N) \leq \exp C_X N$. Notice that the number of *all* X built of N n-dimensional simplices grows super-exponentially, roughly as n^n, and the major difficulty for a given X comes from $\pi_1(X)$ and, possibly (but less likely), from $H_1(X)$, where the issue is to count the number of triangulated manifolds X with a fixed $\pi_1(X)$ or $H_1(X)$.

These questions (coming from physicists working on quantization of gravity) have an (essentially equivalent) combinatorial counterpart (we stumbled upon with Alex Nabutovski): evaluate the number $t_L(N)$ of connected 3-valent (i.e. degree ≤ 3) graphs X with N edges, such that cycles of length $\leq L$ normally generate $\pi_1(X)$ (or, at least, generate $H_1(X)$)? *Is $t_L(N)$ at most exponential in N for a fixed* (say $= 10^{10}$) L? The questions look just great and no idea how to answer them.

Here is a somewhat similar but easier question: *what does a random group* (rather than a space) *look like?* As we shall see the answer is most satisfactory (at least for me): "nothing like we have ever seen before". (No big surprise though: typical objects are usually atypical.)

Random presentation of groups. Given a group F, e.g. the free group F_k with k generators, one may speak of random quotient groups $G = F/[R]$, where $R \subset F$ is a random subset with respect to some probability measure μ on 2^G and $[R]$ standing for the normal subgroup generated by R. The simplest way to make a μ is to choose weights $p(\gamma) \in [0, 1]$ for all $\gamma \in \Gamma$ and take the product measure μ in 2^G, i.e. R is obtained by independent choices of $\gamma \in F$ with probabilities $p(\gamma)$. This is still too general; we specialize to $p(\gamma) = p(|\gamma|)$ where $|\gamma|$ denotes the word length of γ for a given, say finite, system of generators in F. A pretty such p is $p = p_\theta(\gamma) = (\mathrm{card}\{\gamma' \in F \mid |\gamma'| = |\gamma|\})^{-\theta}$, $\theta > 0$. If the "temperature" θ is close to zero, $p_0(\gamma)$ decays slowly and random R is so large that it normally generates all F making $G = \{e\}$ with probability one, provided F is infinite. For example, if $F = F_k$, this happens whenever $p(\gamma) \in \ell_2(F_k)$, i.e. $\sum_\gamma p^2(\gamma) < \infty$. This is easy; but it is not so clear if G may be ever non-trivial for large θ. However, if $F = F_k$ (or a general non-elementary word hyperbolic group), one can show that G is infinite with positive probability for $\theta > \theta_{\mathrm{crit}}(F)$, and $\theta_{\mathrm{crit}}(F_k)$, probably, equals 2, i.e. $p_\theta(\gamma) \notin \ell_2 \Rightarrow \mathrm{card}(G) = \infty$ with non-zero probability (see [G2] for a slightly different $p(\gamma)$, where the critical 2 comes as the Euler characteristic of S^2 via the small cancellation theory).

Random groups G_θ look very different for different θ. It seems that G_{θ_1}

cannot be embedded (even in a most generous sense of the word) into G_{θ_2} for $\theta_2 > \theta_1$ as the "density" of random G_θ decreases with "temperature". Furthermore, generic samples of G_θ for the same (large) θ are, probably, mutually non-isomorphic (not even quasi-isometric) with probability one, yet their "elementary invariants" are likely to be the same. It is clear for all $\theta < \infty$, that G_θ a.s. have no finite factor groups and they may satisfy Kazhdan's property T. (T is more probable for small θ where it is harder for G_θ to be infinite.)

Let us modify the above probability scheme by considering a random homomorphism φ from a fixed group H to F with $G = F/[\varphi(H)]$. To be simple, let $H = \pi_1(\Delta)$ for a (directed) graph Δ and φ be given by random assignment of generators of F to each edge e in Δ, independently for all edges. Denote by $N(L)$ the number of (non-oriented) cycles in Δ of length $\leq L$ and observe that for large $N(L)$ the group G is likely to be trivial. But we care for *infinite G and this can be guaranteed with positive probability if $N(L) \leq \exp L/\beta$ for large $\beta \geq \beta_{\mathrm{crit}}(F)$ for free groups $F = F_k$, $k \geq 2$, and non-elementary hyperbolic groups F in general.* (I have checked this so far under an additional *lacunarity* assumption allowing, for example, Δ being the disjoint union of finite graphs Δ_i of cardinalities d_i, $i = 1, 2, \ldots$, with $d_{i+1} \geq \exp d_i$ and such that the shortest cycle in Δ_i has length $\geq \beta_{cr} \log d_i$ for all i.)

To have an infinite random $G = G(F, \Delta)$ with interesting features, we need a special Δ. We take Δ, such that it contains arbitrarily large λ-*expanders* with a fixed (possibly small) $\lambda > 0$. (Such Δ do exist, in fact random Δ in our category contain such expanders, see [Lu].) Then random groups G a.s. enjoy the following properties.

(A) G are Kazhdan T, i.e. every affine isometric action of such G on the Hilbert space $\mathbb{R}^\infty$ has a fixed point. Furthermore, every action of G on a (possibly infinite dimensional) complete simply connected Riemannian manifold V with $K(V) \leq 0$ has a fixed point (if $\dim V < \infty$, then there is no non-trivial action at all).

(B) For every Lipschitz (for the word metric) map $f : G \to \mathbb{R}^\infty$, there are sequences $g_i, g_i' \in G$ with $\mathrm{dist}_G(g_i, g_i') \to \infty$ and $\mathrm{dist}_{\mathbb{R}^\infty}(f(g_i), f(g_i')) \leq$ const $< \infty$ (and the same remains true for the ℓ_p-spaces for $p < \infty$).

One may think the above "pathologies" are due to the fact that G are *not* finitely presented. But one can show that some "quasi-random" groups among our G are recursively presented and so embed into a *finitely presented* group G' which then automatically satisfies (B) and can be chosen with an *aspherical* presentation by a recent (unpublished) result by Ilia Rips and Mark Sapir (where T can be preserved by adding extra relations). Then, one can arrange $G' = \pi_1(V)$ for a closed aspherical manifold V of a given dimension $n \geq 5$ which, besides (B), has more "nasty" features, arresting, for example, all known (as far as I can tell) arguments for proving (strong) Novikov's conjecture for G'. (See [G8]; but I could not rule out hypersphericity yet.) I feel, random groups altogether may grow up as healthy as random graphs, for example.

There are other possibilities to define random groups, e.g. by following the "symbolic" approach where combinatorial manipulations with finite sets are replaced by parallel constructions in a geometric category. For example, we may give some structure (topology, measure, algebraic geometry) to the generating set B of (future) G with relations being geometric subsets in the Cartesian powers of B. Then, depending on the structure, one may speak of "random" or "generic" groups G (with a possible return to finitely generated groups via a model theoretic reasoning). Looks promising, but I have not arrived at a point of asking questions.

I conclude by thanking Stephanie Alexander and Miriam Hercberg for their help in my struggle with English.

References

[ABCKT] J. Amorós, M. Burger, K. Corlette, D. Kotschick, D. Toledo, Fundamental Groups of Compact Kähler Manifolds, Mathematical Surveys and Monographs, American Mathematical Society, Providence, RI, 1996.

[B]   V. Bangert, Existence of a complex line in tame almost complex tori, Duke Math. J. 94 (1998), 29–40.

[BeEE]   J.K. Beem, P.E. Ehrlich, K.L. Easley, Global Lorentzian Geometry, 2. ed., Monograph + Textbooks in Pure and Appl. Mathematics 202, Marcel Dekker Inc., NY, 1996.

[Ber]   M. Berger, Riemannian geometry during the second half of the twentieth century, Jahrbericht der Deutschen Math. Vereinigung 100 (1998), 45–208.

[Br]   R. Bryant, Recent advances in the theory of holonomy, Sém. N. Bourbaki, vol. 1998-99, juin 1999.

[CC]   J. Cheeger, T. Colding, On the structure of spaces with Ricci curvature bounded below, III, preprint.

[DG]   G. D'Ambra, M. Gromov, Lectures on transformation groups: Geometry and dynamics, Surveys in Differential Geometry 1 (1991), 19–111.

[G1]   M. GROMOV, with Appendices by M. Katz, P. Pansu, and S. Semmes, Metric Structures for Riemannian and Non-Riemannian Spaces, based on "Structures Métriques des Variétés Riemanniennes" (J. LaFontaine, P. Pansu, eds.), English Translation by Sean M. Bates, Birkhäuser, Boston–Basel–Berlin (1999).

[G2]   M. GROMOV, Asymptotic invariants of infinite groups. Geometric group theory, Vol. 2, Proc. Symp. Sussex Univ., Brighton, July 14-19, 1991. London Math. Soc. Lecture Notes 182 (Niblo, Roller, ed.), Cambridge Univ. Press, Cambridge (1993) 1–295.

[G3]   M. GROMOV, Carnot-Caratheodory spaces seen from within sub-Riemannian geometry, Proc. Journées nonholonomes; Géométrie sous-riemannienne, théorie du contrôle, robotique, Paris, June 30–July 1, 1992 (A. Bellaiche, ed.), Birkhäuser, Basel, Prog. Math. 144 (1996), 79–323.

[G4]   M. GROMOV, Endomorphisms of symbolic algebraic varieties, J. Eur. Math. Soc. 1 (1999), 109–197.

[G5]   M. GROMOV, Hyperbolic groups, in "Essays in Group Theory, Mathematical Sciences Research Institute Publications 8 (1978), 75–263, Springer-Verlag.

[G6]   M. GROMOV, Positive curvature, macroscopic dimension, spectral gaps and higher signatures, in "Functional Analysis on the Eve of the 21st Century" (Gindikin, Simon, et al., eds.) Volume II. in honor of the eightieth birthday of I.M. Gelfand, Proc. Conf. Rutgers Univ., New Brunswick, NJ, USA, Oct. 24-27, 1993, Birkhäuser, Basel, Prog. Math. 132 (1996), 1–213,

[G7]   M. GROMOV, Partial Differential Relations, Springer-Verlag (1986), Ergeb. der Math. 3. Folge, Bd. 9.

[G8]   M. GROMOV, Random walk in random groups, in preparation.

[G9]   M. GROMOV, Sign and geometric meaning of curvature, Rend. Sem. Math. Fis. Milano 61 (1991), 9–123.

[G10]   M. GROMOV, Soft and hard symplectic geometry, Proc. ICM-1986 (Berkeley), AMS 1,2 (1987), 81–98.

[G11]   M. GROMOV, Topological invariants of dynamical systems and spaces of holomorphic maps, to appear in Journ. of Geometry and Path. Physics.

[GuS]   V. GUBA, M. SAPIR, Diagram Groups, Memoirs of the AMS, November, 1997.

[K]   P. KANERVA, Sparse Distributed Memory, Cambridge, Mass. MIT Press, 1988.

[Kl]   F. KLEIN, Vorlesungen über die Entwicklung der Mathematik im 19. Jahrhundert, Teil 1, Berlin, Verlag von Julius Springer, 1926.

[LS]   U. LANG, V. SCHROEDER, Kirszbraun's theorem and metric spaces of Bounded Curvature, GAFA 7 (1997), 535–560.

[Lu]   A. LUBOTZKY, Discrete groups, expanding graphs and invariant measures, Birkhäuser Progress in Mathematics 125, 1994).

[MS]　D. McDuff, D. Salamon, J-holomorphic Curves and Quantum Cohomology, Univ. Lect. Series n° 6, A.M.S. Providence, 1994.

[Mc]　M. McQuillan, Holomorphic curves on hyperplane sections of 3-folds, GAFA 9:2 (1999), 370–392.

[Mi]　V.D. Milman, The heritage of P. Lévy in geometrical functional analysis, in "Colloque P. Lévy sur les Processus Stochastiques, Astérisque 157–158 (1988), 273-301.

[N]　J.R. Newman, The World of Mathematics, Vol. 1, Simon and Schuster, NY, 1956.

[P]　G. Perelman, Spaces with curvature bounded below, Proc. ICM (S. Chatterji, ed.), Birkhäuser Verlag, Basel (1995), 517–525.

[Pe]　P. Petersen, Aspects of global Riemannian geometry, BAMS 36:3 (1999), 297–344.

[S]　D. Spring, Convex Integration Theory, Birkhäuser Verlag, Basel–Boston–Berlin, 1998.

[V]　A.V. Vasiliev, Nikolai Ivanovich Lobachevski, Moscow, Nauka, 1992 (in Russian).

Mikhail Gromov, IHES, 35 route de Chartres, 91440 Bures sur Yvette, France; and The Courant Institute Mathematical Sciences, NYU, 251 Mercer Street, NY 10012, USA

GAFA, Geom. funct. anal.
Special Volume – GAFA2000, 162 – 183
1016-443X/00/S10162-22 $ 1.50+0.20/0
© Birkhäuser Verlag, Basel 2000

CLASSIFICATION OF INFINITE-DIMENSIONAL SIMPLE GROUPS OF SUPERSYMMETRIES AND QUANTUM FIELD THEORY

VICTOR G. KAC

Introduction

This work was motivated by two seemingly unrelated problems:

1. Lie's problem of classification of "local continuous transformation groups of a finite-dimensional manifold".
2. The problem of classification of operator product expansions (OPE) of chiral fields in 2-dimensional conformal field theory.

I shall briefly explain in §5 how these problems are related to each other via the theory of conformal algebras [DK], [K4–6]. This connection led to the classification of finite systems of chiral bosonic fields such that in their OPE only linear combinations of these fields and their derivatives occur [DK], which is, basically, what is called a "finite conformal algebra".

It is, of course, well known that a solution to Lie's problem requires quite different methods in the cases of finite- and infinite-dimensional groups. The most important advance in the finite-dimensional case was made by W. Killing and E. Cartan at the end of the 19th century who gave the celebrated classification of simple finite-dimensional Lie algebras over $\mathbb{C}$. The infinite-dimensional case was studied by Cartan in a series of papers written in the beginning of the 20th century, which culminated in his classification of infinite-dimensional "primitive" Lie algebras [Ca].

The advent of supersymmetry in theoretical physics in the 1970's motivated the work on the "super" extension of Lie's problem. In the finite-dimensional case the latter problem was settled in [K2]. However, it took another 20 years before the problem was solved in the infinite-dimensional case [K7], [ChK2,3]. An entertaining account of the historical background of the four classifications mentioned above may be found in the review [St].

A large part of my talk (§§1–4) is devoted to the explanation of the fourth classification, that of simple infinite-dimensional local supergroups of transformations of a finite-dimensional supermanifold. The application of this result to the second problem, that of classification of OPE when fermionic fields are allowed as well, or, equivalently, of finite conformal superalgebras, is explained in § 5.

I am convinced, however, that the classification of infinite-dimensional supergroups may have applications to "real" physics as well. The main reason for this belief is the occurrence in my classification of certain exceptional infinite-dimensional Lie supergroups that are natural extensions of the compact Lie groups $SU_3 \times SU_2 \times U_1$ and SU_5. In §7, I formulate a system of axioms, which, via representation theory of the corresponding Lie superalgebras (see §6), produce precisely all the multiplets of fundamental particles of the Standard model.

I wish to thank K. Gawedzki, D. Freedman, L. Michel, L. Okun, R. Penrose, S. Petcov, A. Smilga, H. Stremnitzer, I. Todorov for discussions and correspondence. I am grateful to P. Littlemann and J. van der Jeugt for providing tables of certain branching rules.

1   Lie's Problem and Cartan's Theorem

In his "Transformation groups" paper [L] published in 1880, Lie argues as follows. Let G be a "local continuous group of transformations" of a finite-dimensional manifold. The manifold decomposes into a union of orbits, so we should first study how the group acts on an orbit and worry later how the orbits are put together. In other words, we should first study transitive actions of G. Furthermore, even for a transitive action it may happen that G leaves invariant a fibration by permuting the fibers. But then we should first study how G acts on fibers and on the quotient manifold and worry later how to put these actions together. We thus arrive at the problem of classifications of transitive primitive (i.e., leaving no invariant fibrations) actions.

Next Lie establishes his famous theorems that relate the action of G on a manifold M in a neighborhood of a point p to the Lie algebra of vector fields in this neighborhood generated by this action. Actually, he talks about formal vector fields in a formal neighborhood of p, hence G gives rise to a Lie algebra L of formal vector fields and its canonical filtration by subalgebras L_j of L consisting of vector fields that vanish at p up to the order $j + 1$. Then transitivity of the action of G is equivalent to the property that $\dim L/L_0 = m := \dim M$, and primitivity is equivalent to the property that L_0 is a maximal subalgebra of L.

The first basic example is the Lie algebra of all formal vector fields in m indeterminates:

$$W_m = \left\{ \sum_{i=1}^{m} P_i(x) \frac{\partial}{\partial x_i} \right\}$$

($P_i(x)$ are formal power series in $x = (x_1, \ldots, x_m)$), endowed with the formal topology. A subalgebra L of W_m is called transitive if $\dim L/L_0 = m$, where $L_0 = (W_m)_0 \cap L$. A rigorous statement of Lie's problem is as follows:

1st formulation. Classify all closed transitive subalgebras L of W_m such that L_0 is a maximal subalgebra of L, up to a continuous automorphism of W_m.

Cartan published a solution to this problem in the infinite-dimensional case in 1909 [Ca]. The result is that a complete list over $\mathbb{C}$ (conjectured by Lie) is as follows ($m \geq 1$):

1. W_m,
2. $S_m = \{X \in W_m \mid \operatorname{div} X = 0\}$ ($m \geq 2$),
2' $CS_m = \{X \in W_m \mid \operatorname{div} X = \text{const}\}$ ($m \geq 2$),
3. $H_m = \{X \in W_m \mid X\omega_s = 0\}$ ($m = 2k$), where $\omega_s = \sum_{i=1}^{k} dx_i \wedge dx_{k+i}$ is a symplectic form,
3' $CH_m = \{X \in W_m \mid X\omega_s = \text{const}\,\omega_s\}$ ($m = 2k$),
4. $K_m = \{X \in W_m \mid X\omega_c = f\omega_c\}$ ($m = 2k + 1$), where $\omega_c = dx_m + \sum_{i=1}^{k} x_i dx_{k+i}$ is a contact form and f is a formal power series (depending on X).

The work of Cartan had been virtually forgotten until the sixties. A resurgence of interest in this area began with the papers [SiS] and [GS], which developed an adequate language and machinery of filtered and graded Lie algebras. The work discussed in the present talk also uses heavily the ideas from [W], [K1] and [G2].

The transitivity of the action implies that L_0 contains no non-zero ideals of L. The pair (L, L_0) is called primitive if L_0 is a (proper) maximal subalgebra of L which contains no non-zero ideals of L. Using the Guillemin-Sternberg realization theorem [GS], [Bl1], it is easy to show [G2] that the 1st formulation of Lie's problem is equivalent to the following, more invariant, formulation:

2nd formulation. Classify all primitive pairs (L, L_0), where L is a linearly compact Lie algebra and L_0 is its open subalgebra.

Recall that a topological Lie algebra is called linearly compact if its underlying space is isomorphic to a topological product of discretely topologized finite-dimensional vector spaces (the basic examples are finite-dimensional spaces with discrete topology and the space of formal power series in x with formal topology).

Any linearly compact Lie algebra L contains an open (hence of finite codimension) subalgebra L_0 [G1]. Hence, if L is simple, choosing any maximal open subalgebra L_0, we get a primitive pair (L, L_0). One can show that there exists a unique such L_0, and this leads to the four series $1, 2, 3$ and 4. The remaining series $2'$ and $3'$ are not simple, they actually are the Lie algebras of derivations of 2 and 3, but the choice of L_0 is again unique.

Using the structure results on general transitive linearly compact Lie algebras [G1], it is not difficult to reduce, in the infinite-dimensional case, the classification of primitive pairs to the classification of simple linearly compact Lie algebras (cf. [G2]). Such a reduction is possible also in the Lie superalgebra case, but it is much more complicated for two reasons:

  (a) a simple linearly compact Lie superalgebra may have several maximal open subalgebras;
  (b) construction of arbitrary primitive pairs in terms of simple primitive pairs is more complicated in the superalgebra case.

In the next sections I will discuss in some detail the classification of infinite-dimensional simple linearly compact Lie superalgebras.

2   Statement of the Main Theorem

The "superization" basically amounts to adding anticommuting indeterminates. In other words, given an algebra (associative or Lie) $\mathcal{A}$ we consider the Grassmann algebra $\mathcal{A}\langle n \rangle$ in n anticommuting indeterminates $\xi_1, \ldots, \xi_n$ over $\mathcal{A}$. This algebra carries a canonical $\mathbb{Z}/2\mathbb{Z}$-gradation, called parity, defined by letting

$$p(\mathcal{A}) = \bar{0}, \quad p(\xi_i) = \bar{1}, \quad \bar{0}, \bar{1} \in \mathbb{Z}/2\mathbb{Z}.$$

For example, $\mathbb{C}\langle n \rangle$ is the Grassmann algebra in n indeterminates over $\mathbb{C}$. If $\mathcal{O}_m$ denotes the algebra of formal power series over $\mathbb{C}$ in m indeterminates, then $\mathcal{O}_m\langle n \rangle$ is the algebra over $\mathbb{C}$ of formal power series in m commuting indeterminates $x = (x_1, \ldots, x_m)$ and n anticommuting indeterminates $\xi = (\xi_1, \ldots, \xi_n)$:

$$x_i x_j = x_j x_i, \quad x_i \xi_j = \xi_j x_i, \quad \xi_i \xi_j = -\xi_j \xi_i.$$

Recall that a derivation D of parity $p(D) \in \mathbb{Z}/2\mathbb{Z}$ of a $\mathbb{Z}/2\mathbb{Z}$-graded algebra is a vector space endomorphism satisfying condition

$$D(ab) = (Da)b + (-1)^{p(D)p(a)} a(Db).$$

Furthermore the sum of the spaces of derivations of parity $\bar{0}$ and $\bar{1}$ is closed under the "super" bracket:

$$[D, D_1] = DD_1 - (-1)^{p(D)p(D_1)} D_1 D.$$

This "super" bracket satisfies "super" analogs of anticommutativity and Jacobi identity, hence defines what is called a Lie superalgebra.

For example, the algebra $\mathcal{A}\langle n \rangle$ has derivations $\partial/\partial\xi_i$ of parity $\bar{1}$ defined by

$$\tfrac{\partial}{\partial\xi_i}(a) = 0 \text{ for } a \in \mathcal{A}, \qquad \tfrac{\partial}{\partial\xi_i}(\xi_j) = \delta_{ij},$$

and these derivations anticommute, so that $[\partial/\partial\xi_i, \partial/\partial\xi_j] = 0$.

The "super" analog of the Lie algebra W_m is the Lie superalgebra, denoted by $W(m|n)$, of all continuous derivations of the $\mathbb{Z}/2\mathbb{Z}$-graded algebra $\mathcal{O}_m\langle n \rangle$, $n \in \mathbb{Z}_+$, with the defined above "super" bracket,

$$W(m|n) = \left\{ \sum_{i=1}^{m} P_i(x,\xi)\frac{\partial}{\partial x_i} + \sum_{j=1}^{n} Q_j(x,\xi)\frac{\partial}{\partial\xi_j} \right\},$$

where $P_i(x,\xi), Q_j(x,\xi) \in \mathcal{O}_m\langle n \rangle$. In a more geometric language, this is the Lie superalgebra of all formal vector fields on a supermanifold of dimension $(m|n)$.

There is a unique way to extend divergence from W_m to $W(m|n)$ such that the divergenceless vector fields form a subalgebra,

$$\operatorname{div}\left(\sum_i P_i \tfrac{\partial}{\partial x_i} + \sum_j Q_j \tfrac{\partial}{\partial\xi_j} \right) = \sum_j \tfrac{\partial P_i}{\partial x_i} + \sum_j (-1)^{p(Q_j)} \tfrac{\partial Q_j}{\partial\xi_j},$$

and the "super" analog of S_m is

$$S(m|n) = \left\{ X \in W(m|n) \mid \operatorname{div} X = 0 \right\}.$$

In order to define "super" analogs of the Hamiltonian and contact Lie algebras H_m and K_m, introduce a "super" analog of the algebra of differential forms [K2]. This is an associative algebra over $\mathcal{O}_m\langle n \rangle$, denoted by $\Omega(m|n)$, on generators $dx_1, \ldots, dx_m$, $d\xi_1, \ldots, d\xi_n$ and defining relations,

$$dx_i dx_j = -dx_j dx_i, \qquad dx_i d\xi_j = d\xi_j dx_i, \qquad d\xi_i d\xi_j = d\xi_j d\xi_i,$$

and the $\mathbb{Z}/2\mathbb{Z}$ gradation defined by

$$p(x_i) = p(d\xi_j) = \bar{0}, \qquad p(\xi_j) = p(dx_i) = \bar{1}.$$

The algebra $\Omega(m|n)$ carries a unique continuous derivation d of parity $\bar{1}$ such that

$$d(x_i) = dx_i, \quad d(\xi_j) = d\xi_j, \quad d(dx_i) = 0, \quad d(d\xi_j) = 0.$$

The operator d has all the usual properties, e.g.,

$$df = \sum_i \tfrac{\partial f}{\partial x_i}\, dx_i + \sum_j \tfrac{\partial f}{\partial \xi_j}\, d\xi_j \text{ for } f \in \mathcal{O}_m\langle n \rangle, \text{ and } d^2 = 0\,.$$

As usual, for any $X \in W(m|n)$ one defines a derivation ι_X (contraction along X) of the algebra $\Omega(m|n)$ by the properties,

$$p(\iota_X) = p(X) + \bar{1}, \quad \iota_X(x_j) = 0, \quad \iota_X(dx_j) = (-1)^{p(X)} X(x_j)\,.$$

The action of any $X \in W(m|n)$ on $\mathcal{O}_m\langle n \rangle$ extends in a unique way to the action by a derivation of $\Omega(m|n)$ such that $[X, d] = 0$. This is called Lie's derivative and is usually denoted by L_X, but we shall write X in place of L_X unless confusion may arise. One has the usual Cartan formula for this action: $L_X = [d, \iota_X]$.

Using this action, one can define super-analogs of the Hamiltonian and contact Lie algebras for any $n \in \mathbb{Z}_+$,

$$H(m|n) = \big\{ X \in W(m|n) \mid X\omega_s = 0 \big\}\,,$$

where $\omega_s = \sum_{i=1}^{k} dx_i \wedge dx_{k+i} + \sum_{j=1}^{n} (d\xi_j)^2$,

$$K(m|n) = \big\{ X \in W(m|n) \mid X\omega_c = f\omega_c \big\}\,,$$

where $\omega_c = dx_m + \sum_{i=1}^{k} x_i dx_{k+i} + \sum_{j=1}^{n} \xi_j\, d\xi_j$, and $f \in \mathcal{O}_m\langle n \rangle$.

Note that $W(0|n)$, $S(0|n)$ and $H(0|n)$ are finite-dimensional Lie superalgebras. The Lie superalgebras $W(0|n)$ and $S(0|n)$ are simple iff $n \geq 2$ and $n \geq 3$, respectively. However, $H(0|n)$ is not simple as its derived algebra $H'(0|n)$ has codimension 1 in $H(0|n)$, but $H'(0|n)$ is simple iff $n \geq 4$. Thus, in the Lie superalgebra case the lists of simple finite- and infinite-dimensional algebras are more closely related than in the Lie algebra case.

The four series of Lie superalgebras are infinite-dimensional if $m \geq 1$, in which case they are simple except for $S(1|n)$. The derived algebra $S'(1|n)$ has codimension 1 in $S(1|n)$, and $S'(1|n)$ is simple iff $n \geq 2$.

In my paper [K2] I conjectured that the four series constructed above exhaust all infinite-dimensional simple linearly compact Lie superalgebras. Remarkably, the situation turned out to be much more exciting.

As was pointed out by several mathematicians, the Schouten bracket [SK] makes the space of polyvector fields on a m-dimensional manifold into a Lie superalgebra. The formal analog of this is the following fifth series of superalgebras, called by physicists the Batalin-Vilkoviski algebra:

$$HO(m|m) = \big\{ X \in W(m|m) \mid X\omega_{os} = 0 \big\}\,,$$

where $\omega_{os} = \sum_{i=1}^{m} dx_i d\xi_i$ is an odd symplectic form. Furthermore, unlike in the $H(m|n)$ case, not all vector fields of $HO(m|n)$ have zero divergence, which gives rise to the sixth series:

$$SHO(m|m) = \{X \in HO(m|m) \mid \operatorname{div} X = 0\}.$$

The seventh series is the odd analog of $K(m|n)$ [ALS]:

$$KO(m|m+1) = \{X \in W(m|m+1) \mid X\omega_{oc} = f\omega_{oc}\},$$

where $\omega_{oc} = d\xi_{m+1} + \sum_{i=1}^{m}(\xi_i \, dx_i + x_i \, d\xi_i)$ is an odd contact form. One can take again the divergence 0 vector fields in $KO(m|m+1)$ in order to construct the eighth series, but the situation is more interesting. It turns out that for each $\beta \in \mathbb{C}$ one can define the deformed divergence $\operatorname{div}_\beta X$ [Ko], [K7], so that $\operatorname{div} = \operatorname{div}_0$ and

$$SKO(m|m+1;\beta) = \{X \in KO(m|m+1) \mid \operatorname{div}_\beta X = 0\}$$

is a subalgebra. The superalgebras $HO(m|m)$ and $KO(m|m+1)$ are simple iff $m \geq 2$ and $m \geq 1$, respectively. The derived algebra $SHO'(m|m)$ has codimension 1 in $SHO(m|m)$, and it is simple iff $m \geq 3$. The derived algebra $SKO'(m|m+1;\beta)$ is simple iff $m \geq 2$, and it coincides with $SKO(m|m+1;\beta)$ unless $\beta = 1$ or $\frac{m-2}{m}$ when it has codimension 1.

Some of the examples described above have simple "filtered deformations", all of which can be obtained by the following simple construction. Let L be a subalgebra of $W(m|n)$, where n is even. Then it happens in three cases that

$$L^\sim := (1 + \Pi_{j=1}^{n}\xi_j)L$$

is different from L, but is closed under bracket. As a result we get the following three series of superalgebras: $S^\sim(0|n)$ [K2], $SHO^\sim(m|m)$ [ChK2] and $SKO^\sim(m|m+1; \frac{m+2}{m})$ [Ko] (the constructions in [Ko] and [ChK2] were more complicated). We thus get the ninth and the tenth series of simple infinite-dimensional Lie superalgebras:

$$SHO^\sim(m|m), \quad m \geq 2, \ m \text{ even},$$

$$SKO^\sim\left(m|m+1; \frac{m+2}{m}\right), \quad m \geq 3, \ m \text{ odd}.$$

It is appropriate to mention here that the four series $W(0|n)$, $S(0|n)$, $S^\sim(0|n)$ and $H'(0|n)$ along with the classical series $s\ell(m|n)$ and $osp(m|n)$, strange series $p(n)$ and $q(n)$, two exceptional superalgebras of dimension 40 and 31 and a family of 17-dimensional exceptional superalgebras along with the marvelous five exceptional Lie algebras, comprise a complete list of simple finite-dimensional Lie superalgebras [K2].

A surprising discovery was made in [Sh1] where the existence of three exceptional simple infinite-dimensional Lie superalgebras was announced. The proof of the existence along with one more exceptional example was given in [Sh2]. An explicit construction of these four examples was given later in [ChK3]. The fifth exceptional example was found in the work on conformal algebras [ChK1] and independently in [Sh2]. (The alleged sixth exceptional example $E(2|2)$ of [K7] turned out to be isomorphic to $SK0(2|3;1)$ [ChK3].)

Now I can state the main theorem.

Theorem 1 [K7]. *The complete list of simple infinite-dimensional linearly compact Lie superalgebras consists of ten series of examples described above and five exceptional examples: $E(1|6)$, $E(3|6)$, $E(3|8)$, $E(4|4)$, and $E(5|10)$.*

It happens that all infinite-dimensional simple linearly compact Lie algebras L have a unique transitive primitive action [G2]. This is certainly false in the Lie superalgebra case. However, if L is a simple linearly compact Lie superalgebra of type $X(m|n)$, it happens that m is minimal such that L acts on a super-manifold of dimension $(m|n)$ (i.e., $L \subset W(m|n)$), n is minimal for this m, and L has a unique action with such minimal $(m|n)$. Incidentally, in all cases the growth of L equals m. (Recall that growth is the minimal m for which $\dim L/L_j$ is bounded by $P(j)$, where P is a polynomial of degree m.)

Let me now describe those linearly compact infinite-dimensional Lie superalgebras L that allow a transitive primitive action. Let S be a simple linearly compact infinite-dimensional Lie superalgebra and let $S\langle n \rangle$ denote, as before, the Grassmann algebra over S with n indeterminates. The Lie superalgebra $\mathrm{Der}(S\langle n \rangle)$ of all derivations of the Lie superalgebra $S\langle n \rangle$ is the following semi-direct sum:

$$\mathrm{Der}(S\langle n \rangle) = (\mathrm{Der}\, S)\langle n \rangle + W(0|n)\,.$$

(For a description of $\mathrm{Der}\, S$ see [K7, Proposition 6.1].) Denote by $\mathcal{L}(S, n)$ the set of all open subalgebras L of $\mathrm{Der}(S\langle n \rangle)$ that contain $S\langle n \rangle$ and have the property that the canonical image of L in $W(0|n)$ is a transitive subalgebra.

Using a description of semi-simple linearly compact Lie superalgebras similar to the one given by Theorem 6 from [K2] (cf. [Ch], [G1] and [Bl2]) and Proposition 4.1 from [G1], it is easy to derive the following result.

PROPOSITION 1. *If a linearly compact infinite-dimensional Lie superalgebra L allows a transitive primitive action, then L is one of the algebras of the sets $\mathcal{L}(S, n)$.*

EXAMPLE.   Consider the semidirect sum $L = S\langle n \rangle + R$, where R is a transitive subalgebra of $W(0|n)$. Then $(L, L_0 = S_0\langle n \rangle + R)$ is a primitive pair if S_0 is a maximal open subalgebra of S, and these are all primitive pairs in the case when $S = \operatorname{Der} S$. One can also replace in this construction S by $\operatorname{Der} S$ and S_0 by a maximal open subalgebra of $\operatorname{Der} S$ having no non-zero ideals of $\operatorname{Der} S$.

3   Explanation of the Proof of Theorem 1

Step 1. Introduce Weisfeiler's filtration [W] of L. For that choose a maximal open subalgebra L_0 of L and a minimal subspace L_{-1} satisfying the properties,
$$L_{-1} \supsetneq L_0, \quad [L_0, L_{-1}] \subset L_{-1}.$$
Geometrically this corresponds to a choice of a primitive action of L and an invariant irreducible differential system. The pair L_{-1}, L_0 can be included in a unique filtration,
$$L = L_{-d} \supset L_{-d+1} \supset \cdots \supset L_{-1} \supset L_0 \supset L_1 \supset \cdots.$$
Of course, if L leaves invariant no non-trivial differential system, then the "depth" $d = 1$ and the Weisfeiler filtration coincide with the canonical filtration. Incidentally, in the Lie algebra case, $d > 1$ only for K_n (when $d = 2$), but in the Lie superalgebra case, $d > 1$ in the majority of cases.

The associated to Weisfeiler's filtration $\mathbb{Z}$-graded Lie superalgebra is of the form $GrL = \Pi_{j \geq -d}\mathfrak{g}_j$, and has the following properties:

(G0) $\dim \mathfrak{g}_j < \infty$ (since $\operatorname{codim} L_0 < \infty$),

(G1) $\mathfrak{g}_{-j} = \mathfrak{g}_{-1}^j$ for $j \geq 1$ (by maximality of L_0),

(G2) $[x, \mathfrak{g}_{-1}] = 0$ for $x \in \mathfrak{g}_j$, $j \geq 0 \Rightarrow x = 0$ (by simplicity of L),

(G3) $\mathfrak{g}_0$-module $\mathfrak{g}_{-1}$ is irreducible (by choice of L_{-1}).

Weisfeiler's idea was that property (G3) is so restrictive, that it should lead to a complete classification of $\mathbb{Z}$-graded Lie algebras satisfying (G0)–(G3). (Incidentally, the infinite-dimensionality of L and hence of GrL, since L is simple, is needed only in order to conclude that $\mathfrak{g}_1 \neq 0$.) This indeed turned out to be the case [K1]. In fact, my idea was to replace the condition of finiteness of the depth by finiteness of the growth, which allowed one to add to the Lie-Cartan list some new Lie algebras, called nowadays affine Kac-Moody algebras.

However, unlike in the Lie algebra case, it is impossible to classify all finite-dimensional irreducible faithful representations of Lie superalgebras.

One needed a new idea to make this approach work.

Step 2. The main new idea is to choose L_0 to be invariant with respect to all inner automorphisms of L (meaning to contain all even adexponentiable elements of L). A non-trivial point is the existence of such L_0. This is proved by making use of the characteristic supervariety, which involves rather difficult arguments of Guillemin [G2], that, unfortunately, I was unable to simplify.

Next, using a normalizer trick of Guillemin [G2], I prove, for the above choice of L_0, the following very powerful restriction on the $\mathfrak{g}_0$-module $\mathfrak{g}_{-1}$:

(G4) $[\mathfrak{g}_0, x] = \mathfrak{g}_{-1}$ for any non-zero even element x of $\mathfrak{g}_{-1}$.

Step 3. Consider a faithful irreducible representation of a Lie superalgebra $\mathfrak{p}$ in a finite-dimensional vector space V. This representation is called strongly transitive if

$$\mathfrak{p} \cdot x = V \text{ for any non-zero even element } x \in V.$$

Note that property (G4) along with (G0), (G2) for $j = 0$ and (G3), shows that the $\mathfrak{g}_0$-module $\mathfrak{g}_{-1}$ is strongly transitive.

In order to demonstrate the power of this restriction, consider first the case when $\mathfrak{p}$ is a Lie algebra and V is purely even. Then the strong transitivity simply means that $V \setminus \{0\}$ is a single orbit of the Lie group P corresponding to $\mathfrak{p}$. It is rather easy to see that the only strongly transitive subalgebras $\mathfrak{p}$ of $g\ell_V$ are $g\ell_V$, $s\ell_V$, sp_V and csp_V. These four cases lead to GrL, where $L = W_n, S_n, H_n$ and K_n, respectively.

In the super case the situation is much more complicated. First we consider the case of "inconsistent gradation", meaning that $\mathfrak{g}_{-1}$ contains a non-zero even element. The classification of such strongly transitive modules is rather long and the answer consists of a dozen series and a half dozen exceptions (see [K7, Theorem 3.1]). Using similar restrictions on $\mathfrak{g}_{-2}, \mathfrak{g}_{-3}, \ldots$, we obtain a complete list of possibilities for

$$GrL_{\leq} := \oplus_{j \leq 0} \mathfrak{g}_j$$

in the case when $\mathfrak{g}_{-1}$ contains non-zero even elements. It turns out that all but one exception are not exceptions at all, but correspond to the beginning members of some series. As a result, only $E(4|4)$ "survives" (but the infamous $E(2|2)$ does not).

Step 4. Next, we turn to the case of a consistent gradation, i.e., when $\mathfrak{g}_{-1}$ is purely odd. But then $\mathfrak{g}_0$ is an "honest" Lie algebra, having a faithful irreducible representation in $\mathfrak{g}_{-1}$ (condition (G4) becomes vacuous). An explicit description of such representations is given by the classical Cartan-Jacobson theorem. In this case I use the "growth" method developed in [K1] and [K2] to determine a complete list of possibilities for $GrL_\leq$. This case produces mainly the (remaining four) exceptions.

Step 5 is rather long and tedious [ChK3]. For each $GrL_\leq$ obtained in Steps 3 and 4 we determine all possible "prolongations", i.e., infinite-dimensional $\mathbb{Z}$-graded Lie superalgebras satisfying (G2), whose negative part is the given $GrL_\leq$.

Step 6. It remains to reconstruct L from GrL, i.e., to find all possible filtered simple linearly compact Lie superalgebras L with given GrL (such an L is called a simple filtered deformation of GrL). Of course, there is a trivial filtered deformation, $GrL := \Pi_{j \geq -d}\mathfrak{g}_j$, which is simple iff GrL is.

It is proved in [ChK2] by a long and tedious calculation that only $SHO(m|m)$ for m even ≥ 2 and $SKO(m|m+1; \frac{m+2}{m})$ for m odd ≥ 3 have a non-trivial simple filtered deformation, which are the ninth and tenth series. It would be nice to have a more conceptual proof. Recall that $SHO(m|m)$ is not simple, though it does have a simple filtered deformation. Note also that in the Lie algebra case all filtered deformations are trivial.

4   Construction of Exceptional Linearly Compact Lie Superalgebras

In order to describe the construction of the exceptional infinite-dimensional Lie superalgebras (given in [ChK3]), I need to make some remarks. Let $\Omega_m = \Omega(m|0)$ be the algebra of differential forms over $\mathcal{O}_m$, let Ω_m^k denote the space of forms of degree k, and $\Omega_{m,cl}^k$ the subspace of closed forms. For any $\lambda \in \mathbb{C}$ the representation of W_m on Ω_m^k can be "twisted" by letting

$$X \mapsto L_X + \lambda \operatorname{div} X, \quad X \in W_m,$$

to get a new W_m-module, denoted by $\Omega_m^k(\lambda)$ (the same can be done for $W_{m,n}$). Obviously, $\Omega_m^k(\lambda) = \Omega_m^k$ when restricted to S_m. Then we have the following obvious W_m-module isomorphisms: $\Omega_m^0 \simeq \Omega_m^m(-1)$ and $\Omega_m^0(1) \simeq \Omega_m^m$. Furthermore, the map $X \mapsto \iota_X(dx_1 \wedge \cdots \wedge dx_m)$ gives the following W_m-module and S_m-module isomorphisms:

$$W_m \simeq \Omega_m^{m-1}(-1), \quad S_m \simeq \Omega_{m,cl}^{m-1}.$$

We shall identify the representation spaces via these isomorphisms.

The simplest is the construction of the largest exceptional Lie superalgebra $E(5|10)$. Its even part is the Lie algebra S_5, its odd part is the space of closed 2-forms $\Omega^2_{5,c\ell}$. The remaining commutators are defined as follows for $X \in S_5$, $\omega, \omega' \in \Omega^2_{5,c\ell}$:

$$[X, \omega] = L_X \omega, \quad [\omega, \omega'] = \omega \wedge \omega' \in \Omega^4_{5,c\ell} = S_5.$$

Each quintuple of integers $(a_1, a_2, \ldots, a_5)$ such that $a = \sum_i a_i$ is even, defines a $\mathbb{Z}$-gradation of $E(5|10)$ by letting

$$\deg x_i = -\frac{\partial}{\partial x_i} = a_i, \quad \deg dx_i = a_i - \tfrac{1}{4}a.$$

The quintuple $(2, 2, \ldots, 2)$ defines the (only) consistent $\mathbb{Z}$-gradation, which has depth 2, $E(5|10) = \Pi_{j \geq -2} \mathfrak{g}_j$, and one has

$$\mathfrak{g}_0 \simeq s\ell_5 \text{ and } \mathfrak{g}_{-1} \simeq \Lambda^2 \mathbb{C}^5, \quad \mathfrak{g}_{-2} \simeq \mathbb{C}^{5*} \text{ as } \mathfrak{g}_0\text{-modules.}$$

Furthermore, $\Pi_{j \geq 0} \mathfrak{g}_j$ is a maximal open subalgebra of $E(5|10)$ (the only one which is invariant with respect to all automorphisms). There are three other maximal open subalgebras in $E(5|10)$, associated to $\mathbb{Z}$-gradations corresponding to quintuples $(1, 1, 1, 1, 2)$, $(2, 2, 2, 1, 1)$ and $(3, 3, 2, 2, 2)$, and one can show that these four are all, up to conjugacy, maximal open subalgebras (cf. [ChK3]).

Another important $\mathbb{Z}$-gradation of $E(5|10)$, which is, unlike the previous four, by infinite-dimensional subspaces, corresponds to the quintuple $(0, 0, 0, 1, 1)$ and has depth 1: $E(5|10) = \Pi_{\lambda \geq -1} \mathfrak{g}^\lambda$. One has $\mathfrak{g}^0 \simeq E(3|6)$ and the $\mathfrak{g}^\lambda$ form an important family of irreducible $E(3|6)$-modules [KRu]. The consistent $\mathbb{Z}$-gradation of $E(5|10)$ induces that of $\mathfrak{g}^0 : E(3|6) = \Pi_{j \geq -2} \mathfrak{a}_j$, where

$$\mathfrak{a}_0 \simeq s\ell_3 \oplus s\ell_2 \oplus g\ell_1, \quad \mathfrak{a}_{-1} \simeq \mathbb{C}^3 \boxtimes \mathbb{C}^2 \boxtimes \mathbb{C} \quad \mathfrak{a}_{-2} \simeq \mathbb{C}^3 \boxtimes \mathbb{C} \boxtimes \mathbb{C}.$$

A more explicit construction of $E(3|6)$ is as follows [ChK3]: the even part is $W_3 + \Omega^0_3 \otimes s\ell_2$, the odd part is $\Omega^1_3(-\tfrac{1}{2}) \otimes \mathbb{C}^2$ with the obvious action of the even part, and the bracket of two odd elements is defined as follows:

$$[\omega \otimes u, \omega' \otimes v] = (\omega \wedge \omega') \otimes (u \wedge v) + (d\omega \wedge \omega' + \omega \wedge d\omega') \otimes (u \cdot v).$$

Here the identifications $\Omega^2_3(-1) = W_3$ and $\Omega^0_3 = \Omega^3_3(-1)$ are used.

The gradation of $E(5|10)$ corresponding to the quintuple $(0, 1, 1, 1, 1)$ has depth 1 and its 0th component is isomorphic to $E(1|6)$ (cf. [ChK3]).

The construction of $E(4|4)$ is also very simple [ChK3]: The even part is W_4, the odd part is $\Omega^1_4(-1/2)$ and the bracket of two odd elements is

$$[\omega, \omega'] = d\omega \wedge \omega' + \omega \wedge d\omega' \in \Omega^3_4(-1) = W_4.$$

The construction of $E(3|8)$ is slightly more complicated, and we refer to [ChK3] for details.

5   Classification of Superconformal Algebras

Superconformal algebras have been playing an important role in superstring theory and in conformal field theory. Here I will explain how to apply Theorem 1 to the classification of "linear" superconformal algebras. By a ("linear") superconformal algebra I mean a Lie superalgebra $\mathfrak{g}$ spanned by coefficients of a finite family F of pairwise local fields such that the following two properties hold:

(1) for $a, b \in F$ the singular part of OPE is finite, i.e.,
$$[a(z), b(w)] = \sum_j c_j(w) \partial_w^j \delta(z - w) \quad \text{(a finite sum)},$$
where all $c_j(w) \in \mathbb{C}[\partial_w]F$,

(2) $\mathfrak{g}$ contains no non-trivial ideals spanned by coefficients of fields from a $\mathbb{C}[\partial_w]$-submodule of $\mathbb{C}[\partial_w]F$.

This problem goes back to the physics paper [RS], some progress in its solution was made in [K6] and a complete solution was stated in [K4,5]. (A complete classification even in the "quadratic" case seems to be a much harder problem, see [FL] for some very interesting examples.) The simplest example is the loop algebra $\widetilde{\mathfrak{g}} = \mathbb{C}[x, x^{-1}] \otimes \mathfrak{g}$ (= centerless affine Kac-Moody (super)algebra), where $\mathfrak{g}$ is a simple finite-dimensional Lie (super)algebra. Then $F = \left\{ a(z) = \sum_{n \in \mathbb{Z}} (x^n \otimes a) z^{-n-1} \right\}_{a \in \mathfrak{g}}$, and $[a(z), b(w)] = [a, b](w) \delta(z - w)$. The next example is the Lie algebra $\text{Vect} \, \mathbb{C}^\times$ of regular vector fields on $\mathbb{C}^\times$ (= centerless Virasoro algebra); F consists of one field, the Virasoro field $L(z) = -\sum_{n \in \mathbb{Z}} \left(x^n \frac{d}{dx} \right) z^{-n-2}$, and $[L(z), L(w)] = \partial_w L(w) \delta(z - w) + 2L(w) \delta'_w(z - w)$.

One of the main theorems of [DK] states that these are all examples in the Lie algebra case. The strategy of the proof is the following. Let $\partial = \partial_z$ and consider the (finitely generated) $\mathbb{C}[\partial]$-module $R = \mathbb{C}[\partial]F$. Define the "$\lambda$-bracket" $R \otimes R \to \mathbb{C}[\lambda] \otimes R$ by the formula
$$[a_\lambda b] = \sum_j \lambda^j c_j.$$

This satisfies the axioms of a conformal (super)algebra (see [DK], [K4]), similar to the Lie (super)algebra axioms:

(i) $[\partial a_\lambda b] = -\lambda[a_\lambda b]$, $[a_\lambda \partial b] = (\partial + \lambda)[a_\lambda b]$,
(ii) $[a_\lambda b] = -[b_{-\lambda - \partial} a]$,
(iii) $[a_\lambda[b_\mu c]] = [[a_\lambda b]_{\lambda + \mu} c] + (-1)^{p(a)p(b)}[b_\mu[a_\lambda c]]$.

The main observation of [DK] is that a conformal (super)algebra is completely determined by the Lie (super)algebra spanned by all coefficients of negative powers of z of the fields $a(z)$, called the annihilation algebra, along with an even locally nilpotent surjective derivation of the annihilation algebra. Furthermore, apart from the case of current algebras, the completed annihilation algebra turns out to be an infinite-dimensional simple linearly compact Lie (super)algebra of growth 1. Since in the Lie algebra case the only such example is W_1, the proof is finished.

In the superalgebra case the situation is much more interesting since there are many infinite-dimensional simple linearly compact Lie superalgebras of growth 1. By Theorem 1, the complete list is as follows:

$$W(1|n), \quad S'(1|n), \quad K(1|n) \text{ and } E(1|6).$$

The corresponding superconformal algebras in the first three cases are defined in the same way, except that we replace $\mathcal{O}_1\langle n\rangle$ by $\mathbb{C}((x))\langle n\rangle$; denote them by $W_{(n)}$, $S_{(n)}$ and $K_{(n)}$, respectively. The superconformal algebras $W_{(n)}$ and $K_{(n)}$ are simple for $n \geq 0$, except for $K_{(4)}$ which should be replaced by its derived algebra $K'_{(4)}$, and $S'_{(n)}$ is simple for $n \geq 2$. The unique superconformal algebra corresponding to $E(1|6)$ is denoted by $CK_{(6)}$. Its construction is more difficult and may be found in [ChK3] or [K6].

However, the superconformal algebra with the annihilation algebra $S'(1|n)$ is not unique since there are two up to conjugacy even surjective locally nilpotent derivations ∂ of $S'(1|n)$. In order to show this, we may assume that both ∂ and $S'(1|n)$ are in $W(1|n)$. Using a change of indeterminates, we may assume that $\partial = \partial/\partial x$, but then the standard volume form v that defines div, may change to $P(\xi)v$ (since it must be annihilated by $\partial/\partial x$). Further change of indeterminates brings this form to v or to $(1 + \xi_1 \ldots \xi_n)v$. This gives us another superconformal algebra with the annihilation algebra $S'(1|n)$ (n even), which is the derived algebra of

$$\widetilde{S}_{(n)} = \left\{X \in W_{(n)} \mid \operatorname{div}\left((1 + \xi_1 \ldots \xi_n)X\right) = 0\right\}.$$

(The situation is more interesting in the case $n = 2$, since the algebra of outer derivations of $S'(1|2)$ is 3-dimensional [P], but this gives no new superconformal algebras.) One argues similarly in the case $K(1|n)$. The case $E(1|6)$ is checked directly. We thus have arrived at the following theorem.

Theorem 2. *A complete list of superconformal algebras consists of loop algebras* $\widetilde{\mathfrak{g}}$, *where* $\mathfrak{g}$ *is a simple finite-dimensional Lie superalgebra, and of the following Lie superalgebras* ($n \in \mathbb{Z}_+$): $W_{(n)}$, $S'_{(n+2)}$, $\widetilde{S}'_{(n+2)}$ (n even),

$K_{(n)}(n \neq 4)$, $K'_{(4)}$, and $CK_{(6)}$.

Note that the first members of the above series are well-known super-algebras; $W_{(0)} \simeq K_{(0)}$ is the Virasoro algebra, $K_{(1)}$ is the Neveu-Schwarz algebra, $K_{(2)} \simeq W_{(1)}$ is the $N = 2$ algebra, $K_{(3)}$ is the $N = 3$ algebra, $S'_{(2)}$ is the $N = 4$ algebra. These algebras, along with $W_{(2)}$ and $CK_{(6)}$ are the only superconformal algebras for which all fields are primary with positive conformal weights [K6]. It is interesting to note that all of them are contained in $CK_{(6)}$, which consists of 32 fields, the even ones are the Virasoro fields and 15 currents that form $\widetilde{so}_6$, and the odd ones are 6 and 10 fields of conformal weight $3/2$ and $1/2$, respectively. Here is the table of inclusions, where in square brackets the number of fields is indicated:

$$CK_{(6)}[32] \quad \supset \quad W_{(2)}[12] \quad \supset W_{(1)} = K_{(2)}[4] \supset K_{(1)}[2] \supset \mathrm{Vir}$$
$$\cup \qquad\qquad \cup$$
$$K_{(3)}[8] \qquad\quad S'_{(2)}[8]$$

All of these Lie superalgebras have a unique non-trivial central extension, except for $CK_{(6)}$ that has none. The only other superalgebras listed by Theorem 2 which have non-trivial central extensions are $\widetilde{S}'_{(2)}$ with one such extension and $K'_{(4)}$ with two inequivalent central extensions. (The presence of a central term is necessary for the construction of an interesting conformal field theory.)

6 Representations of Linearly Compact Lie Superalgebras

By a representation of a linearly compact Lie superalgebra L we shall mean a continuous representation in a vector space V with discrete topology (then the contragredient representation is a continuous representation in a linearly compact space V^*). Fix an open subalgebra L_0 of L. We shall assume that V is locally L_0-finite, meaning that any vector of V is contained in a finite-dimensional L_0-invariant subspace (this property actually often implies that V is continuous). These kinds of representations were studied in the Lie algebra case by Rudakov [Ru].

It is easy to show that such an irreducible L-module V is a quotient of an induced module $\mathrm{Ind}_{L_0}^L U = U(L) \otimes_{U(L_0)} U$, where U is a finite-dimensional irreducible L_0-module, by a (unique in good cases) maximal submodule. The induced module $\mathrm{Ind}_{L_0}^L U$ is called degenerate if it is not irreducible. An irreducible quotient of a degenerate induced module is called a degenerate irreducible module; such a module is called a top module if the corresponding induced module cannot be mapped into another induced module.

One of the most important problems of representation theory is to determine all degenerate representations and among them all top representations. I will state here the result for $L = E(3|6)$ with $L_0 = \Pi_{j \geq 0} \mathfrak{a}_j$ (see §4), so that the finite-dimensional irreducible L_0-modules are actually $\mathfrak{a}_0 = s\ell_3 \oplus s\ell_2 \oplus g\ell_1$-modules (with $\Pi_{j>0} \mathfrak{a}_j$ acting trivially). We shall normalize the generator Y of $g\ell_1$ by the condition that its eigenvalue on $\mathfrak{a}_{-1}$ is $-1/3$. The finite-dimensional irreducible $\mathfrak{a}_0$-modules are labeled by triples (mn, b, Y), where mn (resp. b) are labels of the highest weight of an irreducible representation of $s\ell_3$ (resp. $s\ell_2$), so that $m0$ and $0m$ label $S^m \mathbb{C}^3$ and $S^m \mathbb{C}^{3*}$ (resp. b labels $S^b \mathbb{C}^2$), and Y is the eigenvalue of the central element Y. Since irreducible $E(3|6)$-modules are unique quotients of induced modules, they can be labeled by the above triples as well.

Theorem 3 [KRu]. *The complete list of irreducible degenerate $E(3|6)$-modules is as follows ($m, b \in \mathbb{Z}_+$, $b > 2$ if $m = 0$ in the second series) :*

$$\left(0m, b, -b-\tfrac{2}{3}m-2\right), \left(0m, b, b-\tfrac{2}{3}m\right), \left(m0, b, -b+\tfrac{2}{3}m\right), \left(m0, b, b+\tfrac{2}{3}m+2\right).$$

The top $E(3|6)$-modules are precisely those that are contragredient to the $\mathfrak{g}^{-1}$ and $\mathfrak{g}^0$ from the $\mathbb{Z}$-gradation of depth 1 of $E(5|10)$ described in §4. Their labels are $(00, 1, -1)$ for $\mathfrak{g}^{-1}$ and $(10, 0, 2/3)$ for $\mathfrak{g}^0$.

7   Fundamental Particle Multiplets

In order to explain the connection of representation theory of linearly compact Lie superalgebras to particle physics, let me propose the following axiomatics of fundamental particles:

A. The algebra of symmetries is a linearly compact Lie superalgebra L with an element Y, called the hypercharge operator, such that

   (i) $\operatorname{ad} Y$ is diagonalizable and normalized such that its spectrum is bounded below, $\subset \tfrac{1}{3}\mathbb{Z}$ and $\not\subset \mathbb{Z}$,
   (ii) the centralizer of Y in L is $\mathfrak{a}_0 = s\ell_3 + s\ell_2 + \mathbb{C}Y$ (one may weaken this by requiring $\supset$ in place of $=$).

B. A particle multiplet is an irreducible subrepresentation of $\mathfrak{a}_0$ in a degenerate irreducible representation of L. Particles in a multiplet are linearly independent eigenvectors of the $s\ell_2$ generator $I_3 = \tfrac{1}{2}\left(\begin{smallmatrix} 1 & 0 \\ 0 & -1 \end{smallmatrix}\right)$. Charge Q of a particle is given by the Gell-Mann-Nishijima formula,

$$Q = (I_3 \text{ eigenvalue}) + \tfrac{1}{2} (\text{hypercharge}).$$

C. Fundamental particle multiplet is a particle multiplet such that

   (i) $|Q| \leq 1$ for all particles of the multiplet,
   (ii) only the 1-dimensional, the two fundamental representations or
        the adjoint representation of $s\ell_3$ occur.

Using Theorem 3, it is easy to classify all fundamental multiplets when
the algebra of symmetries $L = E(3|6)$ [KRu]. The answer is given in the
left half of Table 1. The right half contains all the fundamental particles
of the Standard model (see e.g. [Ok]): the upper part is comprised of three
generations of quarks and the middle part of three generations of leptons
(these are all fundamental fermions from which matter is built), and the
lower part is comprised of fundamental bosons (which mediate the strong
and electro-weak interactions). Except for the last line, the match is perfect.

multiplets	charges		particles	
$(01, 1, 1/3)$	$2/3, -1/3$	$\binom{u_L}{d_L}$	$\binom{c_L}{s_L}$	$\binom{t_L}{b_L}$
$(10, 1, -1/3)$	$-2/3, 1/3$	$\binom{\tilde{u}_R}{\tilde{d}_R}$	$\binom{\tilde{c}_R}{\tilde{s}_R}$	$\binom{\tilde{t}_R}{\tilde{b}_R}$
$(10, 0, -4/3)$	$-2/3$	$\tilde{u}_L$	$\tilde{c}_L$	$\tilde{t}_R$
$(01, 0, 4/3)$	$2/3$	u_R	c_R	t_R
$(01, 0, -2/3)$	$-1/3$	d_R	s_R	b_R
$(10, 0, 2/3)$	$1/3$	$\tilde{d}_L$	$\tilde{s}_L$	$\tilde{b}_L$
$(00, 1, -1)$	$0, -1$	$\binom{\nu_L}{e_L}$	$\binom{\nu_{\mu L}}{\mu_L}$	$\binom{\nu_{\tau L}}{\tau_L}$
$(00, 1, 1)$	$0, 1$	$\binom{\tilde{\nu}_R}{\tilde{e}_R}$	$\binom{\tilde{\nu}_{\mu R}}{\tilde{\mu}_R}$	$\binom{\tilde{\nu}_{\tau R}}{\tilde{\tau}_R}$
$(00, 0, 2)$	1	$\tilde{e}_L$	$\tilde{\mu}_L$	$\tilde{\tau}_L$
$(00, 0, -2)$	-1	e_R	μ_R	τ_R
$(11, 0, 0)$	0	gluons		
$(00, 2, 0)$	$1, -1, 0$	W^+, W^-, Z	(gauge bosons)	
$(00, 0, 0)$	0	γ	(photon)	
$(11, 0, \pm 2)$	± 1	$-$		

Table 1

8   Speculations and Visions

As the title of the conference suggests, each speaker is expected to propose his (or her) visions in mathematics for the 21st century. This is an obvious invitation to be irresponsibly speculative. Some of the items proposed below are of this nature, but some others are less so.

1. It is certainly impossible to classify all simple infinite-dimensional Lie algebras or superalgebras. The most popular types of conditions that have emerged in the past 30 years and that I like most are these:

   (a) existence of a gradation by finite-dimensional subspaces and finiteness of growth [K1], [M].
   (b) topological conditions [G2], [K7], §§1–3,
   (c) the condition of locality [DK], [K4,5], §5.

   Problem (a) in the Lie algebra case has been completely solved in [M], but an analogous conjecture in the Lie superalgebra case [KL] is apparently much harder.

   Concerning (b), let me state a concrete problem. Let $L = \mathbb{C}((x))^n$, where $\mathbb{C}((x))$ is the space of formal Laurent series in x with formal topology. Examples of simple topological Lie algebras with the underlying space L are the completed (centerless) affine and Virasoro algebras. Are there any other examples?

   Incidentally, after going to the dual, Theorem 1 gives a complete classification of simple Lie co-superalgebras.

2. In §5 I explained how to use classification of simple linearly compact Lie superalgebras of growth 1 in order to classify simple "linear" OPE of chiral fields in 2-dimensional conformal field theory. Will CK_6 play a role in physics or is it just an exotic animal? Are the linearly compact Lie superalgebras of growth > 1 in any way related to OPE of higher dimensional quantum field theories?

3. Each of the four types W, S, H, K of simple primitive Lie algebras (L, L_0) correspond to the four most important types of geometries of manifolds: all manifolds, oriented manifolds, symplectic and contact manifolds. Since every smooth supermanifold of dimension $(m|n)$ comes from a rank n vector bundle on a m-dimensional manifold, it is natural to expect that each of the simple primitive Lie superalgebras corresponds to one of the most important types of geometries of vector bundles on manifolds. For example, the five exceptional superalgebras have altogether, up to conjugacy, 15 maximal open subalgebras.

They correspond to irreducible $\mathbb{Z}$-gradations listed in [ChK3]: 4 for $E(1|6)$, 3 for $E(3|6)$, 3 for $E(3|8)$, 1 for $E(4|4)$ and 4 for $E(5|10)$ (as Shchepochkina pointed out, we missed two $\mathbb{Z}$-gradations of $E(1|6)$: $(1, 0, 0, 0, 1, 1, 1)$ and $(2, 2, 0, 1, 1, 1, 1)$ in notation of [ChK3]) . There should be therefore 15 exceptional types of geometries of vector bundles on manifolds which are especially important.

4. The main message of §7 of my talk is the following principle:

Nature likes degenerate representations.

There are several theories where this principle works very well. First, is the theory of 2-dimensional statistical lattice models, especially the minimal models of [BPZ], which are (for $0 < c < 1$) nothing else but the top degenerate representations of the Virasoro algebra, and the WZW models, including the case of fractional levels, which are based on degenerate top modules over the affine Kac-Moody algebras (see [K3] for a review on these modules). Second, is the theoretical explanation of the quantum Hall effect by [CTZ] based on degenerate top modules of $W_{1+\infty}$ (see [KR]).

5. In view of the discussion in §§4 and 7, it is natural to suggest that the algebra $su_3 + su_2 + u_1$ of internal symmetries of the Weinberg-Salam-Glashow Standard model extends to $E(3|6)$. I am hopeful that representation theory will shed new light on various features of the Standard model (including the Kobayashi-Maskawa matrix). It turns out [KRu] that all degenerate $E(3|6)$ Verma modules have a unique non-trivial singular vector. This should lead to some canonical differential equations on the correlation functions (cf. [BPZ]).

I find it quite remarkable that the SU_5 Grand unified model of Georgi-Glashow combines the left multiplets of fundamental fermions in precisely the negative part of the consistent gradation of $E(5|10)$ (see §4). This is perhaps an indication of the possibility that an extension from su_5 to $E(5|10)$ algebra of internal symmetries may resolve the difficulties with the proton decay.

One, of course, may try other finite- or infinite-dimensional Lie super-algebras. For example, J. van der Jeugt has tried recently $L = s\ell(3|2)$ and it worked rather nicely, but $osp(6|2)$ has been ruled out.

6. Let me end with the most irresponsible suggestion. Since W_4 is, on the one hand, the algebra of symmetries of Einstein's gravity theory, and, on the other hand, the even part of $E(4|4)$, it is a natural guess that $E(4|4)$ is the algebra of symmetries of a nice super extension of

general relativity. One knows that the algebra of symmetries of the minimal $N = 1$ supergravity theory is $S(4|2)$ [OS].

References

[ALS] D. ALEXSEEVSKI, D. LEITES, I. SHCHEPOCHKINA, Examples of simple Lie superalgebras of vector fields, C.R. Acad. Bul. Sci. 33 (1980), 1187–1190.

[BPZ] A.A. BELAVIN, A.M. POLYAKOV, A.M. ZAMOLODCHIKOV, Infinite conformal symmetry of critical fluctuations in two dimensions, J. Stat. Phys. 34 (1984), 763–774.

[Bl1] R.J. BLATTNER, Induced and produced representations of Lie algebras, Trans. Amer. Math. Soc. 144 (1969), 457-474.

[Bl2] R.J. BLATTNER, A theorem of Cartan and Guillemin, J. Diff. Geom. 5 (1970), 295–305.

[CTZ] A. CAPPELLI, C.A. TRUGENBERGER, G.R. ZEMBA, Stable hierarchal quantum Hall fluids as $W_{1+\infty}$ minimal models, Nucl. Phys. B 448 (1995), 470–511.

[Ca] E. CARTAN, Les groupes des transformations continués, infinis, simples, Ann. Sci. Ecole Norm. Sup. 26 (1909), 93–161.

[Ch] S.-J. CHENG, Differentiably simple Lie superalgebras and representations of semisimple Lie superalgebras, J. Algebra 173 (1995), 1–43.

[ChK1] S.-J. CHENG, V.G. KAC, A new $N = 6$ superconformal algebra, Commun. Math. Phys. 186 (1997), 219–231.

[ChK2] S.-J. CHENG, V.G. KAC, Generalized Spencer cohomology and filtered deformations of $\mathbb{Z}$-graded Lie superalgebras, Adv. Theor. Math. Phys. 2 (1998), 1141–1182.

[ChK3] S.-J. CHENG, V.G. KAC, Structure of some $\mathbb{Z}$-graded Lie superalgebras of vector fields, Transformation Groups 4 (1999), 219–272.

[DK] A. D'ANDREA, V.G. KAC, Structure theory of finite conformal algebras, Selecta Mathematica 4 (1998), 377–418.

[FL] E.S. FRADKIN, V.YA. LINETSKY, Classification of superconformal and quasisuperconformal algebras in two dimensions, Phys. Lett. B 291 (1992), 71–76.

[G1] V.W. GUILLEMIN, A Jordan-Hölder decomposition for a certain class of infinite dimensional Lie algebras, J. Diff. Geom. 2 (1968), 313–345.

[G2] V.W. GUILLEMIN, Infinite-dimensional primitive Lie algebras, J. Diff. Geom. 4 (1970), 257–282.

[GS] V.W. GUILLEMIN, S. STERNBERG, An algebraic model of transitive differential geometry, Bull. Amer. Math. Soc. 70 (1964), 16–47.

[K1] V.G. KAC, Simple irreducible graded Lie algebras of finite growth, Math. USSR-Izvestija 2 (1968), 1271–1311.

[K2]   V.G. KAC, Lie superalgebras, Adv. Math. 26 (1977), 8–96.

[K3]   V.G. KAC, Modular invariance in mathematics and physics, in "Mathematics into the 21st Century", AMS Centennial Publ. (1992), 337–350.

[K4]   V.G. KAC, Vertex Algebras for Beginners, University Lecture Series 10, AMS, Providence, RI, 1996. Second edition 1998.

[K5]   V.G. KAC, The idea of locality, in "Physical Applications and Mathematical Aspects of Geometry, Groups and Algebras" (H.D. Doebner, et al., eds.), World Sci., Singapore (1997), 16–22.

[K6]   V.G. KAC, Superconformal algebras and transitive group actions on quadrics, Comm. Math. Phys. 186 (1997), 233–252.

[K7]   V.G. KAC, Classification of infinite-dimensional simple linearly compact Lie superalgebras, Adv. Math. 139 (1998), 1–55.

[KL]   V.G. KAC, J. VAN DER LEUR, On classification of superconformal algebras, in "Strings 88" (S.J. Gates, et al., eds.), World Sci. (1989), 77–106.

[KR]   V.G. KAC, A.O. RADUL, Representation theory of $W_{1+\infty}$, Transformation Groups 1 (1996), 41–70.

[KRu]   V.G. KAC, A.N. RUDAKOV, Irreducible representations of linearly compact Lie superalgebras, in preparation.

[Ko]   YU. KOCHETKOFF, Déformations de superalgébres de Buttin et quantification, C.R. Acad. Sci. Paris 299, ser I, 14 (1984), 643–645.

[L]   S. LIE, Theorie der transformations Gruppen, Math. Ann. 16 (1880), 441–528.

[M]   O. MATHIEU, Classification of simple graded Lie algebras of finite growth, Invent. Math. 108 (1992), 445–519.

[OS]   V.I. OGIEVETSKII, E.S. SOKACHEV, The simplest group of Einstein supergravity, Sov. J. Nucl. Phys. 31 (1980), 140–164.

[Ok]   L. OKUN, Physics of Elementary Particles, Nauka, 1988 (in Russian).

[P]   E. POLETAEVA, Semi-infinite cohomology and superconformal algebras, preprint.

[RS]   P. RAMOND, J.H. SCHWARZ, Classification of dual model gauge algebras, Phys. Lett. B. 64 (1976), 75–77.

[Ru]   A.N. RUDAKOV, Irreducible representations of infinite-dimensional Lie algebras of Cartan type, Math. USSR-Izvestija 8 (1974), 836–866.

[SK]   J.A. SCHOUTEN, W. VAN DER KULK, Pfaff's Problem and its Generalizations, Clarendon Press, 1949.

[Sh1]   I. SHCHEPOCHKINA, New exceptional simple Lie superalgebras, C.R. Bul. Sci. 36:3 (1983), 313–314.

[Sh2]   I. SHCHEPOCHKINA, The five exceptional simple Lie superalgebras of vector fields, preprint.

[SiS]   I.M. SINGER, S. STERNBERG, On the infinite groups of Lie and Cartan I, J. Analyse Math. 15 (1965), 1–114.

[St]   S. STERNBERG, Featured review of "Classification of infinite-dimensional simple linearly compact Lie superalgebras" by Victor Kac, Math. Reviews 99m:17006, 1999.

[W]   B.Y. WEISFEILER, Infinite-dimensional filtered Lie algebras and their connection with graded Lie algebras, Funct. Anal. Appl. 2 (1968), 88–89.

VICTOR KAC, Department of Mathematics, MIT, Cambridge, MA 02139, USA
kac@math.mit.edu

GAFA, Geom. funct. anal.
Special Volume – GAFA2000, 184 – 187
1016-443X/00/S10184-4 $ 1.50+0.20/0

© Birkhäuser Verlag, Basel 2000

GAFA Geometric And Functional Analysis

GEOMETRIZATION IN REPRESENTATION THEORY

D. Kazhdan

The common theme of the articles "γ-function of Representations and Lifting" and "Geometric and Unipotent Crystals" is the geometrization of familiar objects. In the first article we interpret some special functions appearing in the theory of representations as *materializations* of geometric objects which we call *algebraic-geometric distributions* in the second one we interpret elements of the crystalline basis which parametrize the special basis of finite-dimensional representations of reductive groups as *materializations* of geometric crystals. Moreover these two topics are very much interwoven since the theory of geometric crystals is necessary for the construction of our *algebraic-geometric distributions*.

By definition, an algebraic-geometric distribution Φ on an algebraic variety $\mathbf{X}$ over a field F is quadruple $\Phi = (\mathbf{Y}, \mathbf{p}, \mathbf{f}, \omega)$ where $\mathbf{Y}$ is an F-variety, $\mathbf{p} : \mathbf{Y} \to \mathbf{X}$ is a rational morphism, $\mathbf{f}$ is rational function and ω is a top-form on $\mathbf{Y}$. If F is a local field and ψ is a non-trivial additive character of F then we can associate with an algebraic-geometric distribution $\Phi = (\mathbf{Y}, \mathbf{p}, \mathbf{f}, \omega)$ a distribution $\Phi := p_!(\psi(f)|\omega|)$ on $\mathbf{X}(F)$ where $|\omega|$ is a measure on the set $\mathbf{Y}(F)$ corresponding to the top-form ω and $p_!(\psi(f)|\omega|)$ is the push-forward of the complex-valued measure $\psi(f)|\omega|$.

If W is a finite group acting on $\mathbf{X}$ then *a lift* of this action to Φ is a rational action of W on $\mathbf{Y}$ compatible with $\mathbf{p}$ and preserving $\mathbf{f}$ and ω. Given an action of W on Φ we can define an algebraic-geometric distribution $\Phi/W = (\mathbf{Y}/W, \mathbf{p}, \mathbf{f}, \omega)$ which we call the descent of Φ and therefore a distribution Φ/W on $\mathbf{X}/W(F)$. We call it "the descent of Φ". Of course the descent is not determined by the distribution Φ and an action of W on $\mathbf{X}$ but depends on the choice of the action of W to Φ.

One of the two main ideas of the article "γ-functions of representations and lifting" is to formulate a conjecture that for any reductive group $\mathbf{G}$ over a local field F and a representation ρ of the Langlands dual group $\mathbf{G}^\vee$ of the group $\mathbf{G}$, the distribution $\Phi_\rho^{\mathbf{G}}$ on $\mathbf{G}(F)$ responsible for the lifting l_ρ of ρ can be obtained as a descent of the analogous distribution $\Phi_\rho^{\mathbf{T}}$ for the maximal torus $\mathbf{T}$ of $\mathbf{G}$. It is easy to construct the distribution $\Phi_\rho^{\mathbf{T}}$ as a *materialization* of an algebraic-geometric distribution Φ_ρ^T and therefore a conjectural description of the lifting l_ρ of ρ is reduced to an algebra-geometric question about construction of an action of W on $\Phi_\rho^{\mathbf{T}}$.

The article "Geometric and unipotent crystals" deals with a geometrization of the concept of crystals which we believe is necessary for a construction of an action of W on $\Phi_\rho^{\mathbf{T}}$.

If you wish to construct a representation of a group we have first of all to construct the space V of this representation. But how to construct a vector space? The simplest way is to realize V as a space of complex-valued functions on a set B. In other words one has to find a basis v_b, $b \in B$, of V. It was discovered by Lusztig and Kashiwara ([Lu], [K]) that in the case when V is an irreducible representation of a semisimple group G one can construct a basis v_b of V such that the parameter space B has an interesting *crystalline* structure.

Since the theory of crystals is not widely known I will give with a short introduction to this theory.

Let $\pi_n : SL(2, \mathbb{C}) \to \text{End}(R_n)$, $n \geq 0$, be an irreducible representation of dimension $n+1$. Then we can write any finite-dimensional representation V of $SL(2, \mathbb{C})$ as a direct sum $V = \oplus_{n \geq 0} V_n$ where $V_n = R_n \otimes \text{Hom}(R_n, V)$. Let v_b, $b \in B$, be a basis of V. Define $F_n := \oplus_{0 \leq m \leq n} V_m$. For any $b \in B$ we denote by $\bar{v}_b$ the image of the vector $v_b \in F_n/F_{n-1}$ where $n = n(b)$ is the smallest number such that $v_b \in F_n$. We denote by $R_b \subset F_n/F_{n-1}$ the minimal $SL(2, \mathbb{C})$-invariant subspace of F_n/F_{n-1} containing $\bar{v}_b$. By the construction R_b is a space of a representation of $SL(2, \mathbb{C})$.

We say that a basis v_b is *compatible* with the action of the Lie group $SL(2, \mathbb{C})$ if

a) For any $b \in B$ vector $v_b \subset V$ is a H-eigenvector where $H \subset SL(2, \mathbb{C})$ is the diagonal subgroup. We denote by $\tilde{\gamma}(b)$ the corresponding character of H.

b) For any $n \geq 0$ there exists a subset $B_n \subset B$ such that the subspace F_n is equal to the span of v_b, $b \in B_n$.

c) For any $b \in B_n$ the space R_b is spanned by the images of elements of B_n in F_n/F_{n-1}.

d) For any $b \in B$ the representation of $SL(2, \mathbb{C})$ on R_b is irreducible.

Let $B \subset SL(2, \mathbb{C})$ (resp. $U \subset SL(2, \mathbb{C})$) be the subgroup of upper triangular (resp. unipotent upper triangular) matrices, and let $sl(2)$ be the Lie algebra of $SL(2, \mathbb{C})$. Choose a non-zero element e in the Lie algebra of U.

Given a finite-dimensional representation V of $SL(2, \mathbb{C})$ and a basis v_b,

$b \in B$, of V compatible with the action of $SL(2, \mathbb{C})$ we define a partial bijections $\tilde{e}$ on B in the following way. Given $b \in B$ we say that $\tilde{e}(b)$ is defined if $e(\bar{v}_b) \neq 0$. In such a case it is easy to see that there exists a unique element $\tilde{e}(b) \in B$ such that $v_{\tilde{e}(b)} \in F_{n(b)}$ and that $e(\bar{v}_b)$ is proportional to $\bar{v}_{\tilde{e}(b)}$.

Let G be a semisimple group over $\mathbb{C}$, $T \subset G$ be a maximal torus, $X^*(T)$ the group of characters of T and $B \subset G$ a Borel subgroup of G containing T. We denote by I the set of vertices of the Dynkin diagram of G. For any $i \in I$ we denote by $L_i \subset G$ the subgroup generated by the root subgroups U_i and U_{-i}. As is well known L_i is quotient of $SL(2, \mathbb{C})$.

Let V be a finite-dimensional representation of G and v_b, $b \in B$ be a basis of V. We say that a basis v_b is *compatible* with the action of the Lie group G if for any $i \in I$ the basis v_b is compatible with the action on L_i. As follows from [BG] for any finite-dimensional representation V of G there exists a basis v_b, $b \in B$ of V compatible with the action of G.

Let V be a finite-dimensional representation of G and v_b, $b \in B$, be a basis of V compatible with the action of G. Since the Cartan subgroups $H_i \subset L_i$ generate T for any $b \in B$ the vector v_b is a T-eigenvector. We denote by $\tilde{\gamma}(b) \subset X^*(T)$ the corresponding character of T. By definition, for any $i \in I$, the basis v_b is compatible with the restriction to L_i. We denote by $\tilde{e}_i$ the corresponding partial bijection on B and by $\tilde{f}_i$ the inverse bijection. It is easy to check that for any $b \in B$, $i \in I$, we have $\tilde{\gamma}(\tilde{e}_i(b)) = \tilde{\gamma}(b) + \alpha_i$ were $\alpha_i \in X^*(T)$ is the simple root corresponding to i. The data consisting of $\tilde{\gamma} : B \to X^*(T)$ and partial bijections $\tilde{e}_i$, $i \in I$, on B define a structure of a *pre-crystal* on B (see details in the appendix to "Geometric and Unipotent Crystals"). As follows form the construction in [BG] it is natural to call the date $\tilde{\gamma}, \tilde{e}_i$, $i \in I$ on B a pre-crystal structure associated with the Langlands dual group $\mathbf{G}^\vee$ of G.

It would be very interesting to construct an analog of a crystal associated with infinite-dimensional representations of reductive groups over local fields. We do not know how to find such a construction but we can try to analyze the problem. Finite-dimensional representations V of reductive groups have different sizes and therefore the set $B = B(V)$ parametrizing bases of V are different. On the other hand, the majority of infinite-dimensional representations of reductive groups over local fields are of the same size. Therefore it is natural to look for some "universal" infinite sets $B(\infty)$ with crystal structure. Such infinite "free" crystal appeared earlier in the works of Kashiwara. One could hope that the space of a generic representation of a reductive group $\mathbf{G}(F)$ where F is a local field can be realized in the space of functions on $F^* \otimes B(\infty)$.

The main purpose of the paper "Geometric and unipotent crystals" is a construction of a geometric analog of the notion $\mathbf{G}^\vee\text{-}crystal$. It consists of an algebraic variety X with a rational morphism $\gamma : X \to T$ and a family e_i, $i \in I$, of birational actions of the multiplicative group $\mathbb{G}_m$ on X satisfying some compatibility conditions. The set $B(\infty)$ is isomorphic to the set of connected components of the space $\mathcal{L}(X)$ of formal loops on a particular algebraic variety X and partial bijections $\tilde{e}_i$ come from the actions e_i of the group $\mathbb{G}_m$ on X.

To our surprise these geometric crystals are important for finding the action of the Weyl group W central for the construction of the descent of algebraic-geometric distributions.

References

[BG] A. BRAVERMAN, D. GAITSGORY, Crystals via the affine Grassmannian, preprint math. AG/9909077.

[K] M. KASHIWARA, Crystal bases of modified quantized enveloping algebra, Duke Math. J. 73:2 (1994), 383–413.

[Lu] G. LUSZTIG, Canonical bases arising from quantized enveloping algebras, J. Amer. Math. Soc. 3:2 (1990), 447–498.

DAVID KAZHDAN, Department of Mathematics, Harvard University, Cambridge, MA 02138, USA

GAFA, Geom. funct. anal.
Special Volume – GAFA2000, 188 – 236
1016-443X/00/S10188-49 $ 1.50+0.20/0

© Birkhäuser Verlag, Basel 2000

I GAFA Geometric And Functional Analysis

GEOMETRIC AND UNIPOTENT CRYSTALS

ARKADY BERENSTEIN AND DAVID KAZHDAN

Contents

The research of both authors was supported in part by NSF grants

1 Introduction

Let G be a split semisimple algebraic group over $\mathbb{Q}$, $\mathfrak{g}$ be the Lie algebra of G and $U_q(\mathfrak{g})$ be the corresponding *quantized enveloping algebra*. Lusztig has introduced in [Lu1] *canonical bases* for finite-dimensional $U_q(\mathfrak{g})$-modules. About the same time Kashiwara introduced in [K1] *crystal bases* as a natural framework for parametrizing bases of finite-dimensional $U_q(\mathfrak{g})$-modules. It was shown in [Lu2] that Kashiwara's crystal bases are the limits as $q \to 0$ of Lusztig's canonical bases. Later, in [K2] Kashiwara introduced a new combinatorial concept – *crystals*. Kashiwara's crystals generalize the crystal bases and provide a natural framework for their study.

In this paper we study *geometric crystals* and *unipotent crystals* which are algebro-geometric analogues of respectively the crystals and the crystal bases.

Our approach of the "geometrization" of combinatorial objects comes back to the works of Lusztig. On the one hand, Lusztig constructed for each reduced decomposition $\mathbf{i}$ of the longest element w_0 of the Weyl group W of G a parametrization $\psi^{\mathbf{i}}$ of the canonical basis $\mathbf{B}$ by the cone $(\mathbb{Z}_{\geq 0})^{l_0}$ where l_0 is the length w_0. On the other hand, in his study of the positive elements of $G(\mathbb{R})$, Lusztig constructed for each $\mathbf{i}$ a birational isomorphism $\pi^{\mathbf{i}} : \mathbb{A}^{l_0} \xrightarrow{\sim} U$ where U is the unipotent radical of the Borel subgroup of G. Moreover, Lusztig observed that for any two decompositions $\mathbf{i}, \mathbf{i}'$ of w_0, the piecewise-linear transformation $(\psi^{\mathbf{i}'})^{-1} \circ \psi^{\mathbf{i}}$ of $(\mathbb{Z}_{\geq 0})^{l_0}$ is similar to the birational transformation $(\pi^{\mathbf{i}'})^{-1} \circ \pi^{\mathbf{i}}$ of $\mathbb{A}^{l_0}$ (see [Lu3], [BFZ]).

This analogy was further developed in a series of papers [BFZ], [BS1,2], [FZ]. The main idea of these papers is to introduce the notion of the "tropicalization" – a set of formal rules for the passage from birational isomorphisms to piecewise-linear maps, and to study piecewise-linear automorphisms of **B** using the "tropicalization" of explicit birational isomorphisms $\pi^i : \mathbb{A}^{l_0} \xrightarrow{\sim} U$.

Another geometric approach to constructing crystal bases is suggested in the recent work of Braverman and Gaitsgory [BrG]. Within this approach the crystal bases and, in fact, actual bases in finite-dimensional G-modules are constructed in terms of perverse sheaves on the affine Grassmannian of G.

In the present paper we go in the opposite direction – we study geometric crystals in their own right and "geometrize" some of the piecewise-linear structures of the crystal bases mentioned above.

Let I be the set of vertices of the Dynkin diagram of G and T be a maximal torus of G. The structure of a *geometric crystal* on an algebraic variety X consists of a rational morphism $\gamma : X \rightarrow T$ and a compatible family $e_i : \mathbb{G}_m \times X \rightarrow X$, $i \in I$, of rational actions of the multiplicative group $\mathbb{G}_m$ on X. We show that such a structure induces a rational action of the Weyl group W on X. Surprisingly many interesting rational actions of W come from geometric crystals. For example, the natural action of W on Grothendieck's simultaneous resolution $\tilde{G} \rightarrow G$ comes from a structure of a geometric crystal on $\tilde{G}$. Also all the examples of the action of W in [BrK] come from geometric crystals. Another application of geometric crystals is a construction for any SL_n-crystal built on (X, γ) of a *trivialization* – a W-equivariant isomorphism $X \xrightarrow{\sim} \gamma^{-1}(e) \times T$.

It is also interesting that the Langlands dual group ^LG emerges when we reconstruct Kashiwara's combinatorial crystals out of *positive* geometric crystals (see section 2.4). The presence of ^LG in the "crystal world" has been noticed in [Lu2]. The combinatorial results of [BZ2] also involve ^LG.

A number of statements in the paper are almost immediate. We call them *lemmas*. Proofs of all the lemmas from sections 2, 3 and 4 are left to the reader.

The material of the paper is organized as follows:

• In section 2 we introduce geometric crystals and formulate their main properties.

• In section 3 we introduce *unipotent crystals* as an algebro-geometric analogue of crystal bases for $U_q(^L\mathfrak{g})$-modules, where $^L\mathfrak{g}$ is the Lie algebra of ^LG. A unipotent crystal built on an irreducible algebraic variety X consists of a rational action of U on X and a rational U-equivariant morphism $\mathbf{f} : X \rightarrow G/U$. The unipotent crystals have a number of properties characteristic of the crystal bases. One of the main properties is a natural *product* of unipotent crystals which is an analogue of the tensor product of crystal bases. We prove that the category of unipotent crystals is strictly monoidal with respect to this product. We also define the notion of a *dual unipotent crystal*. One of the main results of the section is a construction for any unipotent crystal on X, of an *induced* geometric crystal $\mathcal{X}$. We also give a closed formula for the action of W on this induced geometric crystal $\mathcal{X}$.

• In section 4 we study the *standard* unipotent crystals which are the U-orbits $BwB/B \subset G/B$. We obtain a decomposition of any standard crystal BwB/B into the product of 1-dimensional crystals corresponding to any reduced decomposition of $w \in W$. This is a geometrization of Kashiwara's results. We also construct W-invariant functions on certain standard unipotent crystals. These functions are important for the study of γ-functions of representations $^L G$ (see [BrK]).

• In section 5 we collect proofs of the results from sections 2, 3 and 4.

• In section 6 we study the restrictions to a standard Levi subgroup $L \subset G$ of standard unipotent G-crystals BwB/B. In particular, for each $w \in W$ we construct a rational morphism $\mathbf{p}_w : BwB/B \to L/(U \cap L)$. In the case when $\mathbf{p}_w$ is a birational isomorphism with its image, we study the direct image of W_L-invariant functions under $\mathbf{p}_w$. In the special case when $G = GL(m+n)$, $L = GL(m) \times GL(n)$ and $w = w_0^L \cdot w_0$, it is possible to write explicitly the corresponding action of the group W_L on the image of $\mathbf{p}_w$, and the W_L-equivariant trivialization. This simplification of the structure of the crystal on BwB/B is used in [BrK] for the proof that Piatetski-Shapiro's γ-function to $GL(m) \times GL(n)$ is equal to the one introduced in [BrK]. We expect that in general the action of W_L on BwB/B, and the trivialization of the corresponding crystal would be easy to describe and that this description can help in the study of γ-functions on L.

• In the Appendix we collect the necessary definitions and results related to Kashiwara's crystals. We refer to these crystals as *combinatorial* in order to distinguish between them and their geometric counterparts.

The notion of a *geometric crystal* was introduced by the first author who noticed that the formulas for the W-action which appear in the definition of γ-functions for the group $GL_2 \times GL_3$ (see [BrK]) can be interpreted as a "geometrization" of the W-action for the free combinatorial crystal defined earlier by the first author. We want to express our gratitude to A. Braverman for his help at different stages of this work and in particular for the definition of the notion of *degree* in section 2.4. We are also grateful to Y. Flicker for numerous remarks about the paper.

2 Definitions and Main Results on Geometric Crystals

2.1   General notation.   Let G be a split reductive algebraic group over $\mathbb{Q}$ and $T \subset G$ a maximal torus. We denote by $\Lambda^\vee$ and Λ the lattices of co-characters and characters of T and by $\langle \cdot, \cdot \rangle$ the evaluation pairing $\Lambda^\vee \times \Lambda \to \mathbb{Z}$.

Let B be a Borel subgroup containing T. Denote by I the set of vertices of the Dynkin diagram of G; for any $i \in I$, denote by $\alpha_i \in \Lambda$ the simple root $\alpha_i : T \to \mathbb{G}_m$, and by $\alpha_i^\vee \in \Lambda^\vee$ the simple coroot $\alpha_i^\vee : \mathbb{G}_m \to T$.

Let $B^- \subset G$ be the Borel subgroup containing T such that $B \cap B^- = T$. Denote by U and U^- respectively the unipotent radicals of B and B^-.

For each $i \in I$ we denote by $U_i \subset U$ and $U_i^- \subset U^-$ the corresponding simple root subgroups and denote by $\xi_i : U \to U_i$, $\xi_i^- : U^- \to U_i^-$ the canonical projections.

We fix a family of isomorphisms $x_i : \mathbb{G}_a \xrightarrow{\sim} U_i$, $y_i : \mathbb{G}_a \xrightarrow{\sim} U_i^-$, $i \in I$, such that

$$x_i(a)y_i(a') = y_i\left(\frac{a'}{1+aa'}\right)\alpha_i^\vee(1+aa')x_i\left(\frac{a}{1+aa'}\right). \qquad (2.1)$$

Clearly, each isomorphism y_i is uniquely determined by x_i.

REMARK.   Each pair x_i, y_i defines a homomorphism $\phi_i : SL_2 \to G$:

$$\phi_i\begin{pmatrix} a & b \\ c & d \end{pmatrix} = y_i\left(\frac{c}{a}\right)\alpha_i^\vee(a)x_i\left(\frac{b}{a}\right).$$

Denote by $\widehat{U}$ and $\widehat{U}^-$ respectively the spaces $\mathrm{Hom}(U, \mathbb{G}_a)$ and $\mathrm{Hom}(U^-, \mathbb{G}_a)$. For each $i \in I$ define $\chi_i \in \widehat{U}$, $\chi_i^- \in \widehat{U}^-$ by

$$\chi_i(u_+) = x_i^{-1}(\xi_i(u_+)), \quad \chi_i^-(u_-) = y_i^{-1}(\xi_i^-(u_-)),$$

for $u_- \in U^-$, $u_+ \in U$. By definition, the family χ_i, $i \in I$, is a basis in the vector space $\widehat{U}$, and the family χ_i^-, $i \in I$, is a basis in the vector space $\widehat{U}^-$. For any $\chi \in \widehat{U}$, and any $\chi^- \in \widehat{U}^-$ define functions $\overline{\chi}, \overline{\chi}^- : U^- \cdot T \cdot U \to \mathbb{G}_a$ by

$$\overline{\chi}^-(u_-tu_+) = \chi^-(u_-), \quad \overline{\chi}(u_-tu_+) = \chi(u_+).$$

The Weyl group $W = \mathrm{Norm}_G(T)/T$ of G is generated by simple reflections $s_i \in W$, $i \in I$. The group W acts on lattices $\Lambda, \Lambda^\vee$ by $s_i(\lambda) = \lambda - \langle \alpha_i^\vee, \lambda \rangle \alpha_i$ for $\lambda \in \Lambda$, $s_i(\lambda^\vee) = \lambda^\vee - \langle \lambda^\vee, \alpha_i \rangle \alpha_i^\vee$ for $\lambda^\vee \in \Lambda^\vee$.

Let $l : W \to \mathbb{Z}_{\geq 0}$ ($w \mapsto l(w)$) be the length function. For any sequence $\mathbf{i} = (i_1, \ldots, i_l) \in I^l$ we write $w(\mathbf{i}) = s_{i_1} \cdots s_{i_l}$. A sequence $\mathbf{i} \in I^l$ is called *reduced* if the length of $w(\mathbf{i})$ is equal to l. For any $w \in W$ we denote by $R(w)$ the set of all reduced sequences $\mathbf{i}$ such that $w(\mathbf{i}) = w$. We denote by $w_0 \in W$ the element of the maximal length in W.

For $i \in I$ define $\bar{s}_i \in G$ by

$$\bar{s}_i = x_i(1)y_i(-1)x_i(1). \tag{2.2}$$

Each $\bar{s}_i$ belongs to $\text{Norm}_G(T)$ and is a representative of $s_i \in W$. It is well known ([Bo]) that the elements $\bar{s}_i$, $i \in I$, satisfy the braid relations. Therefore we can associate to each $w \in W$ its *standard representative* $\bar{w} \in \text{Norm}_G(T)$ in such a way that for any $(i_1, \ldots, i_l) \in R(w)$ we have

$$\bar{w} = \bar{s}_{i_1} \cdots \bar{s}_{i_l}. \tag{2.3}$$

2.2 Geometric pre-crystals and geometric crystals.

Let X and Y be algebraic varieties over $\mathbb{Q}$. Denote by $\mathbf{R}(X, Y)$ the set of all rational morphisms from X to Y.

For any $\mathbf{f} \in \mathbf{R}(X, Y)$ denote by $\text{dom}(f) \subset X$ the maximal open subset of X on which f is defined; denote by $\mathbf{f}_{reg} : \text{dom}(\mathbf{f}) \rightarrow Y$ the corresponding regular morphism. We denote by $\text{ran}(\mathbf{f}) \subset Y$ the closure of the constructible set $\mathbf{f}_{reg}(\text{dom}(\mathbf{f}))$ in Y. For any regular morphism $f : X' \rightarrow Y$ where $X' \subset X$ is a dense subset, we denote by $[f] : X \rightarrow Y$ the corresponding rational morphism. Note that $[\mathbf{f}_{reg}] = \mathbf{f}$ and $\text{dom}(\mathbf{f}_{reg}) = \text{dom}(\mathbf{f})$ for any $\mathbf{f} \in \mathbf{R}(X, Y)$.

It is easy to see that for any irreducible algebraic varieties X, Y, Z and rational morphisms $f : X \rightarrow Y$, $g : Y \rightarrow Z$ such that $\text{dom}(g)$ intersects $\text{ran}(f)$ non-trivially, the composition $(f, g) \mapsto f \circ g$ is well defined and is a rational morphism from X to Z.

We denote by $\mathcal{V}$ the category whose objects are irreducible algebraic varieties and arrows are dominant rational morphisms.

For any algebraic group H we call a rational action $\alpha : H \times X \rightarrow X$ *unital* if $\text{dom}(\alpha) \supset \{e\} \times X$.

DEFINITION. Let $X \in \text{Ob}(\mathcal{V})$ and γ be a rational morphism $X \rightarrow T$. A *geometric G-pre-crystal* (or simply *geometric pre-crystal*) on (X, γ) is a family $e_i : \mathbb{G}_m \times X \rightarrow X$, $i \in I$, of unital rational actions of the multiplicative group $\mathbb{G}_m$: $(c, x) \mapsto e_i^c(x)$ such that $\gamma(e_i^c(x)) = \alpha_i^\vee(c)\gamma(x)$.

REMARK. Geometric pre-crystals are analogues of free combinatorial pre-crystals (see the Appendix). Under this analogy, the variety X corresponds to the set B, the maximal torus T corresponds to the lattice $\Lambda^\vee$, the rational morphism $\gamma : X \rightarrow T$ corresponds to the map $\tilde{\gamma} : B \rightarrow \Lambda^\vee$ and rational actions e_i of $\mathbb{G}_m$ on X correspond to bijections $\tilde{e}_i : B \rightarrow B$. We will make this analogy precise in section 2.4.

Given a geometric pre-crystal $\mathcal{X}$, and a reduced sequence $\mathbf{i}=(i_1,...,i_l)\in I^l$, we define a rational morphism $e_{\mathbf{i}} : T \times X \to X$ by

$$(t, x) \mapsto e_{\mathbf{i}}^t(x) = e_{i_1}^{\alpha^{(1)}(t)} \circ \cdots \circ e_{i_l}^{\alpha^{(l)}(t)}(x), \qquad (2.4)$$

where $\alpha^{(k)} = s_{i_l} s_{i_{l-1}} \cdots s_{i_{k+1}}(\alpha_{i_k})$, $k = 1, \ldots, l$, are the *associated* positive roots.

DEFINITION.   A geometric pre-crystal $\mathcal{X}$ on (X, γ) is called a *geometric crystal* if, for any $w \in W$ and any $\mathbf{i}, \mathbf{i}' \in R(w)$, one has

$$e_{\mathbf{i}} = e_{\mathbf{i}'}. \qquad (2.5)$$

LEMMA 2.1.   *The relations (2.5) are equivalent to the following relations between e_i, e_j for $i, j \in I$:*

$$e_i^{c_1} e_j^{c_2} = e_j^{c_2} e_i^{c_1} \qquad (2.6)$$

if $\langle \alpha_i^{\vee}, \alpha_j \rangle = 0$;

$$e_i^{c_1} e_j^{c_1 c_2} e_i^{c_2} = e_j^{c_2} e_i^{c_1 c_2} e_j^{c_1} \qquad (2.7)$$

if $\langle \alpha_i^{\vee}, \alpha_j \rangle = \langle \alpha_j^{\vee}, \alpha_i \rangle = -1$;

$$e_i^{c_1} e_j^{c_1^2 c_2} e_i^{c_1 c_2} e_j^{c_2} = e_j^{c_2} e_i^{c_1 c_2} e_j^{c_1^2 c_2} e_i^{c_1} \qquad (2.8)$$

if $\langle \alpha_i^{\vee}, \alpha_j \rangle = -2$, $\langle \alpha_j^{\vee}, \alpha_i \rangle = -1$;

$$e_i^{c_1} e_j^{c_1^2 c_2} e_i^{c_1^3 c_2} e_j^{c_1^3 c_2^2} e_i^{c_1 c_2} e_j^{c_2} = e_j^{c_2} e_i^{c_1 c_2} e_j^{c_1^3 c_2^2} e_i^{c_1^3 c_2} e_j^{c_1^2 c_2} e_i^{c_1} \qquad (2.9)$$

if $\langle \alpha_i^{\vee}, \alpha_j \rangle = -3$, $\langle \alpha_j^{\vee}, \alpha_i \rangle = -1$.

REMARKS.   1. The relations (2.6)–(2.9) are multiplicative analogues of the Verma relations in the universal enveloping algebra $U(^L\mathfrak{g})$ (see [Lu3, Proposition 39.3.7]).

2. An analogue of the relations (2.6),(2.7) for a combinatorial GL_n-crystal was considered in [BK, Theorem 1.1].

LEMMA 2.2.   *Let $\mathcal{X}$ be a geometric pre-crystal. For any $\lambda \in \Lambda$, $i \in I$, the formula $x \mapsto e_i^{\lambda(\gamma(x))}(x)$ defines a rational morphism $X \to X$. This is a birational isomorphism $X \widetilde{\to} X$ if and only if $\langle \alpha_i^{\vee}, \lambda \rangle \in \{-2, 0\}$. In the latter case the inverse morphism is given by the formula*

$$x \mapsto \begin{cases} e_i^{\lambda(\gamma(x))}(x), & \text{if } \langle \alpha_i^{\vee}, \lambda \rangle = -2, \\ e_i^{\frac{1}{\lambda(\gamma(x))}}(x), & \text{if } \langle \alpha_i^{\vee}, \lambda \rangle = 0. \end{cases}$$

Let $\mathcal{X}$ be a geometric crystal. For each $w \in W$ we define a rational morphism $e_w : T \times X \to X$ by the formula $e_w := e_{\mathbf{i}}$ for any $\mathbf{i} \in R(w_0)$ (see (2.5)).

PROPOSITION 2.3.   *The correspondence* $W \times X \to X$ *defined by*

$$(w, x) \mapsto w(x) = e_w^{\gamma(x)^{-1}}(x) \tag{2.10}$$

is a rational unital action of W *on* X.

REMARK.   The formula $s_i(x) = e_i^{\frac{1}{\alpha_i(\gamma(x))}}(x)$ is a multiplicative analogue of (7.1) in the Appendix.

2.3   The geometric crystal $\mathcal{X}_{w_0}$.   Let $B_{w_0}^- := U\overline{w_0}U \cap B^-$. The natural inclusion $B_{w_0}^- \hookrightarrow G$ induces the open inclusion $\mathfrak{j}_0 : B_{w_0}^- \hookrightarrow G/B$. Let $\gamma : G/B \to T$ be the rational morphism defined by $\gamma = pr_T \circ [\mathfrak{j}_0]^{-1}$, where $pr_T : B^- \to T = B^-/U^-$ is the natural projection.

For $i \in I$, let $\varphi_i : G/B \to \mathbb{G}_a$ be the rational function given by $\varphi_i := \overline{\chi}_i^- \circ [\mathfrak{j}_0]^{-1}$, where $\overline{\chi}_i^- : B^- \cdot U \to \mathbb{G}_a$ is the regular function defined in section 2.1.

LEMMA 2.4.   *For each* $i \in I$ *we have*

(a) $\varphi_i \not\equiv 0$;
(b) $\frac{1}{\varphi_i(x_i(a)\cdot x)} = \frac{1}{\varphi_i(x)} + a$;
(c) $\gamma(x_i(a) \cdot x) = \alpha_i^\vee(1 + a\varphi_i(x))\gamma(x)$.

For each $i \in I$ define a rational morphism $e_i : \mathbb{G}_m \times G/B \to G/B$ by the formula

$$e_i^c(x) = x_i\left(\frac{c-1}{\varphi_i(x)}\right) \cdot x. \tag{2.11}$$

REMARK.   Lemma 2.4 implies the equality $\varphi_i(e_i^c(x)) = c^{-1}\varphi_i(x)$ for $i \in I$.

Theorem 2.5.   *The morphisms* e_i, $i \in I$, *define a geometric crystal on* $(G/B, \gamma)$.

We denote this crystal by $\mathcal{X}_{w_0}$. This is one of our main examples of geometric crystals.

REMARK.   The geometric crystal $\mathcal{X}_{w_0}$ is an analogue of the free combinatorial W-crystal $\mathcal{B}_{w_0}$, and the rational functions $\varphi_i : G/B \to \mathbb{G}_m$ are analogues of the functions $-\widetilde{\varphi}_i$ on $\mathcal{B}_{w_0}$ (see Appendix).

2.4   Positive geometric crystals and their tropicalization.   Let T' be an algebraic torus over $\mathbb{Q}$. We denote by $X^\star(T')$ and $X_\star(T')$ respectively the lattices of characters and co-characters of T', and by $\langle \cdot, \cdot \rangle$ the canonical pairing $X_\star(T') \times X^\star(T') \to \mathbb{Z}$.

Let $\mathcal{L}(T')$ be the set of *formal loops* $\phi : \mathbb{G}_m \to T'$. In particular $\mathcal{L}(\mathbb{G}_m) = \mathcal{L}(c)$, where $\mathcal{L}(c)$ is the set of all Laurent series in the variable c.

Denote by $\mathcal{L}_0(T')$ the set *formal disks*, that is, the set of all $\phi \in \mathcal{L}_0(T')$ such that for any $\mu \in X^*(T')$ we have $\mu \circ \phi \in \mathcal{L}_0(c)$, where $\mathcal{L}_0(c) \subset \mathcal{L}_0(c)$ is the set of all invertible Talyor series in the variable c.

Clearly, $\mathcal{L}_0(T')$ has a natural structure of an irreducible pro-algebraic variety.

LEMMA 2.6. *The multiplication map $X_*(T') \times \mathcal{L}_0(T') \to \mathcal{L}(T')$ is a bijection.*

This defines a surjective map $\deg_{T'} : \mathcal{L}(T') \to X_*(T')$.

REMARKS. 1. If $T' = \mathbb{G}_m$ then $\deg_{\mathbb{G}_m} : \mathcal{L}(c) \to \mathbb{Z}$ is the valuation map which associates to a non-zero Laurent series $\phi(c)$ its lowest degree. By definition, $\mathcal{L}_0(c) = \deg_{\mathbb{G}_m}^{-1}(0)$.

2. For any co-character $\lambda \in X_*(T')$, the pre-image $\deg_{T'}^{-1}(\lambda) \subset \mathcal{L}(T')$ is naturally isomorphic to $\mathcal{L}_0(T')$. Hence it makes sense to talk about generic points of $\deg_{T'}^{-1}(\lambda)$.

For any $f \in \mathbf{R}(T', T'')$ and any $\lambda \in X_*(T')$, $\mu \in X_*(T'')$, let $U_f(\lambda, \mu)$ be the set of all $\phi \in \deg_{T'}^{-1}(\lambda)$ such that $f \circ \phi \in \deg_{T''}^{-1}(\mu)$.

LEMMA 2.7. *Let $f \in \mathbf{R}(T', T'')$. For any $\lambda \in X_*(T')$ there is a unique $\mu \in X_*(T'')$ such that the set $U_f(\lambda, \mu)$ is dense in $\deg_{T''}^{-1}(\mu)$.*

This allows us to define a map $\deg(f) : X_*(T') \to X_*(T'')$ by $\deg(f)(\lambda) = \mu$, where $\mu \in X_*(T'')$ is determined in Lemma 2.7.

We call this map the *degree* of f. It is easy to check that $\deg(f)$ is a piecewise-linear map. Note that $\deg(f)$ is linear if f is a group homomorphism.

DEFINITION. (a) A rational function f on a torus T' is called *positive* if it can be written as a ratio $f = f'/f''$, where f' and f'' are linear combinations of characters with positive integer coefficients.

(b) For any two algebraic tori T', T'', we call a rational morphism $f \in \mathbf{R}(T', T'')$ *positive* if for any character $\mu : T'' \to \mathbb{G}_m$ the composition $\mu \circ f$ is a positive rational function on $\mathbb{G}_m$.

Denote by $\mathbf{R}_+(T', T'')$ the set of positive rational morphisms from T' to T''.

REMARK. Any homomorphism of algebraic tori is positive.

LEMMA 2.8. *For any two positive morphisms $f \in \mathbf{R}_+(T', T'')$, $g \in \mathbf{R}_+(T'', T''')$ the composition $g \circ f \in \mathbf{R}_+(T', T''')$ is well defined.*

Therefore, we can consider a category $\mathcal{T}_+$ whose objects are algebraic tori, and the arrows are positive rational morphisms.

For any $f \in \mathbf{R}(T', T'')$ and any $\lambda \in X_*(T')$, $\mu \in X_*(T'')$, let $U_{f;\lambda}$ be the set of all $\phi \in \deg_{T'}^{-1}(\lambda)$ such that $f \circ \phi \in \mathcal{L}(T'')$.

We say that a formal loop $\phi \in \mathcal{L}(T')$ is *positive rational* if $\phi \in \mathbf{R}_+(\mathbb{G}_m, T')$.

LEMMA 2.9.   Let $f : T' \to T''$ be a positive morphism. Then for any $\lambda \in X_*(T')$ and any formal positive rational loop $\phi \in \deg_{T'}^{-1}(\lambda)$ we have

$$\deg_{T''} \circ f \circ \phi = \deg(f)(\lambda).$$

COROLLARY 2.10. For any algebraic tori T', T'', T''' and any $f \in \mathbf{R}_+(T', T'')$, $g \in \mathbf{R}_+(T'', T''')$, we have

$$\deg(g \circ f) = \deg(g) \circ \deg(f). \tag{2.12}$$

This implies that is a functor Trop : $\mathcal{T}_+ \to \mathbf{Set}$ such that $\mathrm{Trop}(T') = X_*(T')$ and $\mathrm{Trop}(f : T' \to T'') = (\deg(f) : X_*(T') \to X_*(T''))$.

Following [BFZ], we call the functor Trop *tropicalization*.

REMARK. If $f : T' \to T''$ is not positive then the assertion of Lemma 2.9 is not true even if f is a birational isomorphism. Moreover, one can find a birational isomorphism f, $f : T' \xrightarrow{\sim} T''$, (2.12) does not hold for (f, g), where $g = f^{-1} : T'' \to T'$. Consider, for example, the case when $T' = T'' = \mathbb{G}_m$, $f(c) = c - 1$, $g(c) = c + 1$.

DEFINITION.   Let $\mathcal{X}$ be a geometric pre-crystal on (X, γ). A birational isomorphism $\theta : T' \xrightarrow{\sim} X$ for some algebraic torus T' is called a *positive structure* on $\mathcal{X}$ if the following conditions are satisfied:

1. The rational morphism $\gamma \circ \theta : T' \to T$ is positive.
2. For each $i \in I$ the rational morphism $e_{i;\theta} : \mathbb{G}_m \times T' \to T'$ given by

$$e_{i;\theta}(c, t') = \theta^{-1}(e_i^c(\theta(t'))) \tag{2.13}$$

    is positive.

REMARK.   In all our examples of positive structures θ, the composition $\gamma \circ \theta : T' \to T$ is a group homomorphism.

We say that two positive structures θ, θ' are *equivalent* if the rational morphisms $\theta^{-1} \circ \theta'$ and $\theta'^{-1} \circ \theta$ are positive.

Recall from the Appendix that a combinatorial pre-crystal consists of a set B, a map $\tilde{\gamma} : B \to \Lambda^\vee$ and a compatible collection of partial bijections $\tilde{e}_i : B \to B$.

For any positive structure θ on a geometric pre-crystal $\mathcal{X}$ built on (X, γ) define for $i \in I$ the $\mathbb{Z}$-action $\tilde{e}_i^\bullet : \mathbb{Z} \times X_*(T') \to X_*(T')$ by the formula

$$\tilde{e}_i^\bullet = \mathrm{Trop}(e_{i;\theta}), \tag{2.14}$$

where $e_{i;\theta} : \mathbb{G}_m \times T' \to T'$ is the positive morphism defined by (2.13).

Also denote $\tilde{\gamma}_{trop} := \mathrm{Trop}(\gamma \circ \theta) : X_*(T') \to \Lambda^\vee$.

Obviously, the map $\tilde{e}_i^1 : X_\star(T') \to X_\star(T')$ is a bijection for $i \in I$. The bijections $\tilde{e}_i := \tilde{e}_i^1$, $i \in I$, define a free combinatorial pre-crystal $\mathrm{Trop}_\theta(\mathcal{X})$ on $(X_\star(T'), \tilde{\gamma}_\theta)$.

Theorem 2.11. *Let $\mathcal{X}$ be a geometric crystal, and let θ be a positive structure on X. Then $\mathrm{Trop}_\theta(\mathcal{X})$ is a free combinatorial W-crystal.*

We call this free combinatorial pre-crystal $\mathrm{Trop}_\theta(\mathcal{X})$ the *tropicalization of $\mathcal{X}$ with respect to the positive structure θ.*

For a reduced sequence $\mathbf{i} = (i_1, \ldots, i_l)$ define a morphism $\theta_\mathbf{i} : (\mathbb{G}_m)^l \to G/B$ by

$$\theta_\mathbf{i}(c_1, \ldots, c_l) := x_{i_1}(c_1)\overline{s_{i_1}} \cdots x_{i_l}(c_1)\overline{s_{i_l}} \cdot B. \qquad (2.15)$$

The following theorem was proved in [BZ2] in a slightly different form.

Theorem 2.12. *For each $\mathbf{i} \in R(w_0)$ the morphism $\theta_\mathbf{i}$ is a positive structure on the geometric crystal $\mathcal{X}_{w_0}$, and the tropicalization of $\mathcal{X}_{w_0}$ with respect to $\theta_\mathbf{i}$ is equal to the free combinatorial W-crystal $\mathcal{B}_\mathbf{i}$. All these positive structures $\theta_\mathbf{i}$ are equivalent to each other.*

REMARK.   In [Lu3] Lusztig introduced morphisms $\theta^\mathbf{i} : (\mathbb{G}_m)^{l(w_0)} \to G/B$,

$$\theta^\mathbf{i}(c_1, \ldots, c_l) = y_{i_1}(c_1) \cdots y_{i_l}(c_l) \cdot B, \qquad (2.16)$$

and proved that these morphisms are related to each other in the same way as the corresponding parametrizations of the canonical basis for $U_q(^L\mathfrak{g})$. We discuss similar morphisms $\pi^\mathbf{i}$ in section 4.4.

It was shown in [BZ2] that the morphisms $\theta^\mathbf{i} : (\mathbb{G}_m)^{l(w_0)} \to G/B$, $\mathbf{i} \in R(w_0)$, are also positive structures on $\mathcal{X}_{w_0}$, and, moreover, these positive structures are equivalent to $\theta_\mathbf{i}, \mathbf{i} \in R(w_0)$.

2.5   Trivialization of geometric crystals.   Let $\mathcal{X}$ be a geometric G-crystal built on (X, γ). Without loss of generality we may assume that γ is a regular surjective morphism $X \to T$. Denote by $X_0 = \gamma^{-1}(e)$ the fiber over the unit $e \in T$.

By definition of the action W on X (see Proposition 2.3), $w(x_0) = x_0$ for $x_0 \in X_0$, $w \in W$.

DEFINITION.   A *trivialization* of a geometric crystal $\mathcal{X}$ is a W-invariant rational projection $\tau : X \to X_0$.

It is easy to see that the following formula defines a trivialization for any geometric SL_2-crystal $\mathcal{X}$,

$$\tau(x) = e_i^{\frac{1}{\omega_1(\gamma(x))}}(x),$$

where $I = \{1\}$ and ω_i is the only fundamental weight of T.

In the case when $G = SL_3$ we can construct two different trivializations $\tau, \tau' : X \to X_0$ for any geometric SL_3-crystal $\mathcal{X}$. These trivializations are given by

$$\tau(x) = e_1^{\frac{1}{\omega_2(\gamma(x))}} e_2^{\frac{1}{\omega_2(\gamma(x))}} e_1^{\frac{\omega_2(\gamma(x))}{\omega_1(\gamma(x))}}(x),$$

$$\tau'(x) = e_2^{\frac{1}{\omega_1(\gamma(x))}} e_1^{\frac{1}{\omega_1(\gamma(x))}} e_2^{\frac{\omega_1(\gamma(x))}{\omega_2(\gamma(x))}}(x),$$

where $\omega_1, \omega_2 \in \Lambda$ are the fundamental weights.

Warning. The morphism $X \to X_0$: $x \mapsto e_1^{\frac{1}{\omega_1(\gamma(x))}} e_2^{\frac{1}{\omega_2(\gamma(x))}}(x)$ is not W-invariant.

Actually for any $r > 1$ any geometric $G = SL_{r+1}$-crystal $\mathcal{X}$ we can construct two trivializations $\tau, \tau' : X \to X_0$ of $\mathcal{X}$ in the following way. Let $\omega_1, \ldots, \omega_r \in \Lambda$ be the fundamental weights ordered in the standard way and $\omega_i(x) = \omega_i(\gamma(x))$ for $i = 1, \ldots, r$.

For every geometric SL_{r+1}-crystal $\mathcal{X}$ define a morphism $\tau : X \to X_0$ by the formula

$$\tau(x) = \left(e_1^{\frac{1}{\omega_r(x)}} \cdots e_r^{\frac{1}{\omega_r(x)}} \right) \left(e_1^{\frac{\omega_r(x)}{\omega_{r-1}(x)}} \cdots e_{r-1}^{\frac{\omega_r(x)}{\omega_{r-1}(x)}} \right) \cdots \left(e_1^{\frac{\omega_3(x)}{\omega_2(x)}} e_2^{\frac{\omega_3(x)}{\omega_2(x)}} \right) \left(e_1^{\frac{\omega_2(x)}{\omega_1(x)}} \right)(x).$$

Theorem 2.13. *This morphism $\tau : X \to X_0$ is a trivialization of $\mathcal{X}$.*

The formula for the second trivialization $\tau' : X \to X_0$ is obtained from τ by applying the automorphism of the Dynkin diagram which exchanges i with $r + 1 - i$.

We do not know whether trivializations of geometric G-crystals exist for other reductive groups. We expect that for the *unipotent crystals* one can find a trivialization.

3 Unipotent Crystals

3.1 Definition of unipotent crystals and their product.

As we have seen, geometric crystals are geometric analogues of free combinatorial W-crystals. Our next task is to introduce a geometric analogue of crystal bases.

DEFINITION.   A *U-variety* $\mathbf{X}$ is a pair (X, α), where $X \in \mathrm{Ob}(\mathcal{V})$ and $\alpha : U \times X \to X$ is a rational unital U-action on X such that $e \times X \subset \mathrm{dom}(\alpha)$, where e is the unit of U.

For U-varieties $\mathbf{X}, \mathbf{Y}$ we say that a rational morphism $\mathbf{f} : X \to Y$ is a *U-morphism* if it commutes with the U-actions.

It is well known that the multiplication in G induces a birational iso-morphism $B^- \times U \widetilde{\to} G$. Denote by $\mathbf{g}$ the inverse birational isomorphism:

$$\mathbf{g} : G \widetilde{\to} B^- \times U . \tag{3.1}$$

Let $\pi^- : G \to B^-$ and $\pi : G \to U$ be rational morphisms defined by

$$\pi^- = pr_1 \circ \mathbf{g}, \quad \pi = pr_2 \circ \mathbf{g} .$$

By definition, $\mathrm{dom}(\pi^-) = \mathrm{dom}(\pi) = B^- \cdot U$.

Passing to the quotient, we obtain a birational isomorphism $G/U \widetilde{\to} B^-$. Therefore, the natural left action of U on G/U defines a left U-action $\alpha_{B^-} : U \times B^- \to B^-$. This action satisfies for $u \in U, b \in B^-$:

$$\alpha_{B^-}(u, b) = \pi^-(u \cdot b) = u \cdot b \cdot \left(\pi(u \cdot b)\right)^{-1} . \tag{3.2}$$

In particular, the pair $\mathbf{B}^- := (B^-, \alpha_{B^-})$ is a U-variety.

LEMMA 3.1.   For $u \in U^-$, $t \in T$ one has

$$\pi\left(x_i(a) \cdot u \cdot t\right) = x_i\left((a^{-1} + \chi_i^-(u))^{-1} \cdot \alpha_i(t^{-1})\right) . \tag{3.3}$$

LEMMA 3.2.   (a) For $b = u \cdot t$, $u \in U^-$, $t \in T$, $a \in \mathbb{G}_a$ and $i \in I$ we have

$$\alpha_{B^-}\left(x_i(a), b\right) = x_i(a) \cdot b \cdot x_i\left(-\frac{a}{(1 + a\chi_i^-(u))\alpha_i(t)}\right) . \tag{3.4}$$

(b) Every U-orbit in $\mathbf{B}^-$ is the intersection of B^- with a $U \times U$-orbit in G.

DEFINITION.   A *unipotent crystal* is a pair $(\mathbf{X}, \mathbf{f})$, where $\mathbf{X}$ is a U-variety and $\mathbf{f} : \mathbf{X} \to \mathbf{B}^-$ is a U-morphism.

We denote by U-*Cryst* the category whose objects are unipotent G-crystals and arrows are dominant rational morphisms.

For any $(\mathbf{X}, \mathbf{f}_X), (\mathbf{Y}, \mathbf{f}_Y) \in \mathrm{Ob}(U\text{-Cryst})$ define a rational morphism $\alpha : U \times X \times Y \to X \times Y$ by the formula

$$\alpha\left(u, (x, y)\right) := \left(u(x), (\pi(u \cdot \mathbf{f}_X(x)))(y)\right) , \tag{3.5}$$

where $u(x) = \alpha(u, x)$. We will often write $u(x, y)$ instead of $\alpha(u, (x, y))$.

Theorem 3.3.   (a) The morphism $\alpha : U \times X \times Y \to X \times Y$ defined above is a rational U-action on $X \times Y$.

(b) Let $\mathbf{m} : B^- \times B^- \to B^-$ be the multiplication morphism. Let $\mathbf{f} = \mathbf{f}_{X \times Y} : X \times Y \to B^-$ be the rational morphism defined by $\mathbf{f} = \mathbf{m} \circ (\mathbf{f}_X \times \mathbf{f}_Y)$. Then $\mathbf{f}_{X \times Y}$ is a U-morphism.

We denote the U-variety $(X \times Y, \alpha_{X \times Y})$ by $\mathbf{X} \times_{\mathbf{f}} \mathbf{Y}$. According to the theorem the pair $(\mathbf{X} \times_{\mathbf{f}} \mathbf{Y}, \mathbf{f}_{X \times Y})$ is a unipotent crystal. We call it the *product* of $(\mathbf{X}, \mathbf{f}_X)$ and $(\mathbf{Y}, \mathbf{f}_Y)$ and denote it by $(\mathbf{X}, \mathbf{f}_X) \times (\mathbf{Y}, \mathbf{f}_Y)$.

REMARK. The product of unipotent crystals is analogous to the tensor product of Kashiwara's crystals defined in [K2]. This analogy is made precise in sections 3.3 and 3.4 below.

PROPOSITION 3.4. *The product of unipotent crystals is associative.*

REMARK. The above results define a strictly monoidal structure on the category $U - \mathrm{Cryst}$.

The product of unipotent crystals is not commutative in general. For any family $(\mathbf{X}_k, \mathbf{f}_k) \in \mathrm{Ob}(U\text{-Cryst})$, $k = 1, \ldots, l$ we denote by $\prod_{k=1}^{l}(\mathbf{X}_k, \mathbf{f}_k)$ the product $(\mathbf{X}_1, \mathbf{f}_1) \times \cdots \times (\mathbf{X}_l, \mathbf{f}_l)$.

EXAMPLE. Denote by $\mathbf{U}$ the pair (U, α_U) where $\alpha_U : U \times U \to U$ is the left action. Let $\mathrm{pr}_e^U : U \to B^-$ be the projection on the unit e. Clearly, $(\mathbf{U}, \mathrm{pr}_e^U) \in \mathrm{Ob}(U\text{-Cryst})$. Similarly, denote by $\mathbf{G}$ the pair (G, α_G), where $\alpha_G : U \times G \to U$ is the left action. Clearly, $(\mathbf{G}, \pi^-) \in \mathrm{Ob}(U\text{-Cryst})$.

PROPOSITION 3.5. *The birational isomorphism* $\mathbf{g}$ *induces the isomorphism in U-Cryst,*

$$(\mathbf{G}, \pi^-) \widetilde{\to} (\mathbf{B}^-, \mathrm{id}_{B^-}) \times (\mathbf{U}, \mathrm{pr}_e^U).$$

More generally, taking the right quotient by any subgroup $U' \subset U$, we obtain the birational isomorphism $\mathbf{g}_{U'} : G/U' \widetilde{\to} B^- \times U/U'$ *which induces the isomorphism in U-Cryst,*

$$(\mathbf{G}/\mathbf{U}', \pi_{U'}^-) \widetilde{\to} (\mathbf{B}^-, \mathrm{id}_{B^-}) \times (\mathbf{U}/\mathbf{U}', \mathrm{pr}_e^{U/U'}),$$

where $\pi_{U'}^- = \mathrm{pr}_1 \circ \mathbf{g}_{U'}$.

The following result shows that the unipotent crystal $(\mathbf{G}, \pi^-)$ is universal.

LEMMA 3.6. *For every U-equivariant rational morphism* $\mathbf{f}_U : U \to B^-$ *there is an element* $\tilde{w} \in \mathrm{Norm}_G(T)$ *such that*

$$\mathbf{f}_U(u) = \pi^-(u \cdot \tilde{w})$$

for every $u \in U$.

3.2　From unipotent G-crystals to unipotent L-crystals. Throughout this section we fix a subset J of I. Let $L = L_J$ be the Levi subgroup of G generated by T and by each $U_j, U_j^-, j \in J$.

Let $P = L \cdot U$ and $P^- = U^- \cdot L$. By definition, P and P^- are parabolic subgroups of G such that $P \supset B$, $P^- \supset B^-$, and $P \cap P^- = L$.

Let U_P and U_P^- be respectively the unipotent radicals of P and P^-. Let $U_L = L \cap U$, $U_L^- = L \cap U^-$. Then U_L and U_L^- are the opposite unipotent radicals of L.

Denote $B_L^- := B^- \cap L$, and $B_L = B \cap L$. The open inclusion $B_L^- \hookrightarrow L/U_L$ induces a rational action of U_L on B_L^- which we denote by $\alpha_L : U_L \times B_L^- \to B_L^-$. Let $\mathbf{p}^- = \mathbf{p}_L^- : B^- \to B_L^-$ be the canonical projection. By definition, $\mathbf{p}_L^-$ commutes with the rational action of U_L. In particular, $(\mathbf{B}^-, \mathbf{p}^-) \in \mathrm{Ob}(U_L\text{-Cryst})$.

For any U-variety $\mathbf{X}$ we denote by $\mathbf{X}|_L$ the U_L-variety obtained by the restriction of the U-action $\alpha : U \times X \to X$ to U_L.

LEMMA 3.7. The mapping $(\mathbf{X}, \mathbf{f}_X) \mapsto (\mathbf{X}, \mathbf{f}_X)|_L := (\mathbf{X}|_L, \mathbf{p}_L^- \circ \mathbf{f}_X)$ defines a functor of monoidal categories $|_L : U - \mathcal{C}ryst \longrightarrow U_L - \mathcal{C}ryst$.

We call the unipotent L-crystal $(\mathbf{X}, \mathbf{f}_X)|_L$ restriction of $(\mathbf{X}, \mathbf{f}_X)$ to L.

3.3   From unipotent G-crystals to geometric L-crystals.

For each unipotent G-crystal $(\mathbf{X}, \mathbf{f})$ define a morphism $\gamma = \gamma_X : X \to T$ by $\gamma = \mathrm{pr}_T \circ \mathbf{f}$. For each $i \in I$ define the function $\varphi_i = \varphi_i^X$ by

$$\varphi_i := \overline{\chi}_i^- \circ \mathbf{f}_X . \tag{3.6}$$

Let $\mathrm{supp}(\mathbf{X}, \mathbf{f})$ be the set of all those $i \in I$ for which $\varphi_i^X \not\equiv 0$. We call this set the *support* of the unipotent crystal $(\mathbf{X}, \mathbf{f}_X)$. For $i \in \mathrm{supp}(\mathbf{X}, \mathbf{f})$ define the morphism $e_i : \mathbb{G}_m \times X \to X$ by

$$e_i^c(x) = x_i \left(\frac{c-1}{\varphi_i(x)} \right)(x) . \tag{3.7}$$

It is easy to see that each e_i is a rational action of $\mathbb{G}_m$ on X.

Theorem 3.8.   For any $(\mathbf{X}, \mathbf{f}) \in \mathrm{Ob}(U\text{-Cryst})$ the actions $e_i : \mathbb{G}_m \times X \to X$, $i \in \mathrm{supp}(\mathbf{X}, \mathbf{f})$, define a geometric L_J-crystal on (X, γ_X), where $J = \mathrm{supp}(\mathbf{X}, \mathbf{f})$.

We denote this geometric L_J-crystal by $\mathcal{X}_{\mathrm{ind}}$ and call it *geometric crystal induced by* $(\mathbf{X}, \mathbf{f})$.

REMARK.   For the unipotent crystal $(G/B, [\mathbf{j}_0]^{-1})$ (see section 2.3) Theorem 3.8 specializes to Theorem 2.5.

EXAMPLES.   1. For the unipotent crystal $(\mathbf{B}^-, \mathrm{id}_{B^-})$ the actions $e_i : \mathbb{G}_m \times B^- \to B^-$, $i \in I$, are given by

$$e_i^c(b) = x_i \left(\frac{c-1}{\varphi_i(b)} \right) \cdot b \cdot x_i \left(\frac{c^{-1}-1}{\varphi_i(b)\alpha_i(\gamma(b))} \right) . \tag{3.8}$$

2. For the unipotent crystal $(\mathbf{G}, \pi^-)$ the actions $e_i : \mathbb{G}_m \times G \to G$, $i \in I$, are given by

$$e_i^c(b) = x_i \left(\frac{c-1}{\overline{\chi}_i(g)} \right) \cdot g . \tag{3.9}$$

LEMMA 3.9. *For* $(\mathbf{X}, \mathbf{f}_X), (\mathbf{Y}, \mathbf{f}_Y) \in \mathrm{Ob}(U\text{-}Cryst)$, *put* $(\mathbf{Z}, \mathbf{f}_Z) := (\mathbf{X}, \mathbf{f}_X) \times (\mathbf{Y}, \mathbf{f}_Y)$, *where* $Z = X \times Y$, *and let* $\mathcal{Z}_{ind}$ *be the geometric crystal on* (Z, γ_Z) *induced* $(\mathbf{Z}, \mathbf{f}_Z)$. *We have*

(a) $\gamma_Z = \mathbf{m} \circ (\gamma_X \circ \gamma_Y)$ *where* $\mathbf{m} : T \times T \to T$ *is the multiplication morphism.*

(b) *For each* $i \in I$, $(x, y) \in Z$,

$$\varphi_i^Z(x, y) = \varphi_i^X(x) + \frac{\varphi_i^Y(y)}{\alpha_i(\gamma_X(x))}, \qquad (3.10)$$

which implies that $\mathrm{supp}(\mathbf{Z}, \mathbf{f}_Z) = \mathrm{supp}(\mathbf{X}, \mathbf{f}_X) \cup \mathrm{supp}(\mathbf{Y}, \mathbf{f}_Y)$.

(c) *For any* $i \in \mathrm{supp}(\mathbf{Z}, \mathbf{f}_Z)$ *the action* $e_i : \mathbb{G}_m \times Z \to Z$ *is given by the formula,* $e_i^c(x, y) = (e_i^{c_1}(x), e_i^{c_2}(y))$, *where*

$$c_1 = \frac{c \varphi_i(x) \alpha_i(\gamma(x)) + \varphi_i(y)}{\varphi_i(x) \alpha_i(\gamma(x)) + \varphi_i(y)}, \quad c_2 = \frac{\varphi_i(x) \alpha_i(\gamma(x)) + \varphi_i(y)}{\varphi_i(x) \alpha_i(\gamma(x)) + c^{-1} \varphi_i(y)}. \qquad (3.11)$$

REMARK. The formula (3.11) for the action of e_i on $X \times Y$ is analogous to the formula in [K2, Section 1.3] for the tensor product of Kashiwara's crystals.

3.4 Positive unipotent crystals and duality.

Let $(\mathbf{X}, \mathbf{f}) \in \mathrm{Ob}(U\text{-}Cryst)$, and let T' be an algebraic torus of the same dimension as X. A birational isomorphism $\theta : T' \xrightarrow{\sim} X$ is called a *positive structure* on $(\mathbf{X}, \mathbf{f})$ if the following two conditions are satisfied:

(1) The isomorphism θ is a positive structure on the induced geometric L_J-crystal $\mathcal{X}_{\mathrm{ind}}$, where $J = \mathrm{supp}(\mathbf{X}, \mathbf{f})$.

(2) For any $i \in J$ the function $\varphi_i^X \circ \theta$ on T' is positive.

Theorem 3.10. *Let* $(\mathbf{X}, \mathbf{f}_X), (\mathbf{Y}, \mathbf{f}_Y) \in \mathrm{Ob}(U\text{-}Cryst)$ *and* $\theta_X : T' \to X$, $\theta_Y : T'' \to Y$ *be respectively the positive structures. Then the birational isomorphism* $\theta_{X \times Y} := \theta_X \times \theta_Y$ *is a positive structure on the product* $(\mathbf{X}, \mathbf{f}_X) \times (\mathbf{Y}, \mathbf{f}_Y)$.

For a geometric pre-crystal $\mathcal{X}$ on (X, γ) denote by $\gamma^* : X \to T$ the morphism $\gamma^*(x) = (\gamma(x))^{-1}$ and consider the *dual* geometric pre-crystal $\mathcal{X}^*$ on (X, γ^*) by defining $(e_i^*)^c(x) = e_i^{c^{-1}}(x)$.

Given a geometric pre-crystal $\mathcal{X}$ on (X, γ), for any morphism $\theta : T' \to X$ define a morphism $\theta^* : T' \to X$ as the composition of θ with the inverse $^{-1} : T' \to T'$.

The following fact is obvious.

LEMMA 3.11. *For any geometric crystal* $\mathcal{X}$ *the dual geometric pre-crystal* $\mathcal{X}^*$ *is a geometric crystal. For any positive structure* θ *on* $\mathcal{X}$ *the morphism* θ^* *is a positive structure on* $\mathcal{X}^*$.

REMARK.   Duality in geometric crystals is an analogue of duality in Kashiwara's crystals (see [K2] and the Appendix below). More precisely, the tropicalization of $\mathcal{X}^*$ with respect to the positive structure θ^* is the free combinatorial W-crystal dual to the tropicalization $\mathcal{X}$ with respect to θ.

Let $(\mathbf{X}, \mathbf{f})$ be a unipotent crystal. Define a morphism $\mathbf{f}^* : X \to B$ by $\mathbf{f}^*(x) = (\mathbf{f}(x))^{-1}$ and a rational morphism $\alpha^* : U \times X \to X$ by

$$\alpha^*(u, x) := \pi\big(u \cdot \mathbf{f}^*(x)\big)(x).$$

PROPOSITION 3.12.   The morphism α^* is a rational action of U on X and $\mathbf{f}^*$ is a U-morphism with respect to this action.

Denote the pair (X, α^*) by $\mathbf{X}^*$. Then the pair $(\mathbf{X}^*, \mathbf{f}^*)$ is a unipotent crystal. We call it the *dual unipotent crystal* of $(\mathbf{X}, \mathbf{f})$ and denote it by $(\mathbf{X}, \mathbf{f})^*$.

REMARK.   It follows from the definition that the mapping $(\mathbf{X}, \mathbf{f}) \mapsto (\mathbf{X}, \mathbf{f})^*$ is an involutive functor $* : U\text{-Cryst} \to U\text{-Cryst}$.

Theorem 3.13.   For any unipotent crystals $(\mathbf{X}, \mathbf{f}_X), (\mathbf{Y}, \mathbf{f}_Y)$ the permutation of the factors (12) : $X \times Y \to Y \times X$ induces the isomorphism of unipotent crystals,

$$\big((\mathbf{X}, \mathbf{f}_X) \times (\mathbf{Y}, \mathbf{f}_Y)\big)^* \overset{\sim}{\to} (\mathbf{Y}, \mathbf{f}_Y)^* \times (\mathbf{X}, \mathbf{f}_X)^*. \qquad (3.12)$$

REMARK.   This theorem implies that the functor $* : U\text{-Cryst} \to U\text{-Cryst}$ reverses the monoidal structure on U-Cryst.

Theorem 3.14.   Let $(\mathbf{X}, \mathbf{f})$ be an unipotent crystal and $\mathcal{X}$ be the induced geometric crystal. Then the geometric crystal induced by $(\mathbf{X}, \mathbf{f})^*$ is equal to $\mathcal{X}^*$.

3.5   Diagonalization of the products of unipotent crystals.   Let $\mathbf{v}$ be the regular morphism $Bw_0B \to U$ defined by $\mathbf{v}(ut\overline{w_0}u') = uu'$ for any $u, u' \in U, t \in T$.

By definition, $\mathbf{v}$ is two-sided U-equivariant.

REMARK.   For $G = SL_2$ the map $\mathbf{v} : G \to U$ is given by

$$\mathbf{v}\begin{pmatrix} a & b \\ c & d \end{pmatrix} = \begin{pmatrix} 1 & \frac{a+d}{c} \\ 0 & 1 \end{pmatrix}.$$

We call a unipotent crystal $(\mathbf{X}, \mathbf{f}_X)$ *non-degenerate* if $\mathrm{ran}(\mathbf{f}_X)$ intersects Bw_0B non-trivially.

For any non-degenerate unipotent crystal $(\mathbf{X}, \mathbf{f}_X) \in \mathrm{Ob}(U\text{-Cryst})$ and any $(\mathbf{Y}, \mathbf{f}_Y) \in \mathrm{Ob}(U\text{-Cryst})$ define the rational morphism $F = F_{X \times Y} : X \times Y \to X \times Y$ by $F(x, y) = (x, v_x(y))$, where $v_x := \mathbf{v}(\mathbf{f}_X(x))$. Clearly, F

is a birational isomorphism of $X \times Y \xrightarrow{\sim} X \times Y$, and its inverse is given by $F^{-1}(x, y) = (x, (v_x)^{-1}(y))$.

Denote by $\delta = \delta_{X \times Y} : U \times X \times Y \to X \times Y$ the diagonal action of U,

$$\left(u, (x, y)\right) \mapsto \left(u(x), u(y)\right).$$

PROPOSITION 3.15. Let $(\mathbf{X}, \mathbf{f})$ be a non-degenerate unipotent crystal, and $(\mathbf{Y}, \mathbf{f}_Y)$ be any unipotent crystal. Then

$$F \circ \alpha = \delta \circ (\mathrm{id}_U \times F). \tag{3.13}$$

In other words the action $\alpha = \alpha_{X \times Y} : U \times X \times Y \to X \times Y$ is diagonalized by F.

3.6 Unipotent action of the Weyl group. For each $J \subset I$ let W_J be the subgroup of W generated by s_j, $j \in J$. By definition, W_J is the Weyl group of the standard Levi subgroup $L = L_J$. So we sometimes denote W_J by W_L.

LEMMA 3.16. Let $(\mathbf{X}, \mathbf{f})$ be a unipotent G-crystal, $J = \mathrm{supp}(\mathbf{X}, \mathbf{f})$ and $\mathcal{X}_{\mathrm{ind}}$ be the induced geometric L_J-crystal. Then the formula (2.10) defines rational action of W_J on X.

We call this action of W_L on X the *unipotent action* of W_J on $(\mathbf{X}, \mathbf{f}_X)$ (or, simply, *unipotent action* of W_J on X).

Recall from section 3.1 that $(\mathbf{B}^-, \mathrm{id}_{B^-})$ is a unipotent crystal.

LEMMA 3.17. The unipotent action of W on B^- satisfies, for $i \in I$,

$$s_i(b) = x_i(a) \cdot b \cdot (x_i(a))^{-1}, \tag{3.14}$$

where $a = \frac{1 - \alpha_i(\gamma(b))}{\varphi_i(b)\alpha_i(\gamma(b))}$.

REMARK. This lemma implies that $\mathrm{dom}(s_i) = \{b \in B : \overline{\chi}_i^-(b)\} \neq 0$.

For each $w \in W$ let $\mathrm{supp}(w)$ be the minimal subset $J \subset I$ such that $w \in W_J$. For example, $\mathrm{supp}(s_i) = \{i\}$.

PROPOSITION 3.18. For each $w \in W$ there is a rational morphism $\mathbf{u}_w : B^- \to U$ such that

(a) For any $w' \in W$ such that $\mathrm{supp}(w') \supset \mathrm{supp}(w)$ each rational U-orbit of the form $t \cdot U\overline{w'}U \cap B^-$, $t \in T$, intersects $\mathrm{dom}(\mathbf{u}_w)$ non-trivially.
(b) The unipotent action of W on B^- is given by

$$(w, b) \mapsto w(b) = \mathbf{u}_w(b) \cdot b \cdot (\mathbf{u}_w(b))^{-1}. \tag{3.15}$$

For any unipotent crystal $(\mathbf{X}, \mathbf{f}_X)$ and any $w \in W$ with $\mathrm{supp}(w) \subset \mathrm{supp}(\mathbf{X}, \mathbf{f}_X)$ we define a rational morphism $\mathbf{u}_w^X : X \to U$ by $\mathbf{u}_w^X = \mathbf{u}_w \circ \mathbf{f}_X$.

Theorem 3.19. *Let* $(\mathbf{X}, \mathbf{f}_X) \in \mathrm{Ob}(U\text{-}Cryst)$ *with* $\mathrm{supp}(\mathbf{X}, \mathbf{f}_X) = J$. *Then the crystal action of* W_J *on* X *is given by*

$$w(x) = (\mathbf{u}_w^X(x))(x) \tag{3.16}$$

for $w \in W_J$.

PROPOSITION 3.20. *Let* $(\mathbf{X}_k, \mathbf{f}_{X_k}) \in \mathrm{Ob}(U\text{-}Cryst)$, $k = 1, \ldots, l$, *and let* $J = \cup_k \mathrm{supp}(\mathbf{X}_k, \mathbf{f}_{X_k})$. *Then the unipotent action of* W_J *on* $X = X_1 \times \cdots \times X_l$ *is given by*

$$w(x_1, \ldots, x_l) = \big(u(x_1), \ldots, u(x_l)\big)$$

for $w \in W$, *where* $u = \mathbf{u}_w^X(x_1 \ldots x_l)$ *is as in* (3.16).

4 Examples of Unipotent Crystals

4.1 Standard unipotent crystals.

For $w \in W$ let $\mathcal{O}(w) := U\overline{w}U/U$. Clearly, $\mathcal{O}(w)$ is a left U-orbit in G/U. Define $U(w) := U \cap \overline{w}U\overline{w}^{-1}$.

LEMMA 4.1. *The mapping* $U \to \mathcal{O}(w)$ *defined by* $u \mapsto u\overline{w}U$ *for* $u \in U$ *is left* U-*equivariant and surjective. It induces the isomorphism of homogeneous* U-*varieties,*

$$\bar{\eta}^w : U/U(w) \xrightarrow{\sim} \mathcal{O}(w).$$

REMARK. Under the natural map $G \to G/B$ each orbit $\mathcal{O}(w)$ is identified with the corresponding Schubert cell. In particular, $\dim \mathcal{O}(w) = \dim U/U(w) = l(w)$.

For $w \in W$ define $U^w := U \cap B^-w^{-1}B^-$ and $B_w^- := U\overline{w}U \cap B^-$.

EXAMPLE. For $G = SL_2$, $w = w_0$, the sets U^w and B_w^- consist respectively of the matrices of the form

$$\begin{pmatrix} 1 & c \\ 0 & 1 \end{pmatrix}, \quad \begin{pmatrix} c & 0 \\ 1 & c^{-1} \end{pmatrix},$$

$c \in \mathbb{G}_m$.

Denote by $\mathbf{j}^w : U^w \to U/U(w)$ the morphism induced by the natural inclusion $U^w \hookrightarrow U$, and by $\mathbf{j}_w : B_w^- \to \mathcal{O}(w)$ the morphism induced by the natural inclusion $B_w^- \hookrightarrow U\overline{w}U$.

By definition, the restrictions of π and π^- to $B^- \cdot U$ are the regular projections $B^- \cdot U \to U$ and $B^- \cdot U \to B^-$ respectively. Since $B^-\overline{w}^{-1}B^- \cdot \overline{w} \subset B^- \cdot U$ we can define a regular morphism $\eta^w : U^w \to B^-$ by

$$\eta^w(u) = \pi^-(u\overline{w})$$

for $u \in U^w$.

PROPOSITION 4.2.   (a) *The morphisms* $\mathbf{j}_w$ *and* $\mathbf{j}^w$ *are open inclusions* $B_w^- \hookrightarrow \mathcal{O}(w)$ *and* $U^w \hookrightarrow U/U(w)$ *respectively.*

(b) *The morphism* η^w *is a biregular isomorphism* $U^w \overset{\sim}{\to} B_w^-$. *The inverse isomorphism* $\eta_w = (\eta^w)^{-1} : B_w^- \overset{\sim}{\to} U^w$ *is given by*

$$\eta_w(b) = \pi(\overline{w} \cdot b^{-1})^{-1}. \tag{4.1}$$

(c) *The following diagram is commutative:*

$$
\begin{array}{ccc}
U/U(w) & \xrightarrow{\ \tilde{\eta}^w\ } & \mathcal{O}(w) \\[4pt]
\uparrow{\scriptstyle \mathbf{j}^w} & & \uparrow{\scriptstyle \mathbf{j}_w} \\[4pt]
U^w & \xrightarrow{\ \eta^w\ } & B_w^-
\end{array} \tag{4.2}
$$

The birational isomorphisms $[\mathbf{j}^w]$ and $[\mathbf{j}_w]$ define for any $w \in W$ rational unital U-actions $\alpha^w : U \times U^w \to U^w$ and $\alpha_w : U \times B_w^- \to B^-$ respectively. We denote by $\mathbf{U}^w := (U^w, \alpha^w)$ and $\mathbf{B}^-w := (B_w, \alpha_w)$ the corresponding U-varieties.

Note that $(\mathbf{U}^w, \eta^w)$ and $(\mathbf{B}_w^-, \mathrm{id}_w)$ are unipotent G-crystals. We call these unipotent crystals *standard*.

LEMMA 4.3.   *For each* $w \in W$ *the isomorphism* η^w *induces the isomorphism of unipotent crystals* $(\mathbf{U}^w, \eta^w) \overset{\sim}{\to} (\mathbf{B}_w^-, \mathrm{id}_w)$.

Note that the U-action on $\mathbf{B}_w^-$ is given by (3.2). In order to describe the U-action on $\mathbf{U}^w$ we need to introduce more notation.

By Proposition 4.2, the multiplication morphism $U \times U \to U$ induces the birational isomorphism $U^w \times U(w) \overset{\sim}{\to} U$. Denote by $\pi^w : U \to U(w)$ the composition of the inverse of this isomorphism with pr_2.

By definition, the action $U \times U^w \to U^w \colon (u, x) \mapsto u(x)$ is given by

$$u(x) = u \cdot x \cdot \left(\pi^w(u \cdot x) \right)^{-1}. \tag{4.3}$$

LEMMA 4.4.   *For any* $w, w' \in W$ *such that* $l(ww') = l(w) + l(w')$ *and* $x \in U^{w'}$, $y \in U^w$, $u \in U$ *we have*

$$\pi^{ww'}(u \cdot x \cdot y) = \pi^w \left(\pi^{w'}(u \cdot x) \cdot y \right)$$

or, equivalently, $u(x \cdot y) = u(x)(\pi^{w'}(u \cdot x)(y))$.

Next, we compute the action (4.3) explicitly in terms of the isomorphism η^w.

PROPOSITION 4.5.   *For each* $w \in W$, $u \in U$, *and* $\in U^w$ *we have*

$$\pi^w(u \cdot x) = \pi\big(\mathrm{Ad}\,\overline{w}(\pi(u \cdot b)) \cdot \pi^-(\overline{w} \cdot b^{-1}) \big) \tag{4.4}$$

where $b = \eta^w(x)$.

4.2 Multiplication of standard unipotent crystals. Define a new associative multiplication $\star$ on the set W by

$$w\star w' = ww'$$

if $l(ww') = l(w) + l(w)$ for $w, w' \in W$ and $s_i \star s_i = s_i$ for all i.

Under this new operation W is identified with the standard multiplicative monoid of the degenerate Hecke algebra of W. We denote this monoid by $(W, \star)$.

LEMMA 4.6. *For any $w, w' \in W$ the multiplication morphism $G \times G \to G$ induces a dominant rational morphism*

$$\pi^{w', w} : U^{w'} \times U^w \to U^{w\star w'}. \tag{4.5}$$

This is a birational isomorphism if and only if $w\star w' = ww'$.

It is well known (see e.g., [FZ, Theorems 1.2, 1.3]) that for any $w, w' \in W$ such that $w\star w' = ww'$ the multiplication morphism $G \times G \to G$ induces the open inclusion

$$B_w^- \times B_{w'}^- \to B_{ww'}^-. \tag{4.6}$$

The following result deals with a generalization of (4.6) to any pair w, w'.

PROPOSITION 4.7. *For any $w, w' \in W$,*

(a) *The intersection $B_w^- \cdot B_{w'}^-$ with $B_{w\star w'}^-$ is a dense subset of $B_{w\star w'}^-$.*

(b) *There exists an algebraic sub-torus $\widetilde{T}_{w,w'} \subset T$ such that the restriction of the multiplication morphism $G \times G \to G$ to $B_w^- \times B_{w'}^-$ is a dominant rational morphism*

$$B_w^- \times B_{w'}^- \to B_{w\star w'}^- \cdot \widetilde{T}_{w,w'}. \tag{4.7}$$

This is a birational isomorphism if and only if $l(w\star w') = l(w) + l(w') - \dim \widetilde{T}_{w,w'}$.

(c) *$\widetilde{T}_{w,w'} = \{e\}$ if and only if $w \star w = ww'$.*

(d) *$\widetilde{T}_{s_i,s_i} = \alpha_i^\vee(\mathbb{G}_m)$ for each $i \in I$.*

(e) *For any $w, w', w'' \in W$ one has $\widetilde{T}_{w,w'w''} \cdot \widetilde{T}_{w',w''} = \widetilde{T}_{w\star w',w''} \cdot (w'')^{-1}(\widetilde{T}_{w,w'})$.*

COROLLARY 4.8. *For any $w, w' \in W$ the morphism (4.7) induces the morphism in U-Cryst : $(\mathbf{B}_w^-, \mathrm{id}_w) \times (\mathbf{B}_{w'}^-, \mathrm{id}_{w'}) \to (\mathbf{B}_{w\star w'}^- \cdot \widetilde{T}_{w,w'}, \mathrm{id}_{B_{w\star w'}^- \cdot \widetilde{T}_{w,w'}})$.*

Next, we compute a lower bound for each $\widetilde{T}_{w,w'}$.

For each $w \in W$ let us consider a homomorphism of tori $T \to T$ defined by $t \mapsto w(t) \cdot t^{-1}$. Denote by T_w the image of this homomorphism. Clearly, T_w is a sub-torus of T such that

$$X_*(T_w) = (\Lambda^w)^\perp \bigcap \oplus_i \mathbb{Z}\alpha_i^\vee,$$

where $\Lambda^w = \{\lambda \in \Lambda : w(\lambda) = \lambda\}$. This implies that $X_*(T_w)$ has a $\mathbb{Z}$-basis of certain (not necessarily simple) coroots.

PROPOSITION 4.9.   For any $w, w' \in W$ we have $\widetilde{T}_{w,w'} \supset T_{(w\star w')^{-1} \cdot w \cdot w'}$.

From now on we will freely use the notation of section 3.2. For any standard Levi subgroups $L \subset L' \subset G$ define the elements $w_{L',L}, w_{L,L'} \in W$ by

$$w_{L',L} := w_0^{L'} w_0^L, \qquad w_{L,L'} := w_0^L w_0^{L'}. \tag{4.8}$$

Clearly, $w_{L',L} \cdot w_{L',L} = e$ and $l(w_{L',L}) = l(w_{L,L'}) = l(w_0^{L'}) - l(w_0^L)$.

Note that $U_{L'}(w_{L,L'}) = U_L$ and $U_{L'}(w_{L',L}) = \mathrm{Ad}\,\overline{w_{L',L}}(U_L)$.

It turns out that for $w = w_{L',L}$ the formula (4.4) simplifies when $u \in U_L$.

PROPOSITION 4.10.   Let $L \subset L'$ be standard Levi subgroups of G. Then, for any $u \in U_L$, $x \in U^{w_{L',L}}$, we have

$$\pi^{w_{L',L}}(u \cdot x) = \mathrm{Ad}\,\overline{w_{L',L}}\big(\pi(u \cdot \eta^{w_{L',L}}(x))\big).$$

For each $w \in W$, let $I(w)$ be the set of all $i \in I$ such that $w(\alpha_i) = \alpha_{i'}$ for some $i' \in I$. Note that an element $w' \in W$ belongs to $W_{I(w)}$ if and only if $\mathrm{Ad}\,\overline{w}(U^{w'}) = U^{ww'w^{-1}}$.

LEMMA 4.11.   Let $w \in W$, $w', w'' \in W_{I(w)}$. Then

$$\mathrm{Ad}\,\overline{w} \circ \pi^{w'}\big|_{U^{w''}} = \pi^{ww'w^{-1}}\big|_{U^{ww''w^{-1}}} \circ \mathrm{Ad}\,\overline{w}.$$

Theorem 4.12.   Let $L \subset L'$ be standard Levi subgroups of G, and let $w \in W$ be any element such that $\overline{w_{L',L}}^{-1} \cdot \overline{w}$ centralizes U_L. Then for any $w' \in W_{I(w)}$ we have

(a)   The morphism $f_{w,w'} : U^{w_{L',L}} \times U^{w'} \to U$ given by $(x,y) \mapsto x \cdot \mathrm{Ad}\,\overline{w}(y)$ is a dominant rational morphism $U^{w_{L',L}} \times U^{w'} \to U^{(ww'w^{-1})\star w_{L',L}}$. This is a birational isomorphism if and only if $l(ww'w^{-1}w_{L',L}) = l(w') + l(w_{L',L})$.

(b)   If $f_{w,w'}$ is a birational isomorphism then it induces an isomorphism in the category $U_L - \mathcal{C}ryst$:

$$(\mathbf{U}^{w_{L',L}}, \eta^{w_{L',L}})|_L \times (\mathbf{U}^{w'}, \eta^{w'})|_L \to (\mathbf{U}^{(ww'w^{-1})\star w_{L',L}}, \eta^{(ww'w^{-1})\star w_{L',L}})|_L.$$

Let $G = GL_{m+n}$, $m > 0$, $n > 0$. We use the standard labeling of the Dynkin diagram of type $A_{m+n-1} : I = \{1, 2, \ldots, m+n-1\}$. Let $J = I \backslash \{m\}$. Clearly, $L_J = L_{m,n} = GL_m \times GL_n \subset GL_{m+n}$. Denote $L_m := GL_m \times \{e\}$, $L'_n := \{e\} \times GL_n$, so that $L_{m,n} = L_m \cdot L'_n \cong L_m \times L'_n$. Let $w_{m,n} := w_{L_{m,n},G}$, and $w_{n,m} := w_{G,L_{m,n}} = w_{m,n}^{-1}$.

It is easy to see that

$$w_{m,n} = (s_m s_{m-1} \cdots s_1)(s_{m+1} s_m \cdots s_2) \cdots (s_{m+n-1} s_{m+n-2} \cdots s_n). \tag{4.9}$$

In particular, $\dim U^{w_{n,m}} = mn$. Denote $C_m := s_1 \cdots s_m$ and $C'_n = s_{m+n-1} \cdots s_m$. Note that $(\mathbf{U}^{w_{n,m}}, \eta^{w_{n,m}})$, $(\mathbf{U}^{C_m}, \eta^{C_m})$ and $(\mathbf{U}^{C'_n}, \eta^{C'_n})$ are unipotent GL_{m+n}-crystals.

COROLLARY 4.13. (a) *The restriction* $(\mathbf{U}^{w_{n,m}}, \eta^{w_{n,m}})|_{L_m}$ *is isomorphic in* U_{L_m}-*Cryst to*

$$\prod_{l=1}^{n} (\mathbf{U}^{C_m}, \eta^{C_m})|_{L_m} . \tag{4.10}$$

(b) *The restriction* $(\mathbf{U}^{w_{n,m}}, \eta^{w_{n,m}})|_{L'_n}$ *is isomorphic in* $U_{L'_n}$-*Cryst to*

$$\prod_{k=1}^{m} (\mathbf{U}^{C'_n}, \eta^{C'_n})|_{L'_n} . \tag{4.11}$$

4.3 W-invariant functions on standard unipotent G-crystals. For each $w \in W$ let $\widehat{U}(w)$ be the set of all $\chi \in \widehat{U}$ such that $\chi(\overline{w}^{-1}u\overline{w}) = \chi(u)$ for all $u \in U(w)$.

Clearly, $\widehat{U}(e) = \widehat{U}(w_0) = \widehat{U}$.

For any $\chi \in \widehat{U}$ define a regular function $\chi^w : BwB \to \mathbb{G}_a$ by

$$\chi^w(g) = \overline{\chi}(\overline{w}^{-1}g) . \tag{4.12}$$

Let $\rho^\vee \in X_*(T^{ad})$ be the co-character of $T^{ad} = T/Z(G)$ such that $\langle \rho^\vee, \alpha_i \rangle = 1$ for $i \in I$. Define the anti-automorphism ι of G by $\iota(g) = \text{Ad} \, \rho^\vee(-1)(g^{-1})$.

By definition, $\iota(t) = t^{-1}$ for $t \in T$ and $\iota \circ x_i = x_i$, $\iota \circ y_i = y_i$ for all $i \in I$.

It is easy to see that $\iota(\overline{w}) = \overline{w^{-1}}$ for any $w \in W$ and $\chi \circ \iota = \chi$ for any $\chi \in \widehat{U}$.

For any $\chi, \chi' \in \widehat{U}$ and any $w \in W$ define a regular function $f^w_{\chi,\chi'}$ on BwB by

$$f^w_{\chi,\chi'} = \chi^w + \chi'^{w^{-1}} \circ \iota . \tag{4.13}$$

Let $L = L_J$ be any standard Levi subgroup of G. For any $\chi = \sum_{i \in I} a_i \chi_i \in \widehat{U}$ let $\chi^L \in \widehat{U}$ be defined by

$$\chi^L := \sum_{i \in I \setminus J} a_i \chi_i .$$

By definition, $\chi^L|_{U_L} \equiv 0$.

Theorem 4.14. *For any* $\chi \in \widehat{U}(w_{L,G})$ *and any* $t \in Z(L)$ *the restriction of the function* $f^{w_{G,L}}_{\chi,\chi^L}$ *to* $t \cdot B^-_{w_{L,G}}$ *is invariant under the unipotent action of* W *on* $t \cdot B^-_{w_{L,G}}$.

Next, we describe for each $w \in W$ a basis of the subspace $\widehat{U}(w) \subset \widehat{U}$. For each $w \in W$ define a bijective map $\zeta_w : I(w) \to I(w^{-1})$ by $w(\alpha_i) = \alpha_{\zeta_w(i)}$. This ζ_w is a partial bijection $I \to I$ (see the Appendix) with $\mathrm{dom}(\zeta) = I(w)$ and $\mathrm{ran}(\zeta) = I(w^{-1})$.

For each $i \in I$ we define the set $\mathbf{o}_w(i)$ as follows. For $i \in \mathrm{dom}(\zeta) \cup \mathrm{ran}(\zeta)$ we put

$$\mathbf{o}_w(i) := \{ \ldots, (\zeta_w)^{-2}(i), (\zeta_w)^{-1}(i), i, \zeta_w(i), (\zeta_w)^2(i), \ldots \}$$

(where for each $k \in \mathbb{Z}$ the power ζ_w^k is considered to be a partial bijection $I \to I$). For $i \notin \mathrm{dom}(\zeta) \cup \mathrm{ran}(\zeta)$ we define $\mathbf{o}_w(i) := \{i\}$.

For any $J \subset I$ let $\chi_J \in \widehat{U}$ be defined by $\chi_J = \sum_{j \in J} \chi_j$.

LEMMA 4.15. *For each $i \in I$ the function $\chi_{\mathbf{o}_w(i)}$ belongs to $\widehat{U}(w)$, and the set of all $\chi_{\mathbf{o}_w(i)}$, $i \in I$, is a basis in the vector space $\widehat{U}(w)$.*

Recall that the unipotent action of W_J was definited in section 3.6.

THEOREM 4.16. (a) *For each $\chi \in \widehat{U}(w_{G,L})$ the restriction of χ to $U^{w_{G,L}}$ is invariant under the unipotent action of W_J on the unipotent L-crystal $(\mathbf{U}^{w_{G,L}}, \eta^{w_{G,L}})|_L$.*

(b) *For each $\chi \in \widehat{U}(w_{L,G})$ the restriction of the function $\chi^{w_{L,G}}$ to $B^-_{w_{L,G}}$ is invariant under the unipotent action of W_J on the unipotent L-crystal $(\mathbf{B}^-_{w_{L,G}}, \mathrm{id}_{w_{L,G}})|_L$.*

4.4 Positive structures on standard unipotent crystals.

For any $i \in I$ define the morphism $\pi_i : \mathbb{G}_m \to B^-$ by the formula $\pi_i(c) := y_i(1/c)\alpha_i^\vee(c)$. Clearly, π_i is a biregular isomorphism $\mathbb{G}_m \widetilde{\to} B^-_{s_i}$. The isomorphism $\pi_i : \mathbb{G}_m \widetilde{\to} B^-_{w_i}$ defines a structure of U-variety on $\mathbb{G}_m$ and a unipotent crystal on $\mathbb{G}_m$ which we denote by $(\mathbb{G}_m, \pi_i)$.

For each sequence $\mathbf{i} = (i_1, \ldots, i_l)$ define the regular morphism $\pi_{\mathbf{i}} : (\mathbb{G}_m)^l \to B^-$ by the formula

$$\pi_{\mathbf{i}}(c_1, \ldots, c_l) = \pi_{i_1}(c_1) \cdots \pi_{i_l}(c_l). \qquad (4.14)$$

For any sequence $\mathbf{i} = (i_1, \ldots, i_l) \in I^l$ define $w_*(\mathbf{i}) := s_{i_1} \star \cdots \star s_{i_l}$.

PROPOSITION 4.17. *For any sequence $\mathbf{i} = (i_1, \ldots, i_l) \in I^l$ we have*

(a) *There exists a sub-torus $\widetilde{T}_{\mathbf{i}} \subset T$ such that $[\pi_{\mathbf{i}}]$ is a dominant rational morphism $(\mathbb{G}_m)^l \to B^-_{w_*(\mathbf{i})} \cdot \widetilde{T}_{\mathbf{i}}$. The morphism $[\pi_{\mathbf{i}}]$ is a birational isomorphism if and only if*

$$l = l(w_*(\mathbf{i})) + \dim \widetilde{T}_{\mathbf{i}}. \qquad (4.15)$$

(b) *The rational morphism $[\pi_{\mathbf{i}}]$ induces the morphism in U-Cryst,*

$$(\mathbb{G}_m, \pi_{i_1}) \times \cdots \times (\mathbb{G}_m, \pi_{i_l}) \to \left(\mathbf{B}^-_{w_*(\mathbf{i})} \cdot \widetilde{T}_{\mathbf{i}}, \mathrm{id}_{B^-_{w_*(\mathbf{i})} \cdot \widetilde{T}_{\mathbf{i}}} \right).$$

(c) If the equality (4.15) holds then $[\pi_{\mathbf{i}}]$ is a positive structure on the unipotent crystal $(\mathbf{B}^-_{w_\star(\mathbf{i})} \cdot \widetilde{T}_{\mathbf{i}}, \mathrm{id}_{B^-_{w_\star(\mathbf{i})} \cdot \widetilde{T}_{\mathbf{i}}})$.

(d) There is an inductive formula for the computation of $\widetilde{T}_{\mathbf{i}}$: $\widetilde{T}_i = \alpha_i^\vee(\mathbb{G}_m)$, and for any sequence $\mathbf{i} = (i_1, \dots, i_l) \in I^l$ and any $i \in I$ one has

$$\widetilde{T}_{(\mathbf{i},i)} = \begin{cases} s_i(\widetilde{T}_{\mathbf{i}}), & \text{if } l(w_\star(\mathbf{i})s_i) = l(w_\star(\mathbf{i})) + 1, \\ s_i(\widetilde{T}_{\mathbf{i}}) \cdot \widetilde{T}_i, & \text{if } l(w_\star(\mathbf{i})s_i) = l(w_\star(\mathbf{i})) - 1. \end{cases}$$

REMARKS.   1. For each $w \in W$, $\mathbf{i} \in R(w)$, the tropicalization of the geometric crystal induced by $(\mathbf{B}^-_w, \mathrm{id}_w)$ with respect to $\theta'_{\mathbf{i}}$ is equal to the corresponding free combinatorial crystal $\mathcal{B}_{\mathbf{i}}$ which was constructed in [K2] as a product of 1-dimensional crystals $\mathcal{B}_{i_1}, \dots, \mathcal{B}_{i_l}$. By the construction, $\mathcal{B}_{\mathbf{i}}$ is a free $W_{|\mathbf{i}|}$-crystal, where $|\mathbf{i}| = \{i_1, \dots, i_l\}$.

2. For any dominant $\lambda^\vee \in \Lambda^\vee$ the image of Kashiwara's embedding (7.3) can be described as follows (see e.g. [BZ2]). Let $\chi = \sum_{i \in I} \chi_i \in \widehat{U}$. For any $\mathbf{i} \in R(w_0)$ let $\theta_{\mathbf{i}} = id_T \times [\pi_{\mathbf{i}}] : T \times (\mathbb{G}_m)^{l_0} \widetilde{\to} T \cdot B^-_{w_0}$ be the positive structure on $T \cdot B^-_{w_0}$. Then the image of the embedding $B(V_{\lambda^\vee}) \hookrightarrow B_{\mathbf{i}} = \mathbb{Z}^{l_0}$ is

$$\left\{ b \in B_{\mathbf{i}} : \mathrm{Trop}_{\theta_{\mathbf{i}}}(f^{w_0}_{\chi,\chi})(\lambda^\vee, b) \geq 0 \right\}.$$

For each $\mathbf{i} = (i_1, \dots, i_l) \in I^l$ define the regular morphism $\pi^{\mathbf{i}} : (\mathbb{G}_m)^l \to U$ by

$$\pi^{\mathbf{i}}(c_1, \dots, c_l) = x_{i_1}(c_1) \cdots x_{i_l}(c_l). \tag{4.16}$$

PROPOSITION 4.18.   For any sequence $\mathbf{i} = (i_1, \dots, i_m) \in I^l$, we have

(a) The morphism $\pi^{\mathbf{i}}$ induces a dominant rational morphism $(\mathbb{G}_m)^l \to U^{(w_\star(\mathbf{i}))^{-1}}$. If $\mathbf{i} \in R(w)$ for some $w \in W$ then $\pi^{\mathbf{i}}$ is an open inclusion $(\mathbb{G}_m)^{l(w)} \hookrightarrow U^w$.

(b) If the sequence $\mathbf{i}$ belongs to $R(w)$ for some $w \in W$ then the birational isomorphism $[\pi^{\mathbf{i}}] : (\mathbb{G}_m)^{l(w)} \widetilde{\to} U^{w^{-1}}$ is a positive structure on the unipotent crystal $(\mathbf{U}^{w^{-1}}, \eta^{w^{-1}})$.

4.5   Duality and symmetries for standard unipotent crystals.

It is easy to see that the inverse $\cdot^{-1} : B^- \to B^-$ induces the isomorphism $B^-_w \widetilde{\to} B^-_{w^{-1}}$ for each $w \in W$.

LEMMA 4.19.   The inverse in B^- induces the isomorphism of unipotent crystals

$$(\mathbf{B}^-_w, \mathrm{id}_w)^* \widetilde{\to} (\mathbf{B}^-_{w^{-1}}, \mathrm{id}_{w^{-1}})$$

for each $w \in W$. In particular, the unipotent crystal $(\mathbf{B}^-_w, \mathrm{id}_w)$ is self-dual if $w^2 = e$.

5 Proofs

5.1 Proofs of results in section 2.

Proof of Theorem 2.5. Since Theorem 2.5 is a particular case of Theorem 3.8 we refer to the proof of Theorem 3.8 in section 5.2. □

Proof of Proposition 2.3. Clearly, $s_i : X \to X$ is a birational involution. Thus, it suffices to prove that for any $w_1, w_2 \in W$ satisfying $l(w_1 w_2) = l(w_1) + l(w_2)$ we have

$$w_1(w_2(x)) = (w_1 w_2)(x) .$$

As follows from the definition of $e_w : T \times X \to X$, we have

$$e_{w_1 w_2}^t = e_{w_1}^{w_2(t)} \circ e_{w_2}^t ,$$

for all $w_1, w_2 \in W$ satisfying $l(w_1 w_2) = l(w_1) + l(w_2)$. Note that for any $w \in W$ we have

$$\gamma(e_w^t(x)) = t \cdot w(t^{-1})\gamma(x) .$$

This implies that $\gamma(w(x)) = \gamma(e_w^{\gamma(x)^{-1}}(x)) = w(\gamma(x))$. Therefore, for any $w_1, w_2 \in W$ such that $l(w_1 w_2) = l(w_1) + l(w_2)$ we have

$$
\begin{aligned}
w_1(w_2(x)) &= e_{w_1}^{\gamma(w_2(x))^{-1}}\left(e_{w_2}^{\gamma(x)^{-1}}(x)\right) \\
&= e_{w_1}^{w_2(\gamma(x)^{-1})} e_{w_2}(x)^{\gamma(x)^{-1}} = e_{w_1 w_2}^{\gamma(x)^{-1}}(x) = (w_1 w_2)(x) .
\end{aligned}
$$

Proposition 2.3 is proved. □

Proof of Theorem 2.11. Let $W \times T' \to T'$ be the action of W obtained from the action 2.10 by twisting with θ, that is, $(w, t') \mapsto \theta^{-1}(w(\theta(t')))$. It is easy to see that W acts on T' by a positive birational isomorphism. Therefore, applying the functor Trop to the action $W \times T' \to T'$ we obtain an action of W on $X_*(T')$.

Theorem 2.11 is proved. □

Proof of Theorem 2.13. By definition, for $i, j \in I = \{1, 2, \ldots, r\}$ we have

$$\omega_i(e_j^c(x)) = c^{\delta_{ij}}\omega_i(c) .$$

Thus $\tau(e_1^c(x)) = \tau(x)$ and therefore $\tau(s_1(x)) = \tau(x)$.

Let us compute $\tau(e_j^c(x))$ for $j > 1$. The computation is based on the following obvious statement.

LEMMA 5.1. *For each $j = 2, \ldots, r$ we have*

$$\tau(e_j^c(x)) = (e_{(j;r)}^x \circ \tau_j)(e_j^c(x)) , \tag{5.1}$$

where

$$\tau_j(z) = \left(e_1^{\frac{\omega_{j+1}(z)}{\omega_j(z)}} \cdots e_j^{\frac{\omega_{j+1}(z)}{\omega_j(z)}}\right) \cdots \left(e_1^{\frac{\omega_3(z)}{\omega_2(z)}} e_2^{\frac{\omega_3(z)}{\omega_2(z)}}\right) \left(e_1^{\frac{\omega_2(z)}{\omega_1(z)}}\right)(z),$$

$$e_{(j;r)}^x = \left(e_1^{\frac{\omega_{r+1}(x)}{\omega_r(x)}} \cdots e_r^{\frac{\omega_{r+1}(x)}{\omega_r(x)}}\right) \cdots \left(e_1^{\frac{\omega_{j+2}(x)}{\omega_{j+1}(x)}} \cdots e_{j+1}^{\frac{\omega_{j+2}(x)}{\omega_{j+1}(x)}}\right) \tag{5.2}$$

and we use the convention $\omega_{r+1}(x) \equiv 1$ *and* $e_{(r;r)}^x \equiv 1$.

Substituting $c = \dfrac{1}{\alpha_j(\gamma(x))} = \dfrac{\omega_{j-1}(x)\omega_{j+1}(x)}{\omega_j(x)}$ into (5.1) we obtain

$$\tau(s_j(x)) = e_{(j;r)}(x) \circ \tau_j(s_j(x)).$$

This implies that all we have to do to prove Theorem 2.13 is to check that

$$\tau_j(s_j(x)) = \tau_j(x) \tag{5.3}$$

for $j = 2, \ldots, r$.

Let us first compute $\tau_j(e_j^c(x))$, $j \geq 2$. Substituting $z = e_j^c(x)$ in (5.2) and using the equalities $\omega_k(z) = c^{\delta_{jk}}\omega_j(x)$ and $\tau_{j-2} \circ e_j^c = e_j^c \circ \tau_{j-2}$ we obtain

$$\tau_j(e_j^c(x)) = \left(e_1^{\frac{\omega_{j+1}(x)}{c\omega_j(x)}} \cdots e_j^{\frac{\omega_{j+1}(x)}{c\omega_j(x)}}\right)\left(e_1^{\frac{c\omega_j(x)}{\omega_{j-1}(x)}} \cdots e_{j-1}^{\frac{c\omega_j(x)}{\omega_{j-1}(x)}}\right)\left(e_j^c \circ \tau_{j-2}(x)\right).$$

Substituting $c = \dfrac{\omega_{j-1}(x)\omega_{j+1}(x)}{\omega_j(x)}$ into (5.2), we obtain

$$\tau_j(s_j(x)) = \left(e_1^{\frac{\omega_j(x)}{\omega_{j-1}(x)}} \cdots e_j^{\frac{\omega_j(x)}{\omega_{j-1}(x)}}\right)\left(e_1^{\frac{\omega_{j+1}(x)}{\omega_j(x)}} \cdots e_{j-1}^{\frac{\omega_{j+1}(x)}{\omega_j(x)}}\right)e_j^{\frac{\omega_{j-1}(x)\omega_{j+1}(x)}{\omega_j(x)}}\left(\tau_{j-2}(x)\right).$$

In order to finish the proof of Theorem 2.13 we will use the following result.

LEMMA 5.2. *For any* $j = 1, 2, \ldots, r$ *the following relation holds:*

$$(e_1^c \cdots e_j^c)(e_1^{c'} \cdots e_{j-1}^{c'})e_j^{c'/c} = (e_1^{c'} \cdots e_j^{c'})(e_1^c \cdots e_{j-1}^c). \tag{5.4}$$

Proof. Induction in j. For $j = 1$ the identity $e_1^c e_1^{c'/c} = e_1^{c'}$ is true.

Now let $j > 1$. Using the commutation of e_j with each of $e_1, \ldots, e_{j-2}$ we rewrite the left-hand side of (5.4) as

$$(e_1^c \cdots e_{j-1}^c)(e_1^{c'} \cdots e_{j-2}^{c'})e_j^c e_{j-1}^{c'} e_j^{c'/c}.$$

Applying to the above expression the basic relation (2.7) written in the form $e_j^c e_{j-1}^{c'} e_j^{c'/c} = e_{j-1}^{c'/c} e_j^{c'} e_{j-1}^c$, we see that the left-hand side of (5.4) equals

$$(e_1^{c'} \cdots e_j^{c'})(e_1^c \cdots e_{j-1}^c)e_{j-1}^{c'/c} e_j^{c'} e_{j-1}^c.$$

Finally, applying the inductive hypothesis (5.4) with $j - 1$, we obtain

$$(e_1^{c'} \cdots e_{j-1}^{c'})(e_1^c \cdots e_{j-2}^c)e_j^{c'} e_{j-1}^c = (e_1^{c'} \cdots e_j^{c'})(e_1^c \cdots e_{j-1}^c).$$

The lemma is proved. $\qquad\square$

We see that (5.3) is a special case of Lemma 5.2 with $c = \frac{\omega_j(x)}{\omega_{j-1}(x)}$, $c' = \frac{\omega_{j+1}(x)}{\omega_j(x)}$.

Theorem 2.13 is proved. □

5.2   Proof of the results in section 3.

Proof of Theorem 3.3.   Prove (a). We start with the following result.

LEMMA 5.3.   *For any unipotent crystal* $(\mathbf{X}, \mathbf{f})$ *the following identity holds:*

$$\pi\big(u' \cdot \mathbf{f}(u(x))\big) \cdot \pi\big(u \cdot \mathbf{f}(x)\big) = \pi\big((u'u) \cdot \mathbf{f}(x)\big) \qquad (5.5)$$

for any $u, u' \in U$, $x \in X$.

Proof. Denote $b = \mathbf{f}(x)$. Note that $\mathbf{f}(u(x)) = u(b) = u \cdot b \cdot \pi(u \cdot b)^{-1}$. Then the identity (5.5) can be rewritten as

$$\pi\big(u' \cdot u \cdot b \cdot \pi(u \cdot b)^{-1}\big) \cdot \pi(u \cdot b) = \pi(u'u \cdot b).$$

This identity is true because the morphism $\pi : G \to U$ is right U-equivariant. The lemma is proved. □

One can easily see that (5.5) implies that $(u'u)(x, y) = u'(u(x, y))$ for all $u, u' \in U$, $(x, y) \in X \times Y$.

Thus, the formula (3.5) defines an U-action.

Part (a) is proved. Prove (b). We need the following fact.

LEMMA 5.4.   *For* $u \in U$, $b, b' \in B^-$ *we have*

$$\pi(u \cdot b \cdot b') = \pi\big(\pi(u \cdot b) \cdot b'\big). \qquad (5.6)$$

Proof. It suffices to take $u = x_i(a)$. In this case Lemma 3.1 implies that for $u, u' \in U^-$, $t, t' \in T$, we have

$$\pi\big(x_i(a) \cdot u \cdot tu' \cdot t'\big) = \pi\big(x_i(a) \cdot u \cdot tu't^{-1} \cdot tt'\big)$$
$$= x_i\big((a^{-1} + \chi_i^-(utu't^{-1}))^{-1} \cdot \alpha_i(tt')^{-1}\big)$$
$$= x_i\big((a^{-1} + \chi_i^-(u) + \alpha_i(t^{-1})\chi_i(u'))^{-1} \cdot \alpha_i(t^{-1}) \cdot \alpha_i(t'^{-1})\big)$$
$$= x_i\big(a'^{-1} + \chi_i^-(u')^{-1}\alpha_i(t'^{-1})\big) = \pi\big(x_i(a') \cdot u't'\big)$$

where $a' = (a^{-1} + \chi_i^-(u))^{-1} \cdot \alpha_i(t^{-1})$. The lemma is proved. □

Part (b) is proved, and Theorem 3.3 is also proved. □

Proof of Proposition 3.4.   Let $(\mathbf{X}_1, \mathbf{f}_{X_1})$, $(\mathbf{X}_2, \mathbf{f}_{X_2})$, $(\mathbf{X}_3, \mathbf{f}_{X_3})$ be unipotent crystals and let

$$(\mathbf{X}_{12,3}, \mathbf{f}_{X_{12,3}}) := \big((\mathbf{X}_1, \mathbf{f}_{X_1}) \times (\mathbf{X}_2, \mathbf{f}_{X_2})\big) \times (\mathbf{X}_3, \mathbf{f}_{X_3}),$$
$$(\mathbf{X}_{1,23}, \mathbf{f}_{X_{1,23}}) := (\mathbf{X}_1, \mathbf{f}_{X_1}) \times \big((\mathbf{X}_2, \mathbf{f}_{X_2}) \times (\mathbf{X}_3, \mathbf{f}_{X_3})\big).$$

For any $x_k \in X_k$, $k = 1, 2, 3$ we have

$$\mathbf{f}_{X_{12,3}}\big((x_1, x_2), x_3\big) = \mathbf{f}_{X_1}(x_1) \cdot \mathbf{f}_{X_2}(x_2) \cdot \mathbf{f}_{X_3}(x_3) = \mathbf{f}_{X_{1,23}}\big(x_1, (x_2, x_3)\big),$$

that is, $\mathbf{f}_{X_{12,3}} = \mathbf{f}_{X_{1,23}}$.

It suffices to prove that $\mathbf{X}_{12,3} = \mathbf{X}_{1,23}$, that is, these two U-actions on $X_1 \times X_2 \times X_3$ are equal.

For $x_k \in X_k$, $k = 1, 2, 3$, denote $b_k := \mathbf{f}_{X_k}(x_k)$.

Let us write the U-actions respectively on $\mathbf{X}_{12,3}$, $\mathbf{X}_{1,23}$,

$$\begin{aligned}
u((x_1, x_2), x_3) &= (u(x_1, x_2), \pi(u \cdot b_1 \cdot b_2)(x_3)) \\
&= (u(x_1), \pi(u \cdot b_1)(x_2), \pi(u \cdot b_1 \cdot b_2)(x_3)) \\
u(x_1, (x_2, x_3)) &= (u(x_1), \pi(u \cdot b_1)(x_2, x_3)) \\
&= (u(x_1), \pi(u \cdot b_1)(x_2), \pi(\pi(u \cdot b_1) \cdot b_2)(x_3)).
\end{aligned}$$

It follows from Lemma 5.4 that $u((x_1, x_2), x_3) = u(x_1, (x_2, x_3))$.

This proves Proposition 3.4. $\qquad\square$

Proof of Proposition 3.5. It follows from the definition (3.1) of $\mathbf{g}$ and the definition of π that for $g \in G$, $u \in U$, one has

$$\mathbf{g}(u \cdot g) = \mathbf{f}_G(u \cdot g) \cdot \pi(u \cdot g).$$

Note that $\mathbf{f}_G(u \cdot g) = u \cdot (\mathbf{f}_G(g))$ and $\pi(ug) = \pi(u(\mathbf{f}_G(g))) \cdot \pi(g)$. Thus, Proposition 3.5 follows from Theorem 3.3. $\qquad\square$

Proof of Theorem 3.8. In view of Lemma 2.1, it suffices to prove the relations (2.6)–(2.9).

First, we are going to prove the relations (2.6) and (2.7). For $\varepsilon \in \{0, -1\}$ let

$$(I \times I)_\varepsilon = \{(i, j) \in I \times I : i \neq j \text{ and } \langle \alpha_i^\vee, \alpha_j \rangle = \langle \alpha_j^\vee, \alpha_i \rangle = \varepsilon\}.$$

Note that if G is simply-laced, then $(I \times I)_0 \cup (I \times I)_{-1} \cup \Delta(I) = I \times I$.

LEMMA 5.5. (a) *For any* $(i, j) \in (I \times I)_0$, *we have*

$$x_i(a_1)x_j(a_2) = x_j(a_2)x_i(a_1). \tag{5.7}$$

(b) *For* $(i, j) \in (I \times I)_{-1}$, *we have*

$$x_i(a_1)x_j(a_2)x_i(a_3) = x_j\left(\frac{a_2 a_3}{a_1 + a_3}\right) x_i(a_1 + a_3) x_j\left(\frac{a_1 a_2}{a_1 + a_3}\right). \tag{5.8}$$

Proof. Clear. $\qquad\square$

PROPOSITION 5.6. *Let* $(\mathbf{X}, \mathbf{f})$ *be a unipotent G-crystal and let $\mathcal{X}_{\mathrm{ind}}$ be the induced geometric crystal. For any* $i, j \in (I \times I)_0 \cup (I \times I)_{-1}$ *such that* $\varphi_i \not\equiv 0$, $\varphi_j \not\equiv 0$, *one has*

(a) *If* $(i, j) \in (I \times I)_0$ *then*

$$\varphi_i(e_j^c(x)) = \varphi_i(x), \quad \varphi_i(e_j^c(x)) = \varphi_i(x); \tag{5.9}$$

(b) If $(i, j) \in (I \times I)_{-1}$ then there exist rational functions $\varphi_{ij}, \varphi_{ij}$ on X such that

$$\varphi_i(x)\varphi_j(x) = \varphi_{ij}(x) + \varphi_{ji}(x) \qquad (5.10)$$

and

$$\begin{aligned} \varphi_{ij}(e_j^c(x)) &= \varphi_{ij}(x), \quad \varphi_{ij}(e_i^c(x)) = c^{-1}\varphi_{ij}(x), \\ \varphi_{ji}(e_i^c(x)) &= \varphi_{ji}(x), \quad \varphi_{ji}(e_j^c(x)) = c^{-1}\varphi_{ji}(x). \end{aligned} \qquad (5.11)$$

(c) We have

$$\varphi_j(x)\varphi_i(e_j^c(x)) = c\varphi_{ij}(x) + \varphi_{ji}(x), \quad \varphi_i(x)\varphi_j(e_i^c(x)) = c\varphi_{ji}(x) + \varphi_{ij}(x) \qquad (5.12)$$

$$\begin{aligned} \varphi_i(e_j^{c_1} e_i^{c_2}(x)) &= \varphi_i(x)\frac{c_1 c_2^{-1}\varphi_{ij}(x) + \varphi_{ji}(x)}{c_2\varphi_{ji}(x) + \varphi_{ij}(x)}, \\ \varphi_j(e_i^{c_1} e_j^{c_2}(x)) &= \varphi_j(x)\frac{c_1 c_2^{-1}\varphi_{ji}(x) + \varphi_{ij}(x)}{c_2\varphi_{ij}(x) + \varphi_{ji}(x)}. \end{aligned} \qquad (5.13)$$

Proof. It suffices to prove the lemma in the assumption that $I = \{i, j\}$, that is, when G is semisimple of types $A_1 \times A_1$ and A_2 respectively. Part (a) follows.

Let us prove (b). It suffices to analyze the case when $G = GL_3$. Due to Lemma 3.6 it suffices to prove the statement only for $X = G$, $f_X = \pi^-$. In this case $I = \{i, j\}$, and we set $i := 1$, $j := 2$ in the standard way.

It is easy to see that $\varphi_k(g) = \overline{\chi_k}(g) = \frac{\Delta_k'(g)}{\Delta_k(g)}$ for $k = 1, 2$, where

$$\Delta_1(g) = g_{11}, \quad \Delta_1'(g) = g_{21},$$

$$\Delta_2(g) = \det\begin{pmatrix} g_{11} & g_{12} \\ g_{21} & g_{22} \end{pmatrix}, \quad \Delta_2'(g) = \det\begin{pmatrix} g_{11} & g_{12} \\ g_{31} & g_{32} \end{pmatrix}.$$

Furthermore, is easy to see that the actions $e_1, e_2 : G_m \times B^- \to B^-$ are given by the formula (3.9):

$$e_k^c(g) = x_k\left((c-1)\frac{\Delta_k(g)}{\Delta_k'(g)}\right) \cdot g$$

for $k = 1, 2$.

Define the functions $\varphi_{12}, \varphi_{21}$ on G by

$$\varphi_{12}(g) = \frac{\Delta_1''}{\Delta_1}, \quad \varphi_{21}(g) = \frac{\Delta_2''(g)}{\Delta_2(g)},$$

where

$$\Delta_1''(g) = g_{31}, \quad \Delta_2''(g) = \det\begin{pmatrix} g_{21} & g_{22} \\ g_{31} & g_{32} \end{pmatrix}.$$

It is easy to see that

$$\Delta_1'(g)\Delta_2'(g) = \Delta_1(g)\Delta_2''(g) + \Delta_1''(g)\Delta_2(g).$$

This identity implies (5.10). Furthermore, it is easy to see that

$$\Delta_k(e_l^c(g)) = c^{\delta_{kl}}\Delta_k(g), \quad \Delta_k''(e_l^c(g)) = \Delta_k''(g)$$

for $k, l \in \{1, 2\}$ and $\Delta_k'(e_k^c(g)) = \Delta_k(g)$ for $k = 1, 2$. This implies (5.11). Part (b) is proved.

Part (c) easily follows from (b). $\square$

In order to prove the relations (2.6) for any $(i, j) \in (I \times I)_0$ we compute the left-hand side and the right-hand side of (2.6). By definition

$$e_i^{c_1} e_j^{c_2}(x) = x_i(a_1)x_j(a_2)(x),$$

where $a_1 = \frac{c_1 - 1}{\varphi_i(e_j^{c_2}(x))} = \frac{c_1 - 1}{\varphi_i(x)}$, and $a_2 = \frac{c_2 - 1}{\varphi_j(x)}$. Analogously

$$e_j^{c_2} e_i^{c_1}(x) = x_j(a_2)x_i(a_1)(x).$$

Thus, using the relation (5.7), we see that the left and the right-hand sides of (2.6) are equal. This proves all the relations (2.6).

To prove the relations (2.7) for any $(i, j) \in (I \times I)_{-1}$ we compute the left-hand side and the right-hand side of (2.6). By definition

$$e_i^{c_1} e_j^{c_1 c_2} e_i^{c_2}(x) = x_i(a_1)x_j(a_2)x_i(a_3)(x),$$

where

$$a_3 = \frac{c_2 - 1}{\varphi_i(x)}, \quad a_2 = \frac{c_1 c_2 - 1}{\varphi_j(e_i^{c_2}(x))} = \frac{\varphi_i(x)(c_1 c_2 - 1)}{c_2 \varphi_{ji}(x) + \varphi_{ij}(x)},$$

and

$$a_1 = \frac{c_1 - 1}{\varphi_i(e_j^{c_1 c_2} e_i^{c_2}(x))} = \frac{(c_1 - 1)(c_2 \varphi_{ji}(x) + \varphi_{ij}(x))}{\varphi_i(x)(c_1 \varphi_{ij}(x) + \varphi_{ji}(x))}.$$

Similarly,

$$e_j^{c_2} e_i^{c_1 c_2} e_j^{c_1}(x) = x_j(a_1')x_i(a_2')x_j(a_3')(x),$$

where

$$a_3' = \frac{c_1 - 1}{\varphi_j(x)}, \quad a_2' = \frac{c_1 c_2 - 1}{\varphi_i(e_j^{c_1}(x))} = \frac{\varphi_j(x)(c_1 c_2 - 1)}{c_1 \varphi_{ij}(x) + \varphi_{ji}(x)},$$

and

$$a_1' = \frac{c_2 - 1}{\varphi_j(e_i^{c_1 c_2} e_j^{c_1}(x))} = \frac{(c_2 - 1)(c_1 \varphi_{ij}(x) + \varphi_{ji}(x))}{\varphi_j(x)(c_2 \varphi_{ji}(x) + \varphi_{ij}(x))}.$$

It is easy to see that

$$a_1' a_2' = a_2 a_3, \quad a_2' a_3' = a_1 a_2, \quad a_2' = a_1 + a_3.$$

Thus it follows from (5.8) that

$$e_i^{c_1} e_j^{c_1 c_2} e_i^{c_2}(x) = x_i(a_1)x_j(a_2)x_i(a_3)(x)$$
$$= x_j(a_1')x_i(a_2')x_j(a_3')(x) = e_i^{c_1} e_j^{c_1 c_2} e_i^{c_2}(x).$$

This proves all the relations (2.7).

Thus, we have proved Theorem 3.8 for all simply-laced reductive groups G. In particular, the relations (2.5) hold for such groups.

It remains to prove the relations (2.8) and (2.9). Instead of doing the computations directly, we will prove the relations (2.5) for any group G by deducing these relations from the relations (2.5) for a certain simply-laced group G' containing G.

Without loss of generality we may assume that G is adjoint semisimple.

It is well known that there is an adjoint semisimple simply-laced group G', an outer automorphism $\sigma : G' \to G'$ and an injective group homomorphism $f : G \hookrightarrow G'$ such that $f(G) = (G')^\sigma$. In what follows we identify G with its image $f(G)$.

Moreover, one can always choose an outer automorphism σ in such a way that σ preserves a chosen Borel subgroup $B' \subset G'$ and a maximal torus $T' \subset B'$, and satisfies $B = (B')^\sigma$, $T = (T')^\sigma$. Let $(B')^-$ be the opposite Borel subgroup containing T'. Then $(B')^-$ is also σ-invariant and $B^- = ((B')^-)^\sigma$. Also the automorphism σ induces an injection $\sigma_* : \Lambda^\vee \to (\Lambda')^\vee$.

Let I' be the Dynkin diagram of G'. It is easy to see that for any $i \in I$ there exists a subset $\tau(i) \in I'$ such that $\alpha_i^\vee = \sum_{i' \in \tau(i)} \alpha_{i'}'^\vee$ and that for any $i', j' \in \tau(i)$ the subgroups $U_{i'}, U_{j'}$ commute. This implies that the Weyl group W is the subgroup of the Weyl group W' of G' and its generators s_i, $i \in I$, are given by

$$s_i = \prod_{i' \in \tau(i)} s_{i'}' \, .$$

One can always choose the homomorphisms $x_{i'}' : \mathbb{G}_m \to U_{i'}'$, $y_{i'}' : \mathbb{G}_m \to U_{i'}'$, $i' \in I'$, in such a way that

$$x_i(a) = \prod_{i' \in \tau(i)} x_{i'}'(a) \, , \quad y_i(a) = \prod_{i' \in \tau(i)} y_{i'}'(a) \, ,$$

for any $i \in I$. This implies that $\overline{s}_i = \prod_{i' \in \tau(i)} \overline{s'}_{i'}$.

Due to Lemma 3.6 we may assume without loss of generality that $X = G$. Let $\mathcal{X}_{ind}$ be the geometric G-crystal induced by $(\mathbf{G}, \mathrm{id}_G)$ and let $\mathcal{X}_{ind}'$ be the geometric G'-crystal induced by $(\mathbf{G}', \mathrm{id}_{G'})$. We denote by $e_i : \mathbb{G}_m \times G \to G$, $i \in I$, the actions (3.9) of $\mathbb{G}_m$, and by $e_{i'}' : \mathbb{G}_m \times G' \to G'$, $i' \in I'$ the corresponding actions of $\mathbb{G}_m$. As follows from (2.6), the transformations $e_{i'}'^{\,c}$ commute with $e_{j'}'^{\,c'}$ for any $i \in I$, $i', j' \in \tau(i)$.

For each $i \in I$ let us fix a linear ordering $\overline{\tau}(i)$ of $\tau(i)$. It is easy to see that for any reduced sequence $\mathbf{i} = (i_1, \ldots, i_l) \in I^l$ the sequence $\tau(\mathbf{i}) := (\overline{\tau}(i_1); \ldots; \overline{\tau}(i_l))$ is also reduced.

Let $e_{\mathbf{i}} : T \times G \to G$ and $e_{\overline{\tau}(\mathbf{i})}' : T' \times G' \to G'$ be morphisms as in (2.4).

The following lemma is obvious.

LEMMA 5.7. *For each reduced sequence* $\mathbf{i} = (i_1, \ldots, i_l) \in I^l$ *we have*

(i) $\operatorname{dom}(e_{\mathbf{i}}) = \operatorname{dom}(e'_{\tau(\mathbf{i})}) \cap (T \times G)$.

(ii) $e'_{\overline{\tau}(\mathbf{i})}|_{T \times G} = e_{\mathbf{i}}$.

The lemma implies that for any reduced sequences $\mathbf{i}, \tilde{\mathbf{i}} \in I^l$ satisfying $w(\mathbf{i}) = w(\tilde{\mathbf{i}})$ we have $e_{\mathbf{i}} = e'_{\overline{\tau}(\mathbf{i})}|_{T \times G} = e'_{\overline{\tau}(\tilde{\mathbf{i}})}|_{T \times G} = e_{\tilde{\mathbf{i}}}$.

This proves all the relations (2.5) for geometric G-crystals.

Theorem 3.8 is proved.   □

Proof of Theorem 3.10. It follows from Lemma 3.9 that

(a) The morphism $\gamma_{X \times Y} \circ \theta_{X \times Y} : T' \times T'' \to T$ is positive.
(b) Each function $\phi_i^{X \times Y}$ is positive,
(c) Each action $e_i : \mathbb{G}_m \times X \times Y \to X \times Y$ is positive.

Theorem 3.10 is proved.   □

Proof of Proposition 3.12.   Without loss of generality we may assume that X is a subset of B^- or even $X = B^-$ so that $\mathbf{f}_X$ is the identity map $X \to B^-$. Then it is easy to see that $\alpha^*(u, b) = (u(b^{-1}))^{-1}$. This implies that α^* is an action of U. Furthermore, $\mathbf{f}^*(b) = b^{-1}$. Thus

$$\mathbf{f}^*(\alpha^*(u, b)) = u(b^{-1}) = u(\mathbf{f}^*(b)).$$

Proposition 3.12 is proved.   □

Proof of Proposition 3.15.   It is easy to see that the formula (3.13) is equivalent to the equation

$$u\mathbf{v}(\mathbf{f}(x)) = \mathbf{v}(\mathbf{f}(u(x))\pi(\mathbf{f}(x)))$$

for $x \in B^-, u \in U$. $\mathbf{v}$ is two-sided U-equivariant this identity is equivalent to

$$\mathbf{v}(u \cdot \mathbf{f}(x)) = \mathbf{v}(\mathbf{f}(u(x)) \cdot \pi(\mathbf{f}(u(x)))).$$

This identity is true since

$$\mathbf{f}(u(x))\pi(\mathbf{f}(u(x))) = u(\mathbf{f}(x))\pi(\mathbf{f}(u(x))) = u \cdot \mathbf{f}(x).$$

Proposition 3.15 is proved.   □

Proof of Theorem 3.14. We need the following obvious result.

LEMMA 5.8. *For any unipotent crystal* $(\mathbf{X}, \mathbf{f})$ *let* $\varphi_i^* = \overline{\chi}_i^-((\mathbf{f}_X(x))^{-1})$ *be the function on X defined by* (3.6) *for the dual unipotent crystal* $(\mathbf{X}, \mathbf{f})^*$, $i \in I$. *Then* $\varphi_i^*(x) = -\varphi_i(x)\alpha_i(\gamma(x))$.

Using (3.3), we obtain $\alpha^*(x_i(a), x) = x_i((a^{-1}+\varphi_i^*(x))^{-1}\alpha_i(\gamma^*(x)^{-1}))(x)$.

Substituting $a = \frac{c-1}{\varphi_i^*(x)}$ into this expression for α^*, we obtain the action of the corresponding multiplicative group of the induced dual crystal. On the other hand, simplifying the right-hand side, we obtain

$$\alpha^*\left(x_i\left(\frac{c-1}{\varphi_i^*(x)}\right), x\right) = x_i\left(\frac{c-1}{-c\varphi_i^*(x)\alpha_i(\gamma^*(x))}\right)(x) = x_i\left(\frac{c^{-1}-1}{\varphi_i(x)}\right)(x) = e_i^{c^{-1}}(x)$$

which proves Theorem 3.14.   □

Proof of Proposition 3.18. Denote by T_r the set of regular elements of T and by B_{rss}^- the pre-image pr_T^{-1} which is the set of all regular semisimple elements in B^-. This is an open dense subset. Note that B_{rss}^- intersects non-trivially each $B_w^- t$, $t \in T_r$.

We define the rational morphism $\mathbf{u}_w : B_{rss}^- \to U^w$ inductively. Define $\mathbf{u}_e := \mathrm{pr}_e$,

$$\mathbf{u}_{s_i}(b) := x_i\left(\frac{1 - \alpha_i(\gamma(b))}{\varphi_i(b)\alpha_i(\gamma(b))}\right)$$

for $i \in I$, and for any $w', w'' \in W$ such that $l(w'w'') = l(w') + l(w'')$, define

$$\mathbf{u}_{w'w''}(b) := \mathbf{u}_{w'}\left(\mathbf{u}_{w''}(b) \cdot b \cdot \mathbf{u}_{w''}(b)^{-1}\right) \cdot \mathbf{u}_{w''}(b). \tag{5.14}$$

LEMMA 5.9. *For each $w \in W$ the morphism $\mathbf{u}_w$ does not depend on the choice of the expression (5.14) and satisfies (3.15).*

Proof. It is easy to see that $\mathbf{u}_w$ satisfies (3.15). Since the centralizer in U of any element $b \in B_{rss}^-$ is trivial we see that $\mathbf{u}_w$ does not depend on a choice of the expression (5.14).

The lemma is proved.   □

Furthermore, we can repeat the same definitions for any sub-variety of the form $B_{w'}^- t \cap B_{rss}^-$. In this case for any $i \in \mathrm{supp}(w')$ the rational morphism $\mathbf{u}_{s_i} : B_{w'}^- t \cap B_{rss}^- \to U^{s_i}$ is well defined and is given by the analogous formula.

Proposition 3.18 is proved.   □

5.3   Proof of results in section 4.

Proof of Proposition 4.2. Follows from [BZ2, Theorem 4.7], with $u = e$, $v = w$.   □

Proof of Proposition 4.5. Let us express the action of U on U^w as the conjugation of the action of U on B_w^- with the isomorphism η_w,

$$u(x) = \eta_w(u(b)),$$

where $b = \eta^w(x)$. Using (3.2), (4.1) and the right U-equivariancy of π, we obtain

$$u(x) = \pi\big(\overline{w} \cdot \pi(u \cdot b) \cdot b^{-1} u^{-1}\big)^{-1} = u \cdot \big(\pi(\overline{w} \cdot \pi(u \cdot b) \cdot b^{-1})\big)^{-1}.$$

Using (4.3) in the left-hand side of the above identity, we obtain

$$\pi^w(u \cdot x) = \pi\big(\overline{w} \cdot \pi(u \cdot b) \cdot b^{-1}\big) \cdot x = \pi\big(\overline{w} \cdot \pi(u \cdot b) \cdot b^{-1} \cdot x\big).$$

Finally,

$$b^{-1}x = b^{-1}\eta_w(b) = b^{-1} \cdot \big(\pi(\overline{w} \cdot b^{-1})\big)^{-1} = \overline{w}^{-1} \cdot \pi^-(\overline{w} \cdot b^{-1}).$$

The proposition is proved.   □

Proof of Proposition 4.7. Prove (a) and (b). If $w \star w' = ww'$ then the statement follows from (4.6). The general statement reduces to the case when $w = w' = s_i$ for $i \in I$. Let us use the following identity in SL_2:

$$\begin{pmatrix} c & 0 \\ 1 & c^{-1} \end{pmatrix} \cdot \begin{pmatrix} c' & 0 \\ 1 & c'^{-1} \end{pmatrix} = \begin{pmatrix} d & 0 \\ 1 & d^{-1} \end{pmatrix} \cdot \begin{pmatrix} d' & 0 \\ 0 & d'^{-1} \end{pmatrix},$$

for any $c, c', d, d' \in \mathbb{G}_m$ satisfying $d'c = 1 + dd'$, $dd' = cc' \neq -1$. This implies that $B_{s_i}^- \cdot B_{s_i}^- \equiv B_{s_i}^- \cdot \alpha_i^\vee(\mathbb{G}_m)$, and that $B_{s_i}^- \cdot B_{s_i}^-$ contains the set $B_{s_i}^- \setminus \{\pi_i(-1)\}$, where $\pi_i : \mathbb{G}_m \xrightarrow{\sim} B_{s_i}^-$ is the biregular isomorphism defined in section 4.4.

Parts (a) and (b) are proved. Parts (c)-(e) easily follow.

Proposition 4.7 is proved.   □

Proof of Proposition 4.9. For $w \in W$ define an action of T on G by

$$(t, g) \mapsto \mathrm{Ad}_w t(g) = w(t) \cdot g \cdot t^{-1}. \qquad (5.15)$$

LEMMA 5.10. *For any $w_1, w_2 \in W$ one has $\mathrm{Ad}_{w_1} T(U\overline{w_2}U) = U\overline{w_2}U \cdot T_{w_2^{-1}w_1}$.*

Proof. It suffices to show that $\mathrm{Ad}_{w_1} T(\overline{w_2}) = \overline{w_2} \cdot T_{w_2^{-1}w_1}$. By definition, we have

$$\mathrm{Ad}_{w_1} t(\overline{w_2}) = w_1(t) \cdot \overline{w_2} \cdot t^{-1} = \overline{w_2} \cdot (w_2^{-1}w_1)(t) \cdot t^{-1}$$

for $t \in T$. The lemma is proved.   □

Furthermore, we have $B_w^- \cdot B_{w'}^- \subset U\overline{w}U \cdot U\overline{w'}U$. Thus $\mathrm{Ad}_{w \cdot w'} T(B_w^- \cdot B_{w'}^-) = B_w^- \cdot B_{w'}^-$. On the other hand, $\mathrm{Ad}_{w \cdot w'} T(B_{w \star w'}^-) = B_{w \star w'}^- \cdot T_{(w \star w')^{-1} \cdot w \cdot w'}$. Therefore, by Proposition 4.7(a), $B_{w \star w'}^- \cdot T_{(w \star w')^{-1} \cdot w \cdot w'} \subset B_{w \star w'}^- \cdot \widetilde{T}_{w, w'}$.

Proposition 4.9 is proved.   □

Proof of Proposition 4.10. Denote temporarily $w := w_{L',L}$, $b := \eta^w(x) = \pi^-(x\overline{w})$ and $u' = \mathrm{Ad}\,\overline{w_{L',L}}(\pi(u \cdot b))$.

Furthermore, we have, by (4.4),

$$\pi^w(u \cdot x) = \pi\big(u' \cdot \pi^-(\overline{w} \cdot b^{-1})\big).$$

It is easy to see that $u' \in U$ because $\pi(u \cdot b) \in U_L$. Note that $\overline{w} \cdot b^{-1} \in \overline{w} U \overline{w}^{-1} U$. Thus $\pi^-(\overline{w} \cdot b^{-1}) \in \overline{w} U_{L'} \overline{w}^{-1} \cap U^- = \overline{w}(U_P \cap L') \overline{w}^{-1} \subset \overline{w} U_P \overline{w}^{-1}$. The latter set is the unipotent radical of wPw^{-1}. This implies that $\mathrm{Ad}\, u'(\pi^-(\overline{w} \cdot b^{-1})) \in U^-$ and

$$\pi^w(u \cdot x) = \pi\big(\mathrm{Ad}\, u'(\pi^-(\overline{w} \cdot b^{-1})) \cdot u'\big) = u'.$$

Proposition 4.10 is proved. □

Proof of Theorem 4.12. Part (a) follows from Lemma 4.6. Prove (b). It suffices to show that $f_{w,w'}$ commutes with the action of U_L. By definition, the action of U_L on the product $(\mathbf{U}^{w_{L'},L}, \eta^{w_{L'},L}) \times (\mathbf{U}^w, \eta^w)$ is given by (see 3.5),

$$u(x,y) = \big(u(x), \pi(u \cdot \eta^{w_{L'},L}(x))(y)\big).$$

For $x \in U^{w_{L'},L}$, $y \in U^{w'}$ and $u \in U_L$ we have

$$u\big(f_{w,w'}(x,y)\big) = u(x \cdot y') = u(x) \cdot u'(y'),$$

where $y' = \mathrm{Ad}\,\overline{w}(y)$ and $u' = \pi^{w_{L'},L}(u \cdot x)$. By Proposition 4.10, $u' = \mathrm{Ad}\,\overline{w_{L',L}}(u'')$, where $u'' = \pi(u \cdot \eta^w(x))$.

It is easy to see that

$$u' \cdot y' = \overline{w_{L',L}} \cdot u'' \cdot \overline{w_{L',L}}^{-1} \cdot \overline{w} \cdot y \cdot \overline{w}^{-1} = \mathrm{Ad}\,\overline{w}(u'' \cdot y)$$

because $\overline{w_{L',L}}^{-1} \cdot \overline{w}$ commutes with $u'' \in U_L$.

Furthermore,

$$u'(y') = u' \cdot y' \cdot \pi^{ww'w^{-1}}(u'y')^{-1} = \mathrm{Ad}\,\overline{w}(u''(y))$$

because $\pi^{ww'w^{-1}}(u'y') = \pi^{ww'w^{-1}}(\mathrm{Ad}\,\overline{w}(u''y)) = \mathrm{Ad}\,\overline{w}(\pi^{w'}(u''y))$ by Lemma 4.11 applied with $w'' = w_0^L \star w'$.

Finally,

$$u\big(f_{w,w'}(x,y)\big) = u(x) \cdot \mathrm{Ad}\,\overline{w}(u''(y)) = f_{w,w'}\big(u(x), u''(x)\big)(y) = f_{w,w'}(u(x,y))$$

Theorem 4.12 is proved. □

Proof of Theorem 4.14. For each $w \in W$ denote $V(w) := U \cap \overline{w} U^- \overline{w}^{-1}$. It is easy to see that $U(w) \cdot V(w) = U$, $U(w) \cap V(w) = \{e\}$, and

$$U \overline{w} U = V(w) \overline{w} U \subset \overline{w} U^- \cdot U.$$

LEMMA 5.11. *For any $w \in W$, $\chi \in \widehat{U}$ we have (see (4.12))*

$$\chi^w(u'gu) = \overline{\chi}(\overline{w}^{-1} t^{-1} u' t \overline{w}) + \chi(u) + \chi^w(g) \tag{5.16}$$

for all $g \in t \cdot U \overline{w} U$, $t \in T$, $u' \in \mathrm{Norm}_U(V(w))$ and $u \in U$.

Proof. Note that $\chi^w(u'gu) = \chi^w(u'g) + \chi(u)$ for all u. So we prove the lemma under the assumption that $u = 1$.

Let $g \in U\overline{w}U$. Express g as $g = u_1 t \overline{w} u_2$, where $u_1 \in V(w)$, $u_2 \in U$. Then $\chi^w(g) = \overline{\chi}(\overline{w}^{-1} u_1 t \overline{w} u_2) = \chi(u_2)$ because $\overline{w}^{-1}V(w)\overline{w} \subset U^-$. Furthermore, for each $u' \in \mathrm{Norm}_U(V(w))$ we have $u'g = u'u_1 t \overline{w} u_2 = u'_1 u' t \overline{w} u_2$, where $u'_1 = \mathrm{Ad}\, u'(u_1) \in V(w)$. Finally,

$$\chi^w(u'g) = \overline{\chi}(\overline{w}^{-1}u'g) = \overline{\chi}(\overline{w}^{-1}u'_1 u' t \overline{w} u_2) = \overline{\chi}(\overline{w}^{-1}u't\overline{w}) + \chi(u_2)\,.$$

The lemma is proved.   □

PROPOSITION 5.12.   For any $\chi \in \widehat{U}(w_{L,G})$ and any $t \in Z(L)$ the function $f^{w_{L,G}}_{\chi,\chi^L}$ satisfies

$$f^{w_{L,G}}_{\chi,\chi^L}(u'gu) = \chi(u) + \chi(u') + f^{w_{L,G}}_{\chi,\chi^L}(g)$$

for $u, u' \in U$, $g \in t \cdot U\overline{w_{G,L}}U$. In particular, f_{χ,χ^L} is invariant under the adjoint action of U on $t \cdot U\overline{w_{G,L}}U$.

Proof. Throughout the proof we denote for shortness $w := w_{L,G}$. In this case we have $U(w) = U_L$, $V(w) = U_P$ (where $P = U \cdot B$) which implies that $\mathrm{Norm}_U(V(w_{L,G})) = U$.

The following result is obvious.

LEMMA 5.13.   Let $\chi \in \widehat{U}(w)$. Then for any $u \in U$ and $t \in Z(L)$, we have

$$\overline{\chi^L}\big(\overline{w^{-1}}^{-1} \cdot t^{-1}u \cdot \overline{w^{-1}}\big) = 0\,, \quad \overline{\chi}(\overline{w}^{-1}uw) = \chi(u) - \chi^L(u)\,.$$

Furthermore, let $t \in Z(L)$, $g \in t \cdot U\overline{w}U$, $u', u \in U$. Then

$$f^{w_{L,G}}_{\chi,\chi^L}(u'gu) = \chi^w(u'gu) + (\chi^L)^{w^{-1}}\big(\iota(u)\iota(g)\iota(u')\big)\,.$$

Applying Lemma 5.11 twice, we obtain

$$f^{w_{L,G}}_{\chi,\chi^L}(u'gu) = \overline{\chi}(\overline{w}^{-1}t^{-1}u't\overline{w}) + \chi(u) + \chi^w(g)$$
$$+ \overline{\chi^L}\big(\overline{w^{-1}}^{-1} \cdot t\iota(u)t^{-1} \cdot \overline{w^{-1}}\big) + \chi^L(\iota(u')) + (\chi^L)^{w^{-1}}(\iota(g))\,.$$

By Lemma 5.13, we have $\overline{\chi^L}\big(\overline{w^{-1}}^{-1}\iota(u)\overline{w^{-1}}\big) = 0$ and $\overline{\chi}(\overline{w}^{-1}t^{-1}u't\overline{w}) = \chi(u') - \chi^L(u')$. Taking into account that $\chi^L(\iota(u')) = \chi^L(u')$, we obtain

$$f^{w_{L,G}}_{\chi,\chi^L}(u'gu) = \chi(u') + \chi^w(g) + \chi(u) + \chi^{w^{-1}}(\iota(g))\,.$$

Proposition 5.12 is proved.   □

Theorem 4.14 is proved.   □

Proof of Theorem 4.16. Recall that $W_L = W_J$ denotes the Weyl group of $L = L_J$.

PROPOSITION 5.14.   For any $\chi \in \widehat{U}(w_{L,G})$, $t \in T$, we have

$$\chi^{w_{L,G}}(u'gu) = \chi(t^{-1}u't) + \chi(u) + \chi^{w_{L,G}}(g)$$

for $u' \in U_L$, $u \in U$, $g \in t \cdot U\overline{w_{L,G}}U$. In particular, $\chi^{w_{L,G}}$ is invariant under the adjoint action of U_L on $t \cdot U\overline{w_{L,G}}U$ for $t \in Z(L)$.

Proof. Taking into accout that $\mathrm{Norm}_u(V(w_{L,G})) = U$, we rewrite 5.16:

$$\chi^{w_{L,G}}(u'gu) = \overline{\chi}(\overline{w_{L,G}}^{-1}ut\overline{w_{L,G}}) + \chi(u) + \chi^{w_{L,G}}(g)$$
$$= \chi(t^{-1}u't) + \chi(u) + \chi^{w_{L,G}}(g).$$

Proposition 5.14 is proved.   □

Proof of Theorem 4.16(b). We have $\mathrm{supp}(w_{L,G}) = I$. By (3.14), for each $j \in J$ and generic $b_- \in B^-_{w_{L,G}}$ we have $s_j(b_-) = \mathrm{Ad}\, u(b_-)$ for some $u \in U_i$. This implies that for all $j \in J$ the restriction $\chi^{w_{L,G}}|_{B^-_{w_{L,G}}}$ is s_j-invariant. Theorem 4.16(b) is proved.   □

Recall from section 4.1 that the U-variety $\mathbf{U}^{w_{G,L}}$ equipped with the isomorphism $\eta^{w_{G,L}} : U^{w_{G,L}} \xrightarrow{\sim} B^-_{w_{G,L}}$ is a unipotent G-crystal.

Proof of Theorem 4.16(a). The proof is based on the following property of the biregular isomorphism $\eta_w : B^-_w \xrightarrow{\sim} U^w$ which is defined in section 4.1.

LEMMA 5.15. *For any $\chi \in \widehat{U}$ and any $w \in W$ we have*

$$\chi(\eta_w(b)) = \chi^{w^{-1}}(\iota(b)) \tag{5.17}$$

for $b \in B^-_w$.

Proof. It is easy to see that for any $\chi \in \widehat{U}$ and $u \in U$ we have $\chi(u^{-1}) = -\chi(u)$. It is also clear that for any $\chi \in \widehat{U}$, $g \in U^- \cdot T \cdot U$ we have

$$\overline{\chi}(t_0 g t_0^{-1}) = -\overline{\chi}(g),$$

where $t_0 = \rho^\vee(-1) \in T^{ad}$ as in section 4.3.

Furthermore, by definition of $\overline{w}$, one has

$$t_0 \overline{w} t_0^{-1} = \overline{w^{-1}}^{-1}.$$

Composing $\eta_{w^{-1}}$ with $\overline{\chi}$, we obtain for $b \in B^-_w$, $\chi \in \widehat{U}$,

$$\chi(\eta_w(b)) = \chi(\pi(\overline{w}b^{-1})^{-1}) = -\chi(\pi(\overline{w}b^{-1})) = -\overline{\chi}(\overline{w}b^{-1})$$
$$= \overline{\chi}(t_0 \overline{w} b^{-1} t_0^{-1}) = \chi(\overline{w^{-1}}^{-1} t_0 b^{-1} t_0^{-1}) = \chi^{w^{-1}}(\iota(b)).$$

The lemma is proved.   □

Now we apply Lemma 5.15 with $w = w_{G,L} = w^{-1}_{L,G}$, $\chi \in \widehat{U}(w_{G,L})$. Using Proposition 5.14 and the property of the mapping $b \mapsto \iota(b)$,

$$\iota(U\overline{w_{G,L}}U) = U\overline{w_{L,G}}U, \iota(\mathrm{Ad}\,U_L(g)) = \mathrm{Ad}\,U_L(\iota(g)),$$

we see that the right-hand side of (5.17) is an $\mathrm{Ad}\,U_L$-invariant function on $B^-_{w_{G,L}}$. Therefore, by (3.14), the right-hand side of (5.17) is invariant under the unipotent W_L-action.

Since, by its definition, $\eta_{w_{G,L}}$ is an isomorphism of unipotent G-crystals

$$(\mathbf{B}^-_{w_{G,L}}, \mathrm{id}_{w_{G,L}}) \xrightarrow{\sim} (\mathbf{U}^{w_{G,L}}, \eta_{w_{G,L}}),$$

the restriction $\chi|_{U^w{}_{G,L}}$ is also invariant under the unipotent action of W_J.

Theorem 4.16(a) is proved. □

Proof of Proposition 4.17. Part (a) follows from (4.7), part (b) follows from Corollary 4.8, part (c) follows from Theorem 3.10, and part (d) is clear.

Proposition 4.17 is proved. □

Proof of Proposition 4.18. Part (a) is immediate. So we prove (b). Due to Lemma 4.3 and Proposition 4.17(c), it suffices to prove that for any $\mathbf{i} \in R(w)$ and $\mathbf{i}' \in R(w^{-1})$ the birational isomorphism $[\pi_{\mathbf{i}'}]^{-1} \circ \eta^{w^{-1}} \circ \pi^{\mathbf{i}}$ and its inverse are positive isomorphisms $T' \overset{\sim}{\to} T'$, where $T' := (\mathbb{G}_m)^{l(w)}$. This follows from the results of [BZ2, Sections 4 and 5].

Proposition 4.18 is proved. □

6 Projections of Bruhat Cells to the Parabolic Subgroups

6.1 General facts on projections.
Recall that we have defined in section 4.2 the monoid structure $(W, \star)$ on W. Clearly, for any standard Levi subgroup $L = L_J$, the monoid $(W_L, \star)$ is a sub-monoid of W under the operation $\star$ defined in section 4.1, and this sub-monoid $(W_L, \star)$ is generated by all s_j, $j \in J$.

LEMMA 6.1. *The correspondence*

$$s_i \mapsto [s_i] = \begin{cases} s_i & \text{if } i \in J \\ e & \text{if } i \in I \setminus J \end{cases}$$

extends to the homomorphism of monoids $[\cdot] : (W, \star) \to (W_L, \star)$.

Similarly to the projection $\mathbf{p}_L^- : B^- \to B_L^-$ defined in section 3.2 let $\mathbf{p}^+ = \mathbf{p}_L^+$ be the natural projection $U \to U_L$.

There is a natural equivalence relation between constructible subsets of G. We say that two such subsets $S, S' \subset G$ are *equivalent* if their intersection $S \cap S'$ is dense in each S and S'. In this case we denote $S \equiv S'$.

LEMMA 6.2. *For each* $w \in W$ *one has* $\mathbf{p}^+(U^w) \equiv U^{[w]}$.

Proof. We proceed by the induction in $l(w)$. If $l(w) = 1$, that is, $w = s_i$ for some $i \in I$ then

$$\mathbf{p}^+(U^{s_i}) = \begin{cases} U^{s_i} & \text{if } i \in J \\ \{e\} & \text{if } i \in I \setminus J. \end{cases} \tag{6.1}$$

Let us now assume that $l(w) > 1$. Using Proposition 4.18 and the fact that $\mathbf{p}^+$ is a group homomorphism, we obtain for any $\mathbf{i} = (i_1, \ldots, i_l) \in R(w)$

$$\mathbf{p}^+(U^w) \equiv \mathbf{p}^+(U^{s_{i_l}} U^{s_{i_{l-1}}} \cdots\cdots\cdots U^{s_{i_1}})$$

$$= \mathbf{p}^+(U^{s_{i_l}}) \cdot \mathbf{p}^+(U^{s_{i_{l-1}}}) \cdots\cdots \mathbf{p}^+(U^{s_{i_1}}) \equiv U^{[s_{i_1}] \star \cdots \star [s_{i_l}]} \equiv U^{[w]}.$$

Lemma 6.2 is proved. $\qquad\square$

The computation of $\mathbf{p}_L^-(B_w^-)$ is a little more complicated.

PROPOSITION 6.3.   *For any* $w \in W$ *we have*

(a) *The intersection of* $\mathbf{p}_L^-(B_w^-)$ *with* $B_{[w]}^-$ *is a dense subset of* $B_{[w]}^-$.

(b) *There is an algebraic sub-torus* $\widetilde{T}_w \subset T$ *such that* $\mathbf{p}_L^-(B_w^-) \equiv B_w^- \cdot \widetilde{T}_w$.

(c) *The torus* $\widetilde{T}_w$ *satisfies* $\widetilde{T}_w = \{e\}$ *if and only if* $w \in W_L$, $\widetilde{T}_{s_i} = \alpha_i^\vee(\mathbb{G}_m)$ *for* $i \in I \setminus J$, *and* $\widetilde{T}_{ww'} = [w']^{-1}(\widetilde{T}_w) \cdot \widetilde{T}_{w'} \cdot \widetilde{T}_{[w],[w']}$ *for any* w, w' *such that* $w \star w' = ww'$.

Proof. It follows from Proposition 4.7 that for any $w, w' \in W$ satisfying $w \star w' = ww'$ we have

$$\mathbf{p}_L^-(B_{ww'}^-) \equiv \mathbf{p}_L^-(B_w^- \cdot B_{w'}^-) = \mathbf{p}_L^-(B_w^-) \cdot \mathbf{p}_L^-(B_{w'}^-) \equiv B_{[w]}^- \cdot \widetilde{T}_w \cdot B_{[w']}^- \cdot \widetilde{T}_{w'}$$

$$\equiv B_{[w]}^- \cdot B_{[w']}^- \cdot [w']^{-1}(\widetilde{T}_w) \cdot \widetilde{T}_{w'} \cdot \widetilde{T}_{[w],[w']}.$$

This proves (b) and (c). Let us prove (a) now. Denote $\widetilde{B}_{[w]}^- := \mathbf{p}_L^-(B_w^-) \cap B_{[w]}^-$. We will proceed by induction in $l(w)$. If $w = e$, the statement is obvious. Now let $w \neq e$. Let us express $w = w'w''$, where $w' \neq e$, $w'' \neq e$ and $l(w) = l(w') + l(w'')$. Thus, the inductive hypothesis implies that $\widetilde{B}_{[w']}^-$ is dense in $B_{[w']}^-$, and $\widetilde{B}_{[w'']}^- := \mathbf{p}_L^-(B_{w''}^-) \cap B_{[w'']}^-$ is dense in $B_{[w'']}^-$.

Then (4.6) implies that

$$\mathbf{p}_L^-(B_{w'w''}^-) \supset \mathbf{p}_L^-(B_{w'}^- \cdot B_{w''}^-) \supset \mathbf{p}_L^-(B_{w'}^-) \cdot \mathbf{p}_L^-(B_{w''}^-) \supset \widetilde{B}_{[w']}^- \cdot \widetilde{B}_{[w'']}^-.$$

But $\widetilde{B}_{[w']}^- \cdot \widetilde{B}_{[w'']}^-$ is a dense subset of $B_{[w']}^- \cdot B_{[w'']}^-$. Finally, Proposition 4.7(a) implies that the latter set contains a dense subset of $B_{[w'] \star [w'']}^- = B_{[w'w'']}^-$.

This proves part (a). The proposition is proved. $\qquad\square$

It is well known that the set $G_0 = U_P^- \cdot P$ is open in G. Denote by $\mathbf{p} = \mathbf{p}_L$ the natural projection $G_0 \to P$. By definition, $\mathbf{p}$ commutes with the action of $B_L \times B$, and the restriction of $\mathbf{p}$ to $B^- \subset G_0$ is the natural projection $\mathbf{p}_L^- : B^- \to B_L^-$.

Our next task is to describe the image of each reduced Bruhat cell $U\overline{w}U$ under the projection $\mathbf{p}$.

Recall that in section 4.2 we defined for each $w \in W$ the sub-torus $T_w \subset T$.

PROPOSITION 6.4. *For any $w \in W$, we have*

(a) $U\overline{[w]}U \cdot T_{[w]^{-1}w} \subset \mathbf{p}(U\overline{w}U \cap G_0)$.

(b) *The restriction $\mathbf{p}_w = \mathbf{p}_{L;w}$ of $\mathbf{p}$ to $U\overline{w}U$ is a dominant rational morphism*
$$\mathbf{p}_w : U\overline{w}U \to U\overline{[w]}U \cdot \tilde{T}_w .$$

(c) $T_{[w]^{-1}w} \subset \tilde{T}_w$.

Proof. Prove (b). The multiplication in G induces the open inclusions $B_w^- \times U \hookrightarrow U\overline{w}U \cap G_0$, $B_w^- \times U_L \hookrightarrow U\overline{w}U_L \cap G_0$. Furthermore,
$$\mathbf{p}(B_w^- \cdot U) = \mathbf{p}_L^-(B_w^-) \cdot U , \quad \mathbf{p}(B_w^- \cdot U_L) = \mathbf{p}_L^-(B_w^-) \cdot U_L . \quad (6.2)$$

Therefore, it follows from Lemma 6.2 that
$$\mathbf{p}(U\overline{w}U) \equiv \mathbf{p}_L^-(B_w^-) \cdot U \equiv B_{[w]}^- \cdot \tilde{T}_w \cdot U = B_{[w]}^- \cdot U \cdot \tilde{T}_w \equiv U\overline{[w]}U \cdot \tilde{T}_w .$$

Part (b) is proved.

Let us now prove (a). Recall that $\mathbf{p}$ is $U_L \times U$-equivariant, and $U\overline{[w]}U = U_L\overline{[w]}U$. Note that $\mathbf{p}_L^-(B_w^-)$ intersects $B_{[w]}^-$ non-trivially. Thus
$$U\overline{[w]}U \subset \mathbf{p}(U\overline{w}U \cap G_0)$$

Recall that the action $Ad_w : T \times G \to G$ is defined in the proof of Proposition 4.9. Clearly, both P and the reduced Bruhat cell $U\overline{w}U$ are invariant under this action, and the morphism $\mathbf{p} : G_0 \to P$ commutes with this action. Thus applying Lemma 5.10 to the above identity, we obtain
$$Ad_w\, T(U\overline{[w]}U) = U\overline{[w]}U \cdot T_{[w]^{-1}w} \subset \mathbf{p}(U\overline{w}U \cap G_0) .$$

Part (a) is proved. Part (c) follows.

Proposition 6.4 is proved.   □

Denote $U_{P;w}^- := U_P^- \cap (U\overline{w}U \cdot \overline{[w]}^{-1})$. The set $U_{P;w}^-$ is not empty because $\overline{[w]} \in \mathbf{p}(U\overline{w}U \cap G_0)$ due to Proposition 6.4(a). Note also that $U_{P;w}^-$ is closed in U_P^-. For any $u_- \in U_{P;w}^-$ let $\mathbf{q}_{u_-} : U\overline{[w]}U \cdot T_{[w]^{-1}w} \to U\overline{w}U$ be the morphism defined by
$$\mathbf{q}_{u_-}\big(u' \cdot \overline{[w]} \cdot u \cdot t\big) := u' \cdot u_- \cdot \overline{[w]} \cdot u \cdot t$$
for $u \in U, u' \in V([w])$, and $t \in T_{[w]^{-1}w}$.

For any $w \in W$ let $T^w := \{t \in T : w(t) = t\}$. Clearly, $\dim T^w \cdot T_w = \dim T$.

PROPOSITION 6.5. *For any $w \in W$ we have*

(a) *for each* $u_- \in U^-_{P;w}$ *the morphism* $\mathbf{q}_{u_-}$ *is a section of* $\mathbf{p}_w$, *that is,*

$$\mathbf{p}_w \circ \mathbf{q}_{u_-} = \mathrm{id}_{U\overline{[w]}U \cdot T_{[w]^{-1}w}} \cdot$$

(b) *The variety* $U^-_{P;w}$ *is invariant under the adjoint action of* $T^{[w]^{-1} \cdot w}$ · $U_L([w])$.

(c) *Let* $\mathbf{q}$ *be a morphism* $U^-_{P;w} \times U\overline{[w]}U \cdot T_{[w]^{-1}w} \to U\overline{w}U$ *defined by* $\mathbf{q}(u_-, g) = \mathbf{q}_{u_-}(g)$. *Then* $\mathbf{q}$ *is an inclusion.*

Proof. Prove (a). Indeed, $V([w]) \subset U_L$. Thus $\mathbf{q}_{u_-}$ is a section of $\mathbf{p}_w$ because $\mathbf{p}_w$ is $U_L \times B$-equivariant.

Prove (b). It is easy to see that $U_L([w]) = Norm_{U_L}(\overline{[w]})$. Thus, for $u_- \in U^-_{P;w}$ and any $u \in Norm_{U_L}(\overline{[w]})$, $t \in T^{[w]^{-1} \cdot w}$ we have $Ad_w t(u \cdot u_- \cdot \overline{[w]} \cdot u^{-1}) \in U\overline{w}U$. On the other hand, $Ad_w t(u \cdot u_- \cdot \overline{[w]} \cdot u^{-1}) = (tu) \cdot u_- \cdot (tu)^{-1} \cdot \overline{[w]}$. This implies that $(tu) \cdot u_- \cdot (tu)^{-1} \in U^-_{P;w}$.

Prove (c). We proceed by the contradiction. Assume that $\mathbf{q}$ is not injective. That is, we assume that there are elements $u_-, u'_- \in U^-_{P;w}$ such that $u'_- \neq u_-$ and

$$\mathbf{q}_{u'_-}(\overline{[w]}) = \mathbf{q}_{u_-}(u'\overline{[w]}ut)$$

for some $u' \in V([w])$, $u \in U$ and $t \in T$. In other words,

$$u'_- \cdot \overline{[w]} = u' \cdot u_- \cdot \overline{[w]} \cdot u \cdot t.$$

Denote $\tilde{u}_- := Ad\, u'(u_-)$. Clearly, $\tilde{u}_- \in U^-_P$ because $u' \in U_L$. Furthermore, let us factorize $Ad\, \overline{[w]}(u) = u_+ \cdot u''_-$, where $u_+ \in U(\overline{[w]})$, and $u''_- \in U^-_L$. Thus, the above identity can be rewritten as

$$(\tilde{u}_-)^{-1} \cdot u'_- = u' \cdot u_+ \cdot u''_- \cdot [w](t).$$

This implies that $t = e$, $u''_- = e$, $u' \cdot u_+ = e$, and $u'_- = \tilde{u}_-$. In particular, $u' = (u_+)^{-1} \in V([w]) \cap U([w]) = U_L([w]) = \{e\}$.

In its turn, this implies that $u'_- = u_-$ which contradicts the original assumption.   □

DEFINITION. We call an element $w \in W$ *L-special* (or simply *special*) if $l(w) = l([w]) + \dim T_{[w]^{-1}w}$.

Theorem 6.6. *For any special element* $w \in W$ *we have*

(a) *The set* $U^-_{P;w}$ *consists of a single element* $u_- = u^-_{P;w}$. *This element* u_- *is centralized by the group* $T^{[w]^{-1} \cdot w} \cdot U_L([w])$. *The inclusion* $\mathbf{q}_{u-}$ *is dense.*

(b) $\mathbf{p}_w = [\mathbf{q}_{u_-}]^{-1}$. *In particular,* $\mathbf{p}_w : U\overline{w}U \to U\overline{[w]}U \cdot T_{[w]^{-1}w}$ *is a birational isomorphism.*

(c) $\tilde{T}_w = T_{[w]^{-1}w}$.

(d) *The composition* $\mathbf{p}_L^- \circ \eta^w$ *is a birational isomorphism* $U^w \xrightarrow{\sim} B_{[w]}^- \cdot T_{[w]^{-1}w}.$

Proof. Follows from Proposition 6.5.   □

6.2 Example of the projection for $G = GL_{m+n}$.

Let us fix two integers $m \leq n$. Let $G = GL_{m+n}$, $L = L_{m,n} = GL_m \times GL_n \subset GL_{m+n}$. In this case $P = P_{m,n} = L \cdot B$ is the corresponding maximal parabolic subgroup which consists of those matrices $g \in GL_{m+n}$ which have zeros in the $m \times n$-rectangle in the lower left corner.

We use the standard labeling of the Dynkin diagram of type A_{m+n-1}: $I = \{1, 2, \ldots, m+n-1\}$. Let $J = I \setminus \{m\}$. Then the Weyl group W is naturally identified with the symmetric group S_{m+n} via $s_i \mapsto (i, i+1)$.

Note that the longest element w_0 of W acts on simple roots by $w_0(\alpha_i) = -\alpha_{i^*}$ for $i = 1, \ldots, m+n-1$, where $i^* = m+n-i$.

The corresponding element $w_{L,G} = w_{m,n} \in W$ admits a factorization (4.9).

For $i \leq j$ denote by s_{ij} the reflection about the root $\alpha_i + \cdots + \alpha_j$. Clearly, s_{ij} is the transposition $(i, j+1) \in W = S_{m+n}$.

Denote $\sigma_{m,n} := [w_{m,n}]^{-1} w_{m,n}$. It is easy to see that

$$\sigma_{m,n} = \prod_{i=1}^{m} s_{i,i^*}, \qquad (6.3)$$

that is, $\sigma_{m,n}$ is a product of exactly m commuting reflections in W. This implies that $\dim T_{\sigma_{m,n}} = m = l(w_{m,n}) - l([w_{m,n}])$. Thus $w_{m,n}$ is special.

Let $\mathbf{q}_{m,n}$ be the dense inclusion $U_L\overline{[w_{m,n}]}U \cdot T_{\sigma_{m,n}} \hookrightarrow U\overline{w_{m,n}}U$ prescribed by Theorem 6.6.

Let $\omega_1, \ldots, \omega_{m+n}$ be the fundamental weights for GL_{m+n}. For each $w \in S_{m+n}$ the weight $w(\omega_k)$ is naturally identified with a k-element subset of $\{1, \ldots, m+n\}$.

For any $w_1, w_2 \in S_{m+n}$ let $\Delta_{w_1(\omega_i), w_2(\omega_i)} : G \to \mathbb{A}^1$ be the minor located in the intersection of rows indexed by $w_1(\omega_i)$ and columns indexed by $w_2(\omega_i)$.

PROPOSITION 6.7. *The variety* $U_L\overline{[w_{m,n}]}U_L \cdot T_{\sigma_{m,n}}$ *consists of all* $g \in B_L\overline{[w_{m,n}]}B_L$ *satisfying, for* $i = 1, \ldots, m+n-1$,

$$\Delta_{[w_{m,n}](\omega_i),\omega_i}(g) = \begin{cases} \Delta_{\omega_m,\omega_m}(g), & \text{if } i \in [m; n], \\ \Delta_{[w_{m,n}](\omega_i^*),\omega_i^*}(g), & \text{if } i \notin [m; n]. \end{cases} \qquad (6.4)$$

Proof. Let us describe the sub-torus $T_{\sigma m,n}$ of T. We have

$$T_{\sigma m,n} = \prod_{i=1}^{m} T_{s_{i,i^*}}.$$

That is, $T_{\sigma m,n}$ can be described in T by the equations ($i = 1, \ldots, m+n-1$)

$$\Delta_{\omega_i,\omega_i}(t) = \begin{cases} \Delta_{\omega_m,\omega_m}(t), & \text{if } i \in [m;n], \\ \Delta_{\omega_i^*,\omega_i^*}(t), & \text{if } i \in [m;n]. \end{cases}$$

On the other hand, for any $w \in W_J$, $t \in T$ one has

$$U_L \overline{w} U_L t = \{g \in B_L w B_L : \Delta_{w(\omega_i),\omega_i}(g) = \Delta_{\omega_i,\omega_i}(t), i = 1, \ldots, m+n-1\}.$$

Proposition 6.7 is proved. $\qquad\qquad\qquad\qquad\qquad\qquad\qquad\qquad\qquad\square$

PROPOSITION 6.8. *For any $\chi \in \widehat{U}(w_{m,n})$, $u' \in U_L$, $u \in U$, $t \in T$ we have*

$$\chi^{w_{m,n}}\big(\mathbf{q}_{m,n}(u'\overline{[w_{m,n}]} \cdot \sigma_{m,n}(t) \cdot t^{-1} \cdot u)\big)$$
$$= \chi(u') + \chi(u) - \delta_{m,n}\alpha_m(t) \cdot \chi(x_m(1)). \quad (6.5)$$

Proof. For any $1 \le i \le m \le j$, $w \in W_J$ let $\tilde{s}_{ij} \in \text{Norm}_G(T)$ be the representative of s_{ij} defined by

$$\tilde{s}_{ij} := \overline{w}^{-1} \overline{s}_m \overline{w} \qquad\qquad\qquad (6.6)$$

for any $w \in W_J$ such that $w^{-1} s_m w = s_{ij}$.

We need the following obvious refinement of (6.3).

LEMMA 6.9. $\overline{[w_{m,n}]}^{-1} \overline{w}_{m,n} = \prod_{i=1}^{m} \widetilde{s_{i,i^*}}$.

For any $1 \le i \le m$ choose an element $w_i \in W$ such that $s_{i,i^*} = w_i^{-1} \overline{s}_m w_i$. By (2.2), we have

$$\widetilde{s_{i,i^*}}^{-1} = u_i^+ u_i^- u_i^+,$$

where $u_i^+ = \overline{w}_i^{-1} x_m(-1)\overline{w}_i, u_i^- = \overline{w}_i^{-1} y_m(1)\overline{w}_i$.

Clearly, $u_i^+ \in U_P \cap \text{Ad}\,\overline{w}_{m,n}^{-1}(U_P^-)$, and $u_i^- \in U_P^- \cap \text{Ad}\,\overline{w}_{m,n}^{-1}(U_P)$. It is also clear that u_i^ε commutes with each $u_j^{\varepsilon'}$ whenever $i \ne j$, where $\varepsilon, \varepsilon' \in \{+,-\}$.

Denote $u^\varepsilon := \prod_{i=1}^{m} u_i^\varepsilon$ for $\varepsilon \in \{+,-\}$.

LEMMA 6.10. *The only element of $U_{P,w_{m,n}}$ is equal to $\text{Ad}\,\overline{w}_{m,n}((u^+)^{-1})$, and*

$$u_- \cdot \overline{[w_{m,n}]} = \overline{w}_{m,n} \cdot u^- \cdot u^+.$$

Proof. Let $u_- := Ad\,\overline{w}_{m,n}((u^+)^{-1})$. Clearly, $u_- \in U_P^-$.

Next, let us rewrite the equation from Lemma 6.9: $\overline{w}_{m,n}^{-1}\overline{[w_{m,n}]} = u^+ \cdot u^- \cdot u^+$ or, equivalently, $\overline{[w_{m,n}]} = \overline{w}_{m,n} \cdot u^+ \cdot u^- \cdot u^+ = (u_-)^{-1} \cdot u_1^+ \overline{w}_{m,n} \cdot u^+$, where $u_1^+ = Ad\,\overline{w}_{m,n}(u^-)$. Clearly, $u_1^+ \in U_P$.

Thus, $u_- \in U_P^- \cap U\overline{w_{m,n}}U[\overline{w_{m,n}}]^{-1} = U_{P,w_{m,n}}$.
Lemma 6.10 is proved.   □

Since $\mathbf{q}_{m,n}$ is $U_L \times U$-equivariant it suffices to prove (6.5) in the case when $u = u' = e$. Using Lemma 6.10 we obtain

$$\chi^{w_{m,n}}\big(\mathbf{q}_{m,n}(\mathrm{Ad}_{w_{m,n}} t([\overline{w_{m,n}}]))\big) = \chi^{w_{m,n}}\big(\mathrm{Ad}_{w_{m,n}} t(\mathbf{q}_{m,n}([\overline{w_{m,n}}]))\big)$$

$$\chi^{w_{m,n}}\big(\mathrm{Ad}_{w_{m,n}} t(\overline{w_{m,n}}u^- u^+)\big) = \overline{\chi}(u^- u^+ \cdot t^{-1}) = \chi(tu^+ t^{-1}).$$

Let us compute $\chi(tu^+ t^{-1})$ using the fact that $u_i^+ \in [U, U]$ for $i = 1, \ldots, m-1$,

$$\chi(tu^+ t^{-1}) = \sum_{i=1}^{m} \chi(tu_i^+ t^{-1}) = \chi(tu_m^+ t^{-1}) = -\delta_{m,n}\chi(x_m(t)).$$

The last equality holds because $tu_m^+ t^{-1} \in [U, U]$ for $m \neq n$, and $tu_m^+ t^{-1} = tx_m(-1)t^{-1} = x_m(-\alpha_m(t))$ if $m = n$.

The proposition is proved.   □

Let $L' = L'_{n-m} := (\mathbb{G}_m)^{2m} \times GL_{n-m} \subset L_{m,n}$. By definition, $[w_{m,n}] = w_{L',L_{m,n}}$. It is easy to see that $[w_{m,n}](T_{\sigma_{m,n}}) \subset Z(L')$.

Recall from Proposition 5.12 that for any $\chi \in \widehat{U}$ the function $f_{\chi,\chi^{L'}}^{[w_{m,n}]}$ on $Z(L') \cdot U_L[\overline{w_{m,n}}]U_L$ defined by (4.13) is proper under the action of $U_L \times U_L$.

COROLLARY 6.11. *For any $\chi \in \widehat{U}(w_{m,n}) \cap \widehat{U}([w_{m,n}])$, and $g \in U_L[\overline{w_{m,n}}]U_L \cdot t$, $t \in T_{\sigma_{m,n}}$, one has*

$$\chi^{w_{m,n}}(\mathbf{q}_{m,n}(g)) = f_{\chi,\chi^{L'}}^{[w_{m,n}]}(g) - \delta_{m,n}a_m \cdot t_m^{-1},$$

where a_m is the coefficient of χ_m in the expansion $\chi = \sum_{i=1}^{m+n-1} a_i\chi_i$.

REMARK. It is well known that $U_{P_{m,n}}$ is abelian in the above example. One can expect that for any G and any standard parabolic subgroup $P = L \cdot U_P$ the element $w_{L,G}$ is special if U_P is abelian.

On the other hand, there is an example of a standard parabolic subgroup L in the simple group G of the type D_4 with non-abelian U_P and non-special $w_{L,G}$.

We use the following labeling of the Dynkin diagram of type D_4: $I = \{0, 1, 2, 3\}$ where 0 stands for the "center" of the Dynkin diagram. Let $J = \{1, 2, 3\}$, and $L := L_J$. Then $w_0^L = s_1 s_2 s_3$, and $w_{L,G} = s_0 s_1 s_2 s_3 s_0 s_1 s_2 s_3 s_0$. It is easy to see that $[w_{L,G}] = w_0^L$. Thus, $l(w_{L,G})^{-1} - l([w_{L,G}]) = 9 - 3 = 6 > \dim T_{[w_{L,G}]^{-1}w_{L,G}}$, which implies that $w_{L,G}$ is not special.

Moreover, $\mathbf{p}_w$ is not a birational isomorphism because $\dim U\overline{w_{L,G}}U = 21$, $\dim U_L[\overline{w_{L,G}}]U = 15$, and $\dim \widetilde{T}_{w_{L,G}} \leq 4$.

7　Appendix: Combinatorial Pre-crystals and W-crystals

In this section we recall Kashiwara's framework (see [K2]).

DEFINITION.　A *partial bijection* of sets $f : A \to B$ is a bijection $A' \to B'$ of subsets $A' \subset A$, $B' \subset B$. We denote the subset A' by $\mathrm{dom}(f)$ and the subset B' by $\mathrm{ran}(f)$.

REMARK.　A partial bijection $f : A \to B$ is an embedding $A \hookrightarrow B$ if and only if $\mathrm{dom}(f) = A$.

The inverse f^{-1} of a partial bijection $f : A \to B$ is the inverse bijection $\mathrm{ran}(f) \to \mathrm{dom}(f)$. Composition $g \circ f$ of partial bijections $f : A \to B$, $g : B \to C$ is naturally a partial bijection with $\mathrm{dom}(g \circ f) = \mathrm{dom}(f) \cap f^{-1}(\mathrm{ran}(f) \cap \mathrm{dom}(g))$ and $\mathrm{ran}(g \circ f) = g(\mathrm{ran}(f) \cap \mathrm{dom}(g))$. In particular, for any partial bijection $f : B \to B$ and $n \in \mathbb{Z}$ the n-th power f^n is a partial bijection $B \to B$.

Note that for any partial bijection $f : A \to B$ the composition $f^{-1} \circ f$ is the *partial identity bijection* $id_{\mathrm{dom}(f)} : A \to A$.

DEFINITION.　Let B be a set $\widetilde{\gamma} : B \to \Lambda^\vee$ be a map. We call a family of partial bijections *combinatorial pre-crystal* (or simply *pre-crystal*) on $(B, \widetilde{\gamma})$ if

$$\widetilde{\gamma}(\tilde{e}_i(b)) = \widetilde{\gamma}(b) + \alpha_i^\vee$$

for $b \in \mathrm{dom}(\tilde{e}_i)$, $i \in I$.

With any pre-crystal $\mathcal{B}$ we define partial bijections $\tilde{s}_i : B \to B$, $i \in I$, by

$$\tilde{s}_i(b) = \tilde{e}_i^{-\langle \widetilde{\gamma}(b), \alpha_i \rangle}(b) \,. \tag{7.1}$$

In fact $\tilde{s}_i$ are partial involutions, $\tilde{s}_i^2 = id_{\mathrm{dom}(\tilde{s}_i)}$.

EXAMPLES.　1. Fix $i \in I$. Denote by $\widetilde{\gamma}_i : \mathbb{Z} \to \Lambda^\vee$ the map $n \mapsto n\alpha_i^\vee$. We define a pre-crystal $\mathcal{B}_i$ on $(\mathbb{Z}, \widetilde{\gamma}_i)$ in the following way. We define a partial bijection $\tilde{e}_i : \mathbb{Z} \to \mathbb{Z}$ by $\tilde{e}_i(n) = n + 1$ and for any $j \neq i$ we define $\mathrm{dom}(\tilde{e}_j) = \mathrm{ran}(\tilde{e}_j) = \emptyset$. This defines a pre-crystal on $(\mathbb{Z}, \widetilde{\gamma}_i)$ which we denote by $\mathcal{B}_i$ (see [K2, Example 1.2.6]).

2. The lattice $\Lambda^\vee$ is a pre-crystal with $\widetilde{\gamma} = id|_{\Lambda^\vee}$ and $\tilde{e}_i(\lambda^\vee) = \lambda^\vee + \alpha_i^\vee$ for $\lambda^\vee \in \Lambda^\vee$, $i \in I$.

3. For any pre-crystal $\mathcal{B}$ on $(B, \widetilde{\gamma})$ we denote by $\widetilde{\gamma}'$ the map $\Lambda^\vee \times B \to \Lambda^\vee$ given by $\widetilde{\gamma}'(\lambda^\vee, b) = \lambda^\vee + \widetilde{\gamma}(b)$. One can define a pre-crystal on $(\Lambda^\vee \times B, \widetilde{\gamma}_{\Lambda^\vee})$ by $\tilde{e}_i(\lambda^\vee, b) = (\lambda^\vee, \tilde{e}_i(b))$.

4. For any pre-crystal $\mathcal{B}$ on $(B, \widetilde{\gamma})$ put $\gamma^* := -\gamma$. Define also the partial bijections $\tilde{e}_i^* := \tilde{e}_i^{-1}$. The collection $\tilde{e}_i^*$, $i \in I$, defines the structure of a

pre-crystal on (B, γ^*). This pre-crystal is called *dual* to B and is denoted by B^*.

REMARKS.   1. In [K2] the partial bijection $\tilde{e}_i^{-1}$ was denoted by $\tilde{f}_i$.
   2. Our pre-crystals correspond to Kashiwara's crystals for $^L\mathfrak{g}$.

DEFINITION.   A pre-crystal B is called *free* if each $\tilde{e}_i$ is a bijection $B \to B$.

DEFINITION.   A morphism of pre-crystals $f : B' \to B$ is a partial bijection $f : B' \to B$ such that $\mathrm{dom}(f) = B'$, $\tilde{\gamma} = \tilde{\gamma}' \circ f$, and the partial bijections $\tilde{e}_i' : B' \to B'$, $i \in I$, are obtained from the partial bijections $e_i : B \to B$, $i \in I$, by $\tilde{e}_i' = f^{-1} \circ \tilde{e}_i \circ f$.

DEFINITION.   We say that a pre-crystal B is a *combinatorial W-crystal* (or simply *W-crystal*) if for any $i \in I$ we have $\mathrm{dom}(\tilde{s}_i) = B$, and the involutions $\tilde{s}_i$ satisfy the braid relations.

   It is clear that a structure of a W-crystal on $(B, \tilde{\gamma})$ defines an action of W on B in such a way that, for any $i \in I$, $s_i \in W$ acts by $\tilde{s}_i : B \to B$.

REMARKS.   1. The condition that $\mathrm{dom}(\tilde{s}_i) = B$ is equivalent to

$$\inf \left\{ n : b \in \mathrm{dom}(\tilde{e}_i^n) \right\} \leq -\langle \tilde{\gamma}(b), \alpha_i \rangle \leq \sup \left\{ n : b \in \mathrm{dom}(\tilde{e}_i^n) \right\} \quad (7.2)$$

for all $b \in B, i \in I$.
   2. The pre-crystal on $(\Lambda^\vee, \mathrm{id}_{\Lambda^\vee})$ defined above is a free W-crystal.
   3. For any W-crystal the structure map $\tilde{\gamma} : B \to \Lambda^\vee$ is W-equivariant.

   We denote by *Pre-Cryst* the category such that the objects are pre-crystals and arrows are morphisms of pre-crystals. We denote by *W-Cryst* the full sub-category of *Pre-Cryst* whose objects are W-crystals. Note that each morphism $f : B \to B'$ in *W-Cryst* is an injective W-equivariant map $B \to B'$.

EXAMPLE.   Let V be a finite-dimensional $U_q(^L\mathfrak{g})$-module, and let $B = B(V)$ be a *crystal basis* for V (see [K1]). Denote by $\tilde{\gamma} : B \to \Lambda^\vee$ the weight grading. Then the partial bijections $\tilde{e}_i, \tilde{f}_i : B \to B$ define the structure of a pre-crystal on (B, γ). It is a deep result of [K3] that this is a W-crystal. We denote this W-crystal by $B(V)$.

REMARK.   For $B = B(V)$ all e_i, e_i^{-1} are nilpotent, that is, $\mathrm{dom}(\tilde{e}_i^n) = \mathrm{dom}(\tilde{e}_i^{-n} = \emptyset$ for some $N > 0$, and $-\langle \tilde{\gamma}(b), \alpha_i \rangle = \inf\{n : b \in \mathrm{dom}(\tilde{e}_i^n)\} + \sup\{n : b \in \mathrm{dom}(\tilde{e}_i^n)\}$ for $b \in B(V), i \in I$. Clearly, this identity implies the inequalities (7.2), that is, $\mathrm{dom}(\tilde{s}_i) = B(V)$ for each $i \in I$.

   There are examples of W-crystals for which $l_i, l_i^- \equiv +\infty$, e.g., there are examples of free W-crystals. Among them are the free W-crystals $B_\mathbf{i}$

from [K2, Section 2.2], built on $(\mathbb{Z}^{l(w_0)}, \tilde{\gamma}_\mathbf{i})$, $\mathbf{i} \in R(w_0)$. Kashiwara has constructed each $\mathcal{B}_\mathbf{i}$ as the *tensor product* of $\mathcal{B}_{i_1}, \ldots, \mathcal{B}_{i_l}$, so that $\mathcal{B}_\mathbf{i}$ is equipped with the family of functions $\tilde{\varphi}_i^\mathbf{i} : \mathbb{Z}^{l(w_0)} \to \mathbb{Z}$, $i \in I$, satisfying $\tilde{\varphi}_i^\mathbf{i}(\tilde{e}_i(b)) = \tilde{\varphi}_i^\mathbf{i}(b) + 1$ for $b \in \mathbb{Z}^{l(w_0)}$, $i \in I$.

Let $\mathcal{R}(w_0)$ be the category whose objects are reduced sequences $\mathbf{i} \in R(w_0)$ and for each pair of objects $\mathbf{i}, \mathbf{i}' \in R(w_0)$ there is exactly one arrow $\mathbf{i} \to \mathbf{i}'$.

A functor $F : \mathcal{R}(w_0) \longrightarrow W - Cryst$ such that $F(\mathbf{i}) = \mathcal{B}_\mathbf{i}$ was defined in [BZ2]. Moreover it was shown in [BZ2] that $\tilde{\varphi}_i^\mathbf{i} = \varphi_i^{\mathbf{i}'} \circ F(\mathbf{i} \to \mathbf{i}')$ for any $\mathbf{i}, \mathbf{i}' \in R(w_0)$. Therefore, the free W-crystal $\mathcal{B}_{w_0}$ is well defined and is equipped with the family of functions $\tilde{\varphi} : B_{w_0} \to \mathbb{Z}$, $i \in I$, such that $\tilde{\varphi}(\tilde{e}_i(b)) = \tilde{\varphi}(b) + 1$.

REMARK. Let $V_{\lambda^\vee}$ be the simple $U_q(^L\mathfrak{g})$-module with the lowest weight λ (e.g., $\lambda^\vee$ is an anti-dominant co-weight), and let $\mathcal{B}(V_{\lambda^\vee})$ be the corresponding combinatorial crystal. It follows from the results of Kashiwara that there is a morphism of W-crystals

$$f_{\lambda^\vee} : \mathcal{B}(V_{\lambda^\vee}) \to \Lambda^\vee \times \mathcal{B}_{w_0} \qquad (7.3)$$

such that $\tilde{\varphi}(f_{\lambda^\vee}(b)) = \sup\{n : b \in \mathrm{dom}(\tilde{e}_i^{-n})\}$ for $b \in B(V_{\lambda^\vee})$, $i \in I$.

References

[BFZ] A. BERENSTEIN, S. FOMIN, A. ZELEVINSKY, Parametrizations of canonical bases and totally positive matrices, Adv. in Math. 122 (1996), 49–149.

[BK] A. BERENSTEIN, AN. KIRILLOV, Groups generated by involutions, Gel'fand-Tsetlin patterns, and combinatorics of Young tableaux (in Russian), Algebra i Analiz 7:1 (1995),92–152. Translation in St. Petersburg Math. J. 7:1 (1996), 77–127.

[BZ1] A. BERENSTEIN, A. ZELEVINSKY, Total positivity in Schubert varieties, Comment. Math. Helv. 72 (1997), 128–166.

[BZ2] A. BERENSTEIN, A. ZELEVINSKY, Tensor product multiplicities, canonical bases and totally positive varieties, preprint math. RT/9912012.

[Bo] N. BOURBAKI, Lie groups and Lie algebras. Chapters 1–3. Translated from the French. Elements of Mathematics. Springer-Verlag, Berlin-New York, 1989.

[BrG] A. BRAVERMAN, D. GAITSGORY, Crystals via the affine Grassmannian, preprint math. AG/9909077.

[BrK] A. BRAVERMAN, D. KAZHDAN, γ-functions of representations and lifting, GAFA2000, in this issue.

[FZ] S. FOMIN, A. ZELEVINSKY, Double Bruhat cells and total positivity, J. Amer. Math. Soc. 12:2 (1999), 335–380.

[K1]   M. KASHIWARA, Crystalizing the q-analogue of universal enveloping alge-
       bras, Comm. Math. Phys. 133:2 (1990), 249–260.

[K2]   M. KASHIWARA, The crystal base and Littelmann's refined Demazure
       character formula, Duke Math. J. 71:3 (1993), 839–858.

[K3]   M. KASHIWARA, Crystal bases of modified quantized enveloping algebra,
       Duke Math. J. 73:2 (1994), 383–413.

[L]   P. LITTELMANN, A Littlewood-Richardson rule for symmetrizable Kac-
       Moody algebras, Invent. Math. 116 (1994), 329–346.

[Lu1] G. LUSZTIG, Canonical bases arising from quantized enveloping algebras,
       J. Amer. Math. Soc. 3:2 (1990), 447–498.

[Lu2] G. LUSZTIG, Introduction to quantized enveloping algebras, Progress in
       Math. 105 (J. Tirao, N. Wallach, eds.), Birkhäuser, Boston, 1992, 49–65.

[Lu3] G. LUSZTIG, Introduction to Quantum Groups, Birkhäuser, Boston, 1993.

ARKADY BERENSTEIN, Department of Mathematics, Harvard University, Cambridge, MA 02138, USA   arkadiy@math.harvard.edu

DAVID KAZHDAN, Department of Mathematics, Harvard University, Cambridge,
MA 02138, USA   kazhdan@math.harvard.edu

GAFA, Geom. funct. anal.
Special Volume – GAFA2000, 237 – 278
1016-443X/00/S10237-42 $ 1.50+0.20/0

I GAFA Geometric And Functional Analysis

γ-FUNCTIONS OF REPRESENTATIONS AND LIFTING

A. BRAVERMAN AND D. KAZHDAN

(Appendix by V. VOLOGODSKY)

Abstract

Let F be a local non-Archimedean field and let $\psi : F \to \mathbb{C}^\times$ be a non-trivial additive character of F. To this data one associates a meromorphic function $\gamma_\psi : \pi \mapsto \gamma_\psi(\pi)$ on the set of irreducible representations of the group $\mathbf{GL}(n, F)$ in the following way. Consider the invariant distribution $\Phi_\psi := \psi(\mathrm{tr}(g))|\det(g)|^n|dg|$ on $GL(n, F)$, where $|dg|$ denotes a Haar distribution on $\mathbf{GL}(n, F)$. Although the support of Φ_ψ is not compact, it is well known that for a generic irreducible representation π of $\mathbf{GL}(n, F)$ the action of Φ_ψ in π is well defined and thus it defines a number $\gamma_\psi(\pi)$. These gamma-functions and the associated L-functions were studied by J. Tate for $n = 1$ and by R. Godement and H. Jacquet for arbitrary n.

Now let G be the group of points of an arbitrary quasi-split reductive algebraic group over F and let $\mathbf{G}^\vee$ be the Langlands dual group. Let also $\rho : \mathbf{G}^\vee \to \mathbf{GL}(n, \mathbb{C})$ be a finite-dimensional representation of $\mathbf{G}^\vee$. The local Langlands conjectures predict the existence of a natural map $l_\rho : \mathrm{Irr}(G) \to \mathrm{Irr}(\mathbf{GL}(n, F))$. Assuming that a certain technical condition on ρ (guaranteeing that the image of l_ρ does not lie in the singular set of γ_ψ) is satisfied, one can consider the meromorphic function $\gamma_{\psi,\rho}$ on $\mathrm{Irr}(G)$, setting $\gamma_{\psi,\rho}(\pi) = \gamma_\psi(l_\rho(\pi))$.

The main purpose of this paper is to propose a general framework for an explicit construction of the functions $\gamma_{\psi,\rho}$. Namely, we propose a conjectural scheme for constructing an invariant distribution $\Phi_{\psi,\rho}$ on G, whose action in every $\pi \in \mathrm{Irr}(G)$ is given by multiplication by $\gamma_{\psi,\rho}$. Surprisingly, this turns out to be connected with certain geometric analogs of M. Kashiwara's crystals (cf. [BeK]).

We work out several examples in detail. As a byproduct we obtain a conjectural formula for the lifting $l_\rho(\theta)$ where θ is a character of an arbitrary maximal torus in $\mathbf{GL}(n, F)$. In some of these examples we also give a definition of the corresponding local L-function and state a conjectural ρ-analogue of the Poisson summation formula for Fourier transform. This conjecture implies that the corresponding global L-function has meromorphic continuation and satisfies a functional equation.

1 Introduction and Statement of the Results

1.1 The Langlands program and the lifting. Let F be either a local
or finite field and $\overline{F}$ its separable closure. We will denote algebraic varieties
over F and maps between them by boldface letters. The corresponding
ordinary letters will denote the associated sets of F-points (and the induced
maps between them).

Let $\mathbf{G}$ be a connected reductive algebraic group over F, $G = \mathbf{G}(F)$
its group of F-points, $\mathrm{Irr}(G)$ the set of isomorphism classes of complex
irreducible representations of G. For simplicity we shall assume in the
introduction that $\mathbf{G}$ is split, although most of our results generalize easily
to quasi-split groups. To $\mathbf{G}$ one associates a connected complex algebraic
group $\mathbf{G}^\vee$, called the Langlands dual group.

EXAMPLE. If $\mathbf{G} = \mathbf{GL}(n)$ then $\mathbf{G}^\vee = \mathbf{GL}(n, \mathbb{C})$.

The local Langlands conjectures can now be summarized as follows. To
the field F one associates a pro-algebraic group $\mathfrak{g}_F$ over $\mathbb{C}$, which is very
closely related with the Galois group $\mathrm{Gal}(\overline{F}/F)$ (in the case when F is a
local field it is called the Weil-Deligne group).

The Langlands conjecture says that

1) There exists a finite-to-one map from $\mathrm{Irr}(G)$ to the set of all $\mathbf{G}^\vee$-orbits
   on $\mathrm{Hom}(\mathfrak{g}_F, \mathbf{G}^\vee)$.
2) If $G = \mathbf{GL}(n, F)$ then the above map is a bijection.

REMARK. This conjecture is known for a finite field (due to Lusztig). It
is known for local fields when $\mathbf{G}$ is a torus (local class field theory) and for
$\mathbf{GL}(n)$ (cf. [LaRS], [H] and [HT]).

Now let $\rho : \mathbf{G}^\vee \to \mathbf{GL}(n, \mathbb{C})$ be an algebraic representation of $\mathbf{G}^\vee$.
Then the above conjecture implies that there exists a map $l_\rho : \mathrm{Irr}(G) \to$
$\mathrm{Irr}(\mathbf{GL}(n, F))$.

Our main task is to try to propose a conjecture for an explicit description of the map l_ρ and to formulate a certain ρ-analogue of the Poisson
summation formula.

In the rest of this paper, until section 9 we assume that F is a local
field. Moreover, we assume that the characteristic of F is good with respect
to $\mathbf{G}$ and that the Lie algebra $\mathfrak{g}$ of $\mathbf{G}$ possesses a non-degenerate invariant
form. We denote by $\mathcal{H}(G)$ the Hecke algebra of G. By definition when
F is non-Archimedean (resp. Archimedean) this is the algebra of locally
constant (resp. C^∞) compactly supported distributions on G. We shall
choose a Haar measure $|dg|$ and thus identify $\mathcal{H}(G)$ with the space $C_c^\infty(G)$

of locally constant (resp. C^∞ when F is Archimedean) compactly supported functions on G.

1.2 γ-functions. We will not try to construct explicitly a representation $l_\rho(\pi)$, $\pi \in \mathrm{Irr}(G)$ but give an indirect description of the lifting l_ρ.

For this purpose we define a complex-valued function γ on $\mathrm{Irr}(\mathbf{GL}(n,F))$.

Choose (once and for all) a non-trivial additive character $\psi : F \to \mathbb{C}^\times$. Let $|dg|$ denote the unique Haar measure on $G = \mathbf{GL}(n,F)$ such that the Fourier transform $\phi \mapsto \mathcal{F}(\phi)$ defined by the formula

$$\mathcal{F}(\phi)(y) = \int_G \phi(x)\psi(\mathrm{tr}(xy))|\det(x)|^n|dx| \tag{1.1}$$

is unitary.

Consider the distribution $\Phi_\psi(g) := \psi(\mathrm{tr}(g))|\det(g)|^n|dg|$ on $\mathbf{GL}(n,F)$. Clearly, the distribution $\Phi = \Phi_\psi$ is invariant under the adjoint action. Now let (π, V) be an irreducible representation of G. Consider the integral

$$\pi(\Phi) = \int_G \pi(g)\Phi(g). \tag{1.2}$$

One can show (cf. [GoJ]) that the integral (1.2) is convergent (although not absolutely convergent) for generic $\pi \in \mathrm{Irr}(G)$. Therefore, for generic π the operator $\pi(\Phi)$ is well defined and it is equal to multiplication by a scalar $\gamma(\pi)$. One can consider γ as a meromorphic function on $\mathrm{Irr}(G)$ (cf. section 3.4 for the definition of this notion).

Now let G and ρ be as above and assume that we know the lifting map l_ρ. Then we define a function $\gamma_{\psi,\rho} = \gamma_\rho$ on the set $\mathrm{Irr}(G)$ by $\gamma_\rho = l_\rho^*(\gamma)$. In other words any irreducible representation π of G we define

$$\gamma_\rho(\pi) = \gamma(l_\rho(\pi)). \tag{1.3}$$

The function γ_ρ is a well-defined meromorphic function on $\mathrm{Irr}(G)$ provided that $l_\rho(\pi)$ does not lie in the singular set of γ for generic $\pi \in \mathrm{Irr}(G)$. In order to guarantee this we shall always (except for section 9) assume that ρ satisfies the following condition: there exists a cocharacter $\sigma : \mathbb{G}_m \to Z(\mathbf{G}^\vee)$ of the center $Z(\mathbf{G}^\vee)$ of the group $\mathbf{G}^\vee$ such that $\rho \circ \sigma^\vee = \mathrm{Id}$. Note that σ can be regarded as a character of $\mathbf{G}$.

It is easy to show (assuming that l_ρ satisfies certain natural properties) that there exists an ad-invariant distribution $\Phi_\rho = \Phi_{\rho,G}$ on G such that

$$\gamma_\rho(\pi) = \pi(\Phi_\rho) \cdot \mathrm{Id}. \tag{1.4}$$

We see from (1.4) that a lifting l_ρ of a representation ρ of the dual group $\mathbf{G}^\vee$ determines an invariant distribution Φ_ρ on G. Since the map l_ρ

collapses the L-packets we assume that the function $\Phi_{\rho,G}$ is stable. Our main purpose is to give a conjectural description of the function $\Phi_{\rho,G}$. In general we can only propose a framework for such a description. Sometimes we can make our suggestion precise and in some simplest cases when the lifting is known we can check that our definition is correct.

1.3   The Fourier transform $\mathcal{F}_\rho$.   Assume now that we know the lifting l_ρ and therefore can construct the distribution Φ_ρ on $G = \mathbf{G}(F)$. In section 6 we define certain Fourier-type transform operator $\mathcal{F}_\rho$ acting in the space of functions on G. Namely, for $\phi \in \mathcal{H}(G)$ we define

$$\mathcal{F}_\rho(\phi) = |\sigma|^{-l-1}(\Phi_\rho * {}^\iota\phi)\,, \tag{1.5}$$

where ${}^\iota\phi(g) = \phi(g^{-1})$ and l is the semi-simple rank of $\mathbf{G}$.

In the case when $\mathbf{G} = \mathbf{GL}(n)$ and ρ is the standard representation of $\mathbf{G}^\vee \simeq \mathbf{GL}(n)$ the operator $\mathcal{F}_\rho$ coincides with the usual Fourier transform in the space of functions on the space $\mathbf{M}_n(F)$ of $n \times n$-matrices over F.

We conjecture that for a $\mathcal{F}_\rho$ extends to a unitary operator acting in the space $L^2_\rho(G) = L^2(G, |\sigma|^{l+1}|dg|)$ where $|dg|$ is a Haar measure on G.

Note that in section 8 we give a definition of Φ_ρ (and thus of $\mathcal{F}_\rho$) in certain cases where l_p is not known.

1.4   Local L-functions.   Until now we have discussed only the local lifting. But the only known formulation of the *local lifting conjecture* is to interpret it as a local part of the *global lifting conjecture*. There are two approaches to a proof of the *global lifting conjecture*. The first approach is based on the Trace formula and the second one on the study of L-functions. So it is natural to try to give a direct definition of the L-function $L(l_\rho(\pi), s)$ for $\pi \in \mathrm{Irr}(G)$ in terms of ρ and π. Assume that we know the distribution Φ_ρ. In such a case we present a conjecture for a direct definition of the L-function $L(l_\rho(\pi), s)$. To any representation ρ of $\mathbf{G}^\vee$ as above we associate a certain subspace $\mathcal{S}_\rho$ of the space of functions on G such that $\mathcal{S}_\rho$ contains the space $\mathcal{S}(G)$ of smooth compactly supported functions on G. We conjecture that $\mathcal{S}_\rho$ is invariant under $\mathcal{F}_\rho$. For any representation $\pi \in \mathrm{Irr}(G)$ we define the function $L(l_\rho(\pi), s)$ as the common denominator of rational operator-valued functions $\pi_s(\phi) := \int_{g \in G} \phi(g)|\sigma(g)|^s \pi(g)|dg|$ where $\phi \in \mathcal{S}_\rho$. In the case when $\mathbf{G} = \mathbf{GL}(n)$ and ρ is the standard representation of $\mathbf{G}^\vee \simeq \mathbf{GL}(n)$ the space $\mathcal{S}_\rho$ coincides with the space of Schwartz-Bruhat functions on the space M_n of $n \times n$-matrices with entries in F and our definition of the L-function coincides with the definition from [GoJ].

1.5   Conjectural applications to automorphic L-functions. Now let K denote a global field and let also $\mathbb{A}_K$ be the corresponding adele ring. For a place $\mathfrak{p}$ of K we denote by $K_{\mathfrak{p}}$ the corresponding local completion of K. In [GoJ], Godement and Jacquet defined the L-function $L(\pi, s)$ for every automorphic representation π of $\mathbf{GL}(n, \mathbb{A}_K)$. This L-function is defined as the product of the corresponding local L-functions over all places of K. It is shown in [GoJ] that $L(\pi, s)$ is meromorphic and has a functional equation. The main tools in the construction and the proof are the space $\mathcal{S}(\mathbf{M}_n(F))$ of Schwartz-Bruhat functions on $n \times n$-matrices, the Fourier transform acting in this space and the Poisson summation formula .

Assume now that we are given a group $\mathbf{G}$ and non-singular a representation ρ of $\mathbf{G}^{\vee}$. R. Langlands (cf. [L]) conjectured that one could define an L-function $L_{\rho}(\pi, s)$ attached to every automorphic representation π of $\mathbf{G}(\mathbb{A}_K)$ as a product of local factors. We conjecture that these local factors are equal to ones defined in section 1.4. In section 5 we define a global version $\mathcal{S}_{\rho}(\mathbf{G}(\mathbb{A}_K))$ and define a global ρ-analogue $\mathcal{F}_{\rho}$ of the Fourier transform. By the definition the space $\mathcal{S}_{\rho}(\mathbf{G}(\mathbb{A}_K))$ is the span of functions ρ which are products $\phi = \bigotimes \phi_{\mathfrak{p}}$ of local factors. We conjecture the existence of an $\mathcal{F}_{\rho}$-invariant distribution ε_{ρ} on $\mathcal{S}_{\rho}(\mathbf{G}(\mathbb{A}_K))$ such that

1)

$$\varepsilon_{\rho}(\phi) = \sum_{g \in \mathbf{G}(K)} \phi(g) \tag{1.6}$$

if some local factor $\phi_{\mathfrak{p}}$ has compact support on $G_{\mathfrak{p}}$
2) For any $\phi \in \mathcal{S}_{\rho}(\mathbf{G}(\mathbb{A}_K))$

$$\varepsilon_{\rho}(\phi) = \varepsilon_{\rho}(\mathcal{F}_{\rho}(\phi)) . \tag{1.7}$$

Existence of such ε can be thought of as a ρ-analogue of the Poisson summation formula. In the case when $\mathbf{G} = \mathbf{GL}(n)$ and ρ is the standard representation of $\mathbf{G}^{\vee} \simeq \mathbf{GL}(n)$ our conjecture specializes to the usual Poisson summation formula. The validity of our ρ-analogue of the Poisson summation formula would imply that the automorphic L-function $L_{\rho}(\pi, s)$ has meromorphic continuation with only a finite number of poles and satisfies a functional equation.

1.6   Application to lifting. In section 8 we give a construction of the distribution Φ_{ρ} in the case when $\mathbf{G} = \mathbf{GL}(n) \times \mathbf{GL}(m)$ for arbitrary n and m and when ρ is equal to the tensor product of the standard representations of $\mathbf{GL}(n, \mathbb{C})$ and $\mathbf{GL}(m, \mathbb{C})$. As a byproduct we get a conjectural formula for $l_{\rho}(\theta)$ where θ is a character of an arbitrary maximal torus in $\mathbf{GL}(n, F)$.

1.7   Idea of the construction of Φ_ρ. In section 4 we check that in the case when $\mathbf{G}$ is a split torus $\mathbf{T}$ the lifting and the distribution $\Phi_{\rho,T}$ on T are well defined for any representation ρ_T of $\mathbf{T}^\vee$, satisfying the condition from section 1.2. Moreover, one can construct an algebraic variety $\mathbf{Y}_{\rho,\mathbf{T}}$, a map $\mathbf{p_T} : \mathbf{Y}_{\rho,\mathbf{T}} \to \mathbf{T}$, a function $\mathbf{f_T}$ on $\mathbf{Y}_{\rho,\mathbf{T}}$ and a top-form $\omega_\mathbf{T} \in \Gamma(\Omega^{top}, \mathbf{Y}_{\rho,\mathbf{T}})$ such that the distribution $\Phi_{\rho,T}$ is equal to the push-forward $(p_T)_!(\psi(f_T)|\omega_T|)$ where $|\omega_T|$ is the measure on $Y_{\rho,T}$ associated with the form ω_T (cf. [We]). We say that $\mathbf{\Phi}_{\rho,\mathbf{T}} = (\mathbf{Y}_{\rho,\mathbf{T}}, \mathbf{f}_{\rho,\mathbf{T}}, \omega_{\rho,\mathbf{T}})$ is an *algebro-geometric distribution* representing the distribution $\Phi_{\rho,T}$ and say that the distribution $\Phi_{\rho,T}$ is a *materialization* of an algebro-geometric distribution $\mathbf{\Phi}_{\rho,\mathbf{T}}$.

Now let $\mathbf{G}$ be an arbitrary reductive group. We conjecture that the Ad-invariant distribution Φ_ρ is *stable* and therefore comes from a distribution $\tilde{\Phi}_\rho$ on the set $\mathbf{G}/\mathrm{Ad}(F)$ of F-rational points of the geometric quotient $\mathbf{G}/\mathbf{Ad} = \mathrm{Spec}(\mathcal{O}(\mathbf{G}))^\mathbf{G}$ of $\mathbf{G}$ by the adjoint action. Since $\mathbf{G}/\mathbf{Ad} = \mathbf{T}/W$ where $\mathbf{T} \subset \mathbf{G}$ is a maximal split torus and W is the Weyl group of $\mathbf{G}$ we can consider $\tilde{\Phi}_\rho$ as a distribution on the set $\mathbf{T}/W(F)$. We conjecture that the distribution $\tilde{\Phi}_\rho$ is a *materialization* of an algebro-geometric distribution $\mathbf{\Phi}_\rho$ which is a "descent" of $\mathbf{\Phi}_{\rho_\mathbf{T},\mathbf{T}}$ where $\rho_\mathbf{T}$ is the restriction of ρ on $\mathbf{T}^\vee$. More precisely we conjecture that there exists an action of the Weyl group W on $\mathbf{\Phi}_{\rho,T}$ such that the distribution $\tilde{\Phi}_\rho$ on the set $\mathbf{T}/W(F)$ is the *materialization* of the algebro-geometric distribution $\mathbf{\Phi}_\rho$ on $\mathbf{T}/W$ which is a "descent" of $\mathbf{\Phi}_{\rho_\mathbf{T},\mathbf{T}}$ (we shall explain the "descent" procedure carefully in section 7).

In section 8 we give explicit formulas for this action in a number of cases.

1.8   This paper is organized as follows. In section 2 we formulate the Langlands lifting conjecture and discuss several examples. In section 3 we define γ-functions and formulate their conjectural properties. In section 4 we give an explicit construction of the lifting for the case when our reductive group is a split torus $\mathbf{T}$ and we write an explicit formula for the corresponding distribution $\Phi_{\rho,T}$. Section 5 and section 6 are devoted respectively to the definition of the Schwartz space $\mathcal{S}_\rho$ and an analogue of the Poisson summation formula for the operator $\mathcal{F}_\rho$. In section 7 we develop the notion of an algebro-geometric distribution. Using this notion we reduce (conjecturally) the problem of constructing the distribution Φ_ρ to a purely algebraic question.

Section 8 is devoted to the discussion of this question, in the following case. Let m, n be two positive integers. Define the group

$$\mathbf{G}(m,n) = \big\{(A,B) \in \mathbf{GL}(m) \times \mathbf{GL}(n) \mid \det(A) = \det(B)\big\}. \quad (1.8)$$

The dual group $\mathbf{G}(m,n)^\vee$ is the quotient of $\mathbf{GL}(m,\mathbb{C}) \times \mathbf{GL}(n,\mathbb{C})$ by the subgroup consisting of all pairs of matrices $(t\mathrm{Id}_m, t^{-1}\mathrm{Id}_n)$ for $t \in \mathbb{C}^\times$. Hence the tensor product $\rho_m \otimes \rho_n$ of the standard representations of $\mathbf{GL}(m,\mathbb{C})$ and $\mathbf{GL}(n,\mathbb{C})$ descends to a representation ρ of $\mathbf{G}(m,n)^\vee$. The corresponding γ-function is essentially constructed in [JPS].

In section 8 we construct Φ_ρ in this case. Using [BeK] one can check that our definition gives rise to (almost) the same γ-function as the one defined in [JPS]. As an application we give conjectural formulas for the lifting problem discussed in section 1.6.

In section 9 we explain how to construct Φ_ρ in the case when the field F is finite. More precisely, in this case we construct a perverse sheaf $\boldsymbol{\Phi}_\rho$ on $\mathbf{G}$ such that Φ_ρ is obtained from it by taking traces of the Frobenius morphism in the fibers.

Finally, the Appendix (by V. Vologodsky) contains a proof of a result about non-Archimedean oscillating integrals used throughout the paper.

1.9 Notation. Everywhere, except for section 9, F denotes a local field. If F is non-Archimedean then we denote by $\mathcal{O}_F \subset F$ its ring of integers and by q the number of elements in the residue field of F.

If X is either a totally disconnected topological space or a C^∞-manifold we denote by $C(X)$ the space of $\mathbb{C}$-valued continuous functions on X. We denote by $C_c^\infty(X)$ the space of all compactly supported $\mathbb{C}$-valued functions on X which are

1) locally constant if X is totally disconnected;
2) C^∞ if X is a C^∞-manifold.

We shall denote algebraic varieties over F by boldface letters (e.g. $\mathbf{G}, \mathbf{X}, \dots$). The corresponding ordinary letters $(G, X, \dots)$ will denote the corresponding sets of F-points.

Let $\mathbf{X}$ be a smooth algebraic variety over F and $\omega \in \Gamma(\mathbf{X}, \Omega^{top}(\mathbf{X}))$, where Ω^{top} denotes the sheaf of differential forms of degree $\dim(\mathbf{X})$ on $\mathbf{X}$. According to [We] to ω one can associate a distribution $|\omega|$.

If G is a topological group (resp. a Lie group) then we denote by $\mathrm{Irr}(G)$ the set of isomorphism classes of irreducible algebraic (cf. [BerZ1]) representations of G (resp. the set of equivalence classes of continuous irreducible Banach representations of G with respect to infinitesimal equivalence, cf. [W]).

1.10   Acknowledgments.   We are grateful A. Berenstein, J. Bernstein, V. Drinfeld, D. Gaitsgory, M. Harris, M. Kontsevich, S. Rallis and V. Vologodsky for very helpful and interesting discussions on the subject. We are also grateful to Y. Flicker for numerous remarks about this paper.

2   Lifting

2.1   Unramified lifting.   Let F be a non-Archimedean local field and $\mathbf{G}$ a quasi-split group over $\mathcal{O}_F$, which can be split over an unramified extension of F of degree k (we assume that k is minimal with this property). Then the cyclic group $\mathbb{Z}/k\mathbb{Z}$ acts naturally on $\mathbf{G}^\vee$. We denote the action of the generator of $\mathbb{Z}/k\mathbb{Z}$ by $g \mapsto \kappa(g)$.

The group $\mathbf{G}^\vee$ acts on itself by $g : x \mapsto gx\kappa(g)^{-1}$. Let $S_{\mathbf{G}}$ denote the set of closed orbits on $\mathbf{G}^\vee$ with respect to the above action.

Let $\mathrm{Irr}_{un}(G)$ denote the set of isomorphism classes of unramified representations of G, i.e. the set of irreducible representations of G which have a non-zero $\mathbf{G}(\mathcal{O}_F)$-invariant vector. Then one has the natural identification $\mathrm{Irr}_{un}(G) \simeq S_{\mathbf{G}}$.

Let $\rho : \mathbf{G}^\vee \rtimes \mathbb{Z}/k\mathbb{Z} \to \mathbf{GL}(n)$ be a representation. The map $x \mapsto (x, \kappa)$ gives rise to a well-defined map from $S_{\mathbf{G}}$ to the set of semisimple conjugacy classes in $\mathbf{G}^\vee \rtimes \mathbb{Z}/k\mathbb{Z}$. If we compose this map with ρ we get a map from $S_{\mathbf{G}}$ to $S_{\mathbf{GL}(n)}$.

Consider now the composite map

$$\mathrm{Irr}_{un}(G) \to S_{\mathbf{G}} \to S_{\mathbf{GL}(n)} \to \mathrm{Irr}_{un}(\mathbf{GL}(n, F)). \tag{2.1}$$

We will denote this map by $l_{\rho, un}$ and call it *the unramified lifting*.

2.2   The global lifting.   Let K be a global field and $\mathbb{A}_K$ its ring of adeles. We denote by $\mathcal{P}(K)$ the set of places of K. For every $\mathfrak{p} \in \mathcal{P}(K)$ we let $K_{\mathfrak{p}}$ be the corresponding local completion of K.

Now let $\mathbf{G}$ be a quasi-split reductive group over K, which splits over a finite separable extension of K with Galois group Γ. As is well known any irreducible representation π of the group $\mathbf{G}(\mathbb{A}_K)$ can be uniquely written as a restricted tensor product $\otimes_{\mathfrak{p} \in \mathcal{P}(K)} \pi_{\mathfrak{p}}$ where $\pi_{\mathfrak{p}} \in \mathrm{Irr}(\mathbf{G}(K_{\mathfrak{p}}))$ and $\pi_{\mathfrak{p}} \in \mathrm{Irr}_{un}(\mathbf{G}(K_{\mathfrak{p}}))$ for almost all $\mathfrak{p}$.

We denote by $\mathrm{Aut}(\mathbf{G}, F)$ the set of irreducible automorphic representations of $\mathbf{G}(\mathbb{A}_K)$.

The group Γ acts naturally on $\mathbf{G}^\vee$. Hence we can consider the semidirect product $\mathbf{G}^\vee \rtimes \Gamma$. For almost all $\mathfrak{p} \in \mathcal{P}(K)$ the group $\mathbf{G}$ can be split over an unramified extension of $K_{\mathfrak{p}}$ of degree $k_{\mathfrak{p}}$ and in this case the group $\mathbf{G}^{\vee} \rtimes \mathbb{Z}/k_{\mathfrak{p}}\mathbb{Z}$ is naturally embedded in $\mathbf{G}^{\vee} \rtimes \Gamma$.

Hence every representation $\rho : \mathbf{G}^{\vee} \rtimes \Gamma \to \mathbf{GL}(n)$ defines a representation of $\mathbf{G}^{\vee} \rtimes \mathbb{Z}/k_{\mathfrak{p}}\mathbb{Z}$ for almost all $\mathfrak{p}$.

DEFINITION 2.3. *Let* $\pi = \otimes\pi_{\mathfrak{p}}$ *be an automorphic representation of* $\mathbf{G}(\mathbb{A}_K)$. *The global lifting* $l_{\rho}(\pi)$ *is an automorphic representation* $l_{\rho}(\pi) = \otimes l_{\rho}(\pi)_{\mathfrak{p}}$ *of the group* $\mathbf{GL}(n, \mathbb{A}_K)$ *such that for every place* $\mathfrak{p} \in \mathcal{P}(K)$ *such that* $\pi_{\mathfrak{p}}$ *is unramified one has* $l_{\rho}(\pi)_{\mathfrak{p}} = l_{\rho,un}(\pi_{\mathfrak{p}})$ *where* $l_{\rho,un}$ *is as in section 2.1* .

REMARK. It follows from the strong multiplicity one theorem for $\mathbf{GL}(n)$ (cf. [P]) that if $l_{\rho}(\pi)$ exists then it is unique.

CONJECTURE 2.4. 1. *For every* $\mathbf{G}, \pi$ *and* ρ *as above the lifting* $l_{\rho}(\pi)$ *exists.*

2. *Let* $\mathbf{G}$ *be a connected quasi-split reductive group over a local field* F, *which splits over a normal separable extension* L/F *such that* Gal$(L/F)=\Gamma$. *Then for every representation* $\rho : \mathbf{G}^{\vee} \rtimes \Gamma \to \mathbf{GL}(n, \mathbb{C})$ *there exists a map* $l_{\rho}^F : \mathrm{Irr}(G) \to \mathrm{Irr}(\mathbf{GL}(n, F))$ *such that for any global field* K, *a place* $\mathfrak{p} \in \mathcal{P}(\mathcal{K})$ *such that* $F = K_{\mathfrak{p}}$, *a quasi-split group* $\mathbf{G}'$ *over* K *as in section 2.2 such that* $\mathbf{G}'_{\mathfrak{p}} \simeq \mathbf{G}$ *and every* $\pi \in \mathrm{Aut}(\mathbf{G}', K), \pi = \otimes\pi_{\mathfrak{p}}$ *the* $\mathfrak{p}$-*th component of the automorphic representation* $l_{\rho}(\pi)$ *is equal to* $l_{\rho}^F(\pi_{\mathfrak{p}})$.

We say that ρ is liftable if Conjecture 2 holds for ρ.

REMARK. It is easy to see that the local lifting l_{ρ} is uniquely determined by the conditions of Conjecture 2.

EXAMPLE 2.5. One of the examples that we would like to consider in this paper is the following. Let E/F be a separable extension of F of degree n. Let $\mathbf{T}_E = \mathrm{Res}_{E/F} \, \mathbb{G}_{m,E}$ where $\mathrm{Res}_{E/F}$ denotes the functor of restriction of scalars. Let Γ_F denote the absolute Galois group of F. Γ_F acts naturally on the set of all embeddings of E into $\overline{F}$, which is a finite set with n elements. Hence we get a homomorphism $\alpha_E : \Gamma_F \to S_n$ which is defined uniquely up to S_n-conjugacy. Let $\Gamma = Im(\alpha_E)$. By the definition Γ is a subgroup of S_n which is also a quotient of Γ_F. Moreover, it is clear that the torus $\mathbf{T}_E$ splits over the Galois extension L/F where Gal$(L/F) = \Gamma$.

The dual torus $\mathbf{T}^{\vee}$ is isomorphic to $(\mathbb{C}^{\times})^n$ and the group Γ acts on it by the restriction of the standard action of S_n on $(\mathbb{C}^{\times})^n$ to Γ. Hence the standard embeddings of $(\mathbb{C}^{\times})^n$ and S_n to $\mathbf{GL}(n, \mathbb{C})$ give rise to a homomorphism $\rho_{\mathbf{T}} : \mathbf{T}^{\vee} \rtimes \Gamma \to \mathbf{GL}(n, \mathbb{C})$. One of the purposes of this paper is to give a conjectural construction of l_{ρ} in this case.

3   Gamma-functions and Bernstein's Center

3.1   Bernstein's center.   Let $\mathbf{G}$ be a reductive algebraic group over F, $G = \mathbf{G}(F)$. We denote by $\mathcal{M}(G)$ the category of smooth representations of G.

Recall that the *Bernstein center* $\mathcal{Z}(G)$ of G is the algebra of endomorphisms of the identity functor on the category $\mathcal{M}(G)$. It was shown by Bernstein that $\mathcal{Z}(G)$ is isomorphic to the direct product of algebras of the form $\mathcal{O}(\Omega)$ where Ω is a finite-dimensional irreducible complex algebraic variety. We shall denote the set of all Ω as above by $\mathcal{C}$. By $\mathrm{Spec}(\mathcal{Z}(G))$ we shall mean the disjoint union of all $\Omega \in \mathcal{C}$. It is shown in [BerD] that the natural map $\mathrm{Irr}(G) \to \mathrm{Spec}(\mathcal{Z}(G))$ sending every irreducible representation to its infinitesimal character is finite-to-one and moreover is generically one-to-one.

The following result (also due to J. Bernstein) gives a different interpretation of $\mathcal{Z}(G)$. Let Φ be a distribution on $\mathcal{G}$. We say that Φ is *essentially compact* if for every function $\phi \in \mathcal{H}(G)$ one has $\Phi * \phi \in \mathcal{H}(G)$. Given two essentially compact distributions Φ_1, Φ_2 one defines their convolution $\Phi_1 * \Phi_2$ by setting

$$\Phi_1 * \Phi_2(\phi) = \Phi_1(\Phi_2 * \phi). \tag{3.1}$$

It is easy to see that $\Phi_1 * \Phi_2$ is again an essentially compact distribution.

LEMMA 3.2.   *The algebra $\mathcal{Z}(G)$ is naturally isomorphic to the algebra of invariant essentially compact distributions.*

EXAMPLE 3.3.   Let $\mathbf{G} = \mathbf{GL}(n)$, $G = \mathbf{GL}(n, F)$. Fix any $a \in F^\times$ and a non-trivial additive character $\psi : F \to \mathbb{C}^\times$. Let $G_a = \{g \in G \mid \det(g) = a\}$. Then a choice of a Haar measure on $\mathbf{SL}(n, F)$ defines a measure on G_a for every $a \in F^\times$ (since G_a is a principal homogeneous space over $\mathbf{SL}(n, F)$). We denote this measure by $d_a g$. Define a distribution Φ_a on G supported on G_a by

$$\Phi_a(\phi) = \int_{G_a} \phi \psi(\mathrm{tr}(g)) d_a g \tag{3.2}$$

for any $\phi \in C_c^\infty(G)$.

One can verify that Φ_a is essentially compact.

3.4   Rational functions on $\mathrm{Irr}(G)$.   Let $\mathcal{K}(G)$ denote the "field of fractions" of $\mathcal{Z}(G)$, i.e. the associative algebra $\prod \mathcal{K}(\Omega)$, where Ω runs over connected components of $\mathrm{Spec}(\mathcal{Z}(G))$ as above and $\mathcal{K}(\Omega)$ is the field of fractions of $\mathcal{O}(\Omega)$. Since generically $\mathrm{Spec}(\mathcal{Z}(G))$ and $\mathrm{Irr}(G)$ are identified we shall refer to elements of $\mathcal{K}(G)$ as *rational functions* on $\mathrm{Irr}(G)$.

3.5 σ-regular central elements. For an associative algebra $\mathcal{A}$ we let $\mathcal{M}(G, \mathcal{A})$ denote the category of G-modules endowed with an action of $\mathcal{A}$ by endomorphisms of the G-module structure (we call it the category of $(G, \mathcal{A})$-modules).

Let $\mathcal{A} = \mathbb{C}[t, t^{-1}]$, $\mathcal{B} = \mathbb{C}((t))$. Assume that we are given an F-rational character $\sigma : \mathbf{G} \to \mathbb{G}_m$. Then we can define an action of G on $\mathcal{A}$ and $\mathcal{B}$ by requiring that

$$g \cdot t = |\sigma(g)| t \tag{3.3}$$

for every $g \in G$.

Define the functor $F_\mathcal{A} : \mathcal{M}(G, \mathcal{A}) \to \mathcal{M}(G, \mathcal{B})$ such that $F_\mathcal{A} : V \to V \otimes_\mathcal{A} \mathcal{B}$. Set $\mathcal{Z}_\sigma(G) = \mathrm{End}\, F_\mathcal{A}$ (the End is taken in the category of functors from $\mathcal{M}(G, \mathcal{A})$ to $\mathcal{M}(G, \mathcal{B})$).

We would like now to identify $\mathcal{Z}_\sigma(G)$ with a subspace of $\mathcal{K}(G)$. The assignment $\pi \mapsto \pi \otimes |\sigma|^{\log_q(z)}$ gives rise to an action of $\mathbb{C}^*$ on $\mathrm{Irr}(G)$ (here $z \in \mathbb{C}^*$). It is easy to see that this action descends to a well defined algebraic action of $\mathbb{C}^\times$ on $\mathrm{Spec}(\mathcal{Z}(G))$. For every $f \in \mathcal{K}(G)$ we can write

$$f(z \cdot x) = \sum f_n(x) z^n \tag{3.4}$$

for some $f_n \in \mathcal{K}(G)$.

LEMMA 3.6. *$\mathcal{Z}_\sigma(G)$ is naturally isomorphic to the space of all $f \in \mathcal{K}(G)$ such that $f_n \in \mathcal{Z}(G)$ for every $n \in \mathbb{Z}$.*

Lemma 3.6 explains the term "σ-regular".

3.7 σ-compact distributions. Now let Φ be a distribution on G and let $G_n = \sigma^{-1}(\pi^n \mathcal{O}^*)$. G_n is an open subset of G. For every $n \in \mathbb{Z}$ we define a new distribution Φ_n on G by $\Phi_n = \chi_n \Phi$, where χ_n is the characteristic function of G_n.

DEFINITION 3.8. *We say that Φ is σ-compact if the following two conditions are satisfied:*

1. *Φ_n is essentially compact for every $n \in \mathbb{Z}$.*

2. *For every $\phi \in \mathcal{H}(G)$ there exists a rational function $F_\phi : \mathbb{C} \to \mathcal{H}(G)$ such that for every $g \in G$ one has*

$$\sum_{n \in \mathbb{Z}} z^n (\Phi_n * \phi)(g) = F_\phi(z)(g). \tag{3.5}$$

The following lemma is straightforward from the definitions and Lemma 3.2.

LEMMA 3.9. *The space of all σ-compact distributions is naturally isomorphic to $\mathcal{Z}_\sigma(G)$.*

EXAMPLE 3.10. Let $G = \mathbf{GL}(n, F)$. Choose a non-trivial additive character $\psi : F \to \mathbb{C}^\times$ and set $\Phi_\psi(g) = \psi(\mathrm{tr}(g))|\det(g)|^n|dg|$, where $|dg|$ is a Haar measure on G. Let also $\sigma : \mathbf{GL}(n) \to \mathbb{G}_m$ be given by $\sigma(g) = \det(g)$.

PROPOSITION 3.11. Φ_ψ *is σ-compact.*

Proof. Let $d_a g = |\det|^n|dg|$ be the "additive" measure on G and let also $\mathcal{S}(M_n)$ denote the space of Schwartz-Bruhat functions on $n \times n$-matrices M_n over F. We shall regard $\mathcal{S}(M_n)$ as a subspace of $C(G)$. Consider the Fourier transform $\mathcal{F} : \mathcal{S}(M_n) \to \mathcal{S}(M_n)$ given by

$$\mathcal{F}(\phi)(y) = \int_G \phi(x)\psi(\mathrm{tr}(xy))d_a x. \tag{3.6}$$

It is easy to see that $\mathcal{F}(\phi) = (\Phi_\psi * {}^\iota\phi)|\det|^{-n}$, where ${}^\iota\phi(x) = \phi(x^{-1})$.

Let us now show that condition 1 holds. Without loss of generality we may assume that $\mathrm{supp}\,\phi \subset G_0$. Then $q^{-n}\Phi_n * \iota^\phi = \mathcal{F}(\phi)\chi_{-n}$. Since $\mathcal{F}(\phi) \in \mathcal{S}(M_n)$ it follows that $\mathcal{F}(\phi)\chi_{-n} \in \mathcal{H}(G)$. The verification of the second condition is left to the reader. □

We shall denote by γ_ψ the rational function on $\mathrm{Irr}(\mathbf{GL}(n, F))$ which corresponds to Φ_ψ.

3.12   Lifting and the distributions $\Phi_{\psi,\rho}$. Now let $\rho : \mathbf{G}^\vee \to \mathbf{GL}(n, \mathbb{C})$ be a finite-dimensional representation of the Langlands dual group $\mathbf{G}^\vee$. The local lifting conjecture predicts the existence of the natural map $l_\rho : \mathrm{Irr}(G) \to \mathrm{Irr}(\mathbf{GL}(n, F))$. Assuming that we know l_ρ we can try to define a rational function $\gamma_{\psi,\rho}$ by setting $\gamma_{\psi,\rho}(\pi) = \gamma_\psi(l_\rho(\pi))$. However, it might happen that the image of l_ρ lies entirely inside the singular set of l_ρ. To guarantee that this does not happen we introduce the following notion.

DEFINITION 3.13. *We say that ρ is admissible if the following conditions hold:*

1. *$\mathrm{Ker}(\rho)$ is connected.*
2. *There exists a character $\sigma : \mathbf{G} \to \mathbb{G}_m$ defined over F such that*

$$\rho \circ \sigma = \mathrm{Id} \tag{3.7}$$

where we regard σ as a cocharacter of the center $\mathbf{Z}(\mathbf{G}^\vee)$ of $\mathbf{G}^\vee$ and $I_n \in \mathbf{GL}(n, \mathbb{C})$ denotes the identity matrix.

It is easy to see that if we assume that ρ is an admissible and liftable then $\gamma_{\psi,\rho}$ satisfies the conditions of Lemma 3.6. Thus we can construct the corresponding σ-compact distribution $\Phi_{\psi,\rho,G}$. Our main task will be to try to construct the distribution $\Phi_{\psi,\rho,G}$ explicitly (we will sometimes drop the indices ψ and G, when it does not lead to confusion).

3.14 Compatibility with induction. Let $\mathbf{P} \subset \mathbf{G}$ be a parabolic subgroup of $\mathbf{G}$ defined over F and $\mathbf{M}$ the corresponding Levi subgroup. Let $\delta_{\mathbf{P}}$ be the determinant of the action of $\mathbf{M}$ on the Lie algebra of the unipotent radical $\mathbf{U_P}$ of $\mathbf{P}$.

For every $\phi \in \mathcal{H}(G)$ we can construct a new function $r_P(\phi)$ on G/U_P

$$r_P(\phi)(g) = \int_{U_P} \phi(u) |du| \tag{3.8}$$

Note that M acts on G/U_P on the right.

Now let ρ be a representation of $\mathbf{G}^{\vee}$ as above. Since $\mathbf{M}^{\vee}$ is naturally a subgroup of $\mathbf{G}^{\vee}$ we can regard ρ as a representation of $\mathbf{M}^{\vee}$. We conjecture that the following property holds for the distributions $\Phi_{\psi,\rho,G}$ and $\Phi_{\psi,\rho,M}$.

CONJECTURE 3.15.

$$r_P(\Phi_{\psi,\rho,G} * \phi) = r_P(\phi) * \left(\Phi_{\psi,\rho,M} |\delta_{\mathbf{P}}|^{-1/2}\right). \tag{3.9}$$

3.16   Assume that $\rho = \rho_1 \oplus \rho_2$ is a direct sum of two admissible representations, such that the corresponding character σ is the same for both representations. The following result can be easily deduced from the definitions.

LEMMA 3.17. *One has*

$$\Phi_{\psi,\rho} = \Phi_{\psi,\rho_1} * \Phi_{\psi,\rho_2}. \tag{3.10}$$

3.18 γ-functions determine the lifting. Let $\mathbf{G} = \mathbf{GL}(n-1) \times \mathbf{GL}(n)$ and the character σ be given by $\sigma(g_{n-1}, g_n) = \det(g_{n-1})$. In [GK, Ch. 7] a rational function $\gamma_{n-1,n}$ was defined on the subset of non-degenerate representations of the group G. One can show that $\gamma_{n-1,n}$ extends to a σ-regular central element for G. Hence there exists a σ-regular distribution $\Phi_{n-1,n}$ on the group $\mathbf{GL}(n-1,F) \times \mathbf{GL}(n,F)$ such that $\gamma_{n-1,n} = \gamma_{\Phi_{n-1,n}}$. In [JPS] the analogous function $\gamma_{m,n}$ on $\mathrm{Irr}(\mathbf{GL}(m,F) \times \mathbf{GL}(n,F))$ was defined. It is also easy to see that it comes from a σ-regular distribution $\Phi_{m,n}$ on the group $\mathbf{GL}(m,F) \times \mathbf{GL}(n,F)$.

As shown in [GK] for any non-degenerate representation π of $\mathbf{GL}(n,F)$ the rational function $\gamma_\pi \doteq \gamma_{n-1,n}(\pi, \star)$ on $\mathrm{Irr}\,\mathbf{GL}(n-1,F)$ determines uniquely the representation π. Moreover [GK] contains an explicit recipe for the construction of the representation π in terms of the function γ_π. It is clear from this recipe that we have to know the function γ_π only up to multiplication by a constant.

3.19   Now let $\mathbf{G}$ be arbitrary and $\rho : \mathbf{G}^\vee \to \mathbf{GL}(n, \mathbb{C})$ be an admissible representation. For any positive integer m we denote by $\rho_m :$ $\mathbf{G}^\vee \times \mathbf{GL}(m, \mathbb{C}) \to \mathbf{GL}(mn, \mathbb{C})$ the representation given by $\rho_m(g^\vee, r) = \rho(g^\vee) \otimes r$. Assume that both ρ and ρ_m are *liftable*. For any $\pi \in \mathrm{Irr}(G)$ we define rational functions $\gamma'_\pi, \gamma''_\pi$ on $\mathrm{Irr}\,\mathbf{GL}(m, F)$ by $\gamma'_\pi(\sigma) \doteq \gamma_{\rho_m}(\pi, \sigma)$, $\gamma''_\pi(\sigma) \doteq \gamma_{n,m}(l_\rho(\pi), \sigma)$.

PROPOSITION 3.20.   *For any $\pi \in \mathrm{Irr}(G)$ the two rational functions γ'_π and γ''_π on $\mathrm{Irr}\,\mathbf{GL}(m, F)$ coincide.*

    This proposition follows from the main result of [H].

COROLLARY 3.21.   *For any $\rho : \mathbf{G}^\vee \to \mathbf{GL}(n, \mathbb{C})$ such that the representation ρ_{n-1} is liftable, the lifting l_ρ is determined by the knowledge of the function $\gamma_{\rho_m}(\pi, \star)$ on $\mathrm{Irr}\,\mathbf{GL}(n - 1, F)$. Moreover it is sufficient to know the function $\gamma_{\rho_m}(\pi, \star)$ only up to a multiplication by a constant.*

4   The Case of Split Tori

In this section we assume that F is non-Archimedean.

4.1   Lifting for split tori.   Almost the only case when l_ρ is known for every representation ρ is the case when $\mathbf{G}$ is equal to a split torus $\mathbf{T}$. In this section we show how to carry out our program in this easy case.

    We start with an explicit description of the lifting l_ρ.

    We can write ρ in the form

$$\rho = \bigoplus_{i=1}^n \lambda_i \qquad (4.1)$$

where λ_i are characters of $\mathbf{T}^\vee$. By the definition of $\mathbf{T}^\vee$ we can regard λ_i as cocharacters $\lambda_i : \mathbb{G}_m \to \mathbf{T}$ of $\mathbf{T}$. Let $\Lambda = X_*(\mathbf{T})$ be the group of cocharacters of $\mathbf{T}$. We define $\mathrm{supp}(\rho) \subset \Lambda$ as the union of $\lambda_i, 1 \le i \le n$.

    Let θ be a character of T and $\chi_i = \theta \circ \lambda_i$ be the corresponding characters of $F^\times$. Then for every $i = 1, \ldots, n$ there exists $z_i \in \mathbb{R}_{>0}$ such that $|\chi_i(t)| = z_i^{v_F(t)}$ for every $t \in F^\times$.

    Let us assume that the ordering $\lambda_1, \ldots, \lambda_n$ is chosen in such a way that

$$|z_1| \le |z_2| \le \cdots \le |z_n|. \qquad (4.2)$$

Let B_n be the subgroup of upper triangular matrices in $\mathbf{GL}(n, F)$. We can define the character $\chi : B_n \to \mathbb{C}^\times$ by setting

$$\chi(b) = \prod_{i=1}^{n} \chi_i(b_{ii}) \tag{4.3}$$

where b_{ii} are the diagonal entries of b.

Consider the unitary induced representation $i_{B_n}^{GL(n,F)} \chi$. It is well known (cf. [BerZ2]) that this representation has unique irreducible quotient. We define this quotient to be $l_\rho(\theta)$. As follows from the theory of Eisenstein series this definition satisfies the conditions of Conjecture 2.

4.2 γ-functions for split tori. We will give now an explicit description of the distribution $\Phi_{\rho,T}$. Let $\mathbf{T}_\rho^\vee = \mathbb{G}_m^n$ be the corresponding torus in $\mathbf{GL}(n)$. Thus we get a homomorphism $\mathbf{p}_\rho^\vee : \mathbf{T}^\vee \to \mathbf{T}_\rho^\vee$. Let $\mathbf{T}_\rho$ be the split torus over F dual to $\mathbf{T}_\rho^\vee$ (thus $T_\rho = (F^\times)^n$). We then get an F-rational map $\mathbf{p}_\rho : \mathbf{T}_\rho \to \mathbf{T}$.

Consider a top degree differential form $\omega_\rho := dt_1 \ldots dt_n$ on $\mathbf{T}_\rho$, and a function $\mathbf{f}_\rho : \mathbf{T}_\rho \to \mathbb{A}^1$ where $\mathbf{f}_\rho(t_1, \ldots, t_n) = t_1 + \ldots + t_n$.

For every $r \in \mathbb{R}_{\geq 0}$ we denote by $T_\rho(r)$ the set $\{t \in T_\rho \mid |f_\rho(t)| \leq r\}$.

PROPOSITION 4.3.   1. *For every $r \in \mathbb{R}_{\geq 0}$ and for every open compact subset $C \subset T$ the integral*

$$\Phi_\rho^r(C) = \int_{t \in p_\rho^{-1}(C) \cap T_\rho(r)} \psi(f_\rho(t)) |\omega_\rho| \tag{4.4}$$

is absolutely convergent.

2. *For every C as above the limit*

$$\lim_{r \to \infty} \Phi_\rho^r(C) \tag{4.5}$$

exists. We denote this limit by $\Phi_\rho(C)$. We also denote by Φ_ρ the corresponding distribution on T.

3. *For every character θ of T there exists $s_0(\theta) \in \mathbb{R}$ such that for every $s \in \mathbb{C}$ such that $\mathrm{Re}(s) > s_0(\theta)$ the convolution $\Phi_\rho * \theta|\sigma|^s$ is absolutely convergent and*

$$\Phi_\rho * \theta|\sigma|^s = \gamma_{\rho,T}\theta|\sigma|^s . \tag{4.6}$$

4. *Φ_ρ is σ-regular.*

Our next goal is to describe explicitly the space $\mathcal{S}_\rho$ in the case of a torus.

4.4   The Fourier transform $\mathcal{F}_\rho$.   We assume now that the representation ρ is faithful. This assumption will be kept until the end of section 8.

Consider the map $\mathcal{F}_\rho : C_c^\infty(T) \to C^\infty(T)$ given by

$$\phi \mapsto |\sigma|^{-1} \Phi_\rho * {}^\iota\phi \tag{4.7}$$

where ${}^\iota\phi(x) = \phi(x^{-1})$.

Let also $L_\rho^2(T) = L^2(T, |\sigma|d^*t)$. The following lemma is straightforward.

LEMMA 4.5. $\mathcal{F}_\rho$ extends to a unitary automorphism of $L_\rho^2(T)$.

We now set

$$\mathcal{S}_\rho(T) = C_c^\infty(T) + \mathcal{F}_\rho(C_c^\infty(T)) \subset L_\rho^2(T) \tag{4.8}$$

(the reader should compare this definition with the definition of a Schwartz space $\mathcal{S}(X)$ in [BrK]).

4.6   The function $\mathcal{C}_\rho$.   Let us give an example of a function in $\mathcal{S}_\rho(T)$. Let $T_0 \subset T$ be the maximal compact subgroup of T. Then we can identify the quotient T/T_0 with the lattice Λ of cocharacters of $\mathbf{T}$.

Let

$$\rho = \bigoplus_{i=1}^n \lambda_i \tag{4.9}$$

be an admissible representation of $\mathbf{T}^\vee$. We denote the set of all λ_i above by $\mathrm{supp}(\rho)$.

Let us define a T_0-invariant function $\mathcal{S}_\rho$ on T (i.e. a function on $T/T_0 = \Lambda$) by setting

$$\mathcal{C}_\rho(\mu) = \#\{(a_1, \ldots, a_n) \in \mathbb{Z}_+^n \mid a_1\lambda_1 + \ldots + a_n\lambda_n = \mu\}. \tag{4.10}$$

LEMMA 4.7. $\mathcal{C}_\rho \in \mathcal{S}_\rho(T)$.

4.8   The semigroup $\overline{\mathbf{T}}_\rho$.   We now want to exhibit certain locality properties of the space $\mathcal{S}_\rho(T)$. For this we first have to introduce additional notation. Let Λ_ρ be the sub-semigroup of Λ generated by $\lambda \in \mathrm{supp}(\rho)$, $\mathcal{O}_\rho$ be the group algebra of Λ_ρ over F and $\overline{\mathbf{T}}_\rho := \mathrm{Spec}\,\mathcal{O}_\rho$. It is easy to see that $\overline{\mathbf{T}}_\rho$ is an F-semi-group which contains $\mathbf{T}$ as an open dense subgroup.

PROPOSITION 4.9.   Let ϕ be a function on T. Then $\phi \in \mathcal{S}_\rho(T)$ if and only if ϕ satisfies the following conditions:

1. The closure of $\mathrm{supp}(\phi)$ in $\overline{T}_\rho$ is compact.
2. For every $x \in \overline{T}_\rho$ there exists a neighbourhood U_x of x in $\overline{T}_\rho$ and a function $\phi' \in \mathcal{S}_\rho(T)$ such that

$$\phi|_{U_x} = \phi'|_{U_x}. \tag{4.11}$$

Proposition 4.9 says that one can determine whether a function ϕ lies in $\mathcal{S}_\rho(T)$ by looking at its local behaviour around points of $\overline{T}_\rho$. We will discuss the notion of a local space of functions in more detail in section 5.

For every $x \in \overline{T}_\rho$ we denote by $\mathbf{Z}_x$ the stabilizer of x in $\mathbf{T} \times \mathbf{T}$. If $x \in \overline{T}_\rho$ we define an Z_x-module $\mathcal{S}_{\rho,x}$ as the quotient $\mathcal{S}_\rho(G)/\mathcal{S}^0_{\rho,x}(T)$ where $\mathcal{S}^0_{\rho,x}(T)$ the space of all functions from $\mathcal{S}_\rho(G)$, which vanish in some neighbourhood of x.

LEMMA 4.10. *There exists a unique Z_x-invariant functional $\varepsilon_x : \mathcal{S}_{\rho,x} \to \mathbb{C}$ such that $\varepsilon_x(C_\rho) = 1$.*

Thus given $\phi \in \mathcal{S}_\rho(T)$ we can produce a function ϕ^ε on $\overline{T}$ setting $\phi^\varepsilon(x) = \varepsilon_x(\phi)$. However, the function ϕ^ε is not locally constant in general.

4.11 Mellin transform. We are now going to give yet another (equivalent) definition of $\mathcal{S}_\rho(T)$ using the *Mellin transform* on T. Let $X(T) = \mathrm{Hom}(T, \mathbb{C}^\times)$. Then $X(T)$ has the natural structure of an algebraic variety, whose irreducible components are parametrised by $\mathrm{Hom}(T_0, \mathbb{C}^\times) = \mathrm{Hom}(T_0, S^1)$. Choose a Haar measure d^*t on T.

Let Λ be the lattice of cocharacters of $\mathbf{T}$. As before one can identify Λ with T/T_0 where $T_0 \subset T$ is the maximal compact subgroup of T. We denote by $X^{\mathrm{un}}(T) \subset X(T)$ the subset of unitary characters of T.

We denote by $X_0(T)$ the component of $X(T)$, consisting of characters whose restriction to T_0 is trivial. One has the natural identification $X_0(T) \simeq \mathbf{T}^\vee$ (where $\mathbf{T}^\vee$ denotes the dual torus to $\mathbf{T}$ over $\mathbb{C}$). For any $\lambda \in \mathrm{supp}(\rho)$ and choose a lift λ' of λ to T. Then we denote by $\mathbf{p}_\lambda$ the regular function on $X(T)$ defined by

$$\mathbf{p}_\lambda(\chi) = \begin{cases} \chi(\lambda'), & \text{if } \chi \circ \lambda|_{\mathcal{O}^*} = 1, \\ 0, & \text{otherwise}. \end{cases}$$

Clearly, this definition does not depend on the choice of λ'.

For $\phi \in C_c^\infty(T)$ we define the Mellin transform $M(\phi)$ of ϕ as a function on $X(T)$ given by

$$M(\phi)(\chi) = \int_T \phi(t)\chi(t)d^*t. \tag{4.12}$$

The following lemma is standard.

LEMMA 4.12. *M defines an isomorphism between $C_c^\infty(T)$ and $\mathcal{O}(X(T))$ which extends to an isomorphism between $L^2(T)$ and $L^2(X^{\mathrm{un}}(T))$.*

Let ρ be an admissible representation of $\mathbf{T}^\vee$ and $\mathcal{F}_{\psi,\rho}$ denote the operator defined above. The following result is proved by a straightforward calculation.

Theorem 4.13.   A function $\phi \in L^2_\rho(T)$ lies in $\mathcal{S}_\rho(T)$ if and only if $\prod_{1 \le i \le n}(\mathbf{p}_{\lambda_i} - 1)M(\phi)$ is a regular function on $X(T)$.

Let us now discuss the connection of the space $\mathcal{S}_\rho(T)$ with the corresponding local L-functions.

LEMMA 4.14.   1. For every $\phi \in \mathcal{S}_\rho(T)$ and every character χ of T the integral

$$Z(\phi, \chi, s) = \int_T \phi(t)\chi(g)|\sigma(t)|^s d^*t \qquad (4.13)$$

is absolutely convergent for $\Re e(s) \gg 0$.

2. $Z(\phi, m, s)$ has a meromorphic continuation to the whole of $\mathbb{C}$ and defines a rational function of q^s.

3. The space of all $Z(\phi, m, s)$ as above is a finitely generated non-zero fractional ideal of the ring $\mathbb{C}[q^s, q^{-s}]$. We shall denote this ideal by J_χ. Let also $L_\rho(s, \chi)$ to be the unique generator of J_χ of the form $P(q^{-s})^{-1}$, where P is a polynomial such that $P(0) = 1$.

4. $L(\chi, s) = \prod_{1 \le i \le n} L(\chi \circ \lambda_i, s)$ where $L(\chi \circ \lambda_i, s)$ denotes the corresponding Tate's L-function (cf. [T]).

5   Local Spaces of Functions

We now want to generalize some constructions of the preceding section to the case of arbitrary reductive group G. Let us remember that we assume now that the representation ρ is admissible and faithful.

In this section we assume that we are given an admissible representation ρ of $\mathbf{G}^\vee$ and the corresponding distribution $\Phi_{\psi, \rho}$ which satisfies the assumptions of section 3.

5.1   Saturations.   Let $\overline{X}$ be a topological set and $X \subset \overline{X}$ an open dense subset. Given a space L of functions $\overline{X}$ we say that L is local if there exists a sheaf $\mathcal{L}$ on $\overline{X}$ and an embedding $\mathcal{L} \hookrightarrow i_*(C(X))$ such that $L = \Gamma_c(\overline{X}, \mathcal{L})$ where $C(X)$ is the sheaf of continuous functions on X.

It is easy to see that when the space X is totally disconnected then locality of L is equivalent to the following two conditions:

1) $C_c(X) \subset L \subset C(X)$
2) L is closed under multiplication by elements of $C(\overline{X})$.

Given a subspace $V \subset C(X)$ we denote by $\mathcal{P}_V$ the presheaf on $\overline{X}$ such that $\Gamma(U, \mathcal{P}_V)$ is a subspace of $\phi \in C(U \cap X)$ consisting of all $\phi \in C(U \cap X)$

such that there exists $v \in V$ for which $v|_{U \cap X} = \phi$. Let $\mathcal{L}_V$ be the associated sheaf. It is clear that we have the natural embedding $\mathcal{L}_V \hookrightarrow i_*(C(X))$ We set $\overline{V} = \Gamma_c(\overline{X}, \mathcal{L}_V)$ and call $\overline{V}$ the *saturation* of V with respect to $\overline{X}$.

5.2 C^∞-version. Now let $\overline{X}$ be a closed subset of a C^∞-manifold Y. Let $d(\cdot, \cdot)$ be a metric on $\overline{X}$ coming from a Riemannian metric on Y.

Let $i : X \hookrightarrow \overline{X}$ be an open dense embedding of C^∞-manifold X into $\overline{X}$. Let also $\mathcal{L} \subset i_*C^\infty(X)$ be a subsheaf. We denote by $\Gamma_c^{as}(\overline{X}, \mathcal{L})$ the space of C^∞-functions ϕ on X such that for any $x \in \overline{X}$ and any $N > 0$ there exists an open neighbourhood U of x and a section $l \in \Gamma(U, \mathcal{L})$ such that for any $y \in U$ one has
$$|\phi(y) - l(y)| \leq Cd(x, y)^N$$
where C is a constant (i.e. C does not depend on y).

EXAMPLE.   Let $X = \overline{X} = (0, 1)$ and let $\mathcal{L}$ be the sheaf of polynomial functions on X. Then $\Gamma_c^{as}(X, \mathcal{L}) = C_c^\infty(0, 1)$.

Given a subset V of $C^\infty(X)$ we can define its saturation $\overline{V}$ as $\Gamma_c^{as}(\overline{X}, \mathcal{L}_V)$ where $\mathcal{L}_V$ is defined as in section 5.1.

5.3 The Fourier transform $\mathcal{F}_\rho$. In this subsection we define a certain "twisted" analogue of the Fourier transform in the space of functions of G attached to every ρ as above. As before we denote by $L_\rho^2(G)$ the space of L^2-functions on G with respect to the measure $|\sigma|^{l+1}|dg|$ where l is the semi-simple rank of $\mathbf{G}$ and $|dg|$ is a Haar measure on G.

For every $\phi \in C_c^\infty$ we define a new function $\mathcal{F}_\rho(\phi)$
$$\mathcal{F}_\rho(\phi) = |\sigma|^{-l}(\Phi_{\psi,\rho} *{}^\iota\phi) \tag{5.1}$$
where ${}^\iota\phi(g) = \phi(g^{-1})$. In what follows we will assume the validity of the following conjecture.

CONJECTURE 5.4. $\mathcal{F}_\rho$ extends to a unitary operator on the space $L_\rho^2(G)$ and $\mathcal{S}_\rho$ is $\mathcal{F}_\rho$-invariant.

EXAMPLE.   Let $G = \mathbf{GL}(n, F)$ and ρ be the standard representation of $\mathbf{G}^\vee = \mathbf{GL}(n)$. In this case $L_\rho^2(G)$ is the same as $L^2(M_n)$ with respect to the additive measure on M_n and $\mathcal{F}_\rho$ coincides with the Fourier transform on M_n (where we identify M_n with the dual vector space by means of the form $\langle A, B \rangle = \operatorname{tr}(AB)$).

5.5 The semigroup $\overline{\mathbf{G}}_\rho$. We would like to embed $\mathbf{G}$ as an open subset in a larger affine variety $\overline{\mathbf{G}}_\rho$ in a $\mathbf{G} \times \mathbf{G}$-equivariant way.

The variety $\overline{\mathbf{G}}_\rho$ is an algebraic semigroup, containing $\mathbf{G}$ as an open dense subgroup. It is well known (cf. [V]) that in order to define such a semigroup one needs to exhibit a subcategory of the category of finite-dimensional **G**-modules, closed under subquotients, extensions and tensor products.

We will define now the category of ρ-*positive* representations of **G** which will satisfy the above properties. Let $\lambda : \mathbf{T}^\vee \to \mathbb{G}_m$ be a character, which has non-zero multiplicity in ρ. We can regard λ as a cocharacter of **T**. We say that a **G**-module **V** is ρ-positive if for every λ as above the representation $\rho \circ \lambda$ of $\mathbb{G}_m$ is isomorphic to a direct sum of characters of the form $t \mapsto t^i$ for $i \geq 0$. It is clear that the category of ρ-positive representations satisfies the properties discussed above and thus defines a semi-group $\overline{\mathbf{G}}_\rho$.

5.6   The space $\mathcal{S}_\rho(G)$.   Let

$$V_\rho = C_c^\infty(G) + \mathcal{F}_\rho(C_c^\infty(G)) \subset L_\rho^2(G). \tag{5.2}$$

Unlike in the case of the torus, one can show that the space V_ρ is almost never local (this is not so for example when $\mathbf{G} = \mathbf{GL}(2)$ and ρ is the standard representation of $\mathbf{G}^\vee \simeq \mathbf{GL}(2,\mathbb{C})$). We define $\mathcal{S}_\rho(G)$ to be the saturation of V_ρ (the definition makes sense both for Archimedean and non-Archimedean F).

5.7   The function $\mathcal{C}_\rho$.   Let us now exhibit a certain explicit element in $\mathcal{S}_\rho(T)$. In what follows we choose a square root $q^{1/2}$ of q.

5.7.1   The Satake transform.   Let $K = \mathbf{G}(\mathcal{O})$ and let $\mathcal{H}_K = C_c^\infty(K \backslash G / K)$ be the corresponding Hecke algebra. Recall that there exists the natural isomorphism $S : \mathcal{H}_K \simeq \mathcal{O}(\mathbf{T}^\vee)^W \simeq \mathcal{O}(\mathbf{G}^\vee)^{\mathbf{G}^\vee}$, where $\mathcal{O}(\mathbf{T}^\vee)^W$ is the algebra of W-invariant regular functions on $\mathbf{T}^\vee$. This isomorphism is characterized as follows. One can identify $\mathbf{T}^\vee$ with the group of unramified characters of the torus T. For every $\lambda \in \mathbf{T}^\vee$ let $i(\lambda)$ denote the corresponding representation of G obtained by normalized induction of the character λ from B to G. Let $v_\lambda \in i(\lambda)$ be the (unique up to a constant) non-zero K-invariant vector in $i(\lambda)$. Then for every $\phi \in C_c^\infty(K \backslash G / K)$ one has

$$(\phi|dg|) \cdot v_\lambda = S(\phi)(\lambda) v_\lambda. \tag{5.3}$$

Let us also define the twisted Satake isomorphism $\widetilde{S}$ by requiring that

$$\widetilde{S}(\mathrm{ch}_\theta) = (-q^{1/2})^{\langle \theta, 2\delta_{\mathbf{G}} \rangle} S(\mathrm{ch}_\theta)$$

where ch_θ denotes the character of the irreducible representation of $\mathbf{G}^\vee$ with highest weight θ and $2\delta_{\mathbf{G}}$ is the sum of all positive roots of **G**.

5.7.2   The function $\mathcal{C}_\rho$.   Let ρ be as above.

Set

$$f_{i,\rho}(\lambda) = \mathrm{tr}\left(\lambda, \mathrm{Sym}^i \rho\right) \tag{5.4}$$

for every $i \in \mathbb{Z}_+$.

We now define

$$\mathcal{C}_\rho = \sum_{i=0}^{\infty} \widetilde{S}^{-1}(f_{i,\rho}) . \tag{5.5}$$

It is easy to see that the non-degeneracy assumptions on ρ imply that $\mathcal{C}_\rho$ is a well-defined locally constant function on G, which however does not have compact support.

LEMMA 5.8. $\mathcal{C}_\rho \in \mathcal{S}(G)$ and $\mathcal{F}_\rho(\mathcal{C}) = \mathcal{C}_\rho$.

The lemma follows easily from Conjecture 3.15 of section 3.

CONJECTURE 5.9.   There exists a unique up to a constant Z_x-invariant functional $\varepsilon_x : \mathcal{S}_{\rho,x} \to \mathbb{C}$.

EXAMPLE.   Let $\mathbf{G} = \mathbf{GL}(n)$ and take ρ to be the standard representation of $\mathbf{G}^\vee \simeq \mathbf{GL}(n)$. Then $\overline{\mathbf{G}}_\rho = \mathbf{M}_n$ is the semi-group of $n \times n$-matrices. Also one has $\mathcal{S}_\rho(G) = \mathcal{S}(M_n)$, the space of Schwartz-Bruhat functions on M_n. In this case $\mathcal{S}_{\rho,x}$ is one-dimensional for every x and the functional ε_x is equal to the evaluation of a locally constant function at x.

5.10   Local L-functions.   Here we give a definition of the local L-factor for every admissible non-singular representation ρ.

Let π be an irreducible representation of $\mathbf{G}$. We denote by $M(\pi) \subset C^\infty(G)$ the space of all matrix coefficients of π.

CONJECTURE 5.11.   1. For every $\phi \in \mathcal{S}_\rho(G)$ and every $m \in M(\pi)$ the integral

$$Z(\phi, m, s) = \int_G \phi(g)m(g)|\sigma(g)|^s |dg| \tag{5.6}$$

is absolutely convergent for $\Re e(s) \gg 0$.

2. $Z(\phi, m, s)$ has a meromorphic continuation to the whole of $\mathbb{C}$ and defines a rational function of q^s.

2. The space of all $Z(\phi, m, s)$ as above is a finitely generated non-zero fractional ideal of the ring $\mathbb{C}[q^s, q^{-s}]$. We shall denote this ideal by J_π.

We now define $L_\rho(s, \pi)$ to be the unique generator of J_π of the form $P(q^{-s})^{-1}$, where P is a polynomial such that $P(0) = 1$.

6   The Poisson Summation Formula

As before, let K be a global field and $\mathcal{P}(K)$ be the set of places of K. We also denote by $\mathbb{A}_K$ the adele ring of K.

Let $\mathbf{G}$ be a split reductive algebraic group over K and $\rho : \mathbf{G}^\vee \to \mathbf{GL}(n, \mathbb{C})$ be as before. For every place $\mathfrak{p} \in \mathcal{P}(K)$ we can consider the Schwartz space $\mathcal{S}_{\rho.\mathfrak{p}} = \mathcal{S}_\rho(\mathbf{G}(F_\mathfrak{p}))$ with the distinguished function $C_{\rho,\mathfrak{p}}$ in it. We define the space $\mathcal{S}_\rho(\mathbf{G}(\mathbb{A}_K))$ as the restricted tensor product of the spaces $\mathcal{S}_{\rho,\mathfrak{p}}$ with respect to the functions $C_{\rho,\mathfrak{p}}$.

Choose now a non-trivial character $\psi : \mathbb{A}_K \to \mathbb{C}^\times$ such that $\psi|_F$ is trivial. Then ψ defines an additive character $\psi_\mathfrak{p}$ of $K_\mathfrak{p}$ for every $\mathfrak{p}$. The Fourier transform

$$\mathcal{F}_{\rho,\psi} = \prod_{\mathfrak{p} \in \mathcal{P}(F)} \mathcal{F}_{\rho,\mathfrak{p}} \qquad (6.1)$$

acts on the space $\mathcal{S}_\rho(\mathbf{G}(\mathbb{A}_K))$.

Every $\phi \in \mathcal{S}_\rho(\mathbf{G}(\mathbb{A}_K))$ gives rise to a function on $\mathbf{G}(\mathbb{A}_K)$.

CONJECTURE 6.1.   a)   *There exists a* $\mathbf{G}(K)$-*invariant functional* $\varepsilon :$ $\mathcal{S}_\rho(\mathbf{G}(\mathbb{A}_K)) \to \mathbb{C}$ *such that*

1) $\varepsilon(\phi) = \varepsilon(\mathcal{F}_\rho(\phi))$ *for any* $\phi \in \mathcal{S}_\rho(\mathbf{G}(\mathbb{A}_K))$.
2) *Let* $\phi = \prod \phi_\mathfrak{p} \in \mathcal{S}_\rho(\mathbf{G}(\mathbb{A}_K))$. *Assume that there exists a place* $\mathfrak{p}_0 \in \mathcal{P}(K)$ *such that* $\phi_{\mathfrak{p}_0} \in \mathcal{H}(G)$. *Then*

$$\varepsilon(\phi) = \sum_{g \in \mathbf{G}(K)} \phi(g) \qquad (6.2)$$

b)   *The functional* ε *is supported at* $\overline{\mathbf{G}}_\rho(K)$. *In other words assume that we are given* $\phi \in \mathcal{S}_\rho(\mathbf{G}(\mathbb{A}_K))$ *such that for every* $g \in \overline{\mathbf{G}}_\rho(K)$ *there exists a neigbourhood* U_g *of* g *such that* $\phi(x) = 0$ *for every* $x \in U_g \cap \mathbf{G}(K)$. *Then* $\varepsilon(\phi) = 0$.

When $\mathbf{G} = \mathbf{GL}(n)$ and ρ is the standard representation, the statement of Conjecture 6.1 is the standard Poisson summation formula. In this case ϕ is a smooth function on the space $\mathbf{M}_n(\mathbb{A}_K)$ of adelic $n \times n$ matrices and one has

$$\varepsilon(\phi) = \sum_{g \in \mathbf{M}_n(K)} \phi(g) . \qquad (6.3)$$

We do not know any explicit formula for ε in general.

Now let π be an irreducible automorphic representation of $\mathbf{G}(\mathbb{A}_K)$, and

$\pi_{\mathfrak{p}}$ its component at $\mathfrak{p} \in \mathcal{P}(S)$. One can consider the global L-function

$$L_\rho(\pi, s) = \prod_{\mathfrak{p} \in \mathcal{P}(S)} L_\rho(\pi_{\mathfrak{p}}, s) \,. \tag{6.4}$$

Using the arguments of [GoJ] is easy to see that the validity of the above conjectures implies the meromorphic continuation and functional equation of the corresponding automorphic L-functions.

7 "Algebraic" Integrals over Local Fields

In the next two sections we assume that the local field F has characteristic 0.

Let $\mathbf{X}$ be an algebraic variety over F of dimension d, $X = \mathbf{X}(F)$. Recall that (cf. [We]) any $\omega \in \Gamma(\mathbf{X}, \Omega^{\mathrm{top}})$ defined over F defines a measure $|\omega|$ on X. For a proper morphism $f : \mathbf{X} \to \mathbf{Y}$ and ω as above we denote by $f_!(|\omega|)$ the corresponding push-forward measure on Y.

7.1 Locally integrable functions. We say that a function $\phi : X \to \mathbb{C}$ is locally integrable for every point $x \in X$ and any top-form $\omega \in \Gamma(\mathbf{U}, \Omega^d)$ defined in a compact neighbourhood $\mathbf{U} \subset \mathbf{X}$ of x the integral

$$\int_U |\phi(y)||\omega(y)| \tag{7.1}$$

is convergent.

Theorem 7.2 (V. Vologodsky). *Let* $\mathbf{f} : \mathbf{X} \to \mathbb{A}^1$ *be a proper morphism and let* $\omega \in \Gamma(\mathbf{X}, \Omega^d)$. *Suppose that both are defined over* F. *Then there exist an open compact subgroup* K *of* $\mathcal{O}_F^\times$ *and a* K-*invariant function* $\alpha(x)$ *such that distribution* $f_!(|\omega|)$ *on* F *is equal to* $\alpha(x)dx$ *for* $|x| \gg 0$.

Theorem 7.2 is proved in the Appendix.

COROLLARY 7.3. *Let* $\psi : F \to \mathbb{C}^\times$ *be a non-trivial character. Then the integral*

$$\int_X \psi(f)|\omega| := \int_F f_!(|\omega|) \cdot \psi := \lim_{N \to \infty} \int_{|x| < N} f_!(|\omega|)\psi \tag{7.2}$$

is convergent.

7.4 Algebraic-geometric distributions. Let as before $\mathbf{X}$ be an algebraic variety over F.

DEFINITION 7.5. *By an algebro-geometric distribution on* $\mathbf{X}$ *we mean a quadruple* $\Phi = (\mathbf{Y}, \mathbf{f}, \mathbf{p}, \omega)$ *where*

1) $\mathbf{Y}$ *is a smooth algebraic variety over* F;

2) $\mathbf{p} : \mathbf{Y} \to \mathbf{X}$, $\mathbf{f} : \mathbf{Y} \to \mathbb{A}^1$ are morphisms defined over F;
3) $\omega \in \Gamma(\mathbf{Y}, \Omega_{\mathbf{Y}}^{top})$.

By an isomorphism between two algebro-geometric distributions $\Phi = (\mathbf{Y}, \mathbf{f}, \mathbf{p}, \omega)$ and $\Phi' = (\mathbf{Y}', \mathbf{f}', \mathbf{p}', \omega')$ we shall mean a *birational* isomorphism between $\mathbf{Y}$ and $\mathbf{Y}'$ preserving all the structure.

Given an algebro-geometric distribution Φ we can try to define a usual distribution $\Phi = \Phi_\psi$ on X by writing

$$\Phi = p_!\big(\psi(f)|\omega|\big) = \lim_{N \to \infty} p_!\big(\psi(f)|\omega|\big)\big|_{\{y \in Y \| f(y)| \leq N\}}. \qquad (7.3)$$

Corollary 7.3 guarantees that in a number of cases (7.3) makes sense. For example one can show that the following result holds.

PROPOSITION 7.6. *Let an algebro-geometric distribution Φ as above be given. Assume that there exists an open dominant embedding $\mathbf{Y} \hookrightarrow \widetilde{\mathbf{Y}}$ such that $\widetilde{\mathbf{Y}}$ is smooth and such that $\mathbf{p}, \mathbf{f}$ and ω extend to $\widetilde{\mathbf{Y}}$. Assume furthermore that the morphism $\mathbf{p} \times \mathbf{f} : \widetilde{\mathbf{Y}} \to \mathbf{X} \times \mathbb{A}^1$ is proper. Then Φ given by (7.3) is a well-defined distribution on X.*

We say that an algebro-geometric distribution Φ is *finite* if it satisfies the conditions of Proposition 7.6. Also, we call Φ *weakly finite* if it is finite over the generic point of $\mathbf{X}$. We also say that Φ is *analytically finite* if it is weakly finite and for any finite extension E/F the corresponding distribution Φ defined *a priori* on a dense open subset of $\mathbf{X}(E)$ is locally L^1 on the whole of $\mathbf{X}(E)$. In this case we call Φ the *materialization* of Φ.

LEMMA 7.7. *Let $\mathbf{X} = \mathbf{T}$ be a split torus over F and let ρ be an admissible representation of $\mathbf{T}^\vee$. Let $\Phi_{\rho,\mathbf{T}} = (\mathbf{T}_\rho, \mathbf{f}_\rho, \mathbf{p}_\rho, \omega_\rho)$ be given by the formulas of section 4.2. Then $\Phi_{\rho,\mathbf{T}}$ is analytically finite and the distribution $\Phi_{\rho,\mathbf{T}}$ is the materialization of $\Phi_{\rho,\mathbf{T}}$.*

7.8 Reduction of algebro-geometric distributions. Let $\Phi = (\mathbf{Y}, \mathbf{f}, \mathbf{p}, \omega)$ be an algebro-geometric distribution and let $\mathbf{V}$ be a vector-group defined over F. We say that $\mathbf{V}$ acts on Φ if we are given a free action of $\mathbf{V}$ on $\mathbf{Y}$ such that

1) $\mathbf{p}$ and ω are $\mathbf{V}$-invariant;
2) Let $\mathbf{q} : \mathbf{Y} \to \mathbf{Z} = \mathbf{Y}/\mathbf{V}$ be the quotient map. Then for any $z \in \mathbf{Z}$ the restriction of $\mathbf{f}$ to $\mathbf{q}^{-1}(z)$ is an affine function.

Let $\mathbf{V}^*$ denote the dual vector space to $\mathbf{V}$. We denote by $\mathbf{s} : \mathbf{Z} \to \mathbf{V}^*$ the map sending every $z \in \mathbf{Z}$ to the linear part of $\mathbf{f}|_{\mathbf{q}^{-1}(z)}$. Set

$$\overline{\mathbf{Y}} = \big\{ z \in \mathbf{Z} \mid \mathbf{s}(z) = 0 \big\}. \qquad (7.4)$$

We say that an action of $\mathbf{V}$ on Φ is non-degenerate if for any $\overline{y} \in \overline{\mathbf{Y}}$ the differential $ds : T_{\overline{y}}\overline{\mathbf{Y}} \to \mathbf{V}^*$ is onto. In this case the variety $\overline{\mathbf{Y}}$ is smooth and for any $\overline{y} \in \overline{\mathbf{Y}}$ we get the natural isomorphism $T_{\overline{y}}\mathbf{Z}/T_{\overline{y}}\overline{\mathbf{Y}} \simeq \mathbf{V}^*$. Hence for any $\overline{y} \in \overline{\mathbf{Y}}$ and any $y \in \mathbf{q}^{-1}(\overline{y})$ we have the natural isomorphisms

$$\Lambda^{\text{top}}(T_y\mathbf{Y}) \simeq \Lambda^{\text{top}}(T_{\overline{y}}\mathbf{Z}) \otimes \Lambda^{\text{top}}(\mathbf{V}) \simeq \Lambda^{\text{top}}(T_{\overline{y}}\overline{\mathbf{Y}}) \otimes \Lambda^{\text{top}}(T_{\overline{y}}\mathbf{Z}/T_{\overline{y}}\overline{\mathbf{Y}}) \otimes \Lambda^{\text{top}}(\mathbf{V})$$
$$\simeq \Lambda^{\text{top}}(T_{\overline{y}}\overline{\mathbf{Y}}) \otimes \Lambda^{\text{top}}(\mathbf{V}^*) \otimes \Lambda^{\text{top}}(\mathbf{V}) \simeq \Lambda^{\text{top}}(T_{\overline{y}}\overline{\mathbf{Y}}).$$

Therefore, for any $\overline{y} \in \overline{\mathbf{Y}}$, $y \in \mathbf{q}^{-1}(\overline{y})$ the restriction of ω to $T_y\mathbf{Y}$ defines an element $\overline{\omega}_{\overline{y}} \in \Lambda(T_{\overline{y}}(\overline{\mathbf{Y}}))^*$. Since ω is $\mathbf{V}$-invariant, it follows that $\overline{\omega}_{\overline{y}}$ does not depend on the choice of $y \in \mathbf{q}^{-1}(\overline{y})$. It is easy to see that there exists a top-degree form $\overline{\omega}$ on $\overline{\mathbf{Y}}$ such that for any $\overline{y} \in \overline{\mathbf{Y}}$ one has $(\overline{\omega})_{\overline{y}} = \overline{\omega}_{\overline{y}}$.

By the definition, the restriction of $\mathbf{f}$ to $\mathbf{q}^{-1}(\mathbf{Y})$ is $\mathbf{V}$-invariant. So, $\mathbf{f}|_{\mathbf{q}^{-1}\overline{\mathbf{Y}}} = \overline{\mathbf{f}} \circ \mathbf{q}$ for some function $\overline{\mathbf{f}}$ on $\overline{\mathbf{Y}}$. Also, since $\mathbf{p}$ is $\mathbf{V}$-invariant we have $\mathbf{p}|_{\mathbf{q}^{-1}(\overline{\mathbf{Y}})} = \overline{\mathbf{p}} \circ \mathbf{q}$ for some morphism $\mathbf{p} : \overline{\mathbf{Y}} \to \mathbf{X}$.

We call the algebro-geometric distribution $\overline{\Phi} = (\overline{\mathbf{Y}}, \overline{\mathbf{p}}, \overline{\mathbf{f}}, \overline{\omega})$ the *reduction* of Φ.

Let us see the effect of this reduction procedure on materializations. Assume that $\overline{\Phi}$ is analytically finite and denote by $\overline{\Phi}$ its materialization. For simplicity we shall assume that F is non-Archimedean. Let V_n be a sequence of open compact subgroups of V, such that

$$V = \bigcup_{n=1}^{\infty} V_n. \tag{7.5}$$

Let also Y_n be an increasing covering of Y by V_n-invariant open compact subsets. Consider the sequence Φ_n, $n = 1, 2, \ldots$ of distributions on X where

$$\Phi_n(\phi) = \int_{Y_n} \psi(f) p^*(\phi)|\omega|. \tag{7.6}$$

LEMMA 7.9. *The sequence Φ_n weakly converges to $\overline{\Phi}$.*

When $\dim \mathbf{V} = 1$ this is proved in [GK]($\S1$). The general case is treated similarly.

7.10 W-equivariant distributions. Now let $\mathbf{G}$ be a reductive algebraic group over F and let $\mathbf{T}$ be its Cartan group. The Weyl group W acts naturally on $\mathbf{T}$ and we have the natural F-rational morphism $\mathbf{s} : \mathbf{G} \to \mathbf{T}/W$. Let $\mathbf{G}_r$ denote the set of regular elements in $\mathbf{G}$. Then the restriction of $\mathbf{s}$ to $\mathbf{G}_r$ is a smooth map. In the rest of this section $\mathbf{s}$ will always denote the restriction of $\mathbf{s}$ to $\mathbf{G}_r$.

Now let $\Phi_{\mathbf{T}} = (\mathbf{Y_T}, \mathbf{f_T}, \mathbf{p_T}, \omega_{\mathbf{T}})$ be an algebro-geometric distribution on $\mathbf{T}$. We assume that $\mathbf{p_T}$ is a dominant map. By a W-*equivariant structure* on Φ we shall mean a birational action of W on $\mathbf{T}$ such that

1) $\mathbf{f_T}$ is invariant under this action;
2) $\mathbf{p_T}$ intertwines the action of W on $\mathbf{Y_T}$ and on $\mathbf{T}$;
3) One has $w^*(\omega_{\mathbf{T}}) = (-1)^{l(w)}\omega$, where $l : W \to \mathbb{Z}$ is the length function.

Since W is a finite group we can always find an open dense subset of $\mathbf{Y}$ on which the action W is biregular. We can also assume that this subset is chosen in such a way that it is contained in the set of regular semi-simple elements. Note that birational modifications of $\mathbf{Y_T}$ (such as passing to an open subset) do not change the distribution Φ given by (7.3). We also assume in the sequel that $\mathbf{p_T}$ is a smooth map.

Assume now that we are given a W-equivariant algebro-geometric distribution $\Phi_{\mathbf{T}}$ as above. We assume also that $\omega_{\mathbf{T}} = \sigma \cdot d^\times t$ where $\sigma : \mathbf{G} \to \mathbb{G}_m$ is a character of $\mathbf{G}$ and $d^\times t$ a translation invariant top-degree differential form on $\mathbf{T}$. Then we may construct an algebro-geometric distribution $\Phi = (\mathbf{Y_G}, \mathbf{f_G}, \mathbf{p_G}, \omega_{\mathbf{G}})$ over $\mathbf{G}_r$ in the following way. We may assume that W acts biregularly on $\mathbf{Y_T}$. Then we define

$$\mathbf{Y_G} = \mathbf{G}_r \underset{\mathbf{T}/W}{\times} \mathbf{Y_T}/W. \qquad (7.7)$$

We let $\mathbf{p_G}$ be the projection on the first multiple. Condition 1 above implies that $\mathbf{f_T}$ descends to a function on $\mathbf{Y_T}/W$ and hence defines a function $\mathbf{f_G}$ on $\mathbf{Y_G}$.

The choice of $d^\times t$ gives rise to a choice of a left invariant top degree differential form dg on $\mathbf{G}$. We set $\omega_{\mathbf{G}} = \sigma^{l+1}dg$ where l as before denotes the split rank of G.

CONJECTURE 7.11.   1. *Let* $\mathbf{G}$ *be a split reductive group over* F *and let* $\rho : \mathbf{G}^\vee \to \mathbf{GL}(n, \mathbb{C})$ *be an admissible representation of* $\mathbf{G}^\vee$. *Let* $\Phi_{\rho,\mathbf{T}}$ *be the algebro-geometric distribution as in Lemma 7.7. Then there exists a* W-*equivariant structure on* $\Phi_{\rho,\mathbf{T}}$ *such that the corresponding distribution* $\Phi_{\rho,\mathbf{G}}$ *is analytically finite and the underlying distribution on* G *is equal to* $\Phi_{\rho,G}$ *introduced in section 3.*

Let $\mathbf{P} \subset \mathbf{G}$ *be a parabolic subgroup and let* $\mathbf{L}$ *be the corresponding Levi factor. The representation* ρ *defines canonically a representation* $\rho_{\mathbf{L}}$ *of* $\mathbf{L}^\vee$, *with the same restriction to* $\mathbf{T}^\vee$. *Note that the Weyl group* $W_{\mathbf{L}}$ *is naturally a subgroup in* W.

2. *The restriction of the* W-*action on* $\Phi_{\rho,\mathbf{T}}$ *to* $W_{\mathbf{L}}$ *is equal to the corresponding* $W_{\mathbf{L}}$-*action on* $\Phi_{\rho_{\mathbf{L}},\mathbf{T}}$.

REMARK. We shall see some examples of the above W-equivariant structure in some examples. Surprisingly, the corresponding W-action on $\mathbf{T}_\rho$ turns out to be rather complicated and it is related to the theory of geometric crystals (cf. [BeK]).

8   The Case of $G(m,n)$ and its Applications

Let $\mathbf{G}, \mathbf{T}$ and ρ be as above. In section 4 we have constructed a finite algebro-geometric distribution $\Phi_{\rho,\mathbf{T}} = (\mathbf{T}_\rho, \mathbf{f}_\rho, \mathbf{p}_\rho, \omega_\rho)$ on $\mathbf{T}$. It is explained in section 7 that in order to get from it a finite algebro-geometric distribution on $\Phi_{\rho,\mathbf{G}}$ one has to define a birational action of the Weyl group W on $\mathbf{T}_\rho$ satisfying certain conditions. We do not know how to construct this action in general. In this section we give an explicit construction of this action for the group $\mathbf{G} = \mathbf{G}(n,m)$ (cf. section 3).

8.1   The case $m = 2$. We now want to define the action of W on $\mathbf{T}_\rho$ in the case when $m = 2$. The Weyl group W of $\mathbf{G} = \mathbf{G}(2,n)$ is isomorphic to $\mathbb{Z}_2 \times S_n$. In this subsection we are going to give an explicit formula for the action of the first factor by using part 2 of Conjecture 7.11.

Let $\mathbf{T}$ be as before a maximal torus of $\mathbf{G}$. It can be identified with the variety of all collections $(a_1, a_2, b_1, \dots, b_n)$, where $a_i, b_j \in \mathbb{G}_m$ and $a_1 a_2 = b_1 \dots b_n$.

The torus $\mathbf{T}_\rho$ consists of all matrices $t = (t_{ij})$ where $i = 1, 2$ and $j = 1, \dots, n$. The map $\mathbf{p}_\rho$ is given by

$$\mathbf{p}_\rho((t_{ij})) = (t_{11}t_{12}\dots t_{1n}, t_{21}, t_{22}\dots t_{2n}, t_{11}t_{21}, t_{12}t_{22}, \dots, t_{1n}t_{2n}). \quad (8.1)$$

The function $\mathbf{f}_\rho$ sends the matrix t_{ij} to $\sum_{i,j} t_{ij}$ and the differential form ω_ρ is up to a sign equal to $\prod dt_{ij}$.

To define $\mathbb{Z}_2$-action on $\Phi_{\rho,\mathbf{T}}$ we need to define an involution τ on $\mathbf{T}_\rho$ that preserves $\mathbf{f}_\rho$, sends ω_ρ to $-\omega_\rho$ and has the following two properties:

(1) For every $t \in \mathbf{T}_\rho$ and every $j = 1, \dots, n$ one has $t_{1j}t_{2j} = \tau(t)_{1j}\tau(t)_{2j}$.
(2) For every $t \in \mathbf{T}_\rho$, $t_{11}\dots t_{1n} = \tau(t)_{21}\dots\tau(t)_{2n}$ and $t_{21}\dots t_{2n} = \tau(t)_{11}\dots\tau(t)_{1n}$.

It is easy to see that in the case when $n > 2$ there exist many ways to define an involution on $\mathbf{T}_\rho$ which satisfy the compatibility conditions (1) and (2). We will use the conjectural properties of the distribution $\Phi_{\rho,\mathbf{G}(2,n)}$ to obtain the formula for the involution τ.

Let $\mathbf{T}_n = \mathbb{G}_m^n$ and let $N : \mathbf{T}_n \to \mathbb{G}_m$ be given by $N((t_1, \dots, t_n)) = t_1 \dots t_n$. Define

$$\mathbf{G}_n = \big\{ (g,t) \in \mathbf{GL}(2) \times \mathbf{T}_n \mid \det(g) = N(t) \big\}. \quad (8.2)$$

The group $\mathbf{G}_n$ can be naturally regarded as a Levi subgroup of $\mathbf{G}(2, n)$. Hence the dual group $\mathbf{G}_n^\vee$ is a Levi subgroup of $\mathbf{G}(2, n)^\vee$ and we can restrict the representation ρ to it. This restriction can be explicitly described as follows. The group $\mathbf{G}_n^\vee$ is isomorphic to the quotient of $\mathbf{GL}(2, \mathbb{C}) \times (\mathbb{C}^\times)^n$ by the subgroup consisting of elements of the form $(x \cdot \mathrm{Id}, x^{-1}, x^{-1}, \ldots, x^{-1})$. For every $i = 1, \ldots, n$ we can define a two-dimensional representation ρ_i of $\mathbf{G}_n^\vee$, sending $(g, z_1, \ldots, z_n)$ to $g z_i$. Then it is easy to see that

$$\rho = \bigoplus_{i=1}^n \rho_i. \tag{8.3}$$

The Cartan group of $\mathbf{G}_n$ is naturally isomorphic to the Cartan group $\mathbf{T}$ of $\mathbf{G}(2, n)$ considered in section 8.1. The Weyl group of $\mathbf{G}_n$ is equal to the first factor of the Weyl group W of $\mathbf{G}(2, n)$.

As follows from (3.10) and (8.3) the distribution Φ_{ρ, G_n} is equal to the convolution

$$\Phi_{\rho, G_n}(g, b_1, \ldots, b_n) = (\Phi_{b_1} * \cdots * \Phi_{b_n}) |db_1 \ldots db_n|, \tag{8.4}$$

where distributions $\Phi_b, b \in F^\times$, on $\mathbf{GL}(m, F)$ are defined in section 3.3. In other words, Φ_{ρ, G_n} can be thought of as a materialization of the algebro-geometric distribution $\widetilde{\Phi} = (\widetilde{\mathbf{Y}}, \widetilde{\mathbf{f}}, \widetilde{\mathbf{p}}, \widetilde{\omega})$, where

- $\widetilde{\mathbf{Y}} = \mathbf{GL}(2)^n$;
- $\widetilde{\mathbf{f}}(g_1, \ldots, g_n) = \mathrm{tr}(g_1) + \ldots + \mathrm{tr}(g_n)$;
- $\widetilde{\mathbf{p}}(g_1, \ldots, g_n) = (g_1 \ldots g_n, \det(g_1), \ldots, \det(g_n))$;
- $\widetilde{\omega} = \det(g_1 \ldots g_n) dg_1 \ldots dg_n$;

where dg denotes a translation invariant top degree differential form on $\mathbf{GL}(2)$.

More precisely, for every $N > 0$ define an open subset $\widetilde{Y}(N)$ of $\widetilde{Y}$ by

$$\widetilde{Y}(N) - \{(g_1, \ldots, g_n) \in \mathbf{GL}(2)^n \mid |(g_i)_{\alpha\beta}| \leq N\}. \tag{8.5}$$

Let $\widetilde{\mu}_N$ denote the restriction of the distribution $\psi(\widetilde{f})|\widetilde{\omega}|$ to $\widetilde{Y}(N)$.

LEMMA 8.2.

$$\Phi_{\rho, G_n} = \lim_{N \to \infty} p_!(\widetilde{\mu}_N). \tag{8.6}$$

Fix a maximal unipotent subgroup $\mathbf{U}_+$ in $\mathbf{GL}(2)$. Then $\mathbf{U}_+ \simeq \mathbb{G}_a$. For every $g \in \mathbf{GL}(2)$ the function $u \mapsto \mathrm{tr}(gu) - \mathrm{tr}(g)$ is linear in u. We set $\mathrm{tr}(gu) - \mathrm{tr}(g) = g_- u$ (thus $g \mapsto g_-$ is a function on $\mathbf{GL}(2)$).

Let $\mathbf{V}_+ = \mathbf{U}_+^{n-1}$. Define an action of $\mathbf{V}_+$ on $\widetilde{\mathbf{Y}}$ by setting

$$(u_1, \ldots, u_{n-1})(g_1, \ldots, g_n) = (g_1 u_1^{-1}, u_1 g_2 u_2^{-1}, \ldots, u_{n-1} g_n). \tag{8.7}$$

This action preserves $\widetilde{p}$ and $\widetilde{\omega}$ and for any $(g_1, \ldots, g_n) \in \widetilde{\mathbf{Y}}$ we have

$$\widetilde{\mathbf{f}}(ug) - \widetilde{\mathbf{f}}(g) = \sum_{i=1}^{n-1} u_i \big((g_{i+1})_- - (g_i)_- \big). \tag{8.8}$$

Thus we obtain an action of $\mathbf{V}_+$ on $\widetilde{\Phi} = (\widetilde{\mathbf{Y}}, \widetilde{\mathbf{f}}, \widetilde{p}, \widetilde{\omega})$. Let $\overline{\Phi}_{U_+} = (\overline{\mathbf{Y}}, \overline{p}, \overline{\mathbf{f}}, \overline{\omega})$ be the $\mathbf{U}_+$-reduction of $\widetilde{\Phi}$ (cf. section 7.8). As follows from Lemma 7.9 the materialization $\overline{\Phi}_{U_+}$ is equal to Φ_ρ.

Let $\mathcal{B}$ be the variety of Borel subgroups of $\mathbf{GL}(2)$. For any $g \in \mathbf{GL}(2)$ we denote by $\mathcal{B}^g \subset \mathcal{B}$ the subvariety of Borel subgroups containing g.

LEMMA 8.3. *For any regular* $g \in \mathbf{GL}(2)$ *the centralizer of* g *in* $\mathbf{PGL}(2)$ *acts simply transitively on the variety* $\mathcal{B} - \mathcal{B}^g$.

COROLLARY 8.4. *For any two unipotent subgroups* $\mathbf{U}_+, \mathbf{U}_- \subset \mathbf{GL}(2)$ *we have a canonical isomorphism* $\overline{\Phi}_{\mathbf{U}_+} \simeq \overline{\Phi}_{\mathbf{U}_-}$. *Therefore, we can write* $\overline{\Phi}$ *instead of* $\overline{\Phi}_{\mathbf{U}_+}$

Let

$$\widetilde{\mathbf{G}}_n = \big\{ (\mathbf{B}, g, t) \in \mathcal{B} \times \mathbf{G}_n \mid g \in \mathbf{B} \big\}. \tag{8.9}$$

We have the natural maps $\mathbf{r} : \widetilde{\mathbf{G}}_n \to \mathbf{G}_n$ (sending $(\mathbf{B}, g, t)$ to g) and $\mathbf{m} : \widetilde{\mathbf{G}}_n \to \mathbf{T}$ (sending $(\mathbf{B}, g, t)$ to g mod $\mathbf{U_B}$ where $\mathbf{U_B}$ denotes the unipotent radical of $\mathbf{B}$). Then r is a ramified double covering and for any $(g, t) \in \mathbf{G_n}$ one has $r^{-1}(g, t) = \mathcal{B}^g$. We denote by $\widetilde{\Phi} = (\widetilde{\mathbf{Y}}, \widetilde{p}, \widetilde{\mathbf{f}}, \widetilde{\omega})$ the lift of Φ to $\widetilde{\mathbf{G}}_n$. By the definition $\widetilde{\mathbf{Y}} = \overline{\mathbf{Y}} \times_{\mathbf{G}_n} \widetilde{\mathbf{G}}_n$.

Since the generic fiber of $\mathbf{m}$ carries a canonical top form (see section 7.10) we can define an algebro-geometric distribution $\mathbf{m}^*(\Phi_{\rho, \mathbf{T}})$ on $\widetilde{\mathbf{G}}_n$. We write $\mathbf{m}^*(\Phi_{\rho, \mathbf{T}}) = (\mathbf{Y}'_{\mathbf{T}}, \mathbf{p}''_{\mathbf{T}}, \mathbf{f}''_{\mathbf{T}}, \omega''_{\mathbf{T}})$ where $\mathbf{Y}'_{\mathbf{T}} = \mathbf{Y}_{\rho, \mathbf{T}} \times_{\mathbf{T}} \widetilde{\mathbf{G}}_n$. We will construct now an isomorphism $\mu : \widetilde{\Phi} \to m^*(\Phi_{\rho, \mathbf{T}})$.

For any $\mathbf{B} \in \mathcal{B}$ such that $\mathbf{U}_+ \not\subset \mathbf{B}$ the multiplication map gives rise to birational isomorphisms $\mathbf{B} \times \mathbf{U}_+ \to \mathbf{GL}(2)$ and $\mathbf{U}_+ \times \mathbf{B} \to \mathbf{GL}(2)$. Therefore the subset $\widetilde{\mathbf{Y}}_+ := \mathbf{B}^{n-1} \times \mathbf{GL}(2) \subset \widetilde{\mathbf{Y}}$ is a section of the action of $\mathbf{V}_+$ on $\widetilde{\mathbf{Y}}$. Fix any element $(g, t; \mathbf{B}) \in \widetilde{\mathbf{G}}_n$ such that g is regular and semisimple. Since by definition $g \in \mathbf{B}$ we see that $\widetilde{\mathbf{Y}}_+ \cap \widetilde{p}^{-1}(g, t; \mathbf{B}) \subset \mathbf{B}^n$. So we can describe the variety $\widetilde{p}^{-1}(g, t; \mathbf{B})$ as the preimage $\mathbf{s_B}^{-1}(0)$ where $\mathbf{s_B}$ is the restriction of the map $\mathbf{s} : \widetilde{\mathbf{Y}} \to \mathbf{V}^*$ to $\mathbf{B}^n \cap p^{-1}(g, t)$ where $\mathbf{s}$ is as in section 7.8. In other words we can identify the variety $\widetilde{\mathbf{Y}}$ with $\mathbf{s_B}^{-1}(0)$.

Let $\mathbf{U} \subset \mathbf{B}$ be the unipotent radical of $\mathbf{B}$ and let $\mathbf{T}_g \subset \mathbf{B}$ be the centralizer of g. Then $\mathbf{B} = \mathbf{T}_g \mathbf{U}$ and $\mathbf{T}_g$ is canonically isomorphic to $\mathbb{G}_m^2$. Therefore $\mathbf{T}_g^n$ is canonically isomorphic to $\mathbf{T}_\rho$.

LEMMA 8.5. a) *The natural projection* $\mathbf{B} \to \mathbf{T}_g$ *defines an embedding* $\mu : \mathbf{s}_{\mathbf{B}}^{-1}(0) \hookrightarrow \mathbf{T}_\rho$ *and the image of* i *is equal to* $\mathbf{p}_{\rho,\mathbf{T}}^{-1}(g,t)$.

In other words we have constructed an isomorphism $\mu : \widetilde{\mathbf{Y}} \to \mathbf{Y}'_{\mathbf{T}}$.

b) μ *defines an isomorphism of algebro-geometric distributions.*

The natural involution θ of $\widetilde{\mathbf{G}}_n$ over $\mathbf{G}_n$ induces an involution $\widetilde{\theta}$ of $\widetilde{\Phi}$ and we can define an involution θ' of $\mathbf{m}^*(\Phi_{\rho,\mathbf{T}})$ by $\theta' = \mu \circ \widetilde{\theta} \circ \mu^{-1}$. It is easy to see that the involution θ' is $\mathbf{GL}(2)$-invariant and therefore it is induced by an involution τ on $\Phi_{\rho,\mathbf{T}}$.

Let us give an explicit formula for τ. For every $k = 1, \dots, n$ we define the function Δ_k on $\mathbf{T}_\rho$ by

$$\Delta_k(t) = t_{11} \dots t_{1(k-1)} + t_{11} \dots t_{1(k-2)} t_{2k} + \dots + t_{22} \dots t_{2k} \quad (8.10)$$

(by definition $\Delta_1(t) = 1$).

We also define a rational function η on $\mathbf{T}_\rho$ setting

$$\eta(t) = \frac{t_{11} \dots t_{1n} - t_{21} \dots t_{2n}}{\Delta_n(t)}. \quad (8.11)$$

LEMMA 8.6. *The involution* τ *satisfies*

$$\tau(t)_{21} \tau(t)_{22} \dots \tau(t)_{2k} = t_{21} \dots t_{2k} + \Delta_k(t)\eta(t). \quad (8.12)$$

It is clear that τ is uniquely determined by (8.12) and by the conditions (1) and (2) above.

Assume for example that $n = 2$. Then one can compute the above action explicitly. Namely, in this case we have

$$\mathbf{T} = \left\{ (a_1, b_1, a_2, b_2) \in \mathbb{G}_m^4 \mid a_1 b_1 = a_2 b_2 \right\}. \quad (8.13)$$

Let $w \in W$ be the involution which interchanges a_1 and b_1. Let $\tau = \tau_w$. Then an explicit calculation shows that

$$\tau : \begin{pmatrix} t_{11} & t_{12} \\ t_{21} & t_{22} \end{pmatrix} \mapsto \begin{pmatrix} t_{21} \frac{t_{11}+t_{22}}{t_{12}+t_{21}} & t_{22} \frac{t_{12}+t_{21}}{t_{11}+t_{22}} \\ t_{11} \frac{t_{12}+t_{21}}{t_{11}+t_{22}} & t_{12} \frac{t_{11}+t_{22}}{t_{12}+t_{21}} \end{pmatrix}. \quad (8.14)$$

Moreover, it is easy to see that in this case τ given by (8.14) is the unique birational involution of $\mathbf{T}_\rho$ which satisfies our requirements.

8.7   The case of $\mathbf{G}(n,m)$. Let us now consider the group $\mathbf{G}(n,m)$ for arbitrary n and m. In this case one can identify $\mathbf{T}_\rho$ with the variety $(t_{ij}, i = 1, \dots, m, j = 1, \dots, n)$ of $m \times n$ matrices with non-zero entries. We wish to define a birational action of the Weyl group $W = S_m \times S_n$ on $\mathbf{T}_\rho$. For every $\alpha = 1, \dots, m-1$ we define a birational involution τ_α^1 on $\mathbf{T}_\rho$ in the following way:

1) All rows of $\tau_\alpha^1(t = t_{ij})$ except for the α and $\alpha + 1$st are equal to the corresponding rows of t.

2) The α and $(\alpha + 1)$st rows of $\tau_\alpha^1(t)$ are obtained from those of t by means of (8.12).

Similarly, for every $\beta = 1, \ldots, n - 1$ we define $\tau_\beta^2(t) = \tau_\beta^1(t')$ where t' is the transposed matrix to t. The following lemma is straightforward.

LEMMA 8.8. *The involutions τ_α^1, τ_β^2 commute with $\mathbf{p}_\rho$, preserve $\mathbf{f}_\rho$ and map ω_ρ to $-\omega_\rho$.*

The following theorem is proven in [BeK, (§6.2)].

Theorem 8.9. *The involutions τ_α^1 and τ_β^2 define a birational action of the group $S_m \times S_n$ on $\mathbf{T}_\rho$. In particular, τ_α^1 commutes with τ_β^2 for any α and β.*

Despite the fact the formulas (8.12) are quite explicit we did not manage to prove Theorem 8.9 directly. The proof, given in [BeK] uses the machinery of geometric crystals.

8.10 γ-functions. Let us still assume that $\mathbf{G} = \mathbf{G}(n, m)$. Recall that we have the natural character $\sigma : \mathbf{G} \to \mathbb{G}_m$ (which sends the pair (A, B) of matrices to $\det(A) = \det(B)$). Using Theorem 8.9 we can define an algebro-geometric distribution $\boldsymbol{\Phi}_{\rho, \mathbf{G}}$ over the generic point of $\mathbf{G}$. We would like to check that it gives the "correct" answer.

Let π_m, π_n be generic representations of respectively $GL(m, F)$ and $GL(n, F)$. Then in [JPS] H. Jacquet, I. Piatetskii-Shapiro and J. Shalika define a rational function $\Gamma(\pi_m, \pi_n, s)$ of one complex variable s (which depends on the choice of a character ψ). The following result can be deduced from [BeK].

Theorem 8.11. 1. *The materialization $\Phi_{\rho, G}$ of $\boldsymbol{\Phi}_{\rho, \mathbf{G}}$ defines a σ-compact distribution on the whole of G. We denote by γ_ρ the corresponding rational function on the set* Irr(G).

2. *Assume that $n \geq m$. Let π_m, π_n be as above and let π be an irreducible representation of G such that* Hom$_G(\pi, \pi_m \otimes \pi_n) \neq 0$. *Then*

$$\Gamma(\pi_m, \pi_n, s) = \text{sign}(\pi_n)\gamma_\rho(\pi| \cdot |^s) \tag{8.15}$$

where sign(π) *is the value of the central character of π_n at the matrix $-\text{Id}_n \in GL(n, F)$.*

8.12 Lifting from non-split tori. Let us recall the notation of section 2.5: let E/F be a separable extension of F of degree n. Let $\alpha_E :$ Gal$(\overline{F}/F) \to S_n$ be the corresponding homomorphism. We denote by

$N : \mathbf{T}_E \to \mathbf{G}_m$ the morphism of algebraic groups coming from the norm map $N : E^\times \to F^\times$.

Let $\mathbf{T}_E = \mathrm{Res}_{E/F}\, \mathbb{G}_{m,E}$ where $\mathrm{Res}_{E/F}$ denotes the functor of restriction of scalars. For every $m > 0$ we set

$$\mathbf{G}_{E,m} = \big\{(t,g) \in \mathbf{T}_E \times \mathbf{GL}(m) \mid N(t) = \det(g)\big\}. \qquad (8.16)$$

As is explained in section 3.18, in order to define the lifting $l_E(\theta) \in \mathrm{Irr}(\mathbf{GL}(n,F))$ of a character $\theta : E^\times \to \mathbb{C}^\times$, it is enough to define the γ-function $\gamma(l_E(\theta), \pi_m)$ as a function on $\mathrm{Irr}(\mathbf{GL}(m,F))$. Moreover, it is enough to know this function only up to a constant.

We are now going to give a conjectural definition of $\gamma(l_E(\theta), \pi_m)$ by constructing a weakly finite algebraic geometric distribution $\Phi_{E,m}$ on $\mathbf{G}_{E,m}$ (in fact, we will define it only up to a constant multiple). We conjecture that this distribution is analytically finite and that the corresponding function on $\mathrm{Irr}(E^\times) \times \mathrm{Irr}(\mathbf{GL}(m,F)$ is equal to $\gamma_{E,m}$.

Choose $d \in F^\times$ such that its image in $F^\times/(F^\times)^2$ is equal to the discriminant of E. Let us also choose its square root $\sqrt{d}$ in $\overline{F}$.

Let

$$\mathbf{G}'_{E,m} = \big\{(t_1,\dots,t_n,g) \in \mathbb{G}_m^n \times \mathbf{GL}(m) \mid t_1 \dots t_n = \det(g)\big\}. \qquad (8.17)$$

Using section 8.7 we define an algebro-geometric distribution $\Phi' = (\mathbf{Y}', \mathbf{p}', \mathbf{f}', \omega')$ on $\mathbb{G}_m^n \times \mathbf{GL}(m)$. Moreover, we have the natural S_n-action on $\mathbf{Y}'$, which is compatible with $\mathbf{p}'$ and leaves $\mathbf{f}'$ (resp. ω') invariant (resp. skew-invariant).

Define a new $\mathrm{Gal}(\overline{F}/F)$-action on $\mathbf{Y}'$ by

$$g^{\mathrm{new}}(y) = \alpha_E(g^{\mathrm{old}}(y)). \qquad (8.18)$$

This action defines a new F-rational structure on $\mathbf{Y}'$. We denote by $\mathbf{Y}$ the corresponding F-variety. It is clear that the map $\mathbf{p}'$ gives rise to an F-rational morphism $\mathbf{p} : \mathbf{Y} \to \mathbf{G}_{E,m}$. Also the function $\mathbf{f}'$ and the differential form $\sqrt{d}\omega'$ give rise to a F-rational function $\mathbf{f}$ and a differential form ω on $\mathbf{Y}$. Thus we set $\Phi_{\mathbf{E},m} = (\mathbf{Y}, \mathbf{p}, \mathbf{f}, \omega)$ to be the required algebro-geometric distribution. We denote by $\Phi_{\mathbf{E},m}$ the materialization of $\Phi_{\mathbf{E},m}$.

Clearly, for different choices of $\sqrt{d}$, the resulting materializations $\Phi_{\mathbf{E},m}$ will differ only by a multiplication by $c \in \mathbb{C}^*$. Hence, the above construction suffices in order to determine the lifting uniquely.

Since the local Langlands conjecture is known (see [H], [HT] and [LaRS]) we can ask whether our definition of lifting coincides with one which is implied by [H], [HT] and [LaRS]. More precisely let χ be a character of the group $T = \mathbf{T}(F)$. The local class field theory associates to χ a homomorphism $\theta_\chi : \mathrm{Gal}(\bar{E}/E) \to \mathbb{C}^\times$. Let $\Theta_\chi := \mathrm{Ind}_{\mathrm{Gal}(\bar{E}/E)}^{\mathrm{Gal}(\bar{F}/F)} \theta_\chi$. Then Θ_χ is an n-dimensional representation of the group $\mathrm{Gal}(\bar{F}/F)$. Since the local Langlands conjecture for $\mathbf{G} = \mathbf{GL}(n)$ is known one associates with Θ_χ an irreducible representation $\pi_\chi \in \mathrm{Irr}(\mathbf{GL}(n,F))$. Therefore we can consider a function $\Gamma(\pi_m, \pi_\chi, s)$ on the set $\mathrm{Irr}\,\mathbf{GL}(m,F)$. One can ask whether there exists $c \in \mathbb{C}^\times$ such that for any $\pi_m \in \mathrm{Irr}\,\mathbf{GL}(m,F)$ we have $\gamma(\pi_m, \pi_\chi, 0) = c\chi \otimes \pi_m(\Phi_{\mathbf{E},m})$.

9 The Case of Finite Fields

In this section we shall a give an explicit conjectural construction of $\Phi_{\rho,\mathbf{G}}$ for any $\mathbf{G}$ and for (almost) any ρ in the case when the field F is finite. It turns out that in this case the relevant tool from algebraic geometry which allows us to go from the case of a torus to the case of an arbitrary group is not the language of algebro-geometric distributions, but that of ℓ-adic perverse sheaves. This tool allows us to avoid constructing an action of W on $\mathbf{T}_\rho$.

9.1 Notation. In this section we fix a finite field F with q elements, a prime number l, different from the characteristic of F, and a non-trivial character $\psi : F \to \overline{\mathbb{Q}}_l^\times$. All representations discussed in the note are over the field $\overline{\mathbb{Q}}_l$. We denote by $\mathcal{L}_\psi$ the Artin-Schreier sheaf on $\mathbb{A}_F^1$.

Let $\mathbf{G}$ be a reductive algebraic group over F and let $\mathbf{T}$ be its abstract Cartan group. The torus $\mathbf{T}$ comes equipped with a canonical F-rational structure. In this section we assume for simplicity that $\mathbf{G}$ is split. Then this F-rational structure is split too. We will denote by $Fr : \mathbf{T} \to \mathbf{T}$ the corresponding Frobenius morphism. Let W denote the Weyl group of $\mathbf{G}$. For every $w \in W$ we set $Fr_w = w \circ Fr$. The morphism Fr_w induces a new F-rational structure on $\mathbf{T}$ and we will denote the corresponding algebraic torus over F by $\mathbf{T}_w$. It is well known that there exists an embedding $\mathbf{T}_w \hookrightarrow \mathbf{G}$ and that in this way we get a bijection between conjugacy classes of elements in W and conjugacy classes of F-rational maximal tori in $\mathbf{G}$.

For a character $\theta : T_w \to \overline{\mathbb{Q}}_l^\times$ we denote by $R_{\theta,w}$ the corresponding Deligne-Lusztig representation.

For two ℓ-adic complexes A and B on $\mathbf{G}$ we can define the convolution complex $A \star B$ and by setting

$$A \star B = m_!(A \boxtimes B)[\dim \mathbf{G}], \tag{9.1}$$

where $m : \mathbf{G} \times \mathbf{G} \to \mathbf{G}$ is the multiplication map.

Let $\mathbf{X}$ be an algebraic variety over F and let $\mathcal{F}$ be a complex of ℓ-adic sheaves on it. By a *Weil structure* on $\mathcal{F}$ we shall mean an isomorphism $\xi : Fr^* \mathcal{F} \to \mathcal{F}$ where Fr denotes the geometric Frobenius morphism on $\mathbf{X}$. In this case we can define a function $\mathrm{Tr}(\mathcal{F})$ on $X = \mathbf{X}(F)$ by setting

$$\mathrm{Tr}(\mathcal{F})(x) = \sum_i (-1)^i \, \mathrm{tr} \left(\xi_x : H^i(\mathcal{F}_x) \to H^i(\mathcal{F}_x) \right) \qquad (9.2)$$

(here $H^i(\mathcal{F}_x)$ denotes the i-th cohomology of the fiber of $\mathcal{F}$ at x).

Let us choose a square root $q^{1/2}$ of q. Then for any Weil sheaf $\mathcal{F}$ on $\mathbf{X}$ and any half integer n we can consider the Tate twist $\mathcal{F}(n)$ of $\mathcal{F}$. By the definition one has

$$\mathrm{Tr}(\mathcal{F}(n)) = \mathrm{Tr}(\mathcal{F})q^{-n} . \qquad (9.3)$$

9.2 γ-functions for $\mathbf{GL}(n)$.

9.2.1   The main result.   Let (π, V) be an irreducible representation of $G = GL(n, F)$. Choose a non-trivial additive character $\psi : F \to \overline{\mathbb{Q}}_l^\times$ as above and consider the operator

$$\sum_{g \in G} \psi(\mathrm{tr}(g)) \pi(g) (-1)^n q^{-n^2/2} \in \mathrm{End}_G V . \qquad (9.4)$$

Since π is irreducible, this operator takes the form $\gamma_{G,\psi}(\pi) \cdot \mathrm{Id}_V$ where $\gamma_G(\pi) \in \overline{\mathbb{Q}}_l$ (we will omit the subscripts G and ψ when it does not lead to confusion). The number $\gamma(\pi) = \gamma_{G,\psi}(\pi)$ is called the gamma-function of the representation π. The purpose of this subsection is to compute explicitly the gamma-functions of all irreducible representations of G.

Let $W \simeq S_n$ denote the Weyl group of $\mathbf{G}$.

Fix $w \in W$. For a character $\theta : T_w \to \overline{\mathbb{Q}}_l^\times$ we set

$$\gamma_w(\theta) = (-1)^{n+l(w)} q^{-n/2} \sum_{t \in T_w} \psi(\mathrm{tr}(t)) \theta(t) \in \overline{\mathbb{Q}}_l . \qquad (9.5)$$

EXAMPLE.   Assume that $w \in S_n$ is a cycle of length n. Then $T_w \simeq E^\times$ where E is the (unique up to isomorphism) extension of F of degree n. In this case $\gamma_w(\theta) = \gamma_E(\theta)$ for any character θ of $E^\times$, where by $\gamma_E(\theta)$ we denote the γ-function defined as in (9.4) for the group $\mathbf{GL}(1, E) \simeq E^\times$.

Theorem 9.3.   *Assume that an irreducible representation (π, V) appears in $R_{\theta,w}$ for some w and θ as above. Then*

$$\gamma(\pi) = (-1)^{l(w)} \gamma_w(\theta) \qquad (9.6)$$

where R_+ denotes the set of positive roots of G. In particular, $\gamma(\pi) = \gamma(\pi')$ if π and π' appear in the same virtual representation $R_{\theta,w}$.

The rest of this subsection is occupied with the proof of Theorem 9.3.

9.3.1   Character sheaves.

Let $\mathbf{G}$ be an arbitrary reductive algebraic group over F. Let us recall Lusztig's definition of (some of) the character sheaves. Let $\widetilde{\mathbf{G}}$ denote the variety of all pairs $(\mathbf{B}, g)$, where

- $\mathbf{B}$ is a Borel subgroup of $\mathbf{G}$
- $g \in \mathbf{B}$

One has natural maps $\alpha : \widetilde{\mathbf{G}} \to \mathbf{T}$ and $\pi : \widetilde{\mathbf{G}} \to \mathbf{G}$ defined as follows. First of all, we set $\pi(\mathbf{B}, g) = g$. Now, in order to define α, let us recall that for any Borel subgroup $\mathbf{B}$ of $\mathbf{G}$ one has canonical identification $\mu_{\mathbf{B}} : \mathbf{B}/\mathbf{U_B} \widetilde{\to} \mathbf{T}$, where $\mathbf{U_B}$ denotes the unipotent radical of $\mathbf{B}$ (in fact, this is how the abstract Cartan group T is defined). Now we set $\alpha(\mathbf{B}, g) = \mu_{\mathbf{B}}(g)$.

Let $\mathcal{L}$ be a tame local system on $\mathbf{T}$. We define $\mathcal{K}_{\mathcal{L}} = \pi_! \alpha^*(\mathcal{L})[\dim \mathbf{G}]$. One knows (cf. [Lu], [La]) that the sheaf $\mathcal{K}_{\mathcal{L}}$ is perverse.

Assume now, that for some $w \in W$ there exists an isomorphism $\mathcal{L} \simeq Fr_w^*(\mathcal{L})$ and let us fix it. It was observed by G. Lusztig in [Lu] that fixing such an isomorphism endows $\mathcal{K}_{\mathcal{L}}$ canonically with a Weil structure. Lusztig's definition of this Weil structure was as follows:

Let $j : \mathbf{G}_{rs} \to \mathbf{G}$ denote the open embedding of the variety of regular semisimple elements in $\mathbf{G}$ into $\mathbf{G}$.

LEMMA 9.4.

$$\mathcal{K}_{\mathcal{L}} = j_{!*}(\mathcal{K}_{\mathcal{L}}|_{\mathbf{G}_{rs}}) . \tag{9.7}$$

Here $j_{!*}$ denotes the Goresky-MacPherson (intermediate) extension (cf. [BBD]).

The lemma follows from the fact that the map π is small in the sense of Goresky and McPherson.

The lemma shows that it is enough to construct the Weil structure only on the restriction of $\mathcal{K}_{\mathcal{L}}$ on $\mathbf{G}_{rs}$. The latter now has a particularly simple form. Namely, let $\widetilde{\mathbf{G}}_{rs}$ denote the preimage of G_{rs} under π and let π_{rs} denote the restriction of π to $\widetilde{\mathbf{G}}_{rs}$. Then it is easy to see that $\pi_{rs} : \widetilde{\mathbf{G}}_{rs} \to \mathbf{G}_{rs}$ is an unramified Galois covering with Galois group W. In particular, W acts on $\widetilde{\mathbf{G}}_{rs}$ and this action is compatible with the action of W on T in the sense that the restriction of α on $\widetilde{\mathbf{G}}_{rs}$ is W-equivariant.

Now, an isomorphism $\mathcal{L} \simeq Fr_w^*(\mathcal{L})$ gives rise to an isomorphism

$$\alpha^* \mathcal{L} \simeq (w \circ Fr)^*(\alpha^* \mathcal{L}) \tag{9.8}$$

(here both w and Fr are considered on the variety $\widetilde{\mathbf{G}}$). Since π_{rs} is a Galois covering with Galois group W, it follows that one has canonical identification

$$\pi_{rs!}\big(Fr^*(\alpha^*\mathcal{L})\big) \simeq \pi_{rs!}\big((w \circ Fr)^*(\alpha^*\mathcal{L})\big). \tag{9.9}$$

Hence from (9.8) and (9.9) we get the identifications

$$Fr^*\,\pi_{rs!}(\alpha^*\mathcal{L}) \simeq \pi_{rs!}\big(Fr^*(\alpha^*\mathcal{L})\big) \simeq \pi_{rs!}\big((w \circ Fr)^*(\alpha^*\mathcal{L})\big) \simeq \pi_{rs!}(\alpha^*\mathcal{L}) \tag{9.10}$$

which gives us a Weil structure on $\pi_{rs!}(\alpha^*\mathcal{L}) \simeq \mathcal{K}_\mathcal{L}|_{\mathbf{G}_{rs}}$. Hence we have defined a canonical Weil structure on $\mathcal{K}_\mathcal{L}$.

Now let $\theta : T_w \to \overline{\mathbb{Q}}_l^\times$ be any character. It is well known that one can associate to θ a one-dimensional local system $\mathcal{L}_\theta$ together with an isomorphism $\mathcal{L}_\theta \simeq Fr_w^*\mathcal{L}_\theta$. This local system is constructed as follows: Consider the sheaf $(Fr_w)_*(\overline{\mathbb{Q}}_l)$. It is clear that this sheaf has a natural fiberwise action of the group T_w. Thus we set $\mathcal{L}_\theta$ to be the direct summand of $(Fr_w)_*(\overline{\mathbb{Q}}_l)$ on which T_w acts by means of the character θ. The following result is due to G. Lusztig.

Theorem 9.5.

$$\mathrm{Tr}(\mathcal{K}_{\mathcal{L}_\theta}) = (-1)^{\dim \mathbf{G}}\,ch(R_{\theta,w}). \tag{9.11}$$

9.5.1 Now we assume again that $\mathbf{G}=\mathbf{GL}(n)$. Set $\Phi_\psi = \mathrm{tr}^*\,\mathcal{L}_\psi[n^2](n^2/2)$.

In this case it is well known that for every character $\theta : T_w \to \overline{\mathbb{Q}}_l^\times$ the vector space, spanned by the characters of all irreducible constituents of $R_{\theta,w}$ coincides with the vector space spanned by the functions of the form $\mathrm{Tr}(A)$ where A runs over all direct summands of $\mathcal{K}_{\mathcal{L}_\theta}$. Hence Theorem 9.3 follows from the following result.

PROPOSITION 9.6.   1. *Let* $\mathrm{tr}_{\mathbf{T}} : \mathbf{T} \to \mathbb{A}^1$ *denote the restriction of the trace morphism from* $\mathbf{G}$ *to* $\mathbf{T}$. *Then for every* $\mathcal{L}$ *as above one has*

$$\dim H_c^i\big(\mathrm{tr}_{\mathbf{T}}^*(\mathcal{L}_\psi) \otimes \mathcal{L}\big) = \begin{cases} 0, & \text{if } i \neq n, \\ 1, & \text{if } i = n. \end{cases} \tag{9.12}$$

We set $H_\mathcal{L} := H_c^n(\mathrm{tr}_{\mathbf{T}}^*(\mathcal{L}_\psi) \otimes \mathcal{L})(n/2)$.

2. *Let* $\theta : T_w \to \overline{\mathbb{Q}}_l^\times$ *be a character (for some* $w \in W$*). Then the natural isomorphism* $Fr_w^*\mathcal{L}_\theta \simeq \mathcal{L}_\theta$ *gives rise to a natural endomorphism of* $H_{\mathcal{L}_\theta}$ *which by abuse of language we will also denote by* Fr_w. *Then*

$$Fr_w|_{H_{\mathcal{L}_\theta}} = (-1)^{l(w)}\gamma_w(\theta) \cdot \mathrm{Id}. \tag{9.13}$$

3.   *One has*

$$\Phi_\psi \star \mathcal{K}_\mathcal{L} \simeq H_\mathcal{L} \otimes \mathcal{K}_\mathcal{L}. \tag{9.14}$$

Moreover, if $\mathcal{L}$ is endowed with an isomorphism $Fr_w^* \mathcal{L} \simeq \mathcal{L}$ then (9.14) is Fr_w-equivariant.

The proof follows easily from the standard properties of the Fourier-Deligne transform. The details are left to the reader.

9.7 The case of arbitrary group. Let $\mathbf{G}$ be an arbitrary connected reductive algebraic group over F. For simplicity we will assume that $\mathbf{G}$ is split. Let $\mathbf{G}^\vee$ denote the Langlands dual group of $\mathbf{G}$ (which is a reductive algebraic group over $\overline{\mathbb{Q}}_l$). Let $\rho : \mathbf{G}^\vee \to \mathrm{Aut}(E)$ be a representation of $\mathbf{G}^\vee$, where E is a vector space over $\overline{\mathbb{Q}}_l$ of dimension n. We would like to associate to it a certain function $\pi \mapsto \gamma_\rho(\pi)$ on the set of isomorphism classes of irreducible representations of G. This is done as follows.

The group $\mathbf{G}^\vee$ comes equipped with a canonical maximal torus $\mathbf{T}^\vee$. Let us diagonalize $\rho|_{\mathbf{T}^\vee}$. Let $\lambda_1, \ldots, \lambda_n$ be the corresponding characters of $\mathbf{T}^\vee$ in ρ (with multiplicities). Let also $\mathbf{M}_\rho^\vee$ denote the minimal Levi subgroup containing $\rho(\mathbf{T}^\vee)$ such that the action of $\mathbf{M}_\rho^\vee$ is multiplicity free. We denote by W_ρ' the group $\mathrm{Norm}_{\mathrm{Aut}(E)}(\mathbf{M}_\rho^\vee)/\mathbf{M}_\rho^\vee$. We also denote by W_ρ the Weyl group of $\mathrm{Aut}(E)$.

LEMMA 9.8. *W_ρ' is naturally a subquotient of W_ρ.*

Proof. Standard. $\qquad\qquad\qquad\qquad\qquad\qquad\qquad\qquad\qquad\qquad\quad$ $\square$

Let $\mathbf{T}_\rho^\vee \simeq \mathbb{G}_{m,\overline{\mathbb{Q}}_l}^n$ be the Cartan group of $\mathbf{GL}(n)$. Thus we get a natural map $\mathbf{p}_\rho^\vee : \mathbf{T}^\vee \to \mathbf{T}_\rho^\vee$ sending every t to $(\lambda_1(t), \ldots, \lambda_n(t))$.

Now let $\mathbf{T}_\rho \simeq \mathbb{G}_{m,F}$ denote the dual torus to $\mathbf{T}_\rho^\vee$ over F and let $\mathbf{p}_\rho : \mathbf{T}_\rho \to \mathbf{T}$ denote the map, which is dual to $\mathbf{p}_\rho^\vee$. Explicitly one has

$$\mathbf{p}_\rho(x_1, \ldots, x_n) = \lambda_1(x_1) \ldots \lambda_n(x_n). \qquad (9.15)$$

The representation ρ defines the natural homomorphism from W to W_ρ', which by abuse of the language we will also denote by ρ.

Now let π be an irreducible representation of G. Assume that π appears in some $R_{\theta,w}$ for some $\theta : T_w \to \overline{\mathbb{Q}}_l^\times$. Let w' be any lift of $\rho(w)$ to W_ρ. Then $\mathbf{p}$ induces an F-rational map $\mathbf{p}_w : \mathbf{T}_{\rho,w'} \to \mathbf{T}_w$, hence a homomorphism $p_w : T_{\rho,w'} \to T_w$. Define

$$\gamma_\rho(\pi) := \gamma(\pi'), \qquad (9.16)$$

where π' is any irreducible representation of G_ρ which appears in $R_{p_w^*(\theta),w'}$. By Theorem 9.3 one has

$$\gamma_\rho(\pi) = (-1)^{l(w')} \gamma_{T_{w'}}(p_w^*(\theta)). \qquad (9.17)$$

LEMMA 9.9. *The definition of $\gamma_\rho(\pi)$ does not depend on the choice of w'.*

Now let Φ_ρ denote the unique central function on G such that for every irreducible representation (π, V) of G one has

$$\sum_{g \in G} \Phi_\rho(g)\pi(g) = \gamma_\rho(\pi) \cdot \mathrm{Id}_V \, . \tag{9.18}$$

We would like to compute this function explicitly. We will be able to solve only a slightly weaker problem. Namely, we are going to construct explicitly an ℓ-adic perverse sheaf $\Phi_{\rho,\psi}$ on $\mathbf{G}$ such that

$$\mathrm{Tr}(\Phi_{\rho,\psi}) = \widetilde{\Phi}_\rho \, , \tag{9.19}$$

where $\widetilde{\Phi}_\rho$ is a function which satisfies the following condition:

$$\sum_{g \in G} \widetilde{\Phi}_\rho(g)\pi(g) = \gamma_\rho(\pi) \cdot \mathrm{Id} \tag{9.20}$$

for every irreducible "generic" representation π of G (cf. Theorem 9.11 for the precise statement).

9.9.1 The perverse sheaf $\Phi_{\rho,\psi}$. In what follows we assume that the zero weight does not appear in ρ. Let $\mathrm{tr}_\rho : \mathbf{T}_\rho \to \mathbb{A}^1$ be given by

$$\mathrm{tr}_\rho(x_1, \ldots, x_n) = x_1 + \ldots + x_n \, . \tag{9.21}$$

Consider the complex $A_\rho := (\mathbf{p}_\rho)_! \, \mathrm{tr}_\rho^* \mathcal{L}_\psi[n](n/2)$ (on $\mathbf{T}$). It is easy to see that this complex is perverse. We would like to endow this complex with a W-equivariant structure.

Choose $w \in W$. We need to define an isomorphism $\iota_w : w^*(A_\rho) \widetilde{\to} A_\rho$. Let (as above) w' be any lift of $\rho(w)$ to W_ρ. Then one has

$$\mathbf{p}_\rho(w'(t)) = w(\mathbf{p}_\rho(t)) \, . \tag{9.22}$$

The sheaf $\mathrm{tr}_\rho^* \mathcal{L}_\psi$ is obviously W_ρ-equivariant. This, together with (9.22) gives rise to an isomorphism $\iota'_w : w^*(A_\rho) \widetilde{\to} A_\rho$. We now define $\iota_w := (-1)^{l(w') - l(w)} \iota'_w$.

PROPOSITION 9.10. *The isomorphism ι_w does not depend on the choice of w'. The assignment $w \mapsto \iota_w$ defines a W-equivariant structure on the sheaf A_ρ.*

The above W-equivariant structure on A_ρ gives rise to the sheaf $B_\rho = (\mathbf{q}_! A_\rho)^W$ on the quotient $\mathbf{T}/W$ where $\mathbf{q} : \mathbf{T} \to \mathbf{T}/W$ is the canonical map.

Let $\mathbf{G}_r$ be the set of regular elements of $\mathbf{G}$ and let $\mathbf{i}$ be its embedding to $\mathbf{G}$. Let $\mathbf{s} : \mathbf{G}_r \to \mathbf{T}/W$ be the morphism coming from the identification $\mathbf{T}/W \simeq \mathcal{O}(\mathbf{G})^{\mathbf{G}}$. We define

$$\Phi_{\rho,\psi} := i_{!*} \mathbf{s}^*(B_\rho)[\dim \mathbf{G} - \dim \mathbf{T}] \left(\frac{\dim \mathbf{G} - \dim \mathbf{T}}{2} \right) \, . \tag{9.23}$$

We also define

$$\widetilde{\Phi}_\rho = \mathrm{Tr}(\Phi_{\rho,\psi})\,. \tag{9.24}$$

By the definition the complex $\Phi_{\rho,\psi}$ on $\mathbf{G}$ is $\mathbf{G}$-equivariant with respect to the adjoint action and therefore the function $\widetilde{\Phi}_\rho$ is central.

9.10.1　The action of $\widetilde{\Phi}_\rho$ in $R_{\theta,w}$. Recall that a local system $\mathcal{L}$ on $\mathbf{T}$ is called quasi-regular if for every coroot $\alpha^\vee : \mathbb{G}_m \to \mathbf{T}$ the local system $(\alpha^\vee)^*\mathcal{L}$ is non-trivial. We say that a character $\theta : T_w \to \overline{\mathbb{Q}}_l^\times$ is *quasi-regular* if the local system $\mathcal{L}_\theta$ is quasi-regular.

Theorem 9.11. *Let π be an irreducible representation of G, which appears in the Deligne-Lusztig representation $R_{\theta,w}$ for a quasi-regular character $\theta : T_w \to \overline{\mathbb{Q}}_l^\times$. Then*

$$\sum_{g \in G} \widetilde{\Phi}_\rho(g)\pi(g) = \gamma_\rho(\pi)\,. \tag{9.25}$$

9.11.1　The basic conjecture. In order to proceed we will have to assume that the following conjecture holds.

Choose a maximal unipotent subgroup in $\mathbf{U} \subset \mathbf{G}$. Denote by $\mathbf{B}$ the Borel subgroup of $\mathbf{G}$ which normalizes $\mathbf{U}$. Let $\mathbf{X} = \mathbf{G}/\mathbf{U}$ and let $\mathbf{r} : \mathbf{G} \to \mathbf{X}$ denote the natural projection. By the definition one has canonical identification $\mathbf{T} \simeq \mathbf{B}/\mathbf{U}$. Hence $\mathbf{T}$ is naturally embedded into $\mathbf{X}$.

CONJECTURE 9.12. *The complex $\mathbf{r}_! B_\rho$ vanishes outside of $\mathbf{T}$.*

One can show that Conjecture 9.12 implies that (9.25) holds for any θ.

10　Appendix (by V. Vologodsky)

10.1　Notation. In this appendix we assume that F is a finite extension of $\mathbb{Q}_p$. We denote by $\mathfrak{m}$ the maximal ideal of $\mathcal{O}_F$.

Let $\mathbf{C}$ be a smooth, quasi-projective curve over F and c_0 be a F-point of $\mathbf{C}$. Let $\mathbf{C}' = \mathbf{C}\backslash c_0$.

Let $f : \mathbf{X} \to \mathbf{C}'$ be a smooth, proper morphism of relative dimension n and ω be a differential form $\omega \in \Omega^{\mathrm{top}}_{\mathbf{X}/\mathbf{C}'}$ of the top degree.

For an F-point $c \in \mathbf{C}'(F)$ we denote by $|\omega|$ the measure associated to ω on the space of F-points of the fiber X_c over c. Put $V(c) = \int_{X_c(K)} |\omega|$. Choose a coordinate $\mathbf{t}$ on a neighborhood of c_0 with $\mathbf{t}(c_0) = 0$.

Theorem 10.2. *There exist integers n and r such that $V(tt') = V(t)$ for all $t \in \mathfrak{m}^n$, $t' \in (\mathcal{O}_F^*)^r$.*

This theorem easily implies Theorem 7.2.

Proof. First, we can make use of Nagata's theorem to construct a proper, integral scheme $\overline{\mathbf{X}}$ over $\mathbf{C}$ such that $\overline{\mathbf{X}} \times_{\mathbf{C}} \mathbf{C}' \simeq \mathbf{X}$. By Hironaka's theorem we can blow up $\overline{\mathbf{X}}$ to build $\widetilde{\mathbf{f}}\colon \mathbf{Y} \to \mathbf{C}$ such that the union of the fiber $\mathbf{Y}_{c_0}$ over c_0 and the closure the zero locus of the differential form ω on $\mathbf{Y}\backslash \mathbf{Y}_{c_0}$ is a normal crossing divisor. It follows that for any F-point a of $\mathbf{Y}_{c_0}$ there exist local coordinates $\mathbf{x}_i$ ($i = 0, 1, \ldots, n$) on a neighborhood $\mathbf{Z}_a$ of a satisfying the following conditions:

i) $\mathbf{t}(\widetilde{\mathbf{f}}) = \mathbf{u}\Pi_i\mathbf{x}_i^{\mathbf{n_i}}$, for some function $\mathbf{u}$ on $\mathbf{Z}_a$ with $\mathbf{u}(a) \neq 0$ and $n_i \geq 0$;

ii) $\omega = v\Pi_i x_i^{m_i}dx_0dx_1dx_2 \ldots dx_{n-1}$ on $Z_a\backslash Y_{c_0}$ for some function v on Z_a with $v(a) \neq 0$ and $m_i \in Z$.

It is clear that for a point $c \in \mathbf{C}'(F)$ sufficiently close to c_0 the fibers $\mathbf{X}_c(F)$ and $\mathbf{Y}_c(F)$ have the same measure. Hence we can replace $\mathbf{X}$ by $\mathbf{Y}\backslash\mathbf{Y}_{c_0}$. By the implicit function theorem we can choose a neighborhood Z'_a of a in p-adic topology: $a \in Z'_a \subset Z_a$ such that coordinates x_i give an isomorphism (of sets) $Z'_a \simeq \mathfrak{m}^l \times \mathfrak{m}^l \times \cdots \times \mathfrak{m}^l$, where l is a positive integer. Moreover, if we choose l sufficiently large the function $(u(a)/u)^{n_1^{-1}}$ is well defined on Z'_a, hence, changing $\mathbf{x}_1$ for $(u(a)/\mathbf{u})^{n_1^{-1}}\mathbf{x}_1$ we can suppose that on Z'_a one has $t(\widetilde{f}) = u(a)\Pi_i x_i^{n_i}$. Without loss of generality we can assume that $|v|$ is constant on Z'_a. Since $Y_{c_0}(F)$ is compact it can be covered by finitely many open sets Z'_a satisfying the properties stated above. In fact one can choose them to be disjoint. To prove this we need the following lemma.

LEMMA 10.3. *Let Q be an open, compact subset of the affine space F^n. We claim that Q can be covered by finitely many disjoint balls contained in Q. (A ball is a subset of the form $B_{r,a} = \{(x_1, \ldots x_n) \in K^n \mid |x_i - a_i| < r\}$).*

We apply the lemma to sets $Z'_q \bigcup_{j<q} Z'_{a_j}$ ($q = 1, \ldots, k$). (By construction each of them is identified with an open compact subset of F^{n+1}.)

Proof. Since Q is compact we can cover it by finitely many balls contained in Q. Let r_0 be the smallest radius of these balls. Pick an integer l such that $-\log_q r_0 < l$. Consider the projection $p\colon Q \to (F/m^l)^n$. The preimage of each element of $\operatorname{Im} p$ is a ball. These balls constitute the desirable covering.

Now the proposition follows from the following simple lemma.

LEMMA 10.4. *Let ω stand for differential form*

$$\omega = \prod_i \mathbf{x}_i^{m_i}d\mathbf{x}_0 d\mathbf{x}_1 d\mathbf{x}_2 \ldots d\mathbf{x}_{n-1}$$

where $m_i \in \mathbb{Z}$ *and*

$$V(t) = \int_{\Pi_i x_i^{n_i} = t; x_i \in m} |\omega|.$$

Then $V(t \cdot t') = V(t)$ for any $t \in \mathfrak{m}$, $t' \in (\mathcal{O}_F^*)^{n_1}$.   □

References

[BBD] A. BEILINSON, J. BERNSTEIN, P. DELIGNE, Faisceaux pervers, in "Analysis and Topology on Singular Spaces, I (Luminy, 1981)", Astèrisque 100 (1982), 5–171.

[BeK] A. BERENSTEIN, D. KAZHDAN, Geometric and unipotent crystals, GAFA2000, in this issue.

[BerD] J. BERNSTEIN, P. DELIGNE, Le centre de Bernstein, Travaux en Cours, Representations of Reductive Groups over a Local Field, 1–32, Hermann, Paris, 1984.

[BerZ1] J. BERNSTEIN, A. ZELEVINSKY, Representations of the group $GL(n, F)$, where F is a local non-Archimedean field (in Russian), Uspehi Mat. Nauk 31 (1976), 5–70.

[BerZ2] J. BERNSTEIN, A. ZELEVINSKY, Induced representations of reductive p-adic groups I, Ann. Sci. École Norm. Sup. (4) 10 (1977), 441–472.

[BoJ] A. BOREL, H. JACQUET, Automorphic forms and automorphic representations, with a supplement, "On the notion of an automorphic representation" by R.P. Langlands, Proc. Sympos. Pure Math. XXXIII, Automorphic forms, representations and L-functions (Proc. Sympos. Pure Math., Oregon State Univ., Corvallis, Ore., 1977), Part 1, Amer. Math. Soc., Providence, R.I. (1979), 189–207.

[BrK] A. BRAVERMAN, D. KAZHDAN, On the Schwartz space of the basic affine space, Selecta Math. (N.S.) 5:1 (1999), 1–28.

[GGP] I. GELFAND, M. GRAEV, I. PIATETSKII-SHAPIRO, Representation Theory and Automorphic Functions, Philadelphia, PA-London-Toronto, 1969.

[GK] I.M. GELFAND, D. KAZHDAN, Representations of the group $\mathbf{GL}(n, K)$ where K is a local field, in "Lie Groups and Their Representations" (Proc. Summer School, Bolyai János Math. Soc., Budapest, 1971), Halsted, New York, 1975, 95–118.

[GeK] S. GELFAND, D. KAZHDAN, Conjectural algebraic formulas for representations of $\mathbf{GL}(n)$, in "Sir Michael Atiyah: a Great Mathematician of the Twentieth Century," Asian J. Math. 3:1 (1999), 17–48.

[GoJ] R. GODEMENT, H. JACQUET, Zeta-functions of simple algebras, Springer Lecture Notes in Mathematics 260 (1972).

[H] M. HARRIS, The local Langlands conjecture for $\mathbf{GL}(n)$ over a p-adic field, $n < p$, Invent. Math. 134:1 (1998), 177–210.

[HT]   M. HARRIS, R. TAYLOR, On the geometry of cohomology of some simple Shimura varieties, preprint.

[JPS]   H. JACQUET, I. PIATETSKII-SHAPIRO, J. SHALIKA, Rankin-Selberg convolutions, Amer. J. Math. 105:2 (1983), 367–464.

[K1]   D. KAZHDAN, On lifting, Lie group representations, II (College Park, Md., 1982/1983), Springer Lecture Notes in Math. 1041 (1984), 209–249,

[K2]   D. KAZHDAN, Forms of the principle series for $GL(n)$ in "Functional Analysis on the Eve of the 21st Century, Vol. 1 (New Brunswick, NJ, 1993), Birkhäuser Progr. Math. 131, (1995), 153–171.

[K3]   D. KAZHDAN, An algebraic integration, in "Mathematics: Frontiers and Perspectives" (V. Arnold, M. Atiyah, P. Lax, B. Mazur, eds.), AMS, 2000.

[L]   R.P. LANGLANDS, Problems in the theory of automorphic forms, in "Lectures in Modern Analysis and Applications, III," Springer Lecture Notes in Math. 170 (1970), 18–61.

[La]   G. LAUMON, Faisceaux charactèrs (d'après Lusztig) (in French), Séminaire Bourbaki, Vol. 1988/89. Astérisque 177-178 (1989), 231–260.

[LaRS]   G. LAUMON, M. RAPOPORT, U. STUHLER, $\mathcal{D}$-elliptic sheaves and the Langlands correspondence, Invent. Math. 113:2 (1993), 217–338.

[Lu]   G. LUSZTIG, Character sheaves I, Adv. in Math. 56:3 (1985), 193–237.

[P]   I. PIATETSKII-SHAPIRO, Multiplicity one theorems, in "Automorphic Forms, Representations and L-functions" (Proc. Sympos. Pure Math., Oregon State Univ., Corvallis, Ore., 1977), Part 1, Proc. Sympos. Pure Math. XXXIII, Amer. Math. Soc., Providence, R.I. (1979), 209–212.

[V]   E.B. VINBERG, On reductive algebraic semi-groups, in "Lie Groups and Lie Algebras: E.B. Dynkin's Seminar," Amer. Math. Soc. Transl. Ser. 2, 169 (1995), 145–182.

[T]   J. TATE, Fourier analysis in number fields, and Hecke's zeta-functions, Algebraic Number Theory (Proc. Instructional Conf., Brighton, 1965) (1967), 305–347.

[W]   N. WALLACH, Real Reductive Groups I, Pure and Applied Mathematics, 132, Academic Press, Inc., Boston, MA, 1988.

[We]   A. WEIL, Adéles and Algebraic Groups, Birkhäuser Progress in Mathematics 23, 1982.

A. BRAVERMAN, Department of Mathematics, MIT, 77 Mass. Ave., Cambridge, MA 02139, USA

D. KAZHDAN, Department of Mathematics, Harvard University, Cambridge, MA 02138, USA

V. VOLOGODSKY, Department of Mathematics, Harvard University, Cambridge, MA 02138, USA

GAFA, Geom. funct. anal.
Special Volume – GAFA2000, 279 – 315
1016-443X/00/S10279-37 $ 1.50+0.20/0

I GAFA Geometric And Functional Analysis

PDE AS A UNIFIED SUBJECT

SERGIU KLAINERMAN

Introduction

Given that one of the goals of the conference is to address the issue of the unity of Mathematics, I feel emboldened to talk about a question which has kept bothering me all through my scientific career: Is there really a unified subject of Mathematics which one can call PDE? At first glance this seems easy: we may define PDE as the subject which is concerned with all partial differential equations. According to this view, the goal of the subject is to find a general theory of all, or very general classes of PDE's. This "natural" definition comes dangerously close to what M. Gromov had in mind, I believe, when he warned us, during the conference, that objects, definitions or questions which look natural at first glance may in fact "be stupid". Indeed, it is now recognized by many practitioners of the subject that the general point of view, as a goal in itself, is seriously flawed. That it ever had any credibility is due to the fact that it works quite well for linear PDE's with constant coefficients, in which case the Fourier transform is extremely effective. It has also produced significant results for some general special classes of linear equations with variable coefficients.[1] Its weakness is most evident in connection to nonlinear equations. The only useful general result we have is the Cauchy-Kowalevsky theorem, in the quite boring class of analytic solutions. In the more restrictive frameworks of elliptic, hyperbolic, or parabolic equations, some important local aspects of nonlinear equations can be treated with a considerable degree of generality. It is the passage from local to global properties which forces us to abandon any generality and take full advantage of the special features of the important equations.

The fact is that PDE's, in particular those that are nonlinear, are too subtle to fit into a too general scheme; on the contrary each important

---

[1]Linear equations with variable coefficients appear naturally by linearizing nonlinear equations around specific solutions. They also appear in the study of specific operators on manifolds, in Several Complex Variables, and Quantum Mechanics. The interaction between the $\bar{\partial}$ operator in SCV and its natural boundary value problems have led to very interesting linear equations with exotic features, such as lack of solvability.

PDE seems to be a world in itself. Moreover, general points of view often obscure, through unnecessary technical complications, the main properties of the important special cases. A useful general framework is one which provides a simple and elegant treatment of a particular phenomenon, as is the case of symmetric hyperbolic systems in connection to the phenomenon of finite speed of propagation and the general treatment of local existence for nonlinear hyperbolic equations. Yet even when a general framework is useful, as symmetric hyperbolic systems certainly are, one would be wrong to expand the framework beyond its natural role. Symmetric hyperbolic systems turn out to be simply too general for the study of more refined questions concerning the important examples of hyperbolic equations.

As the general point of view has lost its appeal many of us have adopted a purely pragmatic point of view of our subject; we chose to be concerned only with those PDE's or classes of PDE's which are considered important. And indeed the range of applications of specific PDE's is phenomenal, many of our basic equations being in fact at the heart of fully fledged fields of Mathematics or Physics such as Complex Analysis, Several Complex Variables, Minimal Surfaces, Harmonic Maps, Connections on Principal Bundles, Kahlerian and Einstein Geometry, Geometric Flows, Hydrodynamics, Elasticity, General Relativity, Electrodynamics, Nonrelativistic Quantum Mechanics, etc. Other important subjects of Mathematics, such as Harmonic Analysis, Probability Theory and various areas of Mathematical Physics are intimately tied to elliptic, parabolic, hyperbolic or Schrödinger type equations. Specific geometric equations such as Laplace–Beltrami and Dirac operators on manifolds, Hodge systems, Pseudo-holomorphic curves, Yang–Mills and recently Seiberg–Witten, have proved to be extraordinarily useful in Topology and Symplectic Geometry. The theory of Integrable systems has turned out to have deep applications in Algebraic Geometry; the spectral theory Laplace–Beltrami operators as well as the scattering theory for wave equations are intimately tied to the study of automorphic forms in Number Theory. Finally, Applied Mathematics takes an interest not only in the basic physical equations but also on a large variety of phenomenological PDE's of relevance to engineers, biologists, chemists or economists.

With all its obvious appeal the pragmatic point of view makes it difficult to see PDE as a subject in its own right. The deeper one digs into the study of a specific PDE the more one has to take advantage of the particular features of the equation and therefore the corresponding results may make sense only as contributions to the particular field to which that PDE is relevant. Thus each major equation seems to generate isolated islands of mathematical activity. Moreover, a particular PDE may be studied from largely different points of view by an applied mathematician, a physicist, a geometer or an analyst. As we lose perspective on the common features of our main equations we see PDE less and less as a unified subject. The field of PDE, as a whole, has all but ceased to exist, except in some old fashioned textbooks. What we have instead is a large collection of loosely connected subjects.

In the end I find this view not only somewhat disconcerting but also, intellectually, as unsatisfactory as the first. There exists, after all, an impressive general body of knowledge which would certainly be included under the framework of a unified subject if we only knew what that was. Here are just a few examples of powerful general ideas:[2]

1) *Well-posedness*:   First investigated by Hadamard at the beginning of this century well-posed problems are at the heart of the modern theory of PDE. The issue of well-posedness comes about when we distinguish between analytic and smooth solutions. This is far from being an academic subtlety, without smooth, non-analytic solutions, we cannot talk about *finite speed of propagation*, the distinctive mark of relativistic physics. Problems are said to be well posed if they admit unique solutions for given *smooth* initial or boundary conditions. The corresponding solutions have to depend continuously on the data. This leads to the classification of linear equations into elliptic, hyperbolic and parabolic with their specific boundary value problems. Well-posedness also plays a fundamental role in the study of nonlinear equations, see a detailed discussion in the last section of this paper. The counterpart of well-posedness is also important in many applications. *Ill-posed problems* appear naturally in Control Theory, Inverse Scattering, etc., whenever we have a limited knowledge of the desired solutions. *Unique continuation* of solutions to general classes of PDE's is intimately tied to ill-posedness.

---

[2]I failed to mention, in the few examples given above, the development of topological methods for dealing with global properties of elliptic PDE's as well as some of the important functional analytic tools connected to Hilbert space methods, compactness, the implicit function theorems, etc. I also failed to mention the large body of knowledge with regard to spaces of functions, such as Sobolev, Schauder, BMO and Hardy, etc., or the recent important developments in nonlinear wave and dispersive equations connected to restriction theorems in Fourier Analysis. For a more in depth discussions of many of the ideas mentioned below, and their history, see the recent survey [BreB].

2) *A priori estimates, boot-strap and continuity arguments*:   A priori estimates allow us to derive crucial information about solutions to complicated equations without having to solve the equations. The best known examples are energy estimates, maximum principle or monotonicity type arguments. Carleman type estimates appear in connection to ill-posed problems. The a priori estimates can be used to actually construct the solutions, prove their uniqueness and regularity, and provide other qualitative information. The boot-strap type argument is a powerful general philosophy to derive a priori estimates for nonlinear equations. According to it we start by making assumptions about the solutions we are looking for. This allows us to think of the original nonlinear problem as a linear one whose coefficients satisfy properties consistent with the assumptions. We may then use linear methods, a priori estimates, to try to show that the solutions to the new linear problem behave as well, or better, than we have postulated. A continuity type argument allows us to conclude the original assumptions are in fact true. This "conceptual linearization" of the original nonlinear equation lies at the heart of our most impressive results for nonlinear equations.

3) *Regularity theory for linear elliptic equations*:   We have systematic methods for deriving powerful regularity estimates for linear elliptic equations. The L^∞ estimates are covered by Schauder theory. The more refined L^p theory occupies an important part of modern Real and Harmonic Analysis. The theory of singular integrals and pseudodifferential operators are intimately tied to the development of L^p-regularity theory.

4) *Direct variational methods*:   The simplest example of a direct variational method is the Dirichlet Principle. Though first proposed by Dirichlet as a method of solving the Poisson equation $\Delta \phi = f$ and later used by Riemann in his celebrated proof of the *Riemann Mapping Theorem* in complex analysis, it was only put on a firm mathematical ground in this century. The method has many deep applications to elliptic problems. It allows one to first solve the original problem in a "generalized sense", and then use regularity estimates, to show that the generalized solutions are in fact classical. The ultimate known expression of this second step is embodied in the De Giorgi–Nash method which allows one to derive full regularity estimates for the generalized solutions of nonlinear, scalar, elliptic equations. This provides, in particular the solution to the famous problem of the regularity of minimal hypersurfaces, as graphs over convex, or mean convex, domains, in all dimensions. Other important applications of the De Giorgi–Nash method were found in connection with such diverse situa-

tions as the Calabi problem in Kahler Geometry, R. Hamilton's Ricci flow and free boundary value problem arising in Continuum Mechanics.

5) *Energy type estimates*: The energy estimates provide a very general tool for deriving a priori estimates for hyperbolic and other evolution equations. Together with Sobolev inequalities, which were developed for this reason, they allow us to prove local in time existence, uniqueness and continuous dependence on the initial data for general classes of nonlinear hyperbolic equations, such as symmetric hyperbolic, similar to the classical local existence result for ordinary differential equations. A more general type of energy estimates, based on using the symmetries of the linear part of the equations, allows one to also prove global in time, perturbation results, such as the global stability of the Minkowski space in General Relativity.

6) *Microlocal analysis, parametrices and paradifferential calculus*: One of the fundamental difficulties of hyperbolic equations consists of the interplay between geometric properties, which concern the physical space, and properties intimately tied to oscillations, which are best seen in Fourier space. Microlocal analysis is a general, still developing, philosophy according to which one isolates the main difficulties by careful localizations in physical or Fourier space, or in both. An important application of this point of view is the construction of parametrices, as Fourier integral operators, for linear hyperbolic equations and their use in propagation of singularities results. The paradifferential calculus can be viewed as an extension of this philosophy to nonlinear equations. It allows one to manipulate the form of a nonlinear equation, by taking account of the way large and small frequencies interact, to achieve remarkable technical versatility.

7) *Generalized solutions*: The idea of a generalized solution appears already in the work of D'Alembert (see [Lu]) in connection with the one dimensional wave equation (vibrating string). A systematic and compelling concept of generalized solutions has developed in connection with the Dirichlet principle; more generally via the direct variational method. The construction of fundamental solutions to linear equations led also to various types of such solutions. This and other developments in linear theory led to the introduction of distributions by L. Schwartz. The theory of distributions provides a most satisfactory framework to generalized solutions in linear theory. The question of what is a good concept of a generalized solution in nonlinear equations, though fundamental, is far more murky. For elliptic equations the solutions derived by the direct variational methods have proved very useful. For nonlinear, one dimensional, conser-

vation laws the concept of a generalized solution has been discussed quite early in the works of J.J. Stokes (see [St]), Rankine, Hugoniot, Riemann, etc. For higher dimensional evolution equations the first concept of a *weak solution* was introduced by J. Leray. I call weak a generalized solution for which one cannot prove any type of uniqueness. This unsatisfactory situation may be temporary, due to our technical inabilities, or unavoidable in the sense that the concept itself is flawed. Leray was able to produce, by a compactness method, a weak solution of the initial value problem for the Navier–Stokes equations. The great advantage of the compactness method (and its modern extensions which can, in some cases, cleverly circumvent lack of compactness) is that it produces global solutions for all data. This is particularly important for supercritical, or critical, nonlinear evolution equations where we expect that classical solutions develop finite time singularities. The problem, however, is that one has very little control of these solutions, in particular we don't know how to prove their uniqueness.[3] Similar types of solutions were later introduced for other important nonlinear evolution equations. In most of the interesting cases of supercritical evolution equations, such as Navier–Stokes, the usefulness of the type of weak solutions used so far remains undecided.

8) *Scaling properties and classification of nonlinear equations*:   Essentially all basic nonlinear equations have well-defined scaling properties. The relationship between the nonlinear scaling and the *coercive* a priori estimates[4] of the equations leads to an extremely useful classification between subcritical, critical and supercritical equations. The definition of criticality and its connection to the issue of regularity was first understood in the case of elliptic equations such as Harmonic Maps, the euclidean Yang–Mills or Yamabe problem. The same issue appears in connection with geometric heat flows and nonlinear wave equations.

Given that some PDE's are interesting from a purely mathematical point of view, while others owe their relevance to physical theories, one of the problems we face when trying to view PDE as a coherent subject is that of the fundamental ambiguity of its status; is it part of Mathematics or Physics or both? In the next section I will try to broaden the discussion by considering some aspects of the general relationship between Mathematics

---

[3]Leray was very concerned about this point. Though, like all other researchers after him, he was unable to prove uniqueness of his weak solution; he showed however that it must coincide with a classical one as long as the latter remains smooth.

[4]See the section "The Problem of Breakdown" for a more thorough discussion.

and Physics, relevant to our main concern. I will try to argue that we can redraw the boundaries between the two subjects in a way which allows us to view PDE as a core subject of Mathematics, with an important applied component. In the third section I will attempt to show how some of the basic principles of modern physics can help us organize the immense variety of PDE's into a coherent field. Equally important, in the fourth section, I will attempt to show that our main PDE's are not only related through their derivation; they also share a common fundamental problem, *regularity or breakdown*. I have tried to keep the discussion of the first four sections as general as possible, and have thus avoided giving more than just a few references. I apologize to all those who feel that their contributions, alluded to in my text, should have been properly mentioned. In the last section of the paper I concentrate on a topic of personal research interest, tied to the issue of regularity, concerning the problem of well-posedness for nonlinear wave equations. My main goal here is to discuss three precise conjectures which I feel are important, difficult and accessible to generate future developments in the field. Even in this section, however, I only provide full references to works directly connected to these conjectures.

Many of the important points I make below, such as the unified geometric structure of the main PDE's, the importance of the scaling properties of the equations and its connection to regularity and well-posedness, have been discussed in similar ways before and are shared by many of my friends and collaborators. My only claim to originality in this regard is the form in which I have assembled them. The imperfections, errors and omissions are certainly my own.

I would like to thank my friends H. Brézis, A. Chang, D. Christodoulou, C. Dafermos, P. Deift, Weinan E, G. Huisken, J. Kohn, E. Stein, P. Sarnak, Y. Sinai, M. Struwe, J. Stalker, and my wife Anca for reading previous versions of the paper and suggesting many corrections and improvements.

Between Mathematics and Physics

In search of a unified point of view for our subject it pays to look at the broader problem of Mathematics as a whole. Isn't Mathematics also in danger of becoming a large collection of loosely connected subjects? Our cherished intellectual freedom to pursue whatever problems strike our imagination as worthwhile is a great engine of invention, but, in the absence of unifying goals, it seems to lead to an endless proliferation of subjects.

This is precisely, I believe, what Poincaré [P] had in mind in the following passage, contained in his address to the first International Congress of Mathematicians, more than a hundred years ago.

"... The combinations that can be formed with numbers and symbols are an infinite multitude. In this thicket how shall we choose those that are worthy of our attention? Shall we be guided only by whimsy?...... [This] would undoubtedly carry us far from each other, and we would rapidly cease to understand each other. But that is only the minor side of the problem. Not only will physics perhaps prevent us from getting lost, but it will also protect us from a more fearsome danger ...... turning around forever in circles. History [shows that] physics has not only forced us to choose [from the multitude of problems which arise], but it has also imposed on us directions that would never have been dreamed of otherwise....... What could be more useful!

The full text of [P] is a marvelous analysis of the complex interactions between Mathematics and Physics. Poincaré argues not only that Physics provides us with a great source of inspiration and cohesiveness but that itself, in return, owes its language, sense of beauty and order to Mathematics. Yet Poincaré's viewpoint concerning the importance of close relations with Physics was largely ignored during most of this century by a large segment of the mathematical community. One reason is certainly due to the fact that traditional areas of Mathematics such as Algebra, Number Theory and Topology have, or seemed to have,[5] relatively little to gain from direct interactions with Physics. Another, more subtle, reason may have to do with the remarkable and unexpected effectiveness of pure mathematical structures in the formulation of the major physical theories of the century: Special and General Relativity, Quantum Mechanics and Gauge Theories. This has led to the popular point of view, coined by Wigner [Wi] as "The unreasonable effectiveness of Mathematics," according to which mathematical objects or ideas developed originally without any reference to Physics turn out to be at the heart of solutions to deep physical problems. Einstein, himself, wrote that any important advance in Physics will have to come in the wake of major new developments in Mathematics. This very seductive picture has emboldened us mathematicians to believe that anything we do may turn out, eventually, to have real applications and has thus, paradoxically, contributed to the problem of ignoring the physical world Poincaré

---

[5]The situation has changed dramatically in the last 25 years with the advent of Gauge fields and String Theories.

has warned us against.

But this is only a minor paradox by comparison to the one which seems to arise from the above discussion relative to the remarkable symbiosis between Physics and Mathematics. On one hand, as Poincaré argues very convincingly in [P], Mathematics needs, to keep itself together, unifying goals and principles; Physics, due, I guess, to the perceived unity of the Physical World, is in a perfect position to provide them for us. On the other hand, Physics owes to Mathematics the very tools which makes it possible to uncover and formulate the unified features of physical reality; it is indeed the search for a selfconsistent mathematical formalism which seems to be at the core of the current attempts to find that *unified theory of everything* which, as theoretical physicists often declare, is Physics' ultimate goal. The paradox is due, of course, to the artificial distinctions we make between the two subjects. We imagine them as separated when in fact they have a nontrivial intersection. Can we identify that intersection? The naive picture would be of two sets which intersect in an area, somewhat peripheral to both, which we might call Mathematical Physics. But this picture does not help to solve the paradox we have mentioned above, which concerns the core of both subjects. A central intersection, however, could imply some form of equality or inclusion between the two subjects, which is definitely not the case. Mathematics pursues goals which are not necessarily suggested by the physical sciences. A research direction is deemed important by mathematicians if it leads to elegant developments and unexpected connections. Physics, on the other hand, cannot allow itself the luxury of being carried away by elegant mathematical theories; in the final analysis it has to subject itself to the tough test of real experiments. Moreover the difference between the work practice and professional standards of mathematicians and modern theoretical physicists cannot be more striking. We mathematicians find ourselves constrained by rigor and are often reluctant to proceed without a systematic analysis of all obstacles in our path. In their quest for the ultimate truth theoretical physicists have no time to waste on unexpected hurdles and unpromising territory. Clearly the relationship between the two subjects is far more complex than may seem at first glance.

The task of defining PDE as a unified subject is tied to that of clarifying, somehow, this ambiguous relationship between Mathematics and Physics. The very concept of partial differential equations has its roots in Physics or, more appropriately Mathematical Physics; there were no clear distinctions at the time of D'Alembert, Euler, Poisson, Laplace, between the two subjects. Riemann was the first, I believe, to show how one can use PDE's to attack problems considered pure mathematical in nature, such as conformal mappings in Complex Analysis. The remarkable effectiveness of PDE's as a tool to solve problems in Complex Analysis, Geometry and Topology has been confirmed many times during this century.

One can separate all mathematicians and other scientists concerned with the study of PDE's into four[6] groups, according to their main interests. In the first group I include those developing and using PDE methods to attack problems in Differential Geometry, Complex Analysis, Symplectic Geometry, Topology and Algebraic Geometry. In the second I include those whose main motivation is the development of rigorous mathematical methods to deal with the PDE's arising in the physical theories. In the third group I include mathematicians, physicists or engineers interested in understanding the main consequences of the physical theories, governed by PDE's, using a variety of heuristic, computational or experimental methods. It is only fair to define yet a fourth category[7] which include all those left out of the groups defined above. According to the common preconceptions about the proper delimitations between Mathematics and Physics only the first group belongs unambiguously within Mathematics. The third group is considered, correctly in my view, as belonging either to Applied Mathematics or Applied Physics. The second group however has an ill defined identity. Since the ultimate goals are not directly connected to specific applications to the traditional branches of Mathematics, many view this group as part of either Applied Mathematics or Mathematical Physics. Yet, apart from the original motivation, it is hard to distinguish the second group from the first. Both groups are dedicated to the development of rigorous analytic techniques. They are tied by many similar concerns, concepts and methods. They are both intimately tied to subjects considered pure, mainly Real and Fourier Analysis but also Geometry, Topology and Algebraic Geometry.

In view of the above ambiguities it helps to take a closer look at the role played by Mathematics in developing the consequences of the *established*

---

[6]My classification is mainly rhetorical. There are, of course, many mathematicians who can cross these artificial boundaries. I will in fact argue below that the first two groups should be viewed as one.

[7]This includes, in particular, PDE's appearing in Biology or Economics. Exotic PDE's, not necessarily connected with any specific application, should also be included in this class.

physical theories.[8] I have heard theoretical physicists and also, alas, mathematicians, expressing the view that the consequences of an established physical theory are of lesser importance and may properly be relegated to Engineering or Chemistry. Nothing, in my mind, can be further from the truth. The first successful physical theory, that of space, was written down by Euclid more than two thousand years ago. Undoubtedly Euclidean Geometry was used by engineers to design levers, pulleys and many other marvelous applications, but does anybody view the further development of the subject as Engineering? Geometry is the primary example of a "physical theory" developed for centuries as a pure mathematical discipline, without too much new input from the physical world, which grew to have deep, mysterious, completely unexpected consequences to the point that pre-eminent physicists talk today of a complete "geometrization" of modern physics, see [Ne].

But this is not all; the Principle of Least Action was developed by mathematicians such as Fermat, Leibnitz, Maupertius, the Bernoulli brothers and Euler from the analysis of simple geometric and physical problems (see [HT] for a very good presentation of the early history of the principle). Their work led to a comprehensive reformulation of the laws of Mechanics by Lagrange who showed how to derive them from a simple Variational Principle. Today the Lagrangian point of view, together with its Hamiltonian reformulation and the famous result of E. Noether concerning the relation between the symmetries of the Lagrangian and conservation laws, is a foundational principle for all Physics. Connected to these are the continuous groups of symmetries attached to the name of S. Lie.

Fourier Analysis was initiated in works by D'Alembert, Euler and D. Bernoulli in connection with the study of the initial-boundary value problem for the one dimensional wave equation (vibrating string). Bernoulli's idea of approximating general periodic functions by sums of sines and cosines was later developed by J. Fourier in connection with the Heat Equation. Further mathematical developments made the theory into a fun-

---

[8]I distinguish between the quest to uncover the basic laws of Nature, which defines the core of theoretical Physics, and the scientific activities concerned with deriving the consequences of a given, established, theory which involve applied physicists, engineers, chemists, applied mathematicians and, as I argue below, "pure" mathematicians. Needless to say, mathematicians have often had direct, fundamental, contributions to theoretical Physics. But more often, I believe, the most impressive contributions came from inner developments within Mathematics of subjects with deep roots in the physical world, such as Geometry, Newtonian Mechanics, Electromagnetism, Quantum Mechanics, etc.

damental tool throughout all of Science.

There are plenty of other examples. I suspect that many, if not most, of the examples of the "unreasonable effectiveness of mathematics" are in fact of this type.[9] There are also many other examples of ideas which originate in Mathematical Physics, and turn out to have a deep, mysterious, impact on the traditional subjects of Mathematics, such as Topology, Geometry or even Number Theory.

All this seems to point to the fact that the further development of the established physical theories ought to be viewed as a genuine and central goal of Mathematics itself. In view of this I think we need to reevaluate our current preconception about what subjects we consider as belonging properly within Mathematics. We may gain, consistent with Poincaré's point of view, considerably more unity by enlarging the boundaries of Mathematics to include, on equal footing with all other more traditional fields, physical theories such as Classical and Quantum Mechanics and Relativity Theory, which are expressed in clear and unambiguous mathematical language. We may then develop them, if we wish, on pure mathematical terms asking questions we consider fundamental, which may not coincide, at any given moment, with those physicists are most interested in, and providing full rigor to our proofs. Of course this has happened to a certain extent, Mathematical Physics and parts of Applied Mathematics fulfill precisely this role. Yet their status remains ambiguous and somewhat peripheral. Many mathematicians assume that subjects like Classical General Relativity[10] or

---

[9]A clear example of this type, this century, is the discovery of the soliton and the "integrable method." Though both emerged in connection with simple nonlinear partial differential equations, the integrability method has found deep applications way beyond the original PDE context. There are other examples which do not quite fit into my description. The extraordinary role played by complex numbers in the formulation of Quantum Mechanics is certainly one which has its roots in Algebra rather than Geometry or Mathematical Physics.

[10]The formulation of General Relativity, by A. Einstein, following the work of Gauss and Riemann in Geometry, and that of Lorentz, Poincaré, Einstein and Minkowski on special relativity can be viewed as one of the most impressive triumphs of Mathematics. Following the recent experiments with double pulsars, GR is considered the most accurate of all physical theories. Research in General Relativity involves, in a fundamental way, all aspects of traditional mathematics; Differential Geometry, Analysis, Topology, Group Representation, Dynamical systems, and of course PDE's. Assuming that the further development of the subject is covered by physics departments is misleading; most theoretical physicists view classical GR as a completely understood physical theory, their main goal now is to develop a quantum theory of gravity. Given their lack of interest and the rich mathematical content of the subject, is there any reason why we should not

Quantum Mechanics belong properly to Physics departments while Physicists often consider them as perfectly well understood, closed, subjects. They are indeed closed, or so it seems, in so far as theoretical physicists are concerned. From their perspective Geometry may have become a closed book more than 2000 years ago, with the publication of Euclid's Elements. But they present us, mathematicians, with wonderful, fundamental challenges formulated in the purest mathematical language. Should we relegate subjects such as Classical and Quantum Mechanics or General Relativity to the periphery of Mathematics, despite their well defined and rich mathematical structures, only because they happen to describe important aspects of the physical world? Is it reasonable to hesitate to include General Relativity as a subject of Mathematics simply because it concerns itself with Lorentz rather than Riemannian metrics? Or because it does not seem to have any applications to Topology? (There are in fact proposals to tie GR to the geometrization conjecture of 3D manifolds, see [FM].)

My proposal is not just to accept these disciplines as some applied appendices to pure Mathematics, but to give them the central role[11] they deserve. This would force us to broaden our outlook and would give us fresh energy and cohesion in the spirit envisioned by Poincaré. It would help us, in particular, to clarify the ambiguous status of subjects such as PDE's and Mathematical Physics and their relations with Applied Mathematics. It would also set more natural boundaries between Mathematics and Physics. As theoretical physicists are primarily interested in understanding new physical phenomena, the further mathematical developments of a confirmed physical theory becomes one of our tasks.[12] Though our pure mathematical considerations may lead us into seemingly esoteric directions, we should hold our ground for with time physicists may come to admit, once more, to the unexpected effectiveness of our Science.

Finally I want to distinguish my proposal from another, more radical, point of view, discussed in this conference, according to which Mathematics ought to become fully engaged with the great problems of Chemistry, Biology, Computing, Economics and Engineering. Though I strongly suspect that one day some, still to be discovered, deep mathematical structure

---

take the opportunity and embrace it fully, as our own?

[11] An easy step, which will go a long way in this direction, would be to add, as a requirement for mathematics majors, or graduate students, a course containing a comprehensive discussion of the mathematical structures which underly the main physical theories.

[12] This does not exclude the possibility that the same subject may be pursued, *in different ways*, in both Mathematics and Physics departments.

will help explain some of the important features of complex biological systems, we are very far from that. It is certainly to be hoped that individual mathematicians will make significant contributions to these fields but it is unrealistic to think that Mathematics can fully embrace these areas while maintaining its inner continuity, coherence, and fundamental commitment to rigor. We have to distinguish between the core of Mathematics, where I believe the basic physical theories ought to belong, and various problems of Science and Engineering where mathematicians can play a very useful role.

The Main Equations

To return to PDE, I want to sketch a way of looking at the subject from simple first principles which happen to coincide with some of the underlying geometric principles of modern Physics. It turns out that most of our basic PDE's can be derived in this fashion. Thus the main objects of our subject turn out to be in no way less "pure mathematical" in nature than the other fundamental objects[13] studied by mathematicians: numbers, functions and various types of algebraic and geometric structures. But most importantly, these simple principles provide a unifying framework[14] for our subject and thus help endow it with a sense of purpose and cohesion. It also explains why a very small number of linear differential operators, such as the Laplacian and D'Alembertian, are all pervasive; they are the simplest approximations to equations naturally tied to the two most fundamental geometric structures, Euclidean and Minkowskian. The Heat equation is the simplest paradigm for diffusive phenomena while the Schrödinger equa-

---

[13]Some pure mathematicians distrust the basic physical PDE's, as proper objects of Mathematics, on the spurious notion that they are just imperfect approximations to an ultimate physical reality of which we are still ignorant. On the basis of this analysis groups, C^* algebras, topological vector spaces or the $\bar{\partial}$ operator are perfect mathematical objects, as long as they have no direct relations to Physics, while Hamiltonian systems, the Maxwell, Euler, Schrödinger and Einstein equations are not!

[14]The scheme I present below is only an attempt to show that, in spite of the enormous number of PDE's studied by mathematicians, physicists and engineers, there are nevertheless simple basic principles which unite them. I don't want, by any means, to imply that the equations discussed below are the only ones worthy of our attention. It would be also foolish to presume that we can predict which PDE's are going to lead to the most interesting developments. Certainly, nobody could have predicted 100 years ago the emergence on the scene of the Einstein and Yang–Mills equations, or the remarkable mathematical structure behind the seemingly pedestrian KdV equation.

tion can be viewed as the Newtonian limit of a lower order perturbation of the D'Alembertian. The geometric framework of the former is Galilean space which, itself, is simply the Newtonian limit of the Minkowski space, see [M].

Starting with the Euclidean space $\mathbb{R}^n$, the Laplacian Δ is the simplest differential operator invariant under the group of isometries, or rigid transformations, of $\mathbb{R}^n$. The heat, Schrödinger, and wave operators $\partial_t - \Delta$, $\frac{1}{i}\partial_t - \Delta$ and $\partial_t^2 - \Delta$ are the simplest evolution operators which we can form using Δ. The wave operator $\square = -\partial_t^2 + \Delta$ has a deeper meaning; it is associated to the Minkowski space $\mathbb{R}^{n+1}$ in the same way that Δ is associated to $\mathbb{R}^n$. Moreover, the solutions to the equation $\Delta\phi = 0$ can be viewed as special, time independent solutions, to $\square\phi = 0$. The Schrödinger equation can also be obtained, by a simple limiting procedure, from the Klein–Gordon operator $\square - m^2$. Appropriate, invariant, and local definitions of square roots of Δ and $\square$, or $\square - m^2$, corresponding to spinorial representations of the Lorentz group, lead to the associated Dirac operators.

In the same vein we can associate to every Riemannian, or Lorentzian, manifold (M, g) the operators Δ_g, resp. $\square_g$, or the corresponding Dirac operators. These equations inherit in a straightforward way the symmetries of the spaces on which they are defined. There exists a more general, *unreasonably effective*, scheme of generating equations with prescribed symmetries. The variational Principle allows us to associate to any Lagrangian L a system of partial differential equations, called the Euler–Lagrange equations, which inherit the symmetries *built in L*. In view of Noether's principle, to any continuous symmetry of the Lagrangian there corresponds a conservation law for the associated Euler–Lagrange PDE. Thus, the Variational Principle generates equations with desired conservation laws such as Energy, Linear and Angular Momenta, etc. The general class of Lagrangian equations, plays the same selected role among all PDE's as that played by Hamiltonian systems among ODE's. Calculus of Variations is by itself a venerable and vast subject of Mathematics. The main equations of interest in both Geometry[15] and Physics, however, are not just variational; they

---

[15]There are, however, important geometric problems, such as prescribed curvature and isometric embeddings in Riemannian Geometry or Lewy flat surfaces in Complex Geometry, without an obvious variational structure. The real and complex Monge Ampere equations are typical examples. The Pseudo-holomorphic Curves, used by Gromov in the study of symplectic manifolds, provides another example. Nevertheless these equations have a rich geometric structure and share with the variational PDE's many common characteristics. Moreover on closer inspection they may turn out to have a nontrivial,

are obtained from Lagrangians constructed from simple geometric objects such as:

1) *Lorentz or Riemannian metrics*:   On a Lorentzian manifold (M, g) the Lagrangian given by the scalar curvature $R(g)$ of the metric leads, through variations of the metric, to the Einstein-Vacuum (EV) equations of General Relativity. A similar procedure leads to Einstein metrics in Riemannian geometry. The restriction of the Einstein functional $\int R(g)dv_g$ to a conformal class of metrics leads to the well-known Yamabe equation.

2) *Connections on a principal bundle*:   The quadratic scalar invariant formed by the curvature of a connection defines the Yang–Mills Lagrangian. The Yang–Mills (YM) equations are obtained through variations of the connection. The Maxwell equations correspond to the case of a trivial bundle over the Minkowski space with structure group $U(1)$. The standard model of particle physics corresponds to the group $SU(3) \times SU(2) \times U(1)$. The YM equations used in Topology correspond to Riemannian connections with nonabelian group $SU(2)$.

3) *Scalar equations*:   Are derived for scalar functions $\phi : M \to \mathbb{R}, \mathbb{C}$. The Lagrangian is $L = g^{\mu\nu}\phi_\mu\phi_\nu + V(\phi)$, with $V(\phi) \geq 0$. When $V = 0$ we derive $\Delta_g \phi = 0$, in the Riemannian case, and $\Box_g \phi = 0$ in the Lorentzian case. The case $V(\phi) = \frac{1}{2}m^2|\phi|^2$ corresponds to the Klein–Gordon equation, $V(\phi) = \frac{1}{4}|\phi|^4$ leads to the well-known cubic wave equation. We will refer to this type of equations as nonlinear scalar wave equations (NSWE).

4) *Mappings between two manifolds*:   Consider mappings $\phi : (M, g) \to (N, h)$ between the pseudoriemannian domain manifold M of dimension $d + 1$ and Riemannian target N of dimension n. Let ϕ^*h be the symmetric 2-tensor on M obtained by taking the pull-back of the metric h of N. Let $\lambda_0, \lambda_1, \ldots, \lambda_d$ be the eigenvalues of ϕ^*h relative to the metric g and $S_0, S_1, \ldots, S_d$ the corresponding elementary symmetric polynomials in $\lambda_0, \lambda_1, \ldots, \lambda_d$. Any symmetric function of $\lambda_0, \lambda_1, \ldots, \lambda_d$, or equivalently, any function $L(S_0, S_1, \ldots, S_d)$, can serve as a Lagrangian. By varying the action integral $\int_M Ldv_g$ relative to ϕ, with dv_g the volume element of the metric g, we obtain a vast class of interesting equations. Here are some examples:[16]

---

hidden, variational structure. This is the case, for example, of the Monge Ampere equations, see [Br].

[16]I want to thank D. Christodoulou for his help in the presentation of this section. Most of the examples below, and much more, are discussed in detail in his book [Chr1].

(i) The Harmonic and Wave Maps (WM) are obtained in the particular case $L = \mathrm{tr}_g(\phi^* h)$. The only distinction between them is due to the character, Riemannian respectively Lorentzian, of the metric g.

(ii) The basic equations of Continuum Mechanics are obtained from a general Lagrangian, as described above, in the particular case when g is Lorentzian, $n = d = 3$ and the additional assumptions that ϕ has maximal rank at every point and the curves $\phi^{-1}(p)$ are time-like for all $p \in N$. Since the dimension of N is one less than the dimension of M one of the eigenvalues, say λ_0, is identically zero. Elasticity corresponds to general choices of L as a symmetric function of $\lambda_1, \ldots, \lambda_d$. Fluid Mechanics corresponds to the special case when L depends only on the product $\lambda_1 \cdot \lambda_2 \cdots \cdot \lambda_d$. One can also derive the equations of Magneto-hydrodynamics (MHD) by assuming an additional structure on N given by a 2-form Ω. The 2-form $F = \phi^* \omega$ defines the electromagnetic field on M. The Lagrangian of MHD is obtained by adding the Maxwell Lagrangian $\frac{1}{2} F_{\mu\nu} \cdot F^{\mu\nu}$ to the fluid Lagrangian described above.

(iii) The minimal surface equation is derived from the Lagrangian $L = \sqrt{\det_g \phi^* h} / \sqrt{\det(g)}$ in the case when g is Riemannian and $m = d + 1 < n$. The case when g is Lorentzian leads to a quasilinear wave equation.

5) *Lagrangian leading to higher order equations*: While the main equations of Physics are all first or second order, there is no reason why one should avoid higher order equations for applications to Geometry. It is natural, for example, to consider equations associated to conformally invariant Lagrangians. Many of the known Lagrangians, which lead to second order equations such as Harmonic Maps, are conformally invariant only in dimension 2. To produce a larger class of conformally invariant equations, in even dimensions, it pays to look for higher order theories such as biharmonic maps in 4D, see [CWY]. The variational problem associated to the zeta functional determinant of the Laplace–Beltrami operator, of a higher dimensional Riemannian metric, also leads to higher order equations. Finally the Willmore problem for closed surfaces in $\mathbb{R}^3$ provides another interesting example of a fourth order equation.

6) *Composite Lagrangians*: By adding various Lagrangians we derive other basic equations. This is true, most remarkably, for the gravitational Lagrangian, given by the scalar curvature of the metric. In combination with the Lagrangian of a matter theory, in fact any other relativistic La-

grangians described above, it leads to the famous Einstein Field Equations $R_{\mu\nu} - \frac{1}{2}g_{\mu\nu}R = 8\pi T_{\mu\nu}$, with T the energy momentum tensor of the matter Lagrangian. The Lagrangian of the Seiberg–Witten equations are obtained by coupling the Lagrangian of the Maxwell theory with that of the Dirac equation.

The equations derived by the above geometric constructions are elliptic, if the metric g on M is Riemannian, and hyperbolic if g is Lorentzian. In the hyperbolic case we distinguish between the Field Theories, for which the only characteristics of the corresponding PDE's are given by the Lorentz metric g, and the other equations; Fluids, Continuum Mechanics, MHD, etc., which have additional characteristics. The YM, WM and the EV are all field theories in the sense we have just defined. The EV equations is distinguished from the other field theories by being the only one for which the metric g itself is the solution. This fact gives the EV equations a quasilinear character. For all other field equations, since the metric g is fixed, the equations are semilinear.

With the exceptional case of EV, which does not have local conservation laws, all equations described above have associated to them, a well-defined energy-momentum tensor T which verifies the positive energy condition. I recall that the energy-momentum tensor of a Lagrangian theory is a rank 2 symmetric tensor $T_{\mu\nu}$ verifying the local conservation law $D^\mu T_{\mu\nu} = 0$. We say that T satisfies the positive energy condition if $T(X,Y) \geq 0$ for all time-like future oriented vectorfields X, Y.

Many other familiar equations can be derived from the *fundamental ones* described above by the following procedures:

(a) *Symmetry reductions*: Are obtained by assuming that the solutions we are looking for have certain continuous symmetries. They lead to much simpler equations than the original, often intractable ones. Another, somewhat more general, way of obtaining simpler equations is to look for solutions which verify a certain ansatz, such as stationary, spherically symmetric, equivariant, self-similar, traveling waves, etc.

(b) *The Newtonian approximation and other limits*: We can derive a large class of new equations, from the basic ones described above, by taking one or more characteristic speeds to infinity. The most important one is the Newtonian limit, which is formally obtained by letting the velocity of light go to infinity. At the level of the space-time manifold itself this limit, described in the seminal paper of Minkowski [M], takes a Lorentz manifold to the Galilean space-time of Newtonian mechanics. As we have mentioned above the Schrödinger equation itself can be derived, in this fashion, from the linear Klein–Gordon equation. In the same way we can formally derive the Lagrangian of nonrelativistic Elasticity (see [Z]), Fluids or MHD equations. The formal Newtonian limit of the full Einstein field equations leads to the various continuum mechanics theories in the presence of Newtonian gravity. The Newtonian potential is tied to the lapse function of the original space-time metric.

We should not be surprised that the better known nonrelativistic equations, look more messy than the relativistic ones. The simple geometric structure of the original equations gets lost in the limit. The remarkable simplicity of the relativistic equations is a powerful example of the importance of Relativity as a unifying principle.

Once we are in the familiar world of Newtonian physics we can perform other well-known limits. The famous incompressible Euler equations are obtained by taking the limit of the general nonrelativistic fluid equations as the speed of sound tends to infinity. Various other limits are obtained relative to other characteristic speeds of the system or in connection with specific boundary conditions, such as the boundary layer approximation in fluids. The equations of Elasticity, for example, approach in the limit, when all characteristic speeds tend to infinity, to the familiar equations of a rigid body in Classical Mechanics. Another important type of limit, leading to the well-known Hamilton–Jacobi equations of Classical Mechanics, is the high frequency or the geometric optics approximation.

Many of these very singular limits remain purely formal. While some of them have been rigorously derived, many more present serious analytic difficulties.

(c) *Phenomenological assumptions*:   Even after taking various limits and making symmetry reductions, the equations may still remain unyielding. In various applications it makes sense to assume that certain quantities are small and may be neglected. This leads to simplified equations which could be called *phenomenological*[17] in the sense that they are not derivable from first principles. They are used to illustrate and isolate important physical phenomena present in complicated systems. A typical way of generating interesting phenomenological equations is to try to write down the simplest model equation which describes a particular feature of the original system.

---

[17]I use this term here quite freely, it is typically used in a somewhat different context. Also some of the equations which I call phenomenological below, e.g., dispersive equations, can be given formal asymptotics derivations by Applied Math. techniques.

Thus, the self-focusing, plane wave effects of compressible fluids, or elasticity, can be illustrated by the simple minded Burgers equation $u_t + uu_x = 0$. Nonlinear dispersive phenomena, typical to fluids, can be illustrated by the famous KdV equation $u_t + uu_x + u_{xxx} = 0$. The nonlinear Schrödinger equations provide good model problems for nonlinear dispersive effects in Optics. The Ginzburg–Landau equations provide a simple model equation for symmetry breaking phase transitions. The Maxwell–Vlasov equations is a simplified model for the interactions between Electomagnetic forces and charged particles, used in Plasma Physics.

When well chosen, a model equation leads to basic insights into the original equation itself. For this reason simplified model problems are also essential in the day to day work of the rigorous PDE mathematician. We all test our ideas on such carefully selected model problems. It is crucial to emphasize that good results concerning the basic physical equations are rare; a very large percentage of important rigorous work in PDE deals with simplified equations selected, for technical reasons, to isolate and focus our attention on some specific difficulties present in the basic equations.

It is not at all a surprise that the equations derived by symmetry reductions, various limits and phenomenological assumptions have additional symmetries and therefore additional conservation laws. It is however remarkable that some of them have infinitely many conserved quantities or turn out to be even integrable. The discovery of the integrability of the KdV equation and, later, that of other integrable PDE's is one of the most impressive achievements of the field of PDE's in this century. It remains also the model case of a beneficial interaction between numerical experiments, heuristic applied mathematics arguments algebra and rigorous analysis. Together they have led to the creation of a beautiful mathematical theory with extensive and deep applications outside the field of PDE's where they have originated from. We have to be aware, however, of the obvious limitations of integrable systems; with few exceptions (the KP-I and KP-II equations are, sort of, 2-dimensional) all known integrable evolution equations are restricted to one space dimension.

In all the above discussion we have not mentioned diffusive equations such as the Navier–Stokes. They are in fact not variational and, therefore, do not fit at all in the above description. They provide a link between the microscopic, discrete, world of Newtonian particles and the continuous macroscopic one described by Continuum Mechanics. Passing from discrete to continuous involves some loss of information hence the continuum equations have diffusive features. The best known examples of diffusive effects are the "heat conduction," which appears in connection with the dissipation of energy in compressible fluids, and "viscosity," corresponding to dissipation of momentum, in Fluids. Another example is that of "electrical resistivity" for the electrodynamics of continuum media. The Navier–Stokes equation appears in the incompressible limit. The incompressible Euler equations are the formal limit of the Navier–Stokes equations as the viscosity tends to zero. Because of the loss of information involved in their derivation the diffusive equations have probabilistic interpretations.

Diffusive equations turn out to be also very useful in connection with geometric problems. Geometric flows such as mean curvature, inverse mean curvature, Harmonic Maps, Gauss Curvature and Ricci flows are some of the best known examples. Some can be interpreted as the gradient flow for an associated elliptic variational problem. They can be used to construct nontrivial stationary solutions to the corresponding stationary systems, in the limit as $t \to \infty$, or to produce foliations with remarkable properties, such as that used recently in the proof of the Penrose conjecture.

REMARK. *The equations which are obtained by approximations or by phenomenological assumptions present us with an interesting dilemma. The dynamics of such equations may lead to behavior which is incompatible with the assumptions made in their derivation. Should we continue to trust and study them, nevertheless, for pure mathematical reasons or should we abandon them in favor of the original equations or a better approximation? Whatever one may feel about this in a specific situation it is clear that the problem of understanding, rigorously, the range of validity of various approximations is one of the fundamental problems in PDE.*

The Problem of Breakdown

The most basic mathematical question in PDE is, by far, that of regularity. In the case of elliptic equations, or subelliptic in Complex Analysis, the issue is to determine the regularity of the solutions to a geometric variational problem. In view of the modern way of treating elliptic equations, one first constructs a generalized solution by using the variational character of the equations. The original problem, then, translates to that of showing that the generalized solution has additional regularity. This is a common technique for both linear and nonlinear problems. Moreover the technique can be extended to scalar, fully nonlinear, nonvariational problems, such as

Monge-Ampere equations, with the help of the viscosity method. In linear cases as well as in some famous nonlinear cases, such as the minimal hypersurfaces as graphs over convex domains (or more generally mean convex), one can show that the generalized solutions are smooth. The solutions to the general Plateau problem, however, may have singularities. In this case the main issue becomes the structure of the singular set of nonsmooth solutions. Geometric Measure Theory provides sophisticated analytical tools to deal with this problem. Singularities are also known to occur in the case of higher dimensional harmonic maps, for positively curved target manifolds such as spheres.

In the case of evolution equations the issue is the possible spontaneous, finite time (in view of results concerning local in time existence, the breakdown can only occur after a short time interval), breakdown of solutions, corresponding to perfectly nice initial conditions. This is a typical nonlinear, multidimensional phenomenon.[18] It can be best illustrated in the case of the Burgers equation $u_t + uu_x = 0$. Despite the presence of infinitely many positive conserved quantities, $\int |u(t,x)|^{2k}dx$, $k \in \mathbb{N}$, all solutions, corresponding to smooth, compactly supported, nonzero initial data at $t = 0$, breakdown in finite time. The breakdown corresponds, physically, to the formation of a shock wave. Similar examples of breakdown can be constructed for compressible fluids or Elasticity, see [J], [Si]. Singularities are also known to form, in some special cases, for solutions to the Einstein field equations in General Relativity. Moreover, one expects this to happen, in general, in the presence of strong gravitational fields. It is also widely expected that the general solutions of the incompressible Euler equations in three space dimensions, modeling the behavior of inviscid fluids, breakdown in finite time. Some speculate that the breakdown may have something to do with the onset of turbulence for incompressible fluids with very high Reynolds numbers. These fluids are in fact described by the Navier–Stokes equations. In this case the general consensus is that the evolution of all smooth, finite energy, initial data lead to global in time, smooth, solutions. The problem is still widely open. *It is conceivable that there are in fact plenty of solutions which break down but are unstable, and thus impossible to detect numerically or experimentally.*

Breakdown of solutions is also an essential issue concerning nonlinear

---

[18] For smooth, one dimensional, Hamiltonian systems with positive energy, solutions are automatically global in time. This the case, for example, of the nonlinear harmonic oscillator $\frac{d^2}{dt^2}x + V'(x) = 0$, $V \geq 0$.

geometric flows, such as the mean and inverse mean curvature flows, Ricci flow, etc. As singularities do actually form in many important geometric situations, one is forced to understand the structure of singularities and find ways to continue the flow past them. Useful constructions of generalized flows can lead to the solution of outstanding geometric problems, as in the recent case of the Penrose conjecture [HuI].

The problem of possible breakdown of solutions to interesting, non-linear, geometric and physical systems is not only the most basic problem in PDE; it is also the most conspicuous unifying problem, in that it affects all PDE's. It is intimately tied to the basic mathematical question of understanding what we actually mean by solutions and, from a physical point of view, to the issue of understanding the very limits of validity of the corresponding physical theories. Thus, in the case of the Burgers equation, for example, the problem of singularities can be tackled by extending our concept of solutions to accommodate "shock waves," i.e. solutions discontinuous across curves in the t, x space. One can define a functional space of generalized solutions in which the initial value problem has unique, global solutions. Though the situation for more realistic physical systems is far less clear and far from being satisfactorily solved, the generally held opinion is that shock wave type singularities can be accommodated without breaking the boundaries of the physical theory at hand. The situation of singularities in General Relativity is radically different. The type of singularities expected here is such that no continuation of the solutions is possible without altering the physical theory itself. The prevaling opinion, in this respect, is that only a quantum field theory of Gravity could achieve this.

One can formulate a general philosophy to express our expectations with regard to regularity. To do that we need to classify our main equations according to the strength of their nonlinearities relative to that of the known *coercive* conservation laws or other a priori estimates. Among the basic conservation laws that provided by the Energy is coercive, because it leads to an absolute, local, space-time bound on the size of solutions, or their first derivatives. The others, such as the linear and angular momentum, do not provide any additional information concerning local regularity. For the basic evolution equations, discussed in the previous section, the energy integral provides the best possible a priori estimate and therefore the classification is done relative to it. This raises a question of fundamental importance; *are there other, stronger, local a priori bounds which cannot be derived from Noether's Principle?* There are methods which can rule out the existence of some exact conserved quantities, different from the physical ones, yet there is no reason, I believe, to discount other, more subtle bounds. A well-known *Morawetz multiplier method* leads, for some classes of nonlinear wave equations, to bounded space-time quantities which do not correspond to any conservation law. The Morawetz quantity, however, has the same scaling properties as the energy integral; it only provides additional information in the large.

REMARK. *The discovery of any new bound, stronger than that provided by the energy, for general solutions of any of our basic physical equations would have the significance of a major event.*

In other cases, when there are additional symmetries, one often has better a priori estimates. For many elliptic equations, for example, one can make use of the maximal principle or some monotonicity arguments to derive far more powerful a priori estimates than those given by the energy integral. Integrable equations, such as KdV, also have additional, coercive, conservation laws. As explained above, the Burgers equation has infinitely many positive conserved quantities. The incompressible Euler equations in dimension $n = 2$ have, in addition to the energy, a pointwise a priori estimate for the vorticity. It is for this reason that we can prove global regularity for 2D Euler equations. In all these cases the classification has to be done relative to the optimal available a priori estimate.

In what follows I will restrict myself to the case I find, personally, most interesting, that of the basic evolution equations for which there are no better known, a priori estimates than those provided by the Energy integral. These include all relativistic field theories, Fluids, Continuum Mechanics and Magentofluidynamic, in three space dimensions and the absence of any additional symmetries. In these cases the classification is done by measuring the scaling properties of the energy integral relative to those of the equations. To illustrate how this is done consider the nonlinear scalar equation $\Box \phi - V'(\phi) = 0$ with $V(\phi) = \frac{1}{p+1}|\phi|^{p+1}$. The energy integral is given by $\int \left(\frac{1}{2}|\partial \phi(t, x)|^2 + |\phi|^{p+1}(t, x) \right) dx$. If we assign to the space-time variables the dimension of length, L^1, then $\Box$ has the dimension of L^{-2} and ϕ acquires, from the equation, the dimension $L^{\frac{2}{1-p}}$. Thus the energy integral has the dimension L^e, $e = n - 2 + \frac{4}{1-p}$. We say that the equation is subcritical if $e < 0$, critical for $e = 0$ and supercritical for $e > 0$. The same analysis can be done for all the other basic equations. YM is subcritical for $n \leq 3$, critical for $n = 4$ and supercritical for $n > 4$. WM is subcritical for $n = 1$, critical for $n = 2$, and supercritical for all other dimensions. The same holds true for the Einstein Vacuum equations. Most of our basic equations, such as EV, Euler, Navier–Stokes, Compressible Euler, Elasticity, etc., turn out to be supercritical in the physical dimension $n = 3$. A PDE is said to be regular if all smooth, finite energy, initial conditions lead to global smooth solutions.

The general philosophy is that subcritical equations are regular while supercritical equations may develop singularities. Critical equations are important[19] borderline cases. For the particular case of field theories, as defined in the previous section, one can formulate a more precise conjecture:

GENERAL CONJECTURE. (i) *All basic, subcritical, field theories are regular for all smooth data.*

(ii) *Under well defined restrictions on their geometric set-up the critical field theories are regular for all smooth data.*

(iii) *"Sufficiently small" solutions to the supercritical field theories are regular. There exist solutions, corresponding to large, smooth, finite energy data, which develop singularities in finite time.*

The part (iii) of the Conjecture is the most intriguing. The fact that all small solutions are regular seems to be typical to field theories; it may fail for fluids or the general elasticity equations. The issue of existence of singular solutions for supercritical equations is almost entirely open. In the case of supercritical, defocusing NSWE, $\Box\phi - V'(\phi) = 0$ for positive power law potential V, most analysts, familiar with the problem, expect that global regularity still prevails. Numerical calculations seem to support that view. *It is however entirely possible that singular solutions exist but are unstable and therefore difficult to construct analytically and impossible to detect numerically.* A similar phenomenon may hold true in the case of the 3D Navier–Stokes equations, which would contradict the almost universal assumption that these equations are globally regular.

If this worst case scenario is true, the big challenge for us would be to prove that *almost all* solutions to such equations are globally regular. At the opposite end of possible situations is that for which almost all solutions

---

[19]Some of the most exciting advances in Geometric PDE's in the last twenty five years involve the study of PDE's which are critical relative to the optimal available a priori estimates. This is the case of the Yamabe problem (related to the critical exponent of the Sobolev inequality), Weakly Harmonic Maps in 2D, Yang–Mills connections in 4D, the Wilmore problem in 2D. See [S] for a beautiful survey and [Y] for his updated list of problems in Geometry.

form singularities. The 3D incompressible Euler equations are a good candidate for this situation. Moreover it is not inconceivable that this *most unstable of all known equations* would exhibit the following perverse scenario: *The set of all initial data which lead to global regular solutions has measure zero, yet, it is dense in the set of all regular initial conditions, relative to a reasonable topology.* Such a possibility, which cannot be ruled out, would certainly explain why it is so difficult to make any progress on the 3D Euler equations with our present techniques. It would also explain, in particular, why it is so difficult to produce specific examples, or numerical evidence, of the widely expected finite time breakdown of solutions.

REMARK. *The development of methods which would allow us to prove generic, global, results may be viewed as one of the great challenges for the subject of PDE's in the next century.*

It is expected that the global structure of singularities in General Relativity will have to be phrased in terms of generic conditions (see [AM] and [W] for up to date surveys concerning Cosmic Censorship and recent mathematical progress on it). Understanding the problem of turbulence for the Navier–Stokes equations would almost certainly require a statistical approach. The effectiveness of many geometric flows is hindered by the presence of bad, seemingly nongeneric, type of singularities. So far the subject of nonlinear PDE's has been dominated by methods well suited for the study of individual solutions; we have had very little success in dealing with families of solutions. By comparison in the case of finite dimensional Hamiltonian systems the natural Liouville measure, defined in the space phase, allows one to prove nontrivial generic results[20] such as Poincaré's recurrence theorem.

The Problem of Well-posedness for Nonlinear Equations

With the exception of the a priori estimates derived from conservation laws, or monotonicity and maximum principle for elliptic or parabolic equations, almost all methods currently used to deal with nonlinear PDE's depend on elaborate comparison arguments between solutions to the original system and those of an appropriate linearization of it. It is essential to have very precise estimates for the linear system, in tune with the a priori estimates

---

[20]There exist some interesting generic results in PDE also, based on the construction of Gibbs measures on the space of solutions, see [B1,2]. Unfortunately the class of equations for which such measures can be constructed is extremely limited.

and the scaling properties of the nonlinear equations. In the case of elliptic and parabolic problems we have a large and powerful arsenal of such estimates, almost all developed during the course of this century, see [BreB]. Our knowledge of linear estimates for hyperbolic and dispersive equations is far less satisfactory.

The need for a well adapted linear theory, for evolution equations, can be best understood from the perspective of the problem of optimal well posedness. In what follows I will limit my discussion to field theories such as the nonlinear scalar wave equation (NSWE), Yang–Mills (YM), Wave-Maps (WM) and the Einstein Vacuum (EV) equations. My goal is to write down three specific conjectures, WP1–WP3, which are, I feel, just beyond the boundary of what can be obtained with present day techniques. They are thus both accessible and important to generate interesting mathematics.

The initial value problem for an evolutionary system of equations is said to be well posed (WP) relative to a Banach, functional, space X if, for any data in X, there exist uniquely defined local in time solutions belonging to X for $t \neq 0$, and depending continuously on the data. The problem is said to be strongly WP if the dependence on the data is analytic and weakly WP if the dependence is merely continuous or differentiable. In the case of hyperbolic equations, especially quasilinear, there is a natural, apparently unique, choice for X. Locally, it has to coincide with the Sobolev[21] space $H^s(\mathbb{R}^n)$. This is due to the fact that L^p norms are not preserved by the linear evolution in dimension $n > 1$ while norms defined in Fourier space are meaningless for quasilinear equations. Taking into account the scaling properties of the basic field equations and proceeding in the same manner as in the previous section, one can define the critical WP exponent s_c to be that value of s for which the H^s norm of initial data is dimensionless. With this definition we can formulate the following:

GENERAL WP CONJECTURE. i) *For all basic field theories the initial value problem is locally, strongly well posed for any data in H^s, $s > s_c$.*

ii) *The basic field theories are weakly, globally well posed for all initial data with small H^{s_c} norm.*

iii) *There can be no well defined solutions[22] for $s < s_c$.*

---

[21]We talk of a space H^s rather than a pair H^s, H^{s-1}. Thus, in the case of the IVP for the wave equation $\Box \phi = 0$, $\phi(0) = f$, $\partial_t \phi(0) = g$, $(f, g) \in H^s$ means $f \in H^s$, $g \in H^{s-1}$.

[22]Weak solutions may exist below the s_c threshold but are, completely unstable and have weird properties. In other words weak solutions, corresponding to $s < s_c$, are mathematical "ghosts".

The proof of the WP conjecture for $s \geq s_c$ will provide us with an essential tool for the problem of regularity discussed in the previous section. So far the conjecture was proved only in the case of NSWE (see [K] and [ShS]); it is based entirely on Strichartz type inequalities. Semilinear equations whose nonlinear terms involve derivatives, such as YM and WM, are far more difficult, see discussion below. The case $s < s_c$ is interesting for a philosophical reason. There are supercritical cases (in the case of the supercritical NSWE see [Str], for the case of WM see [Sh], [MüS]), for which one can prove the existence of a weak solution corresponding to any, finite energy, initial conditions. Part iii) of the above conjecture asserts that these solutions are unstable (it is easy in fact to see that they are linearly unstable) and therefore not particularly useful. It is interesting to remark, in this respect, the recent remarkable result of Schaeffer [S], see also [Shn]. Schaeffer has constructed examples of weak solutions for the 2D Euler equations which are compactly supported in space-time! The result is reminiscent of the famous result of Nash [N], see also Kuiper [Ku], on C^1 isometric imbeddings, which turn out to be plentiful, dense in the set of all smooth functions, and a lot more pliant than the more regular ones.[23] Another remarkable example of how bad weak solutions can sometimes be is that of Rivière, concerning weak harmonic maps from a three dimensional space to S^2 with a dense set of singularities [R]. This is in sharp contrast to the case of minimizers [SchU], or stationary solutions [E] for the same equations. I suspect that similar, *unacceptable properties of weak solutions* type results can be proved for solutions to nonlinear wave equations, below the critical regularity. Moreover, short of additional regularity assumptions on the initial data, there may exist no *entropy type conditions* which would stablize the solutions.

In the case of subcritical equations, for which the energy norm is stronger than H^{s_c}, part i) of the conjecture would imply well-posedness in the energy norm, and therefore, by energy conservation, global well-posedness and regularity. In other words the solutions preserve the H^s regularity of the data for any $s > s_c$. This would thus settle the first part of the General Conjecture stated in the previous section.

In the case of critical equations, part ii) of the WP conjecture will imply the following:

SMALL ENERGY CONJECTURE. *For all basic critical field theories all small*

---

[23]This phenomenon has been called the h-Principle and discussed in a very general set-up by M. Gromov, see [Gr].

energy solutions are globally regular.

The small energy conjecture is an essential step in the proof of the general regularity conjecture for critical field theories. In the case of wave equations, whose nonlinearities do not depend on derivatives or in the case of spherical symmetric solutions, one can prove it directly. In the case of equations like YM or WM, with derivatives appearing in the nonlinear terms, it is now believed that the only way to settle the small energy conjecture is to prove the much stronger part ii) of the WP conjecture. In what follows I will give a more precise formulation of it for the special case of the WM and YM equations.

CONJECTURE WP1. *The Wave Maps equation, defined from the Minkowski $\mathbb{R}^{n+1}$ to a complete, Riemannian, target manifold, is globally well posed for small initial data in $H^{n/2}$, $n \geq 2$.*

CONJECTURE WP2. *The Yang–Mills equation, for $SO(N), SU(N)$ structure groups, is globally well posed for small initial data in $H^{\frac{n-2}{2}}$, $n \geq 4$.*

To understand the difficulties involved in WP1, I will summarize below what are the most significant known results in connection to it.

1) The conjecture is true in the case of equivariant wave maps, see [ShZ], in which case the nonlinear terms do not depend on derivatives. In [ChrZ] the small energy conjecture was proved for the special case of spherically symmetric solutions. Their approach avoids the proof of the WP1 conjecture, which is still not known, even in the spherically symmetric case, by proving directly, in this case, the small energy conjecture. In the general case it does not seem possible to prove the small energy conjecture independent of Conjecture WP. This has to do with the lack of any space-time L^p, $p \neq 2$, first derivative estimates (see [Wo1]) for solutions to $\Box \phi = F$.

2) In [KlM3] and [KlS] one proves local well-posedness for all data in H^s, $s > s_c = n/2$, $n \geq 2$ (see also [KeT] for $n = 1$). The result depends heavily on bilinear estimates. This was further improved in [T1], who has established well-posedness (his result is in fact global in time, in view of the scaling properties of the equations) for small data in the Besov space $B_2^{n/2,1}$. Both above mentioned results fail to to take into account the completeness of the target manifold.

3) We know, from simple examples, that we may not have $H^{n/2}$-well-posedness if the target manifold is not complete.

4) The dependence of solutions on the data, with respect to the $H^{n/2}$ norm, cannot be twice differentiable.

The methods which have been used to tackle the case $s > s_c$ depend heavily on an iterative procedure in which one estimates the $H^{s,\delta}$ norm of each iterate, for $s > n/2$, $\delta > 1/2$, in terms of the $H^{s,\delta}$ norms of the previous iterates. These norms, defined with respect to the space-time Fourier transform, are intimately tied to the symbol of $\square$ and to bilinear estimates, see [KlM1,3], [KlS] and [FoK]. Similar norms where introduced by J. Bourgain [B3], see also [KenPV], in connection with nonlinear dispersive equations.

To treat the critical case one needs to overcome two difficulties. The first has to do with improving the estimates at each iterative step, to make them optimal. The second is an important conceptual difficulty, which has to do with the iterative process itself. Any iterative procedure, if successful, would imply not only well posedness but also analytic dependence on the data in the $H^{n/2}$ norm. This is however wrong, according to the observation (3) above. To understand this effect consider the Hilbert space $X = H^{n/2}(\mathbb{R}^n)$, u a function in X, and let $\Phi(t) = e^{itu}$. It is known that $\Phi(t)$ is a C^1 function of t with values in X but, since X is not closed under multiplications, it is not in C^2 (see [KeT]). The reason $e^{itu} \in X$ is due to the fact that the function e^{iu} is bounded, it cannot be guessed by just considering the Taylor expansion $e^{iu} = \sum_{n \geq 0} \frac{1}{n!} (iu)^n$ in which all terms are divergent.

In the case of the WM equations any iterative procedure loses the crucial information about the completeness of the target manifold and therefore leads to logarithmic divergences. To see that consider WM solutions of the form $\phi = \gamma(u)$ where $\square u = 0$ with data in $H^{n/2}$ and γ is a geodesic of the target manifold M. Since the L^∞ norm of u is not controlled, $\gamma(u)$ makes sense only if the geodesic is globally defined. A standard iteration fails to distinguish between complete and incomplete geodesics.

This situation seems to call for a "renormalization" procedure. More precisely, one may hope that by understanding the nature of the logarithmic divergences of each iterate, we can overcome them by a clever regularization and limiting procedure. In view of the simple minded model problem studied in [KlM4] one may hope that such an approach is not impossible.

I will only make a few remarks concerning the WP2 conjecture. The optimal known result, in dimension $n \geq 4$, is small data well-posedness for $s > s_c$, see [KlM5] and [KlT]. In the case $s = s_c$ it can be shown that any iteration procedure leads to logarithmic divergences. The situation seems thus similar to that described in the previous conjecture. In dimension $n = 3$ we have global well posedness in the energy norm $s = 1$, see [KlM2] and the discussion in connection to WP3 below, and local well-posedness for $s > 3/4$. It is not at all known what happens for $s_c = 1/2 < s \leq 3/4$.

The case of the Einstein Vacuum equations is far more difficult than that of WM or YM. Written relative to wave coordinates the EV equations take the form, $g^{\mu\nu}\partial_\mu\partial_\nu g_{\alpha\beta} = N(g, \partial g)$, where g is a Lorentz metric and N is a nonlinear term quadratic in the first derivatives of g. This form of the Einstein equation leads to the study of quasilinear wave equations of the form:

$$\Box_{g(\phi)}\phi = \Gamma(\phi)Qr(\partial\phi, \partial\phi), \tag{1}$$

with $g(\phi)$ a Lorentz metric depending smoothly on ϕ, Γ smooth function of ϕ and $Qr(\partial\phi, \partial\phi)$ quadratic in $\partial\phi$. Other types of quasilinear wave equations, such as those appearing in Elasticity or Compressible Fluids, depending only on $\partial\phi$ can be written as systems of wave equations of type (1).

Using energy estimates and Sobolev inequalities one can prove the "classical local existence" result, or local well-posedness, for H^s initial data with $s > s_c + 1 = \frac{n}{2} + 1$. This result leads, in the case of the EV expressed relative to wave coordinates, to the well-known local existence result of Y.C. Bruhat. (Bruhat's result, see [Bru], requires in fact more derivatives of the data. The optimal $3 + 1$ dimensional result, $s > s_c + 1 = 5/2$ was proved in [FM].)

Getting close to the critical exponent $s = s_c = 3/2$ is entirely out of reach. I believe, however, that the intermediate result, $s = 2$, is both very interesting and accessible.

CONJECTURE WP3. *The Einstein Vacuum equations are strongly, locally, well posed for initial data sets*[24] *(Σ, g, k) for which $Ric(g) \in L^2(\Sigma)$ and $k \in H^1(\Sigma)$.*

The conjecture can be viewed, in a sense, as a far more difficult analogue of the well-posedness result, see [KlM2], for the $3 + 1$ YM equations in the energy norm. Writing the YM in the Lorentz gauge, which is the precise analogue of wave coordinates, one is led to a system of equations of the form

$$\Box\phi = Qr(\phi, \partial\phi) + C(\phi), \tag{2}$$

with Qr quadratic in $\phi, \partial\phi$ and linear in $\partial\phi$ and C cubic in ϕ. In this case the scaling exponent is $s_c = \frac{n-2}{2}$. The classical local existence result, based on energy estimates and the $H^\sigma \subset L^\infty, \sigma > n/2$ Sobolev estimate,

---

[24](Σ, g) is a Riemannian 3D manifold and k a symmetric 2-tensor, verifying the constraint equations.

requires data in H^s, $s > s_c + 1$. One can improve the result to $s > s_c + \frac{1}{2}$ for $n = 3$ and $s = s_c + \frac{1}{2}$ in higher dimensions by using the classical Strichartz[25] type inequalities for solutions to the inhomogeneous standard wave equation $\Box \phi = F$, Moreover one can show, see [L1], that for $n = 3$ the result $s > s_c + \frac{1}{2} = 1$ is optimal for general equations of type (2). Therefore to prove the H^1 well-posedness result for the Yang–Mills equations one needs to take advantage of some additional cancellations present in the nonlinear terms. One can do that by using the "gauge covariance" of the Yang–Mills equations, according to which a solution of YM is a class of equivalence of solutions relative to gauge transformations. In view of this one is free to pick the particular gauge conditions best suited to the problem at hand. In [KlM2] the choice of the Coulomb gauge leads to a coupled system of elliptic-hyperbolic equations which satisfies the "null condition". This means, very roughly, that the hyperbolic part of the YM (Coulomb) system has the form

$$\Box \phi = Q(\phi, \phi) + \text{better behaved terms.}$$

with $Q(\phi, \phi)$ a nonlocal "null" quadratic form. To deal with the cancellations present in the null quadratic forms Q one has developed the so called bilinear estimates, see [KlM1], [FoK].

In trying to implement a similar strategy to EV one encounters fundamental difficulties due the quasilinear character of the Einstein equations. For example, to improve Bruhat's classical local existence result from $s > s_c + 1$ to $s > s_c + \frac{1}{2}$, in wave coordinates, one needs to prove a version of the classical Strichartz estimates for $\Box$ replaced with the wave operator $\Box_g$, where g is a rough (assuming we fix ϕ, the metric $g(\phi)$ will have the "expected" regularity of ϕ) Lorentz metric.

Until recently this seemed to be an intractable problem. In fact it is known that, if the coefficients of a linear wave equation have less regularity than $C^{1,1}$, some of the main Strichartz inequalities may fail, see [SmS1]. H. Smith, see [Sm], was also able to show that all the Strichartz type inequalities hold true if the coefficients are at least $C^{1,1}$ and $n \leq 3$, see [T2, parts II and III] for $n \geq 3$. The $C^{1,1}$ condition, however, is much too strong to apply to nonlinear equations.

Recently J.Y. Chemin and H. Bahouri, see [ChB], have succeeded in deriving the first improvement over the classical result. They have proved

---

[25]The Strichartz type inequalities are intimately tied to restriction results in Fourier Analysis. Together with the more recent bilinear estimates they exemplify the strong, modern, ties between Harmonic Analysis and nonlinear wave equations.

local WP for equations of type 1 provided that $s > s_c + \frac{3}{4}$ for $n \geq 3$ and $s > s_c + \frac{7}{8}$ for $n = 2$. The same result was proved also by D. Tataru [T2] using a somewhat different method. Both Chemin-Bahouri and Tataru have later obtained some further improvements but fall short of the expected optimal result. (The optimal known result, $s > s_c + \frac{2}{3}$ for $n \geq 3$ and $s > s_c + \frac{5}{6}$ is proved in [T2, part III].) In dimension $n = 3$ we also have examples, due to H. Linblad [L2], which show that one cannot have well-posedness, in general, for $s \leq 2$.

Even if the Strichartz based methods initiated by Chemin-Bahouri and Tataru can be made optimal they will still fall short of proving the desired H^2 result, conjectured by WP3. To obtain such a result one needs to take into account the "null structure" of the EV equations. We know, indirectly from the proof of stability of the Minkowski space, [ChrK], that written in appropriate form, i.e. using their general covariance, the equations must exhibit such a structure. Yet the indirect method of [ChrK], based on the Bianchi identities and a careful decomposition of all geometric components appearing in the equation relative to a null frame, cannot be used in this case. One needs instead a method similar to the one we have sketched above for YM. In other words we need a "gauge condition," similar to the Coulomb one in YM, relative to which all quadratic terms of the Einstein equations exhibit a null bilinear structure. Once this is done we need to develop techniques to prove bilinear estimates,[26] similar to those of [KlM1], [FoK], in a quasilinear set-up. A good warm-up problem, in this respect, would be the study of the Minkowski space analogue of the minimal surface equation, for which the null structure, in the sense of [Kl1,2], [Chr2], is obvious.

To summarize, the study of Conjecture WP3 requires:

1) To develop new analytic techniques to improve the results of Chemin-Bahouri to the optimal regularity possible for Strichartz based methods.

2) To investigate quasilinear equations which verify the null condition, and develop bilinear estimates for linear equations with very rough coefficients.

3) To investigate, in a direct way, the null structure of the Einstein equations.

---

[26]The bilinear estimates of [KlM1] have been recently derived, by Smith and Sogge [SmS2], for $C^{1,1}$ coefficients.

References[27]

[AM]   L. ANDERSON, V. MONCRIEF, The global existence problem in general relativity, preprint (1999).

[B1]   J. BOURGAIN, Nonlinear Schrödinger equations, in "Nonlinear Wave Equations and Frequency Interactions", AMS, series 4, Park City, 1999.

[B2]   J. BOURGAIN, Harmonic analysis and nonlinear PDE's, Proceedings of ICM, Zurich (1994).

[B3]   J. BOURGAIN, Fourier transform restriction phenomena for certain lattice subsets and applications to nonlinear wave equations, I: Schrödinger equations; II: The KdV equation, GAFA 3 (1993), 107–156; 209–262.

[Br]   Y. BRENIER, Minimal geodesics on groups of volume preserving maps, Comm. Pure. Appl. Math. 52 (1999), 411–452.

[BreB] H. BREZIS, F. BROWDER, Partial differential equations in the 20th century, Encyclopedia Italiana, in its series on the history of the twentieth century, to appear.

[Bru]   Y.CH. BRUHAT, Theoremes d'existence pour certains systemes d'equations aux derivee partielles nonlineaires, Acta Math. 88 (1952), 141–225.

[CWY]   A. CHANG, L. WANG, P. YANG, A regularity of biharmonic maps, C.P.A.M. LII (1999), 1113–1137.

[ChB]   J.Y. CHEMIN, H. BAHOURI, Equations d'ondes quasilineaires et effect dispersif, AJM, to appear.

[Chr1] D. CHRISTODOULOU, The Action Principle and PDE's, Annals of Math. Studies 146 (1999).

[Chr2] D. CHRISTODOULOU, Global solutions of nonlinear hyperbolic equations for small initial data, C.P.A.M. 39 (1986), 267–282.

[ChrK] D. CHRISTODOULOU, S. KLAINERMAN, The Global Nonlinear Stability of the Minkowski Space, Princeton Mathematical Series, 41 (1993).

[ChrZ] D. CHRISTODOULOU, A.S.T. ZADEH, On the regularity of spherically symmetric wave-maps, Comm. P. Appl. Math. 46 (1993), 1041–1091.

[E]   L.C. EVANS, Partial regularity for harmonic maps into spheres, Arch. Rat. Mech. Anal. 116 (1991), 101–113.

[FM]   A. FISCHER, J. MARSDEN, The Einstein evolution equations as a first order quasilinear, symmetric hyperbolic system, Comm. Math. Phys. 28 (1972), 1–38.

---

[27]With few exceptions most of the references given below are in regard with the last section of the paper, more precisely in connection to the conjectures WP1–WP3. I apologize for giving only a very limited number of references in connection to the first four sections.

[FMo] A. FISCHER, V. MONCRIEF, The Einstein flow, the sigma-constant, and the geometrization of three manifolds, preprint (1999).

[FoK] D. FOSCHI, S. KLAINERMAN, On bilinear estimates for solutions to the wave equation, Annales ENS, to appear.

[GGKM] C.S. GARDNER, J.M. GREEN, M.D. KRUSKAL, R.M. MIURA, Method for solving the KdV equation, Phys. Rev. Lett. 19(1967), 1095–1097.

[Gr] M. GROMOV, Partial Differential Relations, Springer Verlag, Berlin, 1986.

[HT] S. HILDEBRANDT, A. TROMBA, Mathematics and Optimal Form, Scientific American Library, 1984.

[HuI] G. HUISKEN, T. ILMANEN, The inverse mean curvature flow and the Penrose conjecture, JDG, to appear.

[J] F. JOHN, Formation of singularities in elastic waves, Springer Lecture Notes in Phys. 195 (1984), 190–214.

[K] L. KAPITANKY, Global and unique weak solutions of nonlinear wave equations, Math. Res. Lett. 1 (1994), 211-223.

[KeT] M. KEEL, T. TAO, Local and global well-posedness of wave maps on R^{1+1} for rough data, IMRN 21 (1998), 1117–1156.

[KenPV] C. KENIG, G. PONCE, L. VEGA, The Cauchy problem for the KdV equation in Sobolev spaces of negative indices, Duke Math. J. 71 (1994), 1–21.

[Kl1] S. KLAINERMAN, Long time behavior of solutions to nonlinear wave equations, Proc. ICM 1983, Warszawa, 1209–1215.

[Kl2] S. KLAINERMAN, The null condition and global existence to nonlinear wave equations, AMS Lectures in Applied Mathematics 23 (1986), 293–326.

[KlM1] S. KLAINERMAN, M. MACHEDON, Space-time estimates for null forms and the local existence theorem, Comm. Pure Appl. Math. 46 (1993), 1221–1268.

[KlM2] S. KLAINERMAN, M. MACHEDON, Finite energy solutions for the Yang–Mills solutions in $\mathbb{R}^{3+1}$, Annals of Math. 142 (1995), 39–119.

[KlM3] S. KLAINERMAN, M. MACHEDON, Smoothing estimates for null forms and applications, Duke Math. J. 81 (1995), 99–103.

[KlM4] S. KLAINERMAN, M. MACHEDON, On the algebraic properties of the $H_{n/2,1/2}$ spaces, I.M.R.N 15 (1998), 765–774.

[KlM5] S. KLAINERMAN, M. MACHEDON, On the optimal local regularity for Gauge field theories, Diff. Integr. Eqs. 10:6 (1997), 1019–1030.

[KlS] S. KLAINERMAN, S. SELBERG, Remark on the optimal regularity for equations of wave maps type, Comm. P.D.E. 22:5-6 (1997), 901–918.

[KlT] S. KLAINERMAN, D. TATARU, On the optimal local regularity for the Yang–Mills equations, J. AMS 12:1 (1999), 93–116.

[Ku] N.H. KUIPER, On C^1 isometric embeddings, I. Proc. Koninkl. Neder. Ak. Wet A-58 (1955), 545–556.

[L1] H. LINBLAD, Counterexamples to local existence for semilinear wave equations, AJM 118 (1996), 1–16.

[L2]   H. LINDBLAD, Counterexamples to local existence for quasilinear wave equations, MRL, to appear.

[Lu]   J. LUTZEN, The Prehistory of the Theory of Distributions, Springer-Verlag, 1992.

[M]   H. MINKOWSKI, Space and Time, English translation of the original article in "The Meaning of Relativity", Dover, 1952.

[MüS]   S. MÜLLER, M. STRUWE, Global existence for wave maps in $1 + 2$ dimensions for finite energy data, preprint.

[N]   J. NASH, C^1 isometric embeddings, Ann. of Math. 60 (1954), 383–396.

[Ne]   Y. NEEMAN, Pythagorean and Platonic conceptions in XXth Century physics, in this issue.

[P]   H. POINCARÉ, Sur les rapports de l'analyse pure et de la physique mathematique, First Int. Congress of Mathematicians, Zurich, 1897.

[R]   T. RIVIÈRE, Applications harmonique de B^3 dans S^2 partout disconnues, C.R.Acad. Sci. Paris 314 (1992), 719–723.

[S]   V. SCHAEFFER, An inviscid flow with compact support in space-time, Journ. Geom. Anal. 3:4 (1993), 343–401.

[Sc]   R. SCHOEN, A report on some recent progress on nonlinear problems in geometry, Surveys in Differential Geometry 1 (1991), 201–241.

[ScU]   R. SCHOEN, K. UHLENBECK, A regularity theory for harmonic maps, J. Diff. Geom. 17 (1982), 307–335; 18 (1983), 329.

[Sh]   J. SHATAH, Weak solutions and development of singularities in the SU(2) σ model, Comm. Pure Appl. Math. 41 (1988), 459–469.

[ShZ]   J. SHATAH, A.S.T. ZADEH, On the Cauchy problem for equivariant wave maps, C.P.A.M. 47 (1994), 719–754.

[Shn]   A. SHNIRELMAN, On the nonuniqueness of weak solutions for Euler equations, C.P.A.M. 50 (1997), 1261–1286.

[ShS]   J. SHATAH, M. STRUWE, Well posedness in energy space for semilinear wave equations with critical growth, Int. Math. Res. Not. 7 (1994), 303–309.

[Si]   T. SIDERIS, Formation of singularities in three-dimensional Vompressible fluids, Comm. Math. Phys. 101 (1985), 155–185.

[Sm]   H. SMITH, A parametrix construction for wave equations with $C^{1,1}$ coefficients, Annales de L'Institut Fourier 48 (1998), 797–835.

[SmS1] H. SMITH, C. SOGGE, On Strichartz and eigenfunction estimates for low regularity metrics, preprint.

[SmS2] H. SMITH, C. SOGGE, Null form estimates for $(1/2, 1/2)$ symbols and local existence for a quasilinear Dirichlet wave equation, Ann. Sci. ENS, to appear.

[So]   C. SOGGE, Propagation of singularities and maximal functions in the plane, Inv. Mat. 104 (1991), 349–376.

[St] J.J. STOKES, On a difficulty in the theory of sound, Philosophical Magazine 33 (1848), 349–356.

[Str] W. STRAUSS, Weak solutions for nonlinear wave equations, Annais Acad. Brazil Ciencias 42 (1970), 645–651.

[T1] D. TATARU, Local and global results for wave maps I, C.P.D.E 23:9-10 (1998), 1781–1793; part II to appear in AJM.

[T2] D. TATARU, Strichartz estimates for operators with non smooth coefficients and the nonlinear wave equation, AJM, to appear; part II and III, preprints.

[W] R. WALD, Gravitational Collapse and Cosmic Cesorship, 1997, gr-qc/9712055.

[Wi] E. WIGNER, The unreasonable effectiveness of mathematics in the natural sciences, C.P.A.M. 13 (1960), 1–15.

[Wo1] T. WOLFF, Recent Work Connected to the Kakeya Problem, Prospects in Math. AMS, Princeton 1996.

[Wo2] T. WOLFF, A sharp bilinear restriction estimate, IMRM to appear.

[Y] S.T. YAU, Open problems in geometry, preprint.

[Z] A.S.T. ZADEH, Relativistic and nonrelativistic elastodynamics with small shear strain, Ann. Inst. H. Poincaré, Physique Theorique 69 (1998), 275–307.

SERGIU KLAINERMAN, Department of Mathematics, Princeton University, Princeton, NJ 08544, USA

GAFA, Geom. funct. anal.
Special Volume – GAFA2000, 316 – 333
1016-443X/00/S10316-18 $ 1.50+0.20/0

GAFA Geometric And Functional Analysis

LESSONS FOR TURBULENCE

ANTTI KUPIAINEN

1 Navier-Stokes Equations and Turbulence

Turbulence is one of the major unsolved problems of classical physics. The reason for its continued interest among physicists and mathematicians (not to mention engineers) is easy to understand. On the one hand it has enormous practical consequences ranging from airplane and ship design to weather forecasts. On the other hand, it is believed to be a property of solutions of a relatively simple partial differential equation, the Navier-Stokes (NS) equation: the simplicity of the equation stands in contrast with the complexity of the phenomena.

Let us consider a fluid (i.e. a gas or a liquid) confined in a region $\Omega \subset \mathbf{R}^d$ where the spatial dimension d is in applications usually 3 or 2. Denoting the velocity of the fluid as observed at a point $\mathbf{x}$ and at time t by $\mathbf{u}(t, \mathbf{x}) \in \mathbf{R}^d$, the NS equation is

$$\partial_t \mathbf{u} + (\mathbf{u} \cdot \nabla)\mathbf{u} - \nu\nabla^2\mathbf{u} = \tfrac{1}{\rho}(\mathbf{f} - \nabla p) \,, \tag{1}$$

where ν is the viscosity of the fluid, ρ its density, $\mathbf{f}$ is the external force acting on the fluid and p is the pressure. In most applications, one may assume that the fluid is *incompressible*, i.e. that ρ is constant and that $\mathbf{u}$ is divergence free $\nabla \cdot \mathbf{u} = 0$. It follows then by taking the divergence of both sides of Eq. (1) that $\nabla^2 p = -\nabla \cdot (\mathbf{u} \cdot \nabla)\mathbf{u}$ so the pressure is actually a function of velocity field. Its effect is to project the nonlinear term into a space of divergenceless vector fields on Ω.

Eq. (1) is a nonlinear parabolic PDE (actually an integrodifferential equation due to the pressure term). The $\nu\nabla^2\mathbf{u}$ term in the NS equation represents the effect of friction and tends to smoothen and drive down the motion of the fluid. The motion is maintained by the external force $\mathbf{f}$, whereas the function of the nonlinear term is more subtle and will be discussed below.

To have a clean formulation of the problem of turbulence it is important to understand exactly what sort of forces one should consider in (1). In experimental situations the forcing may be produced by boundary conditions like in the case of flow in a pipe or around an obstacle (e.g., airplane wing) where the velocity field vanishes at the boundary and takes a characteristic value far enough away from it (the relative velocity of the air and the plane or the velocity with which the fluid enters the pipe). Another possibility is to use an explicit stirring mechanism, like in recent experiments [PJT] where electric current is led into a solution of water and salt and stirring is achieved by applying a time dependent magnetic field. In these situations the force is applied on a characteristic *lengthscale*: the diameter of the pipe or the wing, the sizes of the magnets. Let us call this scale L. In equation (1) we could, e.g., consider $\mathbf{f}$ whose Fourier transform in $\mathbf{x}$, $\widehat{\mathbf{f}}(t, \xi)$, has compact support near $|\xi| = L^{-1}$ (below we will discuss a more explicit example).

It is intuitively clear that the relative sizes of ν and $\mathbf{f}$ contribute to the strength of the nonlinearity: $\mathbf{f}$ tends to increase $\mathbf{u}$ and ν drive it down. A more precise formulation of this intuition involves the *Reynolds number*, a measure of the strength of the nonlinear effects that takes into account two symmetries of (1): scale and Galilean invariance. The scaling corresponds to simple changes in spatial and temporal scales: setting

$$\mathbf{u}'(t, \mathbf{x}) = \tau s^{-1} \mathbf{u}(\tau t, s \mathbf{x}), \tag{2}$$

then $\mathbf{u}'$ solves the NS equation with viscosity $\nu' = \tau s^{-2} \nu$ and force $\mathbf{f}' = \tau^2 s^{-1} \mathbf{f}(\tau t, s \mathbf{x})$. The new force is on the scale L/s. Thus, a scale invariant measure of the relative strength of the forcing vs viscosity is given by the Reynolds number

$$Re = \frac{L \delta u}{\nu}, \tag{3}$$

where δu is a characteristic size of velocity differences produced by the force. The reason why velocity differences and not absolute velocities matter is that the NS equation is invariant (modulo boundary conditions) under Galilean transformation $\mathbf{u}(t, \mathbf{x}) \to \mathbf{u}(t, \mathbf{x} + \mathbf{v}t) - \mathbf{v}$ where $\mathbf{v}$ is a constant velocity. For the flow in the pipe or around the wing we may take δu as the difference of the velocity at the boundary (zero actually) and far away from it. In practical terms the scale invariance is important: one can substitute say water for air by changing the size of the system and/or strength of the force.

The basic phenomenological facts about hydrodynamic flows are as follows. If $Re \ll 1$, one encounters regular (*"laminar"*) flows. For Re between ~ 1 and $\sim 10^2$, a series of bifurcations occur leading to ever more complicated flows ("transition to chaos"). Finally, for $Re \gg 10^2$, a peculiar chaotic state, turbulence, is reached that seems to exhibit for very large Re ("*developed turbulence*") universal features, i.e. features that are independent on the detailed nature of the forcing. We will discuss these features below, but let us first mention some mathematical developments.

2   The Initial Value Problem

Much of the mathematical work on turbulence has concentrated on the study of the initial value problem for (1), i.e. existence and uniqueness of solutions for all times given $\mathbf{u}(0, \mathbf{x})$. Important role here is played by conservation laws. The kinetic energy of the fluid is given by the L^2-norm of $\mathbf{u}$ (we set for simplicity the density to 1) and it is formally conserved by the $\nu = 0$ flow without forcing. Correspondingly, supposing $\mathbf{u}$ is a smooth solution to (1) vanishing at the boundary of Ω, one gets by integrating the scalar product of (1) with $\mathbf{u}$ over Ω the energy conservation equation:

$$\frac{d}{dt} \frac{1}{2} \int_\Omega \mathbf{u}^2 = -\nu \int_\Omega (\nabla \mathbf{u})^2 + \int_\Omega \mathbf{f} \cdot \mathbf{u}. \tag{4}$$

Physically, the equation states that the rate of change of fluid energy is equal to the energy injection rate $\int \mathbf{f} \cdot \mathbf{u}$ (work of the external forces per unit time) minus the energy dissipation per unit time $\nu \int (\nabla \mathbf{u})^2$ due to the viscous friction. Note that the nonlinear term does not contribute here.

An immediate consequence of (4) is an a priori bound for smooth solutions: the $L^2(\Omega)$-norm of $\mathbf{u}$ stays bounded for all times (say for a bounded force in a compact Ω) as does the $L^2([0, t] \times \Omega)$-norm of $\nabla \mathbf{u}$. This a priori bound is enough to prove the existence of weak, i.e. distributional solutions to the NS equation, as was realized by Leray in 1933 [Ler]. He approximated the equation by a finite dimensional problem for which the apriori bound guarantees a unique solution. By compactness such approximations yield a weak solution to NS.

The uniqueness and smoothness of these weak solutions has ever since been an open problem in the case of spatial dimension 3. Suppose for simplicity Ω is compact and the force is zero (so-called decaying turbulence) and we start with smooth initial data. For the Reynolds number in this case we may take (3) with $\delta u = L(|\Omega|^{-1} \int_\Omega |\nabla \mathbf{u}|^2)^{1/2}$ where L is say the diameter of Ω. Then for small enough Re the viscous term dominates in the NS equation and one may prove uniqueness and smoothness of solutions for all times. As Re of the initial data increases one loses control of the H^1 norm of the solution, i.e. the time t Reynolds number and smoothness can be proved only for a time that tends to zero as Re tends to infinity. The a priori energy conservation implies only that $\nabla \mathbf{u}$ is square integrable in space and time and thus doesnt rule out singularities. Only upper bounds for the Hausdorff dimension of the singularity set can be deduced [S1]. As we will see below, physically one expects the typical (in the sense of ergodic theory) solutions to be smooth and to lose their smoothness only in the limit of infinite Re.

In the case of dimension two the Euler equation has a second conserved quantity, the *enstrophy* which is the L^2-norm of the *vorticity* $\omega = \frac{\partial v_2}{\partial x_1} - \frac{\partial v_1}{\partial x_2}$. The NS equation implies the following equation for ω:

$$\partial_t \omega + (\mathbf{u} \cdot \nabla)\omega - \nu \nabla^2 \omega = \partial_1 f_2 - \partial_2 f_1 \,, \tag{5}$$

which in turn implies a relation analogous to (4) for enstrophy ($g = \frac{\partial f_2}{\partial x_1} - \frac{\partial f_1}{\partial x_2}$):

$$\frac{d}{dt} \frac{1}{2} \int_\Omega \omega^2 = -\nu \int_\Omega (\nabla \omega)^2 + \int_\Omega g\omega \,. \tag{6}$$

(6) yields an a priori bound for the $L^\infty([0,t], H^1(\Omega))$ and $L^2([0,t], H^2(\Omega))$ norms of $\mathbf{u}$. These suffice to yield uniqueness and smoothness of solutions for all times: now we have an apriori bound for the time t Reynolds number.

3   Statistical Hydrodynamics

It is fair to say that the study of the global regularity of the solutions of the NS equation has shed relatively little light on the problem of turbulence. There has been however another line of approach to turbulence, pioneered by A.N. Kolmogorov [K], based on ideas from ergodic theory that has been more successful.

In this approach one views (1) as a dynamical system in some space of vector fields on Ω and asks questions on the behaviour of typical trajectories with respect to an invariant measure. Of course, to really carry out this strategy a necessary prerequisite would be an existence and uniqueness theorem for solutions, but one may also adopt a pragmatic attitude and inquire what can be said about this measure if it exists.

An ergodic invariant measure μ would satisfy

$$\lim_{T \to \infty} T^{-1} \int_0^T F(\mathbf{u}(t))dt = \int F(\mathbf{u})\mu(d\mathbf{u}) \,, \tag{7}$$

i.e. the time average of suitable functions of the velocity field are given by the ensemble averages. Thus the invariant measure can be accessed by taking the average of a time series from an experiment, or an average over "snapshots" of several experiments.

In the context of smooth hyperbolic dynamical systems such a "physical" measure exists. It is the Sinai-Ruelle-Bowen measure and it is selected as the one that is stable under small noise. With this in mind it is reasonable to consider the NS equation with a random force. More precisely, consider a stochastic PDE

$$d\mathbf{u} + ((\mathbf{u} \cdot \nabla)\mathbf{u} - \nu\nabla^2\mathbf{u} + \nabla p)\,dt = d\mathbf{F}\,, \tag{8}$$

where $\mathbf{F}(t,\mathbf{x})$ is a suitable Wiener process. To avoid worrying about the boundaries let us take the spatial region to be the whole $\mathbf{R}^d$ although in two dimensions as we will see this is not expected to be an innocent assumption. We get a model of homogeneous and isotropic turbulence (i.e. invariant under spatial translations and rotations) with forcing at scale L by taking $\mathbf{F}$ with covariance

$$EF_\alpha(t,\mathbf{x})F_\beta(t,\mathbf{y}) = tC_{\alpha\beta}(\mathbf{x} - \mathbf{y}; L)\,,$$

with C having a Fourier transform $\widehat{C}(\xi) = c(|\xi|)$ and $c(k)$ smooth with compact support near $k = L^{-1}$. With good enough control of the solutions of (8) one could then define a Markov chain in a suitable space $\mathcal{V}$ of vector fields on $\mathbf{R}^d$ by the transition probabilities $p(\mathbf{u}, U) = \text{Prob}\{\mathbf{u}(1) \in U \mid \mathbf{u}(0) = \mathbf{u}\}$. The stationary measure(s) μ of this Markov chain would then give the time averages in (7).

The mathematical status of (8) is similar to the nonrandom case. Weak statistical solutions exist [VF] (if we replace $\mathbf{R}^d$ by a compact region) and in two dimensions they are unique. Uniqueness of invariant measure is open in two dimensions (existence is easy) for the kind of forcing considered above. For the nonrandom case existence of a finite dimensional attracting set (two dimensions, compact region) is known, [L], [CoFT], and similar results hold for the Markov chain defined above [BrKL].

4   Universality

The contrast between theoretical understanding of turbulence and experimental data on it is striking. Rather accurate experimental information on the nature of the high Reynolds number invariant measure is available. In particular for large Re another length scale η seems to emerge in the problem in addition to the forcing scale L and the invariant measure seems to possess universal features for scales between η and L. As $Re \to \infty$ the scale $\eta \to 0$ and in this limit the so called *inertial range* $[\eta, L]$ extends all the way to zero. We shall now discuss more explicitly some examples of correlations and the nature of the conjectured universality.

Energy spectrum. The energy conservation equation (4) becomes an energy balance relation for the invariant measure:

$$\nu \int (\nabla \mathbf{u}(\mathbf{x}))^2 \mu(d\mathbf{u}) = \tfrac{1}{2}\mathrm{tr}C(0) \equiv \epsilon . \qquad (9)$$

(To derive (9) note that by an Ito formula (8) implies $\tfrac{1}{2}d(\mathbf{u})^2 = \mathbf{u} \cdot (\nu\nabla^2\mathbf{u} - (\mathbf{u} \cdot \nabla)\mathbf{u} - \nabla p)dt + \mathbf{u} \cdot d\mathbf{F} + \tfrac{1}{2}\mathrm{tr}C(0)$. Taking averages over the invariant measure and noting that by spatial translation invariance the nonlinear term will not contribute the claim follows.) The physical interpretation of (4) is that at the stationary state the energy injection rate $\tfrac{1}{2}\mathrm{tr}C(0)$ is balanced by the energy dissipation rate whose common value we denote by ϵ. The former is a property of the forcing and experimentally controllable, the latter can be deduced from measurements of velocity correlation function. Let

$$\int \mathbf{u}(\mathbf{x}) \cdot \mathbf{u}(\mathbf{y})\mu(d\mathbf{u}) = G(\mathbf{x} - \mathbf{y}) = \int_{\mathbf{R}^d} e^{i\xi(\mathbf{x}-\mathbf{y})}\widehat{G}(\xi)d\xi .$$

Then the average energy density of the fluid is given as

$$\tfrac{1}{2} \int \mathbf{u}(\mathbf{x})^2 \mu(d\mathbf{u}) = \int_{\mathbf{R}^d} \widehat{G}(\xi)d\xi = \int_0^\infty e(k)dk ,$$

where we defined the *energy spectrum* $e(|\xi|) = \widehat{G}(\xi)|\xi|^{d-1}$. In three dimensions $e(k)$ seems to have a power law behaviour for k in the inertial range $[L^{-1}, \eta^{-1}]$, consistent with the conjecture

$$e(k) = Ck^{-\gamma}\big(1 + o(\eta k, (Lk)^{-1})\big) . \qquad (10)$$

The exponent γ is close to

$$\gamma \sim \tfrac{5}{3} .$$

For $k > \eta^{-1}$ $e(k)$ decreases fast, presumably exponentially. From this experimental information we may draw interesting conclusions on the stationary state.

First, writing the energy balance relation (9) in terms of e we have (we suppose the forcing is isotropic, i.e. $\widehat{C}(\xi)$ only depends on $|\xi|$)

$$\nu \int_0^\infty k^2 e(k)dk = \tfrac{1}{2}\mathrm{tr}C(0) = \tfrac{1}{2} \int_0^\infty \mathrm{tr}\widehat{C}(k)k^2 dk . \qquad (11)$$

The LHS gets most of its contribution near $k = \eta^{-1}$ since $k^2 e(k) \sim k^{1/3}$, whereas the integrand on the RHS is supported in a neighbourhood of L^{-1}. This is an indication of the so called *cascade picture* according to which the stationary state is an energy flux state where energy is injected to the system at large scales and transported by the nonlinearity to small scales where it is dissipated by the viscous forces. Note also that the left hand side behaves as $\nu\eta^{-4/3}$ and the right hand side is ν-independent. Thus as the viscosity is taken to zero the viscous scale η tends to zero as $\nu^{3/4}$. It should be stressed, however, that there is no real derivation of the cascade picture from the NS equation.

The exponential decrease of $e(k)$ as $k \to \infty$ would guarantee that for $\nu > 0$ the invariant measure sits on smooth velocities. Hence, if there was finite time blowup in NS, it would have zero probability in the stationary state and thus it would be hard to see numerically. But of course one cannot deduce from experiments for sure the large k asymptotics of $e(k)$.

However, as $\nu \to 0$ the expectation of $(\nabla\mathbf{u}(\mathbf{x}))^2$ diverges. Hence in the inviscid limit the typical solutions are not differentiable, in fact since $E(|\nabla|^\alpha\mathbf{u}(\mathbf{x}))^2 = \int k^{2\alpha}e(k)dk$ we expect only Hölder continuity with exponent approximately $1/3$ in that limit. The inviscid limit of NS is the Euler equation. Thus the typical solutions of the NS equation become at most weak solutions of the Euler equation, solutions that dissipate energy. The observation that solutions of the Euler equation have to have low regularity in order to dissipate energy goes back to Onsager in the late forties [O].

In two dimensions the previous discussion needs interesting modifications due to the additional conserved quantity enstrophy. The conservation law (6) implies for the stationary state

$$\nu \int (\nabla\omega(\mathbf{x}))^2\mu(d\mathbf{u}) = \tfrac{1}{2}\mathrm{tr}\Delta C(0), \tag{12}$$

with the interpretation as equality of enstrophy dissipation rate and enstrophy injection rate. Thus, in addition to (11) e has to satisfy

$$\nu \int_0^\infty k^4 e(k)dk = \int_0^\infty \mathrm{tr}\widehat{C}(k)k^4 dk. \tag{13}$$

Now we seem to have a dilemma: if as before the main contribution to (11) and (13) comes from k near the viscous scale, only the enstrophy dissipation rate can stay nonzero as $\nu \to 0$. Indeed, experiments [PJT] indicate that now $e(k)$ has the scaling behaviour (10) with the exponent gamma

$$\gamma = 3.$$

Hence, enstrophy is dissipated near the viscous scale that now is proportional to $\eta \sim \nu^{1/2}$. Thus the dissipation rate of energy on the interval $k \in [L^{-1}, \eta^{-1}]$ tends to zero as $\nu \to 0$ and (11) cannot hold. To understand what actually happens consider first our infinite volume model system. Starting with say initial condition $\mathbf{u} = 0$, enstrophy flows towards large k, i.e. the k^{-3} spectrum builds up on the interval $[L^{-1}, \eta^{-1}]$. This is called the direct (enstrophy) cascade. On the other hand, energy starts to flow towards small $k < L^{-1}$ in what is called the *inverse cascade*, building a second power law region with the exponent $\gamma = 5/3$. Note that such a power law for $e(k)$ on $[0, L^{-1}]$ would mean infinite total energy (density), i.e. $\mathbf{u}(\mathbf{x})^2$ would have infinite expectation. What should actually happen is that at time t the initial condition has evolved to a measure with total energy proportional to t and energy spectrum having the $5/3$ law down to $k \sim t^{-3/2}$. $\mathbf{u}$ is not a well defined random variable with respect to the invariant measure, only the vorticity ω is. In finite volume the energy can not escape to modes with smaller and smaller k, and will be dissipated. A stationary state then exists also for $\mathbf{u}$. The inverse cascade was postulated in the 60's by Kraichnan and Batchelor [Kr1], [B].

Multiscaling. The discussion above indicates that the turbulent state seems to have scale invariance for length scales much smaller than the forcing scale L. Exact scale invariance would mean that $\mathbf{u}(\mathbf{x}) - \mathbf{u}(y) = \lambda^{-\varsigma}(\mathbf{u}(\lambda\mathbf{x}) - \mathbf{u}(\lambda\mathbf{y}))$ in law and the energy spectrum indicates that $\varsigma \sim 1/3$. Things are more interesting however. One can get hold of the Hölder properties of the velocity by studying so called *structure functions*, namely moments of velocity differences in the invariant measure. The longitudinal structure functions are defined by

$$S_n(r) = \int \left(\tfrac{\mathbf{x}}{|\mathbf{x}|} \cdot (\mathbf{u}(\mathbf{x}+\mathbf{y}) - \mathbf{u}(\mathbf{y})) \right)^n \mu(d\mathbf{u}), \qquad (14)$$

where $r = |\mathbf{x}|$ and translational and rotational invariance of the invariant measure was assumed.

Experimentally the structure functions seem to behave as

$$S_n(r) \propto r^{\varsigma_n},$$

for r in the inertial range $[\eta, L]$. Hölder continuity with exponent $1/3$ would indicate that ς_n were $n/3$. This indeed is the prediction made by Kolmogorov in 1941.

To see how this result could arise, let us consider the model (8). Take for definiteness the large scale forcing covariance as $C(\mathbf{x}; L) = c(\tfrac{\mathbf{x}}{L})$ with c a fixed rapidly decreasing function. S_n depends on the parameters in the equation, namely, ν, L, c. Making the change of variables (2) we get

$$S_n(r, \nu, L, c) = \left(\tfrac{l}{r} \right)^n S_n \left(\tfrac{r}{l}, \tfrac{\tau}{l^2}\nu, \tfrac{L}{l}, \tfrac{\tau^3}{l^2}c \right).$$

Suppose now that the limit $\lim_{L \to \infty} \lim_{\nu \to 0} S_n(r, \nu, L, c) = S_n(r, c)$ exists. Then taking $l = r$ and $\tfrac{\tau^3}{l^2} = \epsilon^{-1}$ (recall, $\epsilon = \tfrac{1}{2}\mathrm{trc}(0)$ is the energy injection rate) we get

$$S_n(r, c) = A_n(\epsilon r)^{n/3} \tag{15}$$

where $A_n = S_n(r, c/\epsilon)$. Kolmogorov made the further assumption that A_n is independent on c and thus the structure functions are universal, depending only on r and on forcing only through the amount of energy pumped in per unit time and volume.

In reality (15) doesn't quite fit with experiments. Only ζ_3 agrees well with the value $\zeta_3 = 1$. The experimental values for the first few exponents are [BenCBR]

$$\zeta_2 = .70\,(.67)\,, \quad \zeta_4 = 1.28\,(1.33)\,, \quad \zeta_5 = 1.53\,(1.67)\,,$$
$$\zeta_6 = 1.77\,(2)\,, \quad \zeta_7 = 2.01\,(2.33)$$

with the Kolmogorov values in the parenthesis for comparison. These values indicate that the limit argument leading to (15) is flawed. Indeed, it is believed that the the $\nu \to 0$ limit should exist but not the $L \to \infty$ limit. Then the right asymptotics should look like

$$S_n(r) = B_n r^{\zeta_n} L^{\frac{n}{3} - \zeta_n} \left(1 + o\left(\tfrac{\eta}{r}, \tfrac{r}{L}\right)\right) \tag{16}$$

with B_n nonuniversal constants, depending on the details of forcing (i.e. c in our model), whereas the ζ_n are believed to be universal. By Hölder inequality, the exponents ζ_n are concave functions of n (for even n). The data indicates that they are strictly concave, not linear. Thus $S_{2n}(r)/S_2^n(r)$ grows with n more and more as $r \to 0$ (this phenomenon is referred to as *intermittency*). The velocity differences behave very differently from gaussian random variables and more so with diminishing distance. The origin of these exponents and derivation of their values is a major theoretical problem of turbulence.

In two dimensions things seem to be simpler. For the direct enstrophy cascade, i.e. distances less than the injection scale, the Kraichnan's k^{-3} energy spectrum seems exact by recent experiments [PJT], and the Kolmogorov's $k^{-5/3}$ spectrum correspondingly for the inverse energy cascade, i.e. distances larger than L, by experiments [PT] as well as numerical simulations [BoCV]. Hence two dimensional turbulence might be more accessible to mathematical understanding than the strongly intermittent three dimensional one.

To summarize, there is very little mathematical understanding on the basic features of turbulence: the energy/enstrophy cascade picture of the stationary state and the phenomenon of intermittency. There is almost no theoretical understanding of the latter as a phenomenon arising from

Navier-Stokes equations in spite of several decades of attempts. Even worse, there doesn't seem to be any reasonable approximation to NS equations where intermittency can be studied. Until very recently there was no other mathematical model where intermittency would allow a mathematical analysis. Much of the recent progress in the field involves a model of advection, originally introduced by Kraichnan [Kr2] in the sixties. It turns out this problem has very similar phenomenology as the NS equation:

- Cascade picture of the stationary state that is supported on weak energy dissipating solutions of the zero diffusivity limit of the equation.
- A range of scales where scaling is present and a Kolmogorov like scaling theory that is violated by the phenomenon of intermittency.

The difference with NS is that in the Kraichnan problem these features can be theoretically understood.

5   Random Advection

Passive scalar. The velocity field $\mathbf{u}(t, \mathbf{x})$ gives rise to a flow, namely diffeomorphisms $\phi_{t,s}$ on Ω, by solving the ODE $\dot{\mathbf{x}}(\tau) = \mathbf{u}(\tau, \mathbf{x}(\tau))$ of so called *Lagrangian trajectories* from time s to time t. Physically $\mathbf{x}(\tau)$ describes motion of "fluid particles". The flow allows us to transport geometrical objects: scalars, densities, vectors. For instance equation (5) for the vorticity ω of the two dimensional fluid tells us that for zero viscosity and force ω is transported by $\mathbf{u}$ as a scalar

$$\omega(t, \mathbf{x}) = \omega\big(0, \phi_{t,0}^{-1}(\mathbf{x})\big). \tag{17}$$

Similarily, the vorticity of a three dimensional fluid is transported as a vector. For nonzero viscosity ω is both transported and diffused. Many physical quantities of interest are transported and diffused by the fluid motion. E.g. the temperature $T(t, \mathbf{x})$ of the fluid at $\mathbf{x}$ at time t depends on the molecular motion and in many situations can be assumed not to affect the fluid motion taking place in much larger scales (the temperature dependence of viscosity can often be neglected). Then T is transported by the fluid velocity and in addition it diffuses via molecular motion. It is an example of a *passive scalar* satisfying the equation

$$\partial_t T + (\mathbf{u} \cdot \nabla)T - \kappa \nabla^2 T = f, \tag{18}$$

where $f(t, \mathbf{x})$ is a source for T.

Let us recall briefly the solution of (18) in terms of particle motion. We suppose first that $\mathbf{u}$ is smooth and proceed in steps.

$\kappa = 0$, $f = 0$. This is just transport and solved by (17) or in other words

$$T(t,\mathbf{x}) = T\big(s, \mathbf{x}_{t,\mathbf{x}}(s)\big),$$

where $\mathbf{x}_{t,\mathbf{x}}(s)$ is the Lagrangian trajectory with $\mathbf{x}_{t,\mathbf{x}}(t) = \mathbf{x}$.

$\kappa \neq 0$, $f = 0$. The diffusion term in (18) arises from adding a Brownian motion to the particle motion. Let β be the standard Brownian motion. Consider the stochastic ODE

$$d\mathbf{x}(s) = \mathbf{u}\big(s, \mathbf{x}(s)\big)ds + (2\kappa)^{1/2}d\beta(s),\tag{19}$$

with $\mathbf{x}(t) = \mathbf{x}$ (note that this equation is solved backward in time). Then T is given as an average over the trajectories

$$T(t,\mathbf{x}) = E_\beta T\big(0, \mathbf{x}(0)\big).\tag{20}$$

In other words, (19) gives rise to a Markov process with state space $\mathbf{R}^d$ and transition probability kernel $P_{t,s}(\mathbf{x}, \mathbf{y} \mid \mathbf{u})$ that is the fundamental solution to the equation (18) (the derivatives act on the variables $\mathbf{x}$ and t) and so (20) reads

$$T(t,\mathbf{x}) = \int P_{t,0}(\mathbf{x}, \mathbf{y} \mid \mathbf{u})T(0, \mathbf{y})d\mathbf{y}$$

$f \neq 0$. The general case is then given by

$$T(t,\mathbf{x}) = \int P_{t,0}(\mathbf{x}, \mathbf{y} \mid \mathbf{u})T(0, \mathbf{y})d\mathbf{y} + \int_0^t \int P_{t,s}(\mathbf{x}, \mathbf{y} \mid \mathbf{u})f(s, \mathbf{y})d\mathbf{y}\, ds.\tag{21}$$

The Kraichnan model. In practical applications the velocity field $\mathbf{u}$ in (18) is a possibly turbulent velocity field satisfying the NS equations. In that case, ideally one would like to study the statistics of T arising from the statistics of $\mathbf{u}$. In the absence of the knowledge of the latter it is still of interest to consider some explicit ensembles of $\mathbf{u}$ and try to deduce statistical properties for T. In the *Kraichnan model* $\mathbf{u}(t, \mathbf{x})$ are taken Gaussian random variables that are Hölder continuous in $\mathbf{x}$. The former assumption is far from true in the case of realistic turbulent velocities, but as we have seen, the latter mimics the expected properties of infinite Reynolds number flows.

More explicitly consider the Wiener process $\mathbf{U}(t, \mathbf{x})$ with covariance

$$EU_\alpha(t,\mathbf{x})U_\beta(t,\mathbf{y}) = tD_{\alpha\beta}(\mathbf{x} - \mathbf{y})$$

where D is Hölder continuous. An example is the Fourier transform of

$$\widehat{D}_{\alpha\beta}(\mathbf{k}) = \Big(\delta_{\alpha\beta} - \tfrac{k_\alpha k_\beta}{|\mathbf{k}|^2}\Big)\chi(L|\mathbf{k}|)|\mathbf{k}|^{-d-\xi}\tag{22}$$

where the first factor assures incompressibility of the velocity and χ is smooth cutoff function that vanishes in a neighbourhood of $k = 0$. The velocity $\mathbf{u}$ is the time derivative of $\mathbf{U}$, i.e. we must now interpret the first term on the RHS of (19) also as a stochastic differential, $\mathbf{u}(t, \mathbf{x})dt = d\mathbf{U}(t, \mathbf{x})$. It is almost surely Hölder continuous in $\mathbf{x}$ with any exponent less than $\frac{1}{2}\xi$. L is the "integral scale" of $\mathbf{u}$. (As we will see below, the product $d\mathbf{U} \cdot \nabla T$ in (18) has to be interpreted in the Stratonowich sense, or equivalently, in Ito sense with the term $-\frac{1}{2}D_{\alpha\beta}(0)\partial_\alpha\partial_\beta$ added.)

To complete the definition of the Kraichnan model, we consider the source f in (18), analogously to what we did with the stochastic NS, to be random and on large scale in $\mathbf{x}$: $f(t, \mathbf{x})dt = dF(t, \mathbf{x})$ with

$$EF(t, \mathbf{x})F(t, \mathbf{y}) = tC\left(\tfrac{\mathbf{x}-\mathbf{y}}{L}\right)$$

(the L here need not be the same as in the velocity covariance but we suppose so for simplicity of notation).

(21) determines the random variables $T(t, \mathbf{x})$ as functions of $\mathbf{u}$ and f. Hence we may inquire about the same questions as in the case of NS, namely the long time behaviour of T and the existence of stationary measure for it. Note that T is a nonlinear function of $\mathbf{u}$, i.e. the statistics of T may be nontrivial even if $\mathbf{u}$ is gaussian. Indeed, that is what happens.

Let us first summarize the results. Let μ denote the invariant measure of T. It satisfies the "energy" balance relation as in NS (we call the L^2-norm of T still by energy):

$$\kappa \int (\nabla T(\mathbf{x}))^2 \mu(dT) = \tfrac{1}{2}C(0)$$

The energy spectrum is defined as in NS and we have

$$e(k) = Ck^{d-3+\xi}\left(1 + o((\ell k), (Lk)^{-1})\right)$$

where the dissipation scale $l \to 0$ as $\kappa \to 0$. Thus there is scaling on the "inertial range" $[\ell, L]$ and one may show [GK1] that the invariant measure exhibits a cascade of "energy".

Moreover, as $\kappa \to 0$, the invariant measure is supported on weak solutions of the $\kappa = 0$ equation and these solutions are only Hölder continuous.

Next, consider the structure functions of T,

$$\int (T(\mathbf{x}) - T(\mathbf{y}))^n \mu(dT) \equiv S_n(|\mathbf{x} - \mathbf{y}|). \tag{23}$$

The following asymptotics holds

$$S_n(r) = A_n r^{\zeta_n} L^{\frac{2-\xi}{2}-\zeta_n}\left(1 + o\left(\tfrac{\ell}{r}, \tfrac{r}{L}\right)\right),$$

with ζ_n strictly concave in n. Thus the passive scalar exhibits just the sort of multiscaling that is expected in the NS turbulence. It should be noted that the "Kolmogorov scaling theory" (here actually due to Corsin) would predict $\zeta_n = n\frac{2-\xi}{2}$.

Turbulence vs. chaos. It is clear from (21) that in order to control the correlation functions of T we need to understand the joint statistics of the probability kernels $P(\mathbf{U})$. For instance, taking $T(0,\mathbf{x}) = 0$ we have for the 2-point function

$$ET(t,\mathbf{x}_1)T(t,\mathbf{x}_2) = \int_0^t \int E\big(P_{t,s}(\mathbf{x}_1,\mathbf{x}_1')P_{t,s}(\mathbf{x}_2,\mathbf{x}_2')\big)C\left(\tfrac{\mathbf{x}_1'-\mathbf{x}_2'}{L}\right)dx_1'dx_2'ds \tag{24}$$

and similar expressions for expectations of higher point functions. The expectation on the RHS has an interesting interpretation in terms of statistics of particle trajectories. Consider the equations for two trajectories

$$d\mathbf{x}_i(s) = d\mathbf{U}\big(s,\mathbf{x}_i(s)\big) + \sqrt{2\kappa}d\beta_i(s)\,, \quad \mathbf{x}_i(t) = \mathbf{x}_i \ \ i = 1,2\,,$$

where β_i are independent Brownian motions and $d\mathbf{U}$ is our velocity field $\mathbf{u}dt$. Then the expectation on the RHS of (24) is the average over the velocity realizations of the probability density of particles moving from $\mathbf{x}_i(t)$ to $\mathbf{x}_i(s)$.

What can one say about the random kernels $P(\mathbf{U})$? Le Jan and Raimond have proved [LeR] that for almost all realizations of $\mathbf{U}$ these kernels define a Markov process, *even as* $\kappa \to 0$. This random Markov process has markedly different behaviour depending on whether $\mathbf{U}$ is Lipschitz or only Hölder. Let us set $\kappa = 0$. By an Ito formula one gets

$$\partial_t EP_{t,s}(\mathbf{x},\mathbf{x}') = \tfrac{1}{2}D_{\alpha\beta}(0)\partial_\alpha\partial_\beta EP_{t,s}(\mathbf{x},\mathbf{x}') \equiv M_1 EP_{t,s}(\mathbf{x},\mathbf{x}')\,.$$

In our case M_1 is just a multiple of Laplacian and thus individual trajectories diffuse. This is no surprise as our velocity field is derivative of a Brownian. However, a similar application of Ito formula yields

$$E\big(P_{t,s}(\mathbf{x}_1,\mathbf{x}_1')P_{t,s}(\mathbf{x}_2,\mathbf{x}_2')\big) = e^{(t-s)M_2}\big((\mathbf{x}_1,\mathbf{x}_2),(\mathbf{x}_1',\mathbf{x}_2')\big)\,,$$

with M_2 the operator

$$M_2 = M_1 \otimes 1 + 1 \otimes M_1 + D_{\alpha\beta}(\mathbf{x}_1 - \mathbf{x}_2)\partial_{x_{1\alpha}}\partial_{x_{2\beta}}\,.$$

Let us consider the *separation* of two particle trajectories, $\mathbf{X}(s) = \mathbf{x}_1(s) - \mathbf{x}_2(s)$. The behaviour of $\mathbf{X}(s)$ depends crucially on the (spatial) smoothness of the velocity. Its transition probability density averaged over the velocity is

$$EP_{t,s}(\mathbf{X},\mathbf{X}') = e^{(t-s)\mathcal{M}_2}(\mathbf{X},\mathbf{X}')\,, \tag{25}$$

with $\mathcal{M}_2$ given by restricting M_2 to the space of functions depending on the difference variable $\mathbf{x}_1 - \mathbf{x}_2$,

$$\mathcal{M}_2 = \big(D_{\alpha\beta}(0) - D_{\alpha\beta}(\mathbf{X})\big)\partial_{X_\alpha}\partial_{X_\beta}\,.$$

Note that the symbol of this elliptic operator is just the velocity 2-point structure function. It vanishes when $\mathbf{X} \to 0$ as $|\mathbf{X}|^\xi$.

For $\xi < 2$ the semigroup (25) behaves very much like the heat kernel of the Laplacian; it gives rise to superdiffusion, $\mathbf{X}(t)^2 \sim t^{\frac{2}{2-\xi}}$. In particular $\mathbf{X}(t)$ (super)diffuses even if $\mathbf{X}(0) = 0$. Thus the relative motion of the particles is not deterministic for a typical realization of the velocity field.

In contrast, for $\xi \geq 2$, determinism holds,

$$e^{t\mathcal{M}_2}(0, \mathbf{X}') = \delta(\mathbf{X}')\,,$$

i.e. identical initial conditions for the particles induce identical trajectories. However, there is sensitive dependence on initial conditions, $\mathbf{X}(t)$ grows exponentially with t. There is a similar picture for the joint probabilities of n particle trajectories,

$$EP^{\otimes n}_{t,s} = e^{(t-s)M_n}\,,$$

where M_n is a singular elliptic operator, giving rise either to a diffusive or deterministic behaviour depending on whether ξ is less or greater or equal to 2.

It should be stressed that the diffusive/deterministic nature of the trajectories holds for a typical realization of the velocity field and depends on whether the field is Lipschitz or just Hölder. Note that this is precisely the distinction that determines whether solutions of ODE's are unique. In our case, a given (typical) Hölder velocity field gives rise not only to nonuniqueness, but a whole diffusion process for the particle separations.

Typical high Reynolds number turbulent velocities are Hölder (for scales larger than the viscous scale) so we expect them to give rise to (super) diffusive particle motion instead of chaotic but deterministic. This indeed is the case as was known experimentally already to L.F. Richardson in 1926, who measured the growth of the separation between balloon probes in the atmosphere to be $|\mathbf{X}(t)| \sim t^{3/2}$. Thus it appears that turbulent transport is very different from chaotic transport.

Let us finally briefly discuss the intermittency. The stationary state of the scalar can be expressed in terms of the operators M_n. E.g. for the two-point function we get from (24)

$$G_2(\mathbf{x}, \mathbf{y}) \equiv \int T(\mathbf{x})T(\mathbf{y})\mu(dT) = \mathcal{M}_2^{-1}C\left(\tfrac{\cdot}{L}\right)(\mathbf{x} - \mathbf{y})\,, \qquad (26)$$

and similar expressions can be written for the n-point case in terms of the operators $\mathcal{M}_m$, $m \leq n$ [GK2]. The various claims on the nature of this state mentioned above can then be explicitly analyzed. For instance the asymptotics of the structure functions (23) results as follows. G_n arises from the long time behaviour of the diffusion process with the generator $\mathcal{M}_n$. This in turn is dominated by the zero-modes of $\mathcal{M}_n$. Since the operator is homogeneous under scaling of x, the zero-modes are also. One obtains an asymptotic expansion

$$\int \prod_{n=1}^{N} T(\mathbf{x}_n)\mu(dT) = B_N \, L^{\frac{N}{2}(2-\xi)-\zeta_N} \varphi_N(\underline{\mathbf{x}}) + \dots , \qquad (27)$$

where $\varphi_N(\underline{\mathbf{x}})$ are homogeneous zero modes of the operators $\mathcal{M}_N$ of degree ζ_N that can be explicitly analyzed for ξ small [GK2] with the result

$$\zeta_N = \tfrac{N}{2}(2-\xi) - \tfrac{N(N-2)}{2(d+2)}\,\xi + \mathcal{O}(\xi^2)\,.$$

We have denoted in (27) only the dominant term in the limit $L \to \infty$ that contributes to the structure function S_n. This is responsible for the behaviour (23).

This small ξ result follows from a non-rigorous perturbative analysis, but it has been confirmed by another formal perturbative analysis in powers of the inverse dimension [CFKL] as well as by numerical simulations [FMV].

Many other phenomena have been uncovered in the passive scalar problem. For instance, if the incompressibility property of the velocity field is relaxed and one introduces a parameter interpolating between divergence-less and potential fields, a phase transition occurs as the parameter is varied to a state where there is an inverse cascade of energy to large distances and no anomalous scaling [CKV, GV].

Lessons and prospects. The main difference between the Kraichnan model discussed above and the NS equation is, that the former is explicitly solvable. The correlation function $G_n(\mathbf{x}_1, \dots, \mathbf{x}_n)$ satisfies the equation

$$\mathcal{M}_n G_n = \sum_{i<j} G_{n-2} C\left(\tfrac{\mathbf{x}_i - \mathbf{x}_j}{L}\right)$$

where the arguments $\mathbf{x}_i, \mathbf{x}_j$ are absent in the G_{n-2}. This allows for the iterative construction of all G_n in terms of C and the Green functions of the operators $\mathcal{M}_n$. The corresponding equation for the stationary state of NS, the so called Hopf equation, involves G_{n+1} on the RHS due to the nonlinear term in NS. Thus the equations don't close and there has been a long and unsuccessful history of finding *closures*, i.e. approximations where only a finite number of equations need to be considered. Hence very little can be learned for solving NS from the actual solution of the Kraichnan model. However, some of the qualitative features uncovered in the passive scalar could have their correspondence in the NS turbulence.

One such feature is the the mechanism for energy dissipation in the infinite Reynolds number limit. In the Kraichnan model the Lagrangian trajectories become stochastic already in a fixed realization of the velocity field, due to the roughness of the latter. Similar phenomenon should occur in more general velocity ensembles thus providing a mechanism for the energy dissipation in weak solutions of the Euler equation. The Euler equation is known to have energy nonconserving solutions [S2], but these are unphysical (the energy may also increase). It would be very interesting to have a natural class of dissipating weak solutions. The recent work by Shnirelman [Sh] and by Duchon and Robert [DR] is building to this direction.

In the Kraichnan model, intermittency is related to nontrivial zero modes of the operators $\mathcal{M}_n$. This can be interpreted in terms of the Lagrangian trajectories as conserved structures in the n-particle motion [BerGK]. This again is a feature that should be studied in more general velocity ensembles.

While the intermittency problem in 3d turbulence probably is beyond current ideas (for some interesting work in that direction however, see [BeLPP]), the two dimensional problem could be accessible. The conjectured (and experimentally and numerically supported) absence of intermittency in both the inverse and the direct cascade is intriguing and an exact analysis is not excluded. In the absence of that, it would still be very interesting to get some mathematical hold of the inverse cascade. For instance, it would be nice to see some growth of norms measuring short wave number activity, in the manner of [Ku, Bou2] who in the case of the nonlinear Schrödinger equation studied the increase of higher Sobolev norms, signaling a direct cascade.

Finally, it would be nice to have a nonlinear model where a mathematical control of the cascade picture could be carried out. A candidate for this is the so called "weak turbulence" phenomenon [ZLF]. A prime example occurs in the nonlinear Schrödinger equation, a Hamiltonian system. This equation is known to have an invariant measure given by the exponential of the Hamiltonian [Bou1]. However, this equilibrium measure is not the one relevant for turbulence, but another one obtained by adding a small dissipative term and a large scale forcing to the equation. Unlike in NS, however the nonlinear Schrödinger equation has a coupling constant multiplying the nonlinearity and the turbulent state is believed to be accessible by perturbation theory around a closure ansatz. It would be very interesting to carry out such an analysis on a more mathematical level.

References

[B]   G.K. BATCHELOR, Computation of the energy spectrum in homogeneous two-dimensional turbulence, Phys. Fluids Suppl. II 12 (1969), 233–239.

[BeLPP] V.I. BELINCHER, V. LVOV, A. POMYALOV, I. PROCACCIA, Computing the scaling exponents in fluid turbulence from first principles: Demonstration of multiscaling, J. of Stat. Phys. 93 (1998), 797–832.

[BenCBR] R. BENZI, S. CILIBERTO, C. BAUDET, G. RUIZ CHAVARIA, On the scaling of three dimensional homogeneous and isotropic turbulence, Physica D 80 (1995), 385–398.

[BerGK] D. BERNARD, K. GAWĘDZKI, A. KUPIAINEN, Slow modes in passive advection, J. Stat. Phys. 90 (1998), 519-569.

[BoCV] G. BOFFETTA, A. CELANI, M. VERGASSOLA, Inverse cascade in two-dimensional turbulence: deviations from Gaussianity, chao-dyn/9906016.

[Bou1] J. BOURGAIN, Invariant measures for the 2D-defocusing nonlinear Schrödinger equation, Commun. Math. Physics 176 (1996), 421–445.

[Bou2] J. BOURGAIN, On growth in time of Sobolev norms of smooth solutions of nonlinear Schrödinger equations, J. dAnalyse Math. 72 (1997), 299–310.

[BrKL] J. BRICMONT, A. KUPIAINEN, R. LEFEVERE, Probabilistic estimates for the two dimensional stochastic Navier-Stokes equations, J. Stat. Phys., to appear.

[CFKL] M. CHERTKOV, G. FALKOVICH, I. KOLOKOLOV, V. LEBEDEV, Normal and anomalous scaling of the fourth-order correlation function of a randomly advected scalar, Phys. Rev. E 52 (1995), 4924–4941.

[CKV]   M. CHERTKOV, I. KOLOKOLOV, M. VERGASSOLA, Inverse versus direct cascades in turbulent advection, Phys. Rev. Lett. 80 (1998), 512–515.

[CoFT] P. CONSTANTIN, C. FOIAS, R. TEMAM, On the dimension of the attractors in two-dimensional turbulence, Phys. D 30:3 (1988), 284–296.

[DR]   J. DUCHON, R. ROBERT, Inertial energy dissipation for weak solutions of incompressible Euler and Navier-Stokes equations, Nonlinearity, to appear; preprint (1999).

[FMV]   U. FRISCH, A. MAZZINO, M. VERGASSOLA, Intermittency in passive scalar advection, Phys. Rev. Lett. 80 (1998), 5532–5535.

[GK1]   K. GAWĘDZKI, A. KUPIAINEN, Universality in turbulence, an exactly solvable model, in "Low Dimensional Models in Statistical Mechanics and Quantum Field Theory" (H. Grosse, L. Pittner, eds.), Springer, Berlin

(1996), 71–105.

[GK2] K. GAWĘDZKI, A. KUPIAINEN, Anomalous scaling of the passive scalar, Phys. Rev. Lett. 75 (1995), 3834–3837.

[GV]   K. GAWĘDZKI, M. VERGASSOLA, Phase transition in the passive scalar advection, cond-mat/9811399, to appear in Physica D.

[K]   A.N. KOLMOGOROV, The local structure of turbulence in incompressible viscous fluid for very large Reynolds' numbers, C.R. Acad. Sci. URSS 30 (1941), 301–305.

[Kr1] R.H. KRAICHNAN, Inertial ranges in two-dimensional turbulence, Phys. Fluids 10 (1967), 1417–1423.

[Kr2] R.H. KRAICHNAN, Small-scale structure of a scalar field convected by turbulence, Phys. Fluids 11 (1968), 945–963.

[Ku]   S. KUKSIN, Spectral properties of solutions for nonlinear PDEs in the turbulent regime, GAFA 9 (1999), 141-184.

[L]   O.A. LADYZHENSKAYA, On dynamical system generated by the Navier-Stokes equation, J. Soviet Math. 3:4 (1975).

[LeR] Y. LE JAN, O. RAIMOND, Solution statistiques fortes des équations différentielles stochastiques, C.R. Acad. Sci. Paris Ser. I Math. 327 (1998), 893–896.

[Ler] J. LERAY, Essai sur le mouvement d'un liquide visqueux remplissant l'espace, Acta Mathematica 63 (1934), 193–248.

[O]   L. ONSAGER, Nuovo Cim. Suppl. 6 (1949), 279–287.

[PJT] J. PARET, M.C. JULLIEN, P. TABELING, Vorticity statistics in the two-dimensional enstrophy cascade, cond-mat/9904044.

[PT]   J. PARET, P. TABELING, Intermittency in the 2D inverse cascade of energy: experimental observations, Phys. Fluids 10 (1998), 3126–3136.

[S1]   V.SCHEFFER, Hausdorff measure and the Navier-Stokes equation, Commun. Math. Phys. 55 (1977), 97–112.

[S2]   V. SCHEFFER, An inviscid flow with compact support in space-time, J. Geom. Anal. 3:4 (1993), 343–401.

[Sh]   A. SHNIRELMAN, Weak solutions with decreasing energy of incompressible Euler equations, IHES/M/99/02, preprint.

[VF]   M.J. VISHIK, A.V. FURSIKOV, Mathematical Problems of Statistical Hydrodynamics, Kluwer, Dordrecht, 1988.

[ZLF] V.E. ZAKHAROV, V.S. LVOV, G. FALKOVICH, Kolmogorov Spectra of Turbulence I, Springer, Berlin, 1992.

ANTTI KUPIAINEN, Mathematics Department, Helsinki University, P.O.Box 4, 00014 Helsinki, Finland

GAFA, Geom. funct. anal.
Special Volume – GAFA2000, 334 – 358
1016-443X/00/S10334-25 $ 1.50+0.20/0

© Birkhäuser Verlag, Basel 2000

GAFA Geometric And Functional Analysis

THE MATHEMATICS OF THE SECOND LAW OF THERMODYNAMICS

ELLIOTT H. LIEB AND JAKOB YNGVASON

Abstract

The essence of the second law is the 'entropy principle' which states that adiabatic processes can be quantified by an entropy function on the space of all equilibrium states, whose increase is a necessary and sufficient condition for such a process to occur. It is one of the few really fundamental physical laws (in the sense that no deviation, however tiny, is permitted) and its consequences are far reaching. Since the entropy principle is independent of models, statistical mechanical or otherwise, it ought to be derivable from a few logical principles without recourse to Carnot cycles, ideal gases and other assumptions about such things as 'heat', 'hot' and 'cold', 'temperature', 'reversible processes', etc., as is usually done. The well known formula of statistical mechanics, $S = -\sum p \log p$, is irrelevant for this problem. In this paper the foundations of the subject and the construction of entropy from a few simple axioms will be presented. The axioms basically are those of a preorder, except for an important additional property called 'the comparison hypothesis', which we analyze in detail and derive from other axioms. It can be said that this theory addresses the question: *'When is a preorder on a set equivalent to a monotone function on the set ?'* As such, it could conceivably be useful in other areas of mathematics. Finally, we consider some open problems and directions for further study.

Foreword

As part of the conference "Visions in Mathematics, Towards 2000", Tel Aviv, 1999, one of us (E.L.) contributed a talk with the above title. It

E.L.'s work partially supported by U.S. National Science Foundation grant PHY98-20650. J.Y.'s work partially supported by the Adalsteinn Kristjansson Foundation, University of Iceland.

was a review of our attempt in [LiY1] at understanding the mathematical structure of one of the most far reaching and exact laws of physics. To some extent we succeeded, but nothing is perfect and open problems about the meaning of entropy remain. The conference organizers' request for a record of the talk and an addendum about open problems could best be met, we believe, by reproducing a summary of [LiY1] that appeared in the AMS Notices [LiY2] and by adding a new section about open problems. Thus, the first part is the Notices article with some modifications. (A shorter summary appears in [LiY3].) The open problems section is highly speculative and is here largely to share the sense that the Second Law is still open to discussion and creative ideas. Some of the problems in part 2 were mentioned in [Li]. A recent scholarly, pedagogical review of various historical approaches to classical thermodynamics and the Second Law is in [U].

1 A Guide to Entropy and the Second Law of Thermodynamics [LiY2]

This article is intended for readers who, like us, were told that the second law of thermodynamics is one of the major achievements of the nineteenth century, that it is a logical, perfect and unbreakable law – but who were unsatisfied with the 'derivations' of the entropy principle as found in text-books and in popular writings.

A glance at the books will inform the reader that the law has 'various formulations' (which is a bit odd, as if to say the ten commandments have various formulations) but they all lead to the existence of an entropy function whose reason for existence is to tell us which processes can occur and which cannot. Contrary to convention, we shall refer to the existence of entropy as *the* second law. This, at least, is unambiguous. The entropy we are talking about is that defined by thermodynamics (and *not* some analytic quantity, usually involving expressions such as $-p \ln p$, that appears in information theory, probability theory and statistical mechanical models).

Why, one might ask, should a mathematician be interested in the second law of thermodynamics which, historically, had something to do with attempts to understand and improve the efficiency of steam engines? The answer, as we perceive it, is that the law is really an interesting mathematical theorem about orderings on sets, with profound physical implications. The axioms that constitute this ordering are somewhat peculiar from the mathematical point of view and might not arise in the ordinary ruminations of abstract thought. They are special, but important, and they are driven by considerations about the world, which is what makes them so interesting. Maybe an ingenious reader will find an application of this same logical structure to another field of science.

Classical thermodynamics, as it is usually presented, is based on three laws (plus one more, due to Nernst, which is mainly used in low temperature physics and is not immutable like the others). In brief, these are:

> *The Zeroth Law,* which expresses the transitivity of equilibrium, and which is often said to imply the existence of temperature as a parametrization of equilibrium states. We use it below but formulate it without mentioning temperature. In fact, temperature makes no appearance here until almost the very end.
>
> *The First Law,* which is conservation of energy. It is a concept from mechanics and provides the connection between mechanics (and things like falling weights) and thermodynamics. We discuss this later on when we introduce simple systems; the crucial usage of this law is that it allows energy to be used as one of the parameters describing the states of a simple system.
>
> *The Second Law.* Three popular formulations of this law are:
>
> > *Clausius:* No process is possible, the sole result of which is that heat is transferred from a body to a hotter one.
> >
> > *Kelvin (and Planck):* No process is possible, the sole result of which is that a body is cooled and work is done.
> >
> > *Carathéodory:* In any neighborhood of any state there are states that cannot be reached from it by an adiabatic process.

All three are supposed to lead to the entropy principle (defined below). These steps can be found in many books and will not be trodden again here. Let us note in passing, however, that the first two use concepts such as hot, cold, heat, cool, that are intuitive but have to be made precise before the statements are truly meaningful. No one has seen 'heat', for example. The last (which uses the term "adiabatic process", to be defined below) presupposes some kind of parametrization of states by points in $\mathbf{R}^n$, and the usual derivation of entropy from it assumes some sort of differentiability;

such assumptions are beside the point as far as understanding the meaning of entropy goes.

The basic input in our analysis of the second law is a certain kind of ordering on a set and denoted by

$$\prec$$

(pronounced 'precedes'). It is transitive and reflexive as in A1, A2 below, but $X \prec Y$ and $Y \prec X$ does not imply $X = Y$, so it is a 'preorder'. The big question is whether $\prec$ can be encoded in an ordinary, real-valued function on the set, denoted by S, such that if X and Y are related by $\prec$, then $S(X) \leq S(Y)$ if and only if $X \prec Y$. The function S is also required to be additive and extensive in a sense that will soon be made precise.

A helpful analogy is the question: When can a vector-field, $V(x)$, on $\mathbf{R}^3$ be encoded in an ordinary function, $f(x)$, whose gradient is V? The well-known answer is that a necessary and sufficient condition is that $\operatorname{curl} V = 0$. Once V is observed to have this property one thing becomes evident and important: It is necessary to measure the integral of V only along some curves – not all curves – in order to deduce the integral along *all* curves. The encoding then has enormous predictive power about the nature of future measurements of V. In the same way, knowledge of the function S has enormous predictive power in the hands of chemists, engineers and others concerned with the ways of the physical world.

Our concern will be the existence and properties of S, starting from certain natural axioms about the relation $\prec$. We present our results without proofs, but full details, and a discussion of related previous work on the foundations of classical thermodynamics, are given in [LiY1]. The literature on this subject is extensive and it is not possible to give even a brief account of it here, except for mentioning that the previous work closest to ours is that of [Gi], and [Bu], (see also [Co], [D] and [RL]). These other approaches are also based on an investigation of the relation $\prec$, but the overlap with our work is only partial. In fact, a major part of our work is the derivation of a certain property (the "comparison hypothesis" below), which is taken as an axiom in the other approaches. It was a remarkable and largely unsung achievement of Giles [Gi] to realize the full power of this property.

Let us begin the story with some basic concepts.

- *Thermodynamic System*: Physically, this consists of certain specified amounts of certain kinds of matter, e.g., a gram of hydrogen in a container with a piston, or a gram of hydrogen and a gram of oxygen in two separate containers, or a gram of hydrogen and two grams of hydrogen in separate containers. The system can be in various states which, physically, are *equilibrium states*. The space of states of the system is usually denoted by a symbol such as Γ and states in Γ by X, Y, Z, etc.

Physical motivation aside, a state-space, mathematically, is just a set – to begin with; later on we will be interested in embedding state-spaces in some convex subset of some $\mathbf{R}^{n+1}$, i.e., we will introduce coordinates. As we said earlier, however, the entropy principle is quite independent of coordinatization, Carathéodory's principle notwithstanding.

2. *Composition and scaling of states*: The notion of Cartesian product, $\Gamma_1 \times \Gamma_2$ corresponds simply to the two (or more) systems being side by side on the laboratory table; mathematically it is just another system (called a *compound system*), and we regard the state space $\Gamma_1 \times \Gamma_2$ as the same as $\Gamma_2 \times \Gamma_1$. Points in $\Gamma_1 \times \Gamma_2$ are denoted by pairs (X, Y), as usual. The subsystems comprising a compound system are physically independent systems, but they are allowed to interact with each other for a period of time and thereby alter each other's state.

The concept of scaling is crucial. It is this concept that makes our thermodynamics inappropriate for microscopic objects like atoms or cosmic objects like stars. For each state-space Γ and number $\lambda > 0$ there is another state-space, denoted by $\Gamma^{(\lambda)}$ with points denoted by λX. This space is called a *scaled copy* of Γ. Of course we identify $\Gamma^{(1)} = \Gamma$ and $1X = X$. We also require $(\Gamma^{(\lambda)})^{(\mu)} = \Gamma^{(\lambda\mu)}$ and $\mu(\lambda X) = (\mu\lambda)X$. The physical interpretation of $\Gamma^{(\lambda)}$ when Γ is the space of one gram of hydrogen, is simply the state-space of λ grams of hydrogen. The state λX is the state of λ grams of hydrogen with the same 'intensive' properties as X, e.g., pressure, while 'extensive' properties like energy, volume, etc., are scaled by a factor λ (by definition).

For any given Γ we can form Cartesian product state spaces of the type $\Gamma^{(\lambda_1)} \times \Gamma^{(\lambda_2)} \times \cdots \times \Gamma^{(\lambda_N)}$. These will be called *multiple scaled copies* of Γ.

The notation $\Gamma^{(\lambda)}$ should be regarded as merely a mnemonic at this point, but later on, with the embedding of Γ into $\mathbf{R}^{n+1}$, it will literally be $\lambda\Gamma = \{\lambda X : X \in \Gamma\}$ in the usual sense.

3. *Adiabatic accessibility*: Now we come to the ordering. We say $X \prec Y$ (with X and Y *possibly in <u>different</u> state-spaces*) if Y is *adiabatically accessible* from X according to the definition below.

What does this mean? Mathematically, we are just given a list of pairs

$X \prec Y$. There is nothing more to be said, except that later on we will assume that this list has certain properties that will lead to interesting theorems about this list, and will lead, in turn, to the existence of an *entropy function*, S characterizing the list.

The physical interpretation is quite another matter. In text books a process taking X to Y is usually called adiabatic if it takes place in 'thermal isolation', which in turn means that 'no heat is exchanged with the surroundings'. Such concepts (heat, thermal etc.) appear insufficiently precise to us and we prefer the following version, which is in the spirit of Planck's formulation of the second law [P] and avoids those concepts. Our definition of adiabatic accessibility might at first sight appear to be less restrictive than the usual one, but as we have shown in [LiY1], in the end anything that we call an adiabatic process (meaning that Y is adiabatically accessible from X) can also be accomplished in 'thermal isolation' as the concept is usually understood. Our definition has the great virtue (as discovered by Planck) that it avoids having to distinguish between work and heat – or even having to define the concept of heat. We emphasize, however, that the theorems do not require agreement with our physical definition of adiabatic accessibility; other definitions are conceivably possible. We emphasize also that we do not care about the temporal development involved in the state change; we only care about the net result for the system and the rest of the universe.

A state Y is adiabatically accessible from a state X, in symbols $X \prec Y$, if it is possible to change the state from X to Y by means of an interaction with some device consisting of some auxiliary system and a weight, in such a way that the auxiliary system returns to its initial state at the end of the process whereas the weight may have risen or fallen.

The role of the 'weight' in this definition is merely to provide a particularly simple source (or sink) of mechanical energy. Note that an adiabatic process, physically, does not have to be gentle, or 'static' or anything of the kind. It can be arbitrarily violent! The 'device' need not be a well-defined mechanical contraption. It can be another thermodynamic system, and even a gorilla jumping up and down on the system, or a combination of these – as long as the device returns to its initial state.

An example might be useful here. Take a pound of hydrogen in a container with a piston. The states are describable by two numbers, energy and volume, the latter being determined by the position of the piston.

Starting from some state, X, we can take our hand off the piston and let the volume increase explosively to a larger one. After things have calmed down, call the new equilibrium state Y. Then $X \prec Y$. Question: Is $Y \prec X$ true? Answer: No. To get from Y to X adiabatically we would have to use some machinery and a weight, with the machinery returning to its initial state, and there is no way this can be done. Using a weight we can, indeed, recompress the gas to its original volume, but we will find that the energy is then larger than its original value.

Let us write

$$X \prec\prec Y \quad \text{if} \quad X \prec Y \quad \text{but not} \quad Y \prec X \ (\text{written } Y \not\prec X).$$

In this case we say that we can go from X to Y by an *irreversible adiabatic process*. If $X \prec Y$ and $Y \prec X$ we say that X and Y are *adiabatically equivalent* and write

$$X \overset{\text{A}}{\sim} Y.$$

Equivalence classes under $\overset{\text{A}}{\sim}$ are called *adiabats*.

4. *Comparability*: Given two states X and Y in two (same or different) state-spaces, we say that they are comparable if $X \prec Y$ or $Y \prec X$ (or both). This turns out to be a crucial notion. Two states are not always comparable; a necessary condition is that they have the same material composition in terms of the chemical elements. Example: Since water is H_2O and the atomic weights of hydrogen and oxygen are 1 and 16 respectively, the states in the compound system of 2 gram of hydrogen and 16 grams of oxygen are comparable with states in a system consisting of 18 grams of water (but not with 11 grams of water or 18 grams of oxygen).

Actually, the classification of states into various state-spaces is done mainly for conceptual convenience. The second law deals only with states, and the only thing we really have to know about any two of them is whether or not they are comparable. Given the relation $\prec$ for all possible states of all possible systems, we can ask whether this relation can be encoded in an entropy function according to the following:

Entropy principle. *There is a real-valued function on all states of all systems (including compound systems), called* **entropy** *and denoted by S such that*

a) MONOTONICITY: *When X and Y are comparable states then*

$$X \prec Y \quad \text{if and only if} \quad S(X) \leq S(Y). \tag{1}$$

b) ADDITIVITY AND EXTENSIVITY: *If X and Y are states of some (possibly different) systems and if (X,Y) denotes the corresponding state in the compound system, then the entropy is additive for these states, i.e.,*

$$S(X,Y) = S(X) + S(Y).\tag{2}$$

S is also extensive, i.e., for or each $\lambda > 0$ and each state X and its scaled copy $\lambda X \in \Gamma^{(\lambda)}$, (defined in 2. above)

$$S(\lambda X) = \lambda S(X).\tag{3}$$

A formulation logically equivalent to a), not using the word 'comparable', is the following pair of statements:

$$X \overset{A}{\sim} Y \Longrightarrow S(X) = S(Y) \quad \text{and}$$
$$X \prec\prec Y \Longrightarrow S(X) < S(Y).\tag{4}$$

The last line is especially noteworthy. It says that entropy must increase in an irreversible adiabatic process.

The additivity of entropy in compound systems is often just taken for granted, but it is one of the startling conclusions of thermodynamics. First of all, the content of additivity, (2), is considerably more far reaching than one might think from the simplicity of the notation. Consider four states X, X', Y, Y' and suppose that $X \prec Y$ and $X' \prec Y'$. One of our axioms, A3, will be that then $(X, X') \prec (Y, Y')$, and (2) contains nothing new or exciting. On the other hand, the compound system can well have an adiabatic process in which $(X, X') \prec (Y, Y')$ but $X \not\prec Y$. In this case, (2) conveys much information. Indeed, by monotonicity, there will be many cases of this kind because the inequality $S(X) + S(X') \leq S(Y) + S(Y')$ certainly does not imply that $S(X) \leq S(Y)$. The fact that the inequality $S(X) + S(X') \leq S(Y) + S(Y')$ tells us *exactly* which adiabatic processes are allowed in the compound system (among comparable states), independent of any detailed knowledge of the manner in which the two systems interact, is astonishing and is at the *heart of thermodynamics*. The second reason that (2) is startling is this: From (1) alone, restricted to one system, the function S can be replaced by $29S$ and still do its job, i.e., satisfy (1). However, (2) says that it is possible to calibrate the entropies of all systems (i.e., simultaneously adjust all the undetermined multiplicative constants) so that the entropy $S_{1,2}$ for a compound $\Gamma_1 \times \Gamma_2$ is $S_{1,2}(X,Y) = S_1(X) + S_2(Y)$, even though systems 1 and 2 are totally unrelated!

We are now ready to ask some basic questions:

Q1: Which properties of the relation $\prec$ ensure existence and (essential) uniqueness of S?

Q2: Can these properties be derived from simple physical premises?

Q3: Which convexity and smoothness properties of S follow from the premises?

Q4: Can temperature (and hence an ordering of states by "hotness" and "coldness") be defined from S and what are its properties?

The answer to question Q1 can be given in the form of six axioms that are reasonable, simple, 'obvious' and unexceptionable. An additional, crucial assumption is also needed, but we call it a 'hypothesis' instead of an axiom because we show later how it can be derived from some other axioms, thereby answering question Q2.

A1. Reflexivity. $X \overset{\text{A}}{\sim} X$.

A2. Transitivity. If $X \prec Y$ and $Y \prec Z$, then $X \prec Z$.

A3. Consistency. If $X \prec X'$ and $Y \prec Y'$, then $(X, Y) \prec (X', Y')$.

A4. Scaling Invariance. If $\lambda > 0$ and $X \prec Y$, then $\lambda X \prec \lambda Y$.

A5. Splitting and Recombination. $X \overset{\text{A}}{\sim} ((1 - \lambda)X, \lambda X)$ for all $0 < \lambda < 1$. Note that the two state-spaces are different. If $X \in \Gamma$, then the state space on the right side is $\Gamma^{(1-\lambda)} \times \Gamma^{(\lambda)}$.

A6. Stability. If $(X, \varepsilon Z_0) \prec (Y, \varepsilon Z_1)$ for some Z_0, Z_1 and a sequence of ε's tending to zero, then $X \prec Y$. This axiom is a substitute for continuity, which we cannot assume because there is no topology yet. It says that 'a grain of dust cannot influence the set of adiabatic processes'.

An important lemma is that (A1)–(A6) imply the *cancellation law*, which is used in many proofs. It says that for any three states X, Y, Z

$$(X, Z) \prec (Y, Z) \Longrightarrow X \prec Y . \tag{5}$$

The next concept plays a key role in our treatment.

CH. Definition: We say that the *Comparison Hypothesis*, (CH), holds for a state-space Γ if all pairs of states in Γ are comparable.

Note that A3, A4 and A5 automatically extend comparability from a space Γ to certain other cases, e.g., $X \prec ((1 - \lambda)Y, \lambda Z)$ for all $0 \leq \lambda \leq 1$ if $X \prec Y$ and $X \prec Z$. On the other hand, comparability on Γ alone does not allow us to conclude that X is comparable to $((1 - \lambda)Y, \lambda Z)$ if $X \prec Y$ but $Z \prec X$. For this, one needs CH on the product space $\Gamma^{(1-\lambda)} \times \Gamma^{(\lambda)}$, which is not implied by CH on Γ.

The significance of A1–A6 and CH is borne out by the following theorem:

Theorem 1 (Equivalence of entropy and A1-A6, given CH). *The following are equivalent for a state-space* Γ:

(i) *The relation* $\prec$ *between states in (possibly different) multiple scaled copies of* Γ *e.g.,* $\Gamma^{(\lambda_1)} \times \Gamma^{(\lambda_2)} \times \cdots \times \Gamma^{(\lambda_N)}$, *is characterized by an entropy function, S, on* Γ *in the sense that*

$$(\lambda_1 X_1, \lambda_2 X_2, \dots) \prec (\lambda_1' X_1', \lambda_2' X_2', \dots) \tag{6}$$

is equivalent to the condition that

$$\sum_i \lambda_i S(X_i) \leq \sum_j \lambda_j' S(X_j') \tag{7}$$

whenever

$$\sum_i \lambda_i = \sum_j \lambda_j'. \tag{8}$$

(ii) *The relation* $\prec$ *satisfies conditions (A1)–(A6), and (CH) holds for* <u>every</u> *multiple scaled copy of* Γ.

This entropy function on Γ *is unique up to affine equivalence, i.e.,* $S(X) \rightarrow aS(X) + B$, *with* $a > 0$.

That (i) $\Rightarrow$ (ii) is obvious. The proof of (ii) $\Rightarrow$ (i) is carried out by an explicit construction of the entropy function on Γ – reminiscent of an old definition of heat by Laplace and Lavoisier in terms of the amount of ice that a body can melt.

Basic construction of S. Pick two reference points X_0 and X_1 in Γ with $X_0 \prec\prec X_1$. (If such points do not exist then S is the constant function.) Then define for $X \in \Gamma$

$$S(X) := \sup \left\{ \lambda : ((1 - \lambda)X_0, \lambda X_1) \prec X \right\}. \tag{9}$$

REMARKS. As in axiom A5, two state-spaces are involved in (9). By axiom A5, $X \overset{A}{\sim} ((1-\lambda)X, \lambda X)$, and hence, by CH in the space $\Gamma^{(1-\lambda)} \times \Gamma^{(\lambda)}$, X is comparable to $((1-\lambda)X_0, \lambda X_1)$. In (9) we allow $\lambda \leq 0$ and $\lambda \geq 1$ by using the convention that $(X, -Y) \prec Z$ means that $X \prec (Y, Z)$ and $(X, 0Y) = X$. For (9) we only need to know that CH holds in two-fold scaled products of Γ with itself. CH will then automatically be true for all products. In (9) the reference points X_0, X_1 are fixed and the supremum is over λ. One can ask how S changes if we change the two points X_0, X_1. The answer is that the change is affine, i.e., $S(X) \rightarrow aS(X) + B$, with $a > 0$.

Theorem 1 extends to products of multiple scaled copies of different systems, i.e. to general *compound* systems. This extension is an immediate consequence of the following theorem, which is proved by applying

Theorem 1 to the product of the system under consideration with some standard reference system.

Theorem 2 (Consistent entropy scales). *Assume that CH holds for* <u>all</u> *compound systems. For each system* Γ *let* S_Γ *be some definite entropy function on* Γ *in the sense of Theorem 1. Then there are constants* a_Γ *and* $B(\Gamma)$ *such that the function* S, *defined for all states of all systems by*

$$S(X) = a_\Gamma S_\Gamma(X) + B(\Gamma) \tag{10}$$

for $X \in \Gamma$, *satisfies additivity* (2), *extensivity* (3), *and monotonicity* (1) *in the sense that whenever* X *and* Y *are in the same state space then*

$$X \prec Y \qquad \text{if and only if} \qquad S(X) \leq S(Y). \tag{11}$$

Theorem 2 is what we need, except for the question of mixing and chemical reactions, which is treated at the end and which can be put aside at a first reading. In other words, as long as we do not consider adiabatic processes in which systems are converted into each other (e.g., a compound system consisting of vessel of hydrogen and a vessel of oxygen is converted into a vessel of water), the entropy principle has been verified. If that is so, what remains to be done, the reader may justifiably ask? The answer is twofold: First, Theorem 2 requires that CH holds for *all* systems, and we are not content to take this as an axiom. Second, important notions of thermodynamics such as 'thermal equilibrium' (which will eventually lead to a precise definition of 'temperature') have not appeared so far. We shall see that these two points (i.e., thermal equilibrium and CH) are not unrelated.

As for CH, other authors, [Gi], [Bu], [Co] and [RL] essentially *postulate* that it holds for all systems by making it axiomatic that comparable states fall into equivalence classes. (This means that the conditions $X \prec Z$ and $Y \prec Z$ always imply that X and Y are comparable: likewise, they must be comparable if $Z \prec X$ and $Z \prec Y$.) By identifying a 'state-space' with an equivalence class, the comparison hypothesis then holds in these other approaches *by assumption* for all state-spaces. We, in contrast, would like to derive CH from something that we consider more basic. Two ingredients will be needed: The analysis of certain special, but commonplace systems called 'simple systems' and some assumptions about thermal contact (the 'Zeroth law') that will act as a kind of glue holding the parts of a compound systems in harmony with each other.

A **Simple System** is one whose state-space can be identified with some open convex subset of some $\mathbf{R}^{n+1}$ with a distinguished coordinate denoted by U, called the *energy*, and additional coordinates $V \in \mathbf{R}^n$, called *work coordinates*. The energy coordinate is the way in which thermodynamics makes contact with mechanics, where the concept of energy arises and is precisely defined. The fact that the amount of energy in a state is independent of the manner in which the state was arrived at is, in reality, the first law of thermodynamics. A typical (and often the only) work coordinate is the volume of a fluid or gas (controlled by a piston); other examples are deformation coordinates of a solid or magnetization of a paramagnetic substance.

Our goal is to show, with the addition of a few more axioms, that CH holds for simple systems and their scaled products. In the process, we will introduce more structure, which will capture the intuitive notions of thermodynamics; thermal equilibrium is one.

First, there is an axiom about convexity:

A7. Convex combination. If X and Y are states of a simple system and $t \in [0, 1]$ then

$$\left(tX, (1 - t)Y \right) \prec tX + (1 - t)Y,$$

in the sense of ordinary convex addition of points in $\mathbf{R}^{n+1}$. A straightforward consequence of this axiom (and A5) is that the *forward sectors*

$$A_X := \{ Y \in \Gamma : X \prec Y \} \tag{12}$$

of states X in a simple system Γ are *convex* sets.

Another consequence is a connection between the existence of irreversible processes and Carathéodory's principle ([C], [B]) mentioned above.

LEMMA 1. *Assume (A1)–(A7) for $\Gamma \subset R^N$ and consider the following statements:*

(a) Existence of irreversible processes: *For every $X \in \Gamma$ there is a $Y \in \Gamma$ with $X \prec\prec Y$.*

(b) Carathéodory's principle: *In every neighborhood of every $X \in \Gamma$ there is a $Z \in \Gamma$ with $X \not\prec Z$.*

Then (a) $\Rightarrow$ (b) *always. If the forward sectors in Γ have interior points, then* (b) $\Rightarrow$ (a).

We need three more axioms for simple systems, which will take us into an analytic detour. The first of these establishes (a) above.

A8. Irreversibility. For each $X \in \Gamma$ there is a point $Y \in \Gamma$ such that $X \prec\prec Y$. (This axiom is implied by A14 below, but is stated here separately because important conclusions can be drawn from it alone.)

A9. Lipschitz tangent planes. For each $X \in \Gamma$ the *forward sector* $A_X = \{Y \in \Gamma : X \prec Y\}$ has a *unique* support plane at X (i.e., A_X has a *tangent plane* at X). The tangent plane is assumed to be a *locally Lipschitz continuous* function of X, in the sense explained below.

A10. Connectedness of the boundary. The boundary ∂A_X (relative to the open set Γ) of every forward sector $A_X \subset \Gamma$ is connected. (This is technical and conceivably can be replaced by something else.)

Axiom A8 plus Lemma 1 asserts that every X lies on the boundary ∂A_X of its forward sector. Although axiom A9 asserts that the convex set, A_X, has a true tangent at X only, it is an easy consequence of axiom A2 that A_X has a true tangent everywhere on its boundary. To say that this tangent plane is locally Lipschitz continuous means that if $X = (U^0, V^0)$ then this plane is given by

$$U - U^0 + \sum_1^n P_i(X)(V_i - V_i^0) = 0 . \tag{13}$$

with locally Lipschitz continuous functions P_i. The function P_i is called the generalized *pressure* conjugate to the work coordinate V_i . (When V_i is the volume, P_i is the ordinary pressure.)

Lipschitz continuity and connectedness is a well known guarantee that the coupled differential equations

$$\frac{\partial U}{\partial V_j}(V) = -P_j\big(U(V), V\big) \quad \text{for } j = 1, \ldots, n , \tag{14}$$

not only have a solution (since we know that the surface ∂A_X exists) but this solution must be unique. Thus, if $Y \in \partial A_X$ then $X \in \partial A_Y$. In short, the surfaces ∂A_X foliate the state-space Γ. What is less obvious, but very important because it instantly gives us the comparison hypothesis for Γ, is the following.

Theorem 3 (Forward sectors are nested). *If A_X and A_Y are two forward sectors in the state-space, Γ, of a simple system then exactly one of the following holds.*

(a) $A_X = A_Y$, i.e., $X \overset{\mathrm{A}}{\sim} Y$.
(b) $A_X \subset \text{Interior}(A_Y)$, i.e., $Y \prec\prec X$
(c) $A_Y \subset \text{Interior}(A_X)$, i.e., $X \prec\prec Y$.

It can also be shown from our axioms that the orientation of forward sectors w.r.t. the energy axis is the same for *all* simple systems. By convention we choose the direction of the energy axis so that the the energy always *increases* in adiabatic processes at fixed work coordinates. When temperature is defined later, this will imply that temperature is always positive.

Theorem 3 implies that Y is on the boundary of A_X if and only if X is on the boundary of A_Y. Thus the adiabats, i.e., the $\overset{A}{\sim}$ equivalence classes, consist of these boundaries.

Before leaving the subject of simple systems let us remark on the connection with Carathéodory's development. The point of contact is the fact that $X \in \partial A_X$. We assume that A_X is convex and use transitivity and Lipschitz continuity to arrive, eventually, at Theorem 3. Carathéodory uses Frobenius's theorem, plus assumptions about differentiability to conclude the existence – locally – of a surface containing X. Important *global* information, such as Theorem 3, are then not easy to obtain without further assumptions, as discussed, e.g., in [B].

The next topic is *thermal contact* and the zeroth law, which entails the very special assumptions about $\prec$ that we mentioned earlier. It will enable us to establish CH for products of several systems, and thereby show, via Theorem 2, that entropy exists and is additive. Although we have established CH for a simple system, Γ, we have not yet established CH even for a product of two copies of Γ. This is needed in the definition of S given in (9). The S in (9) is determined up to an affine shift and we want to be able to calibrate the entropies (i.e., adjust the multiplicative and additive constants) of all systems so that they work together to form a global S satisfying the entropy principle. We need five more axioms. They might look a bit abstract, so a few words of introduction might be helpful.

In order to relate systems to each other, in the hope of establishing CH for compounds, and thereby an additive entropy function, some way must be found to put them into contact with each other. Heuristically, we imagine two simple systems (the same or different) side by side, and fix the work coordinates (e.g., the volume) of each. Connect them with a "copper thread" and wait for equilibrium to be established. The total energy U will not change but the individual energies, U_1 and U_2 will adjust to values that depend on U and the work coordinates. This new system (with the thread permanently connected) then behaves like a simple system (with one energy coordinate) but with several work coordinates (the union of the two work coordinates). Thus, if we start initially with $X_1 = (U_1, V_1)$ for system 1 and $X_2 = (U_2, V_2)$ for system 2, and if we end up with $X = (U, V_1, V_2)$ for the new system, we can say that $(X_1, X_2) \prec X$. This holds for every choice of U_1 and U_2 whose sum is U. Moreover, after thermal equilibrium is reached, the two systems can be disconnected, if we wish, and once more form a compound system, whose component parts we say are in thermal equilibrium. That this is transitive is the zeroth law.

Thus, we cannot only make compound systems consisting of independent subsystems (which can interact, but separate again), we can also make a new simple system out of two simple systems. To do this an energy coordinate has to disappear, and thermal contact does this for us. All of this is formalized in the following three axioms.

A11. Thermal contact. For any two simple systems with state-spaces Γ_1 and Γ_2, there is another *simple* system, called the *thermal join* of Γ_1 and Γ_2, with state-space

$$\Delta_{12} = \left\{ (U, V_1, V_2) : U = U_1 + U_2 \text{ with } (U_1, V_1) \in \Gamma_1 , \ (U_2, V_2) \in \Gamma_2 \right\}. \tag{15}$$

Moreover,

$$\Gamma_1 \times \Gamma_2 \ni \big((U_1, V_1), (U_2, V_2)\big) \prec (U_1 + U_2, V_1, V_2) \in \Delta_{12}. \tag{16}$$

A12. Thermal splitting. For any point $(U, V_1, V_2) \in \Delta_{12}$ there is at least one pair of states, $(U_1, V_1) \in \Gamma_1$, $(U_2, V_2) \in \Gamma_2$, with $U = U_1 + U_2$, such that

$$(U, V_1, V_2) \overset{A}{\sim} \big((U_1, V_1), (U_2, V_2)\big). \tag{17}$$

If $(U, V_1, V_2) \overset{A}{\sim} \big((U_1, V_1), (U_2, V_2)\big)$ we say that the states $X = (U_1, V_1)$ and $Y = (U_2, V_2)$ are in *thermal equilibrium* and write

$$X \overset{\text{T}}{\sim} Y.$$

A13. Zeroth law of thermodynamics. If $X \overset{\text{T}}{\sim} Y$ and if $Y \overset{\text{T}}{\sim} Z$ then $X \overset{\text{T}}{\sim} Z$.

A11 and A12 together say that for each choice of the individual work coordinates there is a way to divide up the energy U between the two systems in a stable manner. A12 is the stability statement, for it says that joining is reversible, i.e., once the equilibrium has been established, one can cut the copper thread and retrieve the two systems back again, but with a special partition of the energies.

This reversibility allows us to think of the thermal join, which is a simple system in its own right, as a special subset of the product system, $\Gamma_1 \times \Gamma_2$, which we call the *thermal diagonal*. In particular, A12 allows us to prove easily that $X \overset{\text{T}}{\sim} \lambda X$ for all X and all $\lambda > 0$.

A13 is the famous zeroth law, which says that the thermal equilibrium is transitive, and hence an equivalence relation. Often this law is taken to mean that there the equivalence classes can be labeled by an 'empirical' temperature, but we do not want to mention temperature at all at this point. It will appear later.

Two more axioms are needed.

A14 requires that for every adiabat (i.e., an equivalence class w.r.t. $\overset{A}{\sim}$) there exists at least one isotherm (i.e., an equivalence class w.r.t. $\overset{T}{\sim}$), containing points on both sides of the adiabat. Note that, for each given X, only two points in the entire state space Γ are required to have the stated property. This assumption essentially prevents a state-space from breaking up into two pieces that do not communicate with each other. Without it, counterexamples to CH for compound systems can be constructed. A14 implies A8, but we listed A8 separately in order not to confuse the discussion of simple systems with thermal equilibrium.

A15 is a technical and perhaps can be eliminated. Its physical motivation is that a sufficiently large copy of a system can act as a heat bath for other systems. When temperature is introduced later, A15 will have the meaning that all systems have the same temperature range. This postulate is needed if we want to be able to bring every system into thermal equilibrium with every other system.

A14. Transversality. If Γ is the state space of a simple system and if $X \in \Gamma$, then there exist states $X_0 \overset{T}{\sim} X_1$ with $X_0 \prec\prec X \prec\prec X_1$.

A15. Universal temperature range. If Γ_1 and Γ_2 are state spaces of simple systems then, for every $X \in \Gamma_1$ and every V belonging to the projection of Γ_2 onto the space of its work coordinates, there is a $Y \in \Gamma_2$ with work coordinates V such that $X \overset{T}{\sim} Y$.

The reader should note that the concept 'thermal contact' has appeared, but not temperature or hot and cold or anything resembling the Clausius or Kelvin-Planck formulations of the second law. Nevertheless, we come to the main achievement of our approach: *With these axioms we can establish CH for products of simple systems* (each of which satisfies CH, as we already know). First, the thermal join establishes CH for the (scaled) product of a simple system with itself. The basic idea here is that the points in the product that lie on the thermal diagonal are comparable, since points in a simple system are comparable. In particular, with X, X_0, X_1 as in A14, the states $((1 - \lambda)X_0, \lambda X_1)$ and $((1 - \lambda)X, \lambda X)$ can be regarded as states of the *same* simple system and are, therefore, comparable. *This is the key point needed for the construction of S, according to* (9). The importance of transversality is thus brought into focus.

With some more work we can establish CH for multiple scaled copies of a simple system. Thus, we have established S within the context of one system and copies of the system, i.e. condition (ii) of Theorem 1. As long as we stay within such a group of systems there is no way to determine the unknown multiplicative or additive entropy constants. The next task is to show that the multiplicative constants can be adjusted to give a universal entropy valid for copies of *different* systems, i.e. to establish the hypothesis of Theorem 2. This is based on the following.

LEMMA 4 (Existence of calibrators). *If Γ_1 and Γ_2 are simple systems, then there exist states $X_0, X_1 \in \Gamma_1$ and $Y_0, Y_1 \in \Gamma_2$ such that*

$$X_0 \prec\prec X_1 \qquad and \qquad Y_0 \prec\prec Y_1$$

and

$$(X_0, Y_1) \overset{A}{\sim} (X_1, Y_0).$$

The significance of Lemma 4 is that it allows us to fix the *multiplicative* constants by the condition

$$S_1(X_0) + S_2(Y_1) = S_1(X_1) + S_2(Y_0). \tag{18}$$

The proof of Lemma 4 is complicated and really uses all the axioms A1 to A14. With its aid we arrive at our chief goal, which is CH for compound systems.

Theorem 4 (Entropy principle in products of simple systems). *The comparison hypothesis CH is valid in arbitrary scaled products of simple systems. Hence, by Theorem 2, the relation $\prec$ among states in such statespaces is characterized by an entropy function S. The entropy function is <u>unique</u>, up to an overall multiplicative constant and one additive constant for each simple system under consideration.*

At last, we are now ready to define *temperature*. Concavity of S (implied by A7), Lipschitz continuity of the pressure and the transversality condition, together with some real analysis, play key roles in the following, which answers questions Q3 and Q4 posed at the beginning.

Theorem 5 (Entropy defines temperature). *The entropy, S, is a concave and continuously differentiable function on the state space of a simple system. If the function T is defined by*

$$\frac{1}{T} := \left(\frac{\partial S}{\partial U}\right)_V \tag{19}$$

then $T > 0$ and T characterizes the relation $\overset{T}{\sim}$ in the sense that $X \overset{T}{\sim} Y$ if and only if $T(X) = T(Y)$. Moreover, if two systems are brought into

*thermal contact with fixed work coordinates then, since the total entropy
cannot decrease, the energy flows from the system with the higher T to the
system with the lower T.*

The temperature need not be a strictly monotone function of U; indeed,
it is not so in a 'multiphase region'. It follows that T is not always capable of specifying a state, and this fact can cause some pain in traditional
discussions of the second law – if it is recognized, which usually it is not.

Mixing and chemical reactions.   The core results of our analysis have
now been presented and readers satisfied with the entropy principle in the
form of Theorem 4 may wish to stop at this point. Nevertheless, a nagging
doubt will occur to some, because there are important adiabatic processes in
which systems are not conserved, and these processes are not yet covered
in the theory. A critical study of the usual textbook treatments should
convince the reader that this subject is not easy, but in view of the manifold
applications of thermodynamics to chemistry and biology it is important
to tell the whole story and not ignore such processes.

One can formulate the problem as the determination of the additive
constants $B(\Gamma)$ of Theorem 2. As long as we consider only adiabatic processes that preserve the amount of each simple system (i.e., such that Eqs.
(6) and (8) hold), these constants are indeterminate. This is no longer
the case, however, if we consider mixing processes and chemical reactions
(which are not really different, as far as thermodynamics is concerned.) It
then becomes a nontrivial question whether the additive constants can be
chosen in such a way that the entropy principle holds. Oddly, this determination turns out to be far more complex, mathematically and physically
than the determination of the multiplicative constants (Theorem 2). In traditional treatments one usually resorts to *gedanken* experiments involving
strange, nonexistent objects called 'semipermeable' membranes and 'van
t'Hofft boxes'. We present here a general and rigorous approach which
avoids all this.

What we already know is that every system has a well-defined entropy
function, e.g., for each Γ there is S_Γ, and we know from Theorem 2 that
the multiplicative constants a_Γ can been determined in such a way that the
sum of the entropies increases in any adiabatic process in any compound
space $\Gamma_1 \times \Gamma_2 \times \ldots$. Thus, if $X_i \in \Gamma_i$ and $Y_i \in \Gamma_i$ then

$$(X_1, X_2, \ldots) \prec (Y_1, Y_2, \ldots) \quad \text{if and only if} \quad \sum_i S_i(X_i) \le \sum_j S_j(Y_j) . \quad (20)$$

where we have denoted S_{Γ_i} by S_i for short. The additive entropy constants do not matter here since each function S_i appears on both sides of this inequality. It is important to note that this applies even to processes that, in intermediate steps, take one system into another, provided the total compound system is the same at the beginning and at the end of the process.

The task is to find constants $B(\Gamma)$, one for each state space Γ, in such a way that the entropy defined by

$$S(X) := S_\Gamma(X) + B(\Gamma) \qquad \text{for} \qquad X \in \Gamma \tag{21}$$

satisfies

$$S(X) \le S(Y) \tag{22}$$

whenever

$$X \prec Y \qquad \text{with} \qquad X \in \Gamma, \ Y \in \Gamma'.$$

Additionally, we require that the newly defined entropy satisfies scaling and additivity under composition. Since the initial entropies $S_\Gamma(X)$ already satisfy them, these requirements become conditions on the additive constants $B(\Gamma)$:

$$B(\Gamma_1^{(\lambda_1)} \times \Gamma_2^{(\lambda_2)}) = \lambda_1 B(\Gamma_1) + \lambda_2 B(\Gamma_2) \tag{23}$$

for all state spaces Γ_1, Γ_2 under considerations and $\lambda_1, \lambda_2 > 0$. Some reflection shows us that consistency in the definition of the entropy constants $B(\Gamma)$ requires us to consider all possible chains of adiabatic processes leading from one space to another via intermediate steps. Moreover, the additivity requirement leads us to allow the use of a 'catalyst' in these processes, i.e., an auxiliary system, that is recovered at the end, although a state change *within* this system might take place. With this in mind we define quantities $F(\Gamma, \Gamma')$ that incorporate the entropy differences in all such chains leading from Γ to Γ'. These are built up from simpler quantities $D(\Gamma, \Gamma')$, which measure the entropy differences in one-step processes, and $E(\Gamma, \Gamma')$, where the 'catalyst' is absent. The precise definitions are as follows. First,

$$D(\Gamma, \Gamma') := \inf \left\{ S_{\Gamma'}(Y) - S_\Gamma(X) : X \in \Gamma, \ Y \in \Gamma', \ X \prec Y \right\}. \tag{24}$$

If there is no adiabatic process leading from Γ to Γ' we put $D(\Gamma, \Gamma') = \infty$. Next, for any given Γ and Γ' we consider all finite chains of state spaces, $\Gamma = \Gamma_1, \Gamma_2, \ldots, \Gamma_N = \Gamma'$ such that $D(\Gamma_i, \Gamma_{i+1}) < \infty$ for all i, and we define

$$E(\Gamma, \Gamma') := \inf \left\{ D(\Gamma_1, \Gamma_2) + \cdots + D(\Gamma_{N-1}, \Gamma_N) \right\}, \tag{25}$$

where the infimum is taken over all such chains linking Γ with Γ'. Finally we define

$$F(\Gamma, \Gamma') := \inf \left\{ E(\Gamma \times \Gamma_0, \Gamma' \times \Gamma_0) \right\}, \tag{26}$$

where the infimum is taken over all state spaces Γ_0. (These are the 'catalysts'.)

The importance of the F's for the determination of the additive constants is made clear in the following theorem:

Theorem 6 (Constant entropy differences). *If Γ and Γ' are two state spaces then for any two states $X \in \Gamma$ and $Y \in \Gamma'$*

$$X \prec Y \quad \text{if and only if} \quad S_\Gamma(X) + F(\Gamma, \Gamma') \leq S_{\Gamma'}(Y). \tag{27}$$

An essential ingredient for the proof of this theorem is Eq. (20).

According to Theorem 6 the determination of the entropy constants $B(\Gamma)$ amounts to satisfying the inequalities

$$-F(\Gamma', \Gamma) \leq B(\Gamma) - B(\Gamma') \leq F(\Gamma, \Gamma') \tag{28}$$

together with the linearity condition (23). It is clear that (28) can only be satisfied with finite constants $B(\Gamma)$ and $B(\Gamma')$, if $F(\Gamma, \Gamma') > -\infty$. To exclude the pathological case $F(\Gamma, \Gamma') = -\infty$ we introduce our last axiom A16, whose statement requires the following definition.

DEFINITION.   A state-space, Γ is said to be *connected* to another state-space Γ' if there are states $X \in \Gamma$ and $Y \in \Gamma'$, and state spaces $\Gamma_1, \ldots, \Gamma_N$ with states $X_i, Y_i \in \Gamma_i$, $i = 1, \ldots, N$, and a state space Γ_0 with states $X_0, Y_0 \in \Gamma_0$, such that

$$(X, X_0) \prec Y_1, \quad X_i \prec Y_{i+1}, \quad i = 1, \ldots, N-1, \quad X_N \prec (Y, Y_0).$$

A16. (ABSENCE OF SINKS): If Γ is connected to Γ' then Γ' is connected to Γ.

This axiom excludes $F(\Gamma, \Gamma') = -\infty$ because, on general grounds, one always has

$$-F(\Gamma', \Gamma) \leq F(\Gamma, \Gamma'). \tag{29}$$

Hence $F(\Gamma, \Gamma') = -\infty$ (which means, in particular, that Γ is connected to Γ') would imply $F(\Gamma', \Gamma) = \infty$, i.e., that there is no way back from Γ' to Γ. This is excluded by Axiom 16.

The quantities $F(\Gamma, \Gamma')$ have simple subadditivity properties that allow us to use the Hahn-Banach theorem to satisfy the inequalities (28), with constants $B(\Gamma)$ that depend linearly on Γ, in the sense of Eq. (23). Hence we arrive at

Theorem 7 (Universal entropy). *The additive entropy constants of all systems can be calibrated in such a way that the entropy is additive and extensive, and $X \prec Y$ implies $S(X) \leq S(Y)$, even when X and Y do not belong to the same state space.*

Our final remark concerns the remaining non-uniqueness of the constants $B(\Gamma)$. This indeterminacy can be traced back to the non-uniqueness of a linear functional lying between $-F(\Gamma',\Gamma)$ and $F(\Gamma,\Gamma')$ and has two possible sources: One is that some pairs of state-spaces Γ and Γ' may not be connected, i.e., $F(\Gamma,\Gamma')$ may be infinite (in which case $F(\Gamma',\Gamma)$ is also infinite by axiom A16). The other is that there might be a true gap, i.e.,

$$-F(\Gamma',\Gamma) \; < \; F(\Gamma,\Gamma') \tag{32}$$

might hold for some state spaces, even if both sides are finite.

In nature only states containing the same amount of the chemical elements can be transformed into each other. Hence $F(\Gamma,\Gamma') = +\infty$ for many pairs of state spaces, in particular, for those that contain different amounts of some chemical element. The constants $B(\Gamma)$ are, therefore, never unique: For each equivalence class of state spaces (with respect to the relation of connectedness) one can define a constant that is arbitrary except for the proviso that the constants should be additive and extensive under composition and scaling of systems. In our world there are 92 chemical elements (or, strictly speaking, a somewhat larger number, N, since one should count different isotopes as different elements), and this leaves us with at least 92 free constants that specify the entropy of one gram of each of the chemical elements in some specific state.

The other possible source of non-uniqueness, a nontrivial gap (32) for systems with the same composition in terms of the chemical elements is, as far as we know, not realized in nature. (Note that this assertion can be tested experimentally without invoking semipermeable membranes.) Hence, once the entropy constants for the chemical elements have been fixed and a temperature unit has been chosen (to fix the multiplicative constants) the universal entropy is completely fixed.

We are indebted to many people for helpful discussions, including Fred Almgren, Thor Bak, Bernard Baumgartner, Pierluigi Contucci, Roy Jackson, Anthony Knapp, Martin Kruskal, Mary Beth Ruskai and Jan Philip Solovej.

2　Some Speculations and Open Problems

1.　As we have stressed, the purpose of the entropy function is to quantify the list of equilibrium states that can evolve from other equilibrium states. The evolution can be arbitrarily violent, but always $S(X) \le S(Y)$ if $X \prec Y$. Indeed, the early thermodynamicists understood the meaning of entropy as defined for equilibrium states. Of course, in the real world, one is often close to equilibrium without actually being there, and it makes sense to talk about entropy as a function of time, and even space, for situations close to equilibrium. We do a similar thing with respect to temperature, which has the same problem that temperature is only strictly defined for a homogeneous system in equilibrium. At some point the thought arose (and we confess our ignorance about how it arose and by whom) that it ought to be possible to define an entropy function rigorously for manifestly *non*-equilibrium states in such a way that the numerical value of this function will increase with time as a system goes from one equilibrim state to another.

Despite the fact that most physicists believe in such a non-equilibrium entropy it has so far proved to be impossible to define it in a clearly satisfactory way. (For example Boltzmann's famous H-Theorem shows the steady increase of a certain function called H. This, however, is not the whole story, as Boltzmann himself knew; for one thing, $H \neq S$ in equilibrium (except for ideal gases), and, for another, no one has so far proved the increase without making severe assumptions, and then only for a short time interval (cf. [L]).) Even today, there is no universal agreement about what, precisely, one should try to prove (as an example of the ongoing discussion, see [LePR]).

It is not clear if entropy can be consistently extended to non-equilibrium situations in the desired way. After a century and a half of thought, including the rise of the science of statistical mechanics as a paradigm (which was not available to the early thermodynamicists and, therefore, outside their thoughts), we are far from success. It has to be added, however, that a great deal of progress in understanding the problem has been made recently (e.g., [G]).

If such a concept can be formulated precisely, will it have to involve the notion of atoms and statistical mechanical concepts, or can it be defined irrespective of models, as we have done for the entropy of equilibrium states? This is the question we pose.

There are several major problems to be overcome, and we list two of them.

a. The problem of time reversibility: If the concept is going to depend upon mechanical models, we have to account for the fact that both classical and quantum mechanics are time reversible. This makes it difficult to construct a mechanical quantity that can only increase under classical or quantum mechanical time evolution. Indeed, this problem usually occupies center stage in most discussions of the subject, but it might, ultimately, not be the most difficult problem after all.

b. In our view of the subject, a key role is played by the idea of a more or less arbitrary (peaceful or violent) interaction of the system under discussion with the rest of the universe whose final result is a change of the system, the change in height of a weight, and nothing more. How can one model such an arbitrary interaction in a mechanical way? By means of an additional term in the Hamiltonian? That hardly seems like a reasonable way to model a sledgehammer that happens to fall on the system or a gorilla jumping up and down. All discussions of entropy increase refer to the evolution of *isolated* systems subject to Hamiltonian dynamical evolution. This can hardly even cope with describing a steam engine, much less a random, violent external force.

As a matter of fact, most people would recognize a) as the important problem, now and in the past. In b) we interject a new note, which, to us, is possibly more difficult. There are several proposals for a resolution of the irreversibility problem, such as the large number ($10^{23} \approx \infty$) of atoms involved, or the 'sensitive dependence on initial conditions' (one can shoot a ball out of a cannon, but it is very difficult to shoot it back into the cannon's mouth). Problem b), in contrast, has not received nearly as much attention.

2. An essential role in our story was played by axioms A4 and A5, which require the possiblity of having arbitrarily small samples of a given material and that these small samples behave in exactly the same way as a 1 kilogram sample. While this assumption is made in everyone's formulation of the second law, we have to recognize that absurdities will arise if we push the concept to its extreme. Eventually the atomic nature of matter will reveal itself and entropy will cease to have a clear meaning. What protects us is the huge power of ten (e.g., 10^{23}) that separates macroscopic physics and the realm of atoms.

Likewise, a huge power of ten separates time scales that make physical sense in the ordinary macroscopic world and time scales (such as 10^{25} seconds $= 10^7$ times the age of the universe) which are needed for atomic fluctuations to upset the time evolution one would obtain from macroscopic dynamics. One might say that one of the hidden assumptions in our (and everyone else's) analysis is that $\prec$ *is reproducible*, i.e., $X \prec Y$ either holds or it does not, and there are no hidden stochastic or probabilistic mechanisms that would make the list of pairs $X \prec Y$ 'fuzzy'.

One of the burgeoning area of physics research is 'mesoscopics', which deals with the interesting properties of tiny pieces of matter that might contain only a million atoms (= a cube of 100 atoms on a side) or less. At some point the second law has to get fuzzy and a significant open problem is to formulate a fuzzy version of what we have done in [LiY1]. Of course, no amount of ingenuity with mesoscopic systems is allowed to violate the second law on the macroscopic level, and this will have to be taken into account. One possibility could be that an entropy function can still be defined for mesoscopic systems but that $\prec$ is fuzzy, with the consequence that entropy increases only on 'the average', but in a totally unpredictable way – so that the occasional decrease of entropy cannot be utilized to violate the second law on the macroscopic level?

There are other problems as well. A simple system, such as a container of hydrogen gas, has states described by energy and volume. For a mesoscopic quantity of matter, this may not suffice to describe an equilibrium state. Another problem is the meaning of equilibrium and the implicit assumption we made that after the (violent) adiabatic process is over the system will eventually come to some equilibrium state in a time scale that is short compared to the age of the universe. On the mesoscopic level, the achievement of equilibrium may be more delicate because a mesoscopic system might never settle down to a state with insignificant fluctuations that one would be pleased to call an equilibrium state.

To summarize, we have listed two (and there are surely more) areas in which more thought, both mathematical and physical, is needed: the extension of the second law and the entropy concept to 1) non-equilibrium situations and 2) mesoscopic and even atomic situations. One might object that the problems cannot be solved until the 'rules of the game' are made clear, but discovering the rules is part of the problem. That is sometimes inherent in mathematical physics, and that is one of the intellectual challenges of the field.

References

[B]   J.G. BOYLING, An axiomatic approach to classical thermodynamics, Proc. Roy. Soc. London A329 (1972), 35–70.

[Bu]   H.A. BUCHDAHL, The Concepts of Classical Thermodynamics, Cambridge University Press, Cambridge, 1966.

[C]   C. CARATHÉODORY, Untersuchung über die Grundlagen der Thermodynamik, Math. Annalen 67 (1909), 355–386.

[Co]   J.L.B. COOPER, The foundations of thermodynamics, Jour. Math. Anal. and Appl. 17 (1967), 172–193.

[D]   J.J. DUISTERMAAT, Energy and entropy as real morphisms for addition and order, Synthese 18 (1968), 327–393.

[G]   G. GALLAVOTTI, Statistical Mechanics; A Short Treatise, Springer Texts and Monographs in Physics, (1999).

[Gi]   R. GILES, Mathematical Foundations of Thermodynamics, Pergamon, Oxford, 1964.

[L]   O.E. LANFORD III, Time evolution of large classical systems, Springer Lecture Notes in Physics (J. Moser, ed.) 38 (1975), 1–111.

[LePR] J.L. LEBOWITZ, I. PRIGOGINE, D. RUELLE, Round table on irreversibility, in "Statistical Physics XX", North Holland (1999), 516-527, 528-539, 540-544.

[Li]   E.H. LIEB, Some problems in statistical mechanics that I would like to see solved, 1998 IUPAP Boltzmann Prize Lecture, Physica A 263 (1999), 491-499.

[LiY1] E.H. LIEB, J. YNGVASON, The physics and mathematics of the Second Law of Thermodynamics, Physics Reports 310(1999), 1–96; Austin Math. Phys. archive 97–457; Los Alamos archive cond-mat/9708200.

[LiY2] E.H. LIEB, J. YNGVASON, A guide to entropy and the Second Law of Thermodynamics, Notices of the Amer. Math. Soc. 45 (1998), 571–581; Austin Math. Phys. archive 98–339; Los Alamos archive cond-mat/9805005.

[LiY3] E.H. LIEB, J. YNGVASON, A fresh look at entropy and the Second Law of Thermodynamics, Physics Today 53 (2000), 32–37; mp_arc 00-123, arXiv math-ph/0003028.

[P]   M. PLANCK, Über die Begrundung des zweiten Hauptsatzes der Thermodynamik, Sitzungsber. Preuss. Akad. Wiss., Phys. Math. Kl (1926), 453–463.

[RL]   F.S. ROBERTS, R.D. LUCE, Axiomatic thermodynamics and extensive measurement, Synthese 18 (1968), 311–326.

[U]   J. UFFINK, Bluff your way in the second law of thermodynamics, preprint 1999.

ELLIOT H. LIEB, Depts. of Mathematics and Physics, Princeton University, Jadwin Hall, P.O. Box 708, Princeton, NJ 08544, USA

JAKOB YNGVASON, Institut für Theoretische Physik, Universität Wien, Boltzmanngasse 5, A 1090 Vienna, Austria

GAFA, Geom. funct. anal.
Special Volume – GAFA2000, 359 – 382
1016-443X/00/S10359-24 $ 1.50+0.20/0

▌GAFA Geometric And Functional Analysis

DISCRETE AND CONTINUOUS:
TWO SIDES OF THE SAME?

László Lovász

Abstract

How deep is the dividing line between discrete and continuous mathematics? Basic structures and methods of both sides of our science are quite different. But on a deeper level, there is a more solid connection than meets the eye.

1 Introduction

There is a rather clear dividing line between discrete and continuous mathematics. Continuous mathematics is classical, well established, with a rich variety of applications. Discrete mathematics grew out of puzzles and then is often identified with specific application areas like computer science or operations research. Basic structures and methods of both sides of our science are quite different (continuous mathematicians use limits; discrete mathematicians use induction).

This state of mathematics at this particular time, the turn of the millennium, is a product of strong intrinsic logic, but also of historical coincidence.

The main external source of mathematical problems is science, in particular physics. The traditional view is that space and time are continuous, and that the laws of nature are described by differential equations. There are, of course, exceptions: chemistry, statistical mechanics, and quantum physics are based on at least a certain degree of discreteness. But these discrete objects (molecules, atoms, elementary particles, Feynman graphs) live in the continuum of space and time. String theory tries to explain these discrete objects as singularities of a higher dimensional continuum.

Accordingly, the mathematics used in most applications is analysis, the real hard core of our science. But especially in newly developing sciences, discrete models are becoming more and more important.

One might observe that there is a finite number of events in any finite domain of space-time. Is there a physical meaning of the rest of a four (or ten) dimensional manifold? Does "the point in time half way between two consecutive interactions of an elementary particle" make sense? Should the answer to this question be yes or no, discrete features of subatomic world are undeniable. How far would a combinatorial description of the world take us? Or could it happen that the descriptions of the world as a continuum, or as a discrete (but terribly huge) structure, are equivalent, emphasizing two sides of the same reality?

But let me escape from these unanswerable questions and look around elsewhere in science. Biology tries to understand the genetic code: a gigantic task, which is the key to understanding life and, ultimately, ourselves. The genetic code is discrete: simple basic questions like finding matching patterns, or tracing consequences of flipping over substrings, sound more familiar to the graph theorist than to the researcher of differential equations. Questions about the information content, redundancy, or stability of the code may sound too vague to a classical mathematician but a theoretical computer scientist will immediately see at least some tools to formalize them (even if to find the answer may be too difficult at the moment).

Economics is a heavy user of mathematics — and much of its need is not part of the traditional applied mathematics toolbox. Perhaps the most successful tool in economics and operations research is linear programming, which lives on the boundary of discrete and continuous. The applicability of linear programming in these areas, however, depends on conditions of convexity and unlimited divisibility; taking indivisibilities into account (for example, logical decisions, or individual agents) leads to integer programming and other models of combinatorial nature.

Finally, the world of computers is essentially discrete, and it is not surprising that so many discrete mathematicians are working in this science. Electronic communication and computation provides a vast array of well-formulated, difficult, and important mathematical problems on algorithms, data bases, formal languages, cryptography and computer security, VLSI layout, and much more.

In all these areas, the real understanding involves, I believe, a synthesis of the discrete and continuous, and it is an intellectually most challenging goal to develop these mathematical tools.

There are different levels of interaction between discrete and continuous mathematics, and I treat them (I believe) in the order of increasing significance.

1. We often use the finite to approximate the infinite. To discretize a complicated continuous structure has always been a basic method—from the definition of the Riemann integral through triangulating a manifold in (say) homology theory to numerically solving a partial differential equation on a grid.

It is a slightly more subtle thought that the infinite is often (or perhaps always?) an approximation of the large finite. Continuous structures are often cleaner, more symmetric, and richer than their discrete counterparts (for example, a planar grid has a much smaller degree of symmetry than the whole euclidean plane). It is a natural and powerful method to study discrete structures by "embedding" them in the continuous world.

2. Sometimes, the key step in the proof of a purely "discrete" result is the application of a purely "continuous" theorem, or vice versa. We all have our favorite examples; I describe two in section 2.

3. In some areas of discrete mathematics, key progress has been achieved through the use of more and more sophisticated methods from analysis. This is illustrated in section 3 by describing two powerful methods in discrete optimization. (I could not find any area of "continuous" mathematics where progress would be achieved at a similar scale through the introduction of discrete methods. Perhaps algebraic topology comes closest.)

4. Connections between discrete and continuous may be the subject of mathematical study on their own right. *Numerical analysis* may be thought of this way, but *discrepancy theory* is the best example. In this article, we have to restrict ourselves to discussing two classical results in this blooming field (section 4); we refer to the book of Beck and Chen [BeC], the expository article of Beck and Sós [BeS], and the recent book of Matoušek [M].

5. The most significant level of interaction is when one and the same phenomenon appears in both the continuous and discrete setting. In such cases, intuition and insight gained from considering one of these may be extremely useful in the other. A well-known example is the connection between sequences and analytic functions, provided by the power series expansion. In this case, there is a "dictionary" between combinatorial aspects (recurrences, asymptotics) of the sequence and analytic properties (differential equations, singularities) of its generating function.

In section 5, I discuss in detail the discrete and continuous notion of "Laplacian", connected with a variety of dispersion-type processes. This notion connects topics from the heat equation to Brownian motion to random walks on graphs to linkless embeddings of graphs.

An exciting but not well understood further parallelism is connected with the fact that the iteration of very simple steps results in very complex structures. In continuous mathematics, this idea comes up in the theory of dynamical systems and numerical analysis. In discrete mathematics, it is in a sense the basis of many algorithms, random number generation, etc. A synthesis of ideas from both sides could bring spectacular developments here.

One might mention many other areas with substantial discrete and continuous components: groups, probability, geometry... Indeed, my starting observation about this division becomes questionable if we think of some of the recent developments in these areas. I believe that further development will make it totally meaningless.

Acknowledgement. I am indebted to several of my colleagues, in particular to Alan Connes and Mike Freedman, for their valuable comments on this paper.

2   Discrete in Continuous and Continuous in Discrete

2.1   Marriage and measures. A striking application of graph theory to measure theory is the construction of the Haar measure on compact topological groups. This application was mentioned by Rota and Harper [RoH], who elaborated upon the idea in the example of the construction of a translation invariant integral for almost periodic functions on an arbitrary group. The proof is also related to Weil's proof of the existence of Haar measure. (There are other, perhaps more significant, applications of matchings to measures that could be mentioned here, for example the theorem of Ornstein [O] on the isomorphism of Bernoulli shifts; cf. also [Du], [LoM], [LoP], [St].)

Here we shall describe the construction of a translation invariant integral for continuous functions on a compact topological group, equivalent to the existence of the Haar measure [LoP]. An *invariant integration* for continuous functions on a compact topological group is a linear functional L defined on the continuous real-valued functions on G, with the following properties:

(a) $L(\alpha f + \beta g) = \alpha L(f) + \beta L(g)$ (linearity)
(b) If $f \geq 0$ then $L(f) \geq 0$ (monotonicity)
(c) If l denotes the identity function, then $L(l) = 1$ (normalization)
(d) If s and t are in G and f and g are two continuous functions such that $g(x) = f(sxt)$ for every x, then $L(g) = L(f)$ (double translation invariance).

Theorem 1. *For every compact topological group there exists an invariant integration for continuous functions.*

Proof. Let f be the function we want to integrate. The idea is to approximate the integral by the average

$$f(A) = \tfrac{1}{|A|} \sum_{a \in A} f(a) \,,$$

where A is a "uniformly dense" finite subset of group G. The question is, how to find an appropriate set A?

The key definition is the following. Let U be an open non-empty subset of G (think of it as "small"). A subset $A \subset G$ is called a U-*net* if $A \cap sUt \neq \emptyset$ for every $s, t \in G$. It follows from the compactness of G that there exists a finite U-net.

Of course, a U-net may be very unevenly distributed over the group, and accordingly, the average over it may not approximate the integral. What we show is that the simple trick of restricting ourselves to U-nets with minimum cardinality (minimum U-nets, for short), we get a sufficiently uniformly dense finite set.

To measure this uniformity we define

$$\delta(U) = \sup \left\{ |f(x)\, f(y)| \mid x, y \in sUt \text{ for some } s, t \in G \right\}.$$

The compactness of G implies that if f is continuous then it is also uniformly continuous in the sense that for every $\epsilon > 0$ there exists a non-empty open set U such that $\delta(U) < \epsilon$.

Let A and B be minimum U-nets. The following inequality is the heart of the proof:

$$\left| f(A) - f(B) \right| \leq \delta(U). \tag{1}$$

The combinatorial core of the construction lies in the proof of this inequality. Define a bipartite graph H with bipartition (A, B) by connecting $x \in A$ to $y \in B$ if and only if there exists $s, t \in G$ such that $x, y \in sUt$. We use the Marriage theorem to show that this bipartite graph has a perfect matching. We have to verify two things:

(I) $|A| = |B|$. This is trivial.

(II) Every set $X \subseteq A$ has at least $|X|$ neighbors in B. Indeed, let Y be the set of neighbors of X. We show that $T = Y \cup (A \setminus X)$ is a U-net. Let $s, t \in G$, we show that T intersects sUt. Since A is a U-net, there exists an element $x \in A \cap sUt$. If $x \notin X$ then $x \in T$ and we are done. Otherwise, we use that there exists an element $y \in B \cap sUt$. Then xy is an line of H and so $y \in Y$, and we are done again.

Thus T is a U-net. Since A is a U-net with minimum cardinality, we must have $|T| \geq |A|$, which is equivalent to $|Y| \geq |X|$.

Thus H has a perfect matching $\{a_1b_1, \ldots, a_nb_n\}$. Then

$$|f(A) - f(B)| = \frac{1}{n}\left|\sum_{i=1}^{n}(f(a_i) - f(b_i))\right|$$

$$\leq \frac{1}{n}\sum_{i=1}^{n}|f(a_i) - f(b_i)| \leq \frac{1}{n}\big(n\delta(U)\big) = \delta(U).$$

The rest of the proof is rather routine. First, we need to show that averaging over minimum U-nets for different U's gives approximately the same result. More exactly, let A be a minimum U-net and B a minimum V-net, then

$$|f(A) - f(B)| \leq \delta(U) + \delta(V). \tag{2}$$

Indeed, for every $b \in B$, Ab is also a U-net with minimum cardinality, so by (1),

$$|f(A) - f(Ab)| \leq \delta(U).$$

Hence

$$|f(A) - f(AB)| = \left|f(A) - \frac{1}{|B|}\sum_{b \in B}f(Ab)\right| \leq \frac{1}{|B|}\sum_{b \in B}|f(A) - f(Ab)| \leq \delta(U).$$

Similarly,

$$|f(B) - f(AB)| \leq \delta(V),$$

whence (2) follows.

Now choose a sequence U_n of open sets such that $\delta(U_n) \to 0$, and let A_n be a minimum U_n-net. Then (2) implies that the sequence $f(A_n)$ tends to a limit $L(f)$, which is independent of the choice of U_n and A_n, and so it is well defined. Conditions (a)–(d) are trivial to verify. $\square$

2.2 Disjoint subsets and topology. Now we turn to examples demonstrating the converse direction: applications of continuous methods to a purely combinatorial problems. Our first example is a result where algebraic topology and geometry are the essential tools in the proof of a combinatorial theorem.

Theorem 2. Let us partition the k-element subsets of an n-element set into $n - 2k + 1$ classes. Then one of the classes contains two disjoint k-subsets.

This result was conjectured by Kneser [K], and proved in [Lo1]. The proof was simplified by Bárány [B], and we describe his version here.

Proof. We first invoke a geometric construction due to Gale *There exists a set of $2k + d$ vectors on the d-sphere S^d such that every open hemisphere contains at least k of them.* (For $d = 1$, take the vertices of a regular $(2k + 1)$-gon.)

Choosing $d = n - 2k$, we thus get a set S of n points. Suppose that all the k-subsets are partitioned into $d + 1 = n - 2k + 1$ classes $\mathcal{P}_0, \mathcal{P}_1, \ldots, \mathcal{P}_d$. Let A_i be the set of those unit vectors $h \in S^d$ for which the open hemisphere centered at h contains a k-subset of S which belongs to $\mathcal{P}_i$. Clearly the sets $A_0, A_1, \ldots, A_d$ are open, and by the definition of S, they cover the whole sphere.

Now we turn from geometry to topology: by the Borsuk-Ulam theorem, one of the sets A_i contains two antipodal points h and $-h$. Thus the hemispheres about h and $-h$ both contain a k-subset from $\mathcal{P}_i$. Since these hemispheres are disjoint, these two k-subsets are disjoint.   □

There are numerous other examples where methods from algebraic topology have been used to prove purely combinatorial statements (see [Bj] for a survey).

3   Optimization: Discrete, Linear, Semidefinite

Our next example illustrates how a whole area of important applied mathematical problems, namely discrete optimization, relies on ideas bringing more and more sophisticated tools from more traditional continuous optimization.

In a typical optimization problem, we are given a set S of *feasible solutions*, and a function $f : S \to \mathbf{R}$, called the *objective function*. The goal is to find the element of S which maximizes (or minimizes) the objective function.

In a traditional optimization problem, S is a decent continuum in a euclidean space, and f is a smooth function. In this case the optimum can be found by considering the equation $\nabla f = 0$, or by an iterative method using some version of steepest descent.

Both of these depend on the continuous structure in a neighborhood of the optimizing point, and therefore fail for a discrete (or combinatorial) optimization problem, when S is finite. This case is trivial from a classical point of view: "in principle" one could evaluate f for all elements in S, and choose the best. But in most cases of interest, S is very large in comparison with the number of data needed to specify the instance of the problem. For example, S may be the set of spanning trees, or the set of perfect matchings, of a graph; or the set of all possible schedules of all trains in a country; or the set of states of a spin glass. In such cases, we have to use the implicit structure of S, rather than brute force, to find the optimizing element.

In some cases, totally combinatorial methods enable us to solve such problems. For example, let S be the set of all spanning trees of a graph G, and assume that each edge of G has a non-negative "length" associated with it. In this case a spanning tree with minimum length can be found by the *greedy algorithm* (due to Borøuvka and Kruskal): we repeatedly choose the shortest edge that does not form a cycle with the edges chosen previously, until a tree is obtained.

In many (in a sense most) cases, such a direct combinatorial algorithm is not available. A general approach is to embed the set S into a continuum S' of solutions, and also extend the objective function f to a function f' defined on S'. The problem of minimizing f' over S' is called a *relaxation* of the original problem. If we do this right (say, S' is a convex set and f' is a convex function), then the minimum of f' over S' can be found by classical tools (differentiation, steepest descent, etc.). If we are really lucky (or clever), the minimizing element of S' will belong to S, and then we have solved the original discrete problem. (If not, we may still use the solution of the relaxation to obtain a bound on the solution of the original, or to obtain an approximate solution. See later.)

3.1 Polyhedral combinatorics.
The first successful realization of this scheme was worked out in the 60's and 70's, where techniques of linear programming were applied to combinatorics. Let us assume that our combinatorial optimization problem can be formulated so that S is a set of $0-1$ vectors and the objective function is linear. The set S may be specified by a variety of logical and other constraints; in most cases, it is quite easy to translate these into linear inequalities:

$$S = \left\{ x \in \{0,1\}^n : a_1^\mathsf{T} x \le b_1, \ldots, a_m^\mathsf{T} x \le b_m \right\}. \tag{3}$$

The objective function is

$$f(x) = c^\mathsf{T} x = \sum_{i=1}^{n} c_i x_i, \tag{4}$$

where the c_i are given real numbers. Such a formulation is typically easy to find.

Thus we have translated our combinatorial optimization problem into a linear program with integrality conditions. It is quite easy to solve this, if we disregard the integrality conditions; the real game is to find ways to write up these linear programs in such a way that disregarding integrality conditions is justified.

A nice example is matching in bipartite graphs. Suppose that G is a bipartite graph, with node set $\{u_1, \ldots, u_n, v_1, \ldots, v_n\}$, where every edge connects a node u_i to a node v_j. We want to find a perfect matching, i.e., a set of edges covering each node exactly once.

Suppose that G has a perfect matching M and let

$$x_{ij} = \begin{cases} 1, & \text{if } u_i v_j \in M, \\ 0, & \text{otherwise}. \end{cases}$$

Then the defining property of perfect matchings can be expressed as follows:

$$\sum_{i=1}^{n} x_{ij} = 1 \quad \text{for all } j, \qquad \sum_{j=1}^{n} x_{ij} = 1 \quad \text{for all } i. \qquad (5)$$

Conversely, we find a solution of this system of linear equations with every $x_{ij} = 0$ or 1, then we have a perfect matching.

Unfortunately, the solvability of a system of linear equations (even of a single equation) is NP-hard. What we can do is to replace the condition $x_{ij} = 0$ or 1 by the weaker condition

$$x_{ij} \geq 0. \qquad (6)$$

To solve a linear system like (5) in non-negative real numbers is still not trivial, but doable efficiently (in polynomial time) using linear programming.

We are not done, of course, since if we find that (5)–(6) has a solution, this solution may not be in integers and hence it may not "mean" a perfect matching. There are various ways to conclude, extracting a perfect matching from a fractional solution of (5)–(6). The most elegant is the following. The set of all solutions forms a convex polytope. Now *every vertex of this convex polytope is integral.* (The proof of this fact is amusing, using Cramer's Rule and basic determinant calculations. See, e.g. [LoP].)

So we run a linear programming algorithm to see if (5)-(6) has a solution in real numbers. If not, then the graph has no perfect matching. If yes, most linear programming algorithms automatically give a basic solution, i.e., a vertex. This is an integral solution of (5)–(6), and hence corresponds to a perfect matching.

Many variations of this idea have been developed both in theory and practice. There are ways to automatically generate new constraints and add them to (3) if these fail; there are more subtle, more efficient methods to generate new constraints in for special problems; there are ways to handle the system (3) even if it gets too large to be written up explicitly.

3.2   Semidefinite optimization.   This new technique of producing relaxations of discrete optimization problems makes the problems continuous in a more sophisticated way. We illustrate it by the Maximum Cut problem. Let $G = (V, E)$ be a graph. We want to find a partition $(S, \bar{S})$ of V for which the number M of edges connecting S to $\bar{S}$ is maximum.

This problem is NP-hard, so we cannot hope to find an efficient (polynomial time) algorithm that solves it. Instead, we have to settle for less: we try to find a cut that is close to being optimal.

It is easy to find a cut that picks up half of the edges. Just process the nodes one by one, placing them in the set S or $\bar{S}$, whichever gives the larger number of new edges going between the two sets. Since no cut can have more than all edges, this simple algorithm obtains an approximation of the maximum cut with at most 50% relative error.

It turned out to be quite difficult to improve upon this simple fact, until Goemans and Williamson [GW] combined semidefinite optimization with a randomized rounding technique to obtain an approximation algorithm with a relative error of about 12%. On the other hand, it was proved by Håstad [H] (tightening earlier results of Arora et al. [ArLMSS]) that no polynomial time algorithm can produce an approximation with a relative error less than (about) 6%, unless $P = NP$.

We sketch the Goemans–Williamson algorithm. Let us describe any partition $V_1 \cup V_2$ of V by a vector $x \in \{-1, 1\}^V$ by letting $x_i = -1$ iff $i \in V_1$. Then the size of the cut corresponding to x is $(1/4) \sum_{i,j} (x_i - x_j)^2$. Hence the MAX CUT problem can be formulated as maximizing the quadratic function

$$\tfrac{1}{4} \sum_{i,j} (x_i - x_j)^2 \tag{7}$$

over all $x \in \{-1, 1\}^V$. The condition on x can also be expressed by quadratic constraints:

$$x_i^2 = 1 \qquad (i \in V). \tag{8}$$

Reducing a discrete optimization problem to the problem of maximizing a quadratic function subject to a system of quadratic equations may sound great, and one might try to use Lagrange multiplicators and other classical techniques. But these won't help; in fact the new problem is very difficult (NP-hard), and it is not clear that we gain anything.

The next trick is to linearize, by introducing new variables $y_{ij} = x_i x_j$ $(1 \leq i, j \leq n)$. The objective function and the constraints become linear in these new variables y_{ij}:

$$\text{maximize} \quad \tfrac{1}{4} \sum_{i,j} (y_{ii} + y_{jj} - 2y_{ij}) \tag{9}$$

$$\text{subject to} \quad y_{11} = \cdots = y_{nn} = 1. \tag{10}$$

Of course, there is a catch: the variables y_{ij} are not independent! Introducing the symmetric matrix $Y = (y_{ij})_{i,j=0}^n$, we can note that

$$Y \text{ is positive semidefinite}, \tag{11}$$

and

$$Y \text{ has rank } 1. \tag{12}$$

The problem of solving (9)–(10) with the additional constraints (11) and (12) is equivalent to the original problem, and thus NP-hard in general. But if we drop (12), then we get a tractable relaxation! Indeed, what we get is a **semidefinite program**: *maximize a linear function of the entries of a positive semidefinite matrix, subject to linear constraints on the matrix entries.*

Since positive semidefiniteness of a matrix Y can be translated into (infinitely many) linear inequalities involving the entries of Y:

$$v^{\mathsf{T}} Y v \geq 0 \qquad \text{for all } v \in \mathbf{R}^{n+1},$$

semidefinite programs can be viewed as linear programs with infinitely many constraints. However, they behave much nicer than one would expect. Among others, there is a duality theory for them (see, e.g. Wolkowitz [W]). It is also important that semidefinite programs are polynomial time solvable (up to an arbitrarily small error; note that the optimum solution may not be rational). In fact, the ellipsoid method [GrLS] and, more importantly from a practical point of view, interior point methods [Al] extend to semidefinite programs.

Coming back to the Maximum Cut problem, let Y be an optimal solution of (9)-(10)-(11), and let M^* be the optimum value of the objective function. Since every cut defines a solution, we have $M^* \geq M$.

The second trick is to observe that since Y is positive semidefinite, we can write it as a Gram matrix, i.e., there exist vectors u_i if $\mathbf{R}^n$ such that $u_i^T u_j = Y_{ij}$ for all i and j. In particular, we get that $|u_i|^2 = 1$, and

$$\sum_{i,j} (u_i - u_j)^2 = 4M^* . \tag{13}$$

Now choose a uniformly distributed random hyperplane H through the origin. This divides the u_i into two classes, and thereby defines a cut in G. The expected weight of this cut is

$$\sum_{i,j} C_{ij} \mathsf{P} \quad (H \text{ separates } u_i \text{ and } u_j) .$$

But it is trivial that the probability that H separates u_i and u_j is $(1/\pi)$ times the angle between u_i and u_j. In turn, the angle between u_i and u_j is at least $0.21964(u_i - u_j)^2$, which is easily seen by elementary calculus. Hence the expected size of the cut obtained is at least

$$\sum_{i,j} 0.21964(u_i - u_j)^2 = 0.87856M^* \geq 0.87856M .$$

Thus we get a cut which is at least about .88 percent of the optimal.

Except for the last part, this technique is entirely general and has been used in a number of proofs and algorithms in combinatorial optimization [LoS1].

4   Discrepancy Theory

In section 3, we started with a discrete problem, and tried to find a good continuous approximation (relaxation) of it. Let us discuss a reverse problem a bit (only to the extent of a couple of classical examples). Suppose that we have a set with a measure; how well can it be approximated by a discrete measure?

In our first example, we consider the Lebesgue measure λ on the unit square $[0,1]^2$. Suppose that we want to approximate this by the uniform measure on a finite subset T. Of course, we have to specify how the error of approximation is measured: here we define it as the maximum error on axis-parallel rectangles:

$$\Delta(T) = \sup_R \big| |T \cap R| - \lambda(R)|T| \big| ,$$

where R ranges over all sets of the form $R = [a,b] \times [c,d]$ (we scaled up by $|T|$ for convenience). We are interested in finding the best set T, and in determining the "discrepancy"

$$\Delta_n = \inf_{|T|=n} \Delta(T).$$

The question can be raised for any family of "test sets" instead of rectangles: circles, ellipses, triangles, etc. The answer depends in an intricate
way on the geometric structure of the test family; see [BeC] for an exposition of these beautiful results.

The question can also be raised in other dimensions. The 1-dimensional
case is easy, since the obvious choice of $T = \{1/(n+1), \ldots, n/(n+1)\}$ is
optimal. But for dimensions larger than 2, the order of magnitude of Δ_n
is not known.

Returning to dimension 2, the obvious first choice is to try a $\sqrt{n} \times \sqrt{n}$
grid for T. This leaves out a rectangle of size about $1/\sqrt{n}$, and so it
has $\Delta(T) \approx \sqrt{n}$. There are many constructions that do better; the most
elementary is the following. Let $n = 2^k$. Take all points (x, y) where both
x and y are multiples of 2^{-k}, and expanding them to k bits in binary, we
get these bits in reverse order: $x = 0.b_1 b_2 \ldots b_k$ and $y = 0.b_n b_{n-1} \ldots b_1$. It
is a nice exercise to verify that this set has discrepancy $\Delta(T) \approx k = \log_2 n$.

It is hard to prove that one cannot do better; even to prove that $\Delta_n \to$
∞ was difficult [A]. The lower bound matching the above construction was
finally proved by Schmidt [S]:

$$\Delta_n = \Theta(\log n).$$

This fundamental result has many applications.

Our second example can be introduced through a statistical motivation.
Suppose that we are given a $0-1$ sequence $x = x_1 a_2 \ldots x_n$ of length n. We
want to test whether it comes from independent coin flips.

One approach (related to von Mises's proposal for the definition of a random sequence) could be to count 0's and 1's; their numbers should be about
the same. This should also be true if we count bits in the even positions,
or odd positions, or more generally, in any fixed arithmetic progression of
indices.

So we consider the quantity

$$\Delta(x) = \max_A \left| \sum_{i \in A} x_i - \tfrac{1}{2}|A| \right|,$$

where A ranges through all arithmetic progressions in $\{1, \ldots, n\}$. Elementary probability theory tells us that if x is generated by independent coin
flips, then $\Delta(x) \approx \sqrt{n}/2$ with high probability. So if $\Delta(x)$ is larger than

this (which is the case of most non-random sequences one thinks of), then we can conclude that it is not generated by independent coin flips.

But can x fail this test in the other direction, being "too smooth"? In other words, what is

$$\Delta_n = \min_x \Delta(x),$$

where x ranges over all $0 - 1$ sequences of length n? We can think of the question as a problem of measure approximation: we are given the measure on $\{1, \ldots, n\}$ in which each atom has measure $1/2$. This is a discrete measure but not integral valued; we want to approximate it by an integral valued measure. The test family defining the error of approximation consists of arithmetic progressions.

It follows from considering random sequences that $\Delta_n = O(\sqrt{n})$ The first result in the opposite direction was proved by Roth [Rot], who showed that $\Delta_n = \Omega(n^{1/4})$. It was expected that the bound would be improved to $\sqrt{n}$ eventually, showing that random sequences are the extreme, at least in the order of magnitude (in many similar instances of combinatorial extremal problems, for example in Ramsey Theory, random choice is the best). But surprisingly, Sárközy showed that $\Delta_n = O(n^{1/3}$, which was improved by Beck [Be] to the almost sharp $\Delta_n = O(n^{1/4} \log n)$, and even the logarithmic factor was recently removed by Matoušek and Spencer [MS]. Thus there are sequences which simulate random sequences too well!

5   The Laplacian

The Laplacian, as we learn it in school, is the differential operator

$$\sum_i \frac{\partial^2}{\partial x_i^2}.$$

What could be more tied to the continuity of euclidean spaces, or at least of differential manifolds, than a differential operator? In this section we show that the Laplacian makes sense in graph theory, and in fact it is a basic tool. Moreover, the study of the discrete and continuous versions interact in a variety of ways, so that the use of one or the other is almost a matter of convenience in some cases.

5.1   Random walks.   One of the fundamental general algorithmic problems is sampling: *generate a random element from a given distribution over a set V*. In non-trivial cases, the set V is either infinite or finite but very large, and often only implicitly described. In most cases, we want to generate an element from the uniform distribution, and this special case will be enough for the purpose of our discussions.

I'll use as examples two sampling tasks:

    (a) Given a convex body K in $\mathbf{R}^n$, generate a uniformly distributed random point of K.

    (b) Given a graph G, generate a uniformly distributed random perfect matching in G.

Problem (a) comes up in many applications (volume computation, Monte-Carlo integration, optimization, etc.). Problem (b) is related to the Ising model in statistical mechanics.

In simple cases one may find elegant special methods for sampling. For example, if we need a uniformly distributed random point in the unit ball in $\mathbf{R}^n$, then we can generate n independent coordinates from the standard normal distribution, and "normalize" the obtained vector appropriately.

But in general, sampling from a large finite set, or an infinite set, is a difficult problem, and the only general approach known is the use of *Markov chains*. We define (using the structure of S) an ergodic Markov chain whose state space is S and whose stationary distribution is uniform. In the case of a finite set S, it may be easier to think of this as a connected regular non-bipartite graph G with node set V. Starting at an arbitrary node (state), we take a random walk on the graph: from a node i, we step to a next node selected uniformly from the set of neighbors of i. After a sufficiently large number T of steps, we stop: the distribution of the last node is approximately uniform.

In the case of a convex body, a rather natural Markov chain to consider is the following: starting at a convenient point in K, we move at each step to a uniformly selected random point in a ball of radius δ about the current point (if the new point is outside K, we stay where we were, and consider the step "wasted"). The step-size δ will be chosen appropriately, but typically it is about $1/\sqrt{n}$.

If we want to sample from perfect matchings in a graph G, the random walk to consider is not so obvious. Jerrum and Sinclair consider a random walk on perfect and near-perfect matchings (matchings that leave just two nodes unmatched). If we generate such a matching, it will be perfect with a small but non-negligible probability ($1/n^{\text{const}}$ if the graph is dense). So we just iterate until we get a perfect matching.

One step of the random walk is generated by picking a random edge e of G. If the current matching is perfect and contains e, then we delete e; if the current matching is near-perfect and e connects the two unmatched nodes, we add it; if e connects an unmatched node to a matched node, we add e to the matching and delete the edge it intersects.

Another way of looking at this is do a random walk on a "big" graph H whose nodes are the perfect and near-perfect matchings of the "small" graph G. Two nodes of H are adjacent if the corresponding matchings arise from each other by changing (deleting, adding, or replacing) a single edge. Loops are added to make H regular. Note that we do not want to construct H explicitly; its size may be exponentially large compared with G. The point is that the random walk on H can be implemented using only this implicit definition.

It is generally not hard to achieve that the stationary distribution of the chain is uniform, and that the chain is ergodic. The crucial question is: what is the *mixing time*, i.e., how long do we have to walk? This question leads to estimating the mixing time of Markov chains (the number of steps before the chain becomes essentially stationary). From the point of view of practical applications, it is natural to consider finite Markov chains – a computation in a computer is necessarily finite. But in the analysis, it depends on the particular application whether one prefers to use a finite, or a general measurable, state space. All the essential (and very interesting) connections that have been discovered hold in both models. In fact, the general mathematical issue is *dispersion*: we might be interested in dispersion of heat in a material, or dispersion of probability during a random walk, or many other related questions.

A Markov chain can be described by its transition matrix $M = (p_{ij})$, where p_{ij} is the probability of stepping to j, given that we are at i. In the special case of random walks on a regular undirected graph we have $M = (1/d)A$, where A is the usual adjacency matrix of the graph G. Note that the all-1 vector $\mathbf{1}$ is an eigenvector of M belonging to the eigenvalue 1.

If σ is the starting distribution (which can be viewed as a vector in $\mathbf{R}^V$), then the distribution after t steps is $M^t\sigma$. From this it is easy to see that the speed of convergence to the stationary distribution (which is the eigenvector $(1/n)\mathbf{1}$), depends on the difference between the largest eigenvalue $\mathbf{1}$ and the second largest eigenvalue λ (the *spectral gap*). We could also define the spectral gap as the smallest positive eigenvalue of the matrix $L = M - I$.

We call this matrix L the *Laplacian* of the graph. For any vector $f \in \mathbf{R}^V$, the value of Lf at node i is the average of f over the neighbors of i, minus f_i. This property shows that the Laplacian is indeed a discrete analog of the classical Laplace operator.

Some properties of the Laplace operator depend on the fine structure of differentiable manifolds, but many basic properties can be generalized easily to any graph. One can define harmonic functions, the heat kernel, prove identities like the analogue Green's formula:

$$\sum_i \left(f_i(Lg)_i - g_i(Lf)_i \right) = 0$$

and so on. Many properties of the "continuous" Laplace operator can be derived using such easy combinatorial formulas on a grid and then taking limits. See Chung [Chu] for an exposition of some of these connections.

But more significantly, the discrete Laplacian is an important tool in the study of various graph theoretic properties. As a first illustration of this fact, let us return to random walks.

Often information about the spectral gap is difficult to obtain (this is the case in both of our introductory examples). One possible remedy is to relate the dispersion speed to *isoperimetric inequalities* in the state space. To be more exact, define the *conductance* of the chain as

$$\Phi = \max_{\emptyset \subset S \subset V} \sum_{i \in S, j \in V \setminus S} \frac{\pi_i p_{ij}}{\pi(S)\pi(V \setminus S)}.$$

(The numerator can be viewed as the probability that choosing a random node from the stationary distribution, and then making one step, the first node is in S and the second is in $V \setminus S$. The denominator is the same probability for two independent random nodes from the stationary distribution.) The following inequality was proved by Jerrum and Sinclair [JS]:

Theorem 3.

$$\tfrac{\Phi^2}{8} \leq 1 - \lambda \leq \Phi.$$

This means that the mixing time (which we have to use informally here, since the exact definition depends on how we start, how we measure convergence, etc.) is between $1/\Phi$ and $1/\Phi^2$.

In the case of random walks in a convex body, the conductance is very closely related to the following isoperimetric theorem [LoSi], [DyF]:

Theorem 4. *Let K be a convex body in $\mathbf{R}^n$, of diameter D. Let the surface F divide K into two parts K_1 and K_2. Then*

$$\mathrm{vol}_{n-1}(F) \geq \frac{2}{D}\frac{\mathrm{vol}(K_1)\mathrm{vol}(K_2)}{\mathrm{vol}(K)} .$$

(A long thin cylinder shows that the bound is sharp.)

From Theorem 4, it follows that (after appropriate preprocessing and other technical difficulties which are now swept under the carpet) one can generate a sample point in an n-dimensional convex body in $O^*(n^3)$ steps.

In the case of matchings, Jerrum and Sinclair prove that for dense graphs, the conductance of the chain described is bounded from below by $1/n^{\mathrm{const}}$, and hence the mixing time is bounded by n^{const}.

How to establish isoperimetric inequalities? The generic method is to construct, explicitly or implicitly, *flows* or *multicommodity flows*. Suppose that for each subset $\emptyset \subset S \subset V$ we can construct a flow through the graph so that each node $i \in S$ is a source that produces an amount of $\pi_i\pi(V \setminus S)$ of flow, and each $j \in V \setminus S$ is a sink that consumes $\pi_j\pi(S)$ amount of flow. Suppose further that each edge ij carries at most $K\pi_i p_{ij}$ flow. Then a simple computation shows that the conductance satisfies

$$\Phi \geq \tfrac{1}{K} .$$

Instead of constructing a flow for each subset S of nodes, one might prefer to construct a *multicommodity flow*, i.e., a flow of value $\pi_i\pi_j$ for each i and j. If the total flow through each edge ij is at most $K\pi_i p_{ij}$, then the conductance is at least $1/K$.

How good is the bound on the conductance obtained by the multicommodity flow method? An important result of Leighton and Rao [LR] implies that for the best choice of the flows, it gets within a factor of $O(\log n)$. One of the reasons I bring this up is that this result, in turn, can be derived from the theorem of Bourgain [Bo] about the embedability of finite metric spaces in ℓ_1-spaces with small distortion. Cf. also [LiLR], [DeL].

To construct the "best" flows or multicommodity flows is often the main mathematical difficulty. In the case of the proof of Theorem 4, it can be done by a method used earlier by Payne and Weinberger [PW] in the theory of partial differential equations. We only give a rough sketch.

Suppose that we want to construct a "flow" from K_1 to K_2. By the "Ham–Sandwich" theorem, there exists a hyperplane that cuts both K_1 and K_2 into two parts with equal volume. We can separately construct the flows on both side of this hyperplane. Repeating this procedure, we can cut up K into "needles" so that each needle is split by the partition $(S, V \setminus S)$ in the same proportion, and hence it suffices to construct a flow between K_1 and K_2 inside each needle. This is easily done using the Brun–Minkowski theorem.

For the random walk on matchings, the construction of the multicommodity flow (called *canonical paths* by Jerrum and Sinclair) also goes back to one of the oldest methods in matching theory. Given (say) a perfect matching M and a near-perfect matching M', we form the union. This is a subgraph of the "small" graph G that decomposes into a path (with edges alternating between M and M'), some cycles (again alternating between M and M') and the common edges of M and M'. Now it is easy to transform M into M' by walking along each cycle and the path, and replacing the edges one by one. What this amounts to is a path in the "big" graph H, connecting nodes M and M'. If we use this path to carry the flow, then it can be shown that no edge of the graph H is overloaded, provided the graph G is dense.

5.2 The Cage theorem and conformal mappings. It was proved by Steinitz that every 3-connected planar graph can be represented as the skeleton of a (3-dimensional) polytope. In fact, there is a lot of freedom in choosing the geometry of this representing polytope. Among various extensions, the most interesting for us is the classical construction going back to Koebe [Ko] (first proved by Andre'ev [An]; cf. also [T]):

Theorem 5 (The Cage theorem). *Let H be a 3-connected planar graph. Then H can be represented as the skeleton of a 3-dimensional polytope, all whose edges touch the unit sphere.*

We may add that the representing polytope is unique up to a projective transformation of the space that preserves the unit sphere. By considering the "horizon" from each vertex of the polytope, we obtain a representation of the nodes of the graph by openly disjoint circular disks in the plane so that adjacent nodes correspond to touching circles.

The Cage theorem may be considered as a discrete form of the Riemann mapping theorem, in the sense that it implies the Riemann mapping theorem. Indeed, suppose that we want to construct a conformal mapping of a simply connected domain K in the complex plane onto the unit disk D. For simplicity, assume that K is bounded. Consider a triangular grid in the plane (an infinite 6-regular graph), and let G be the graph obtained by identifying all gridpoints outside K into a single node s. Consider the Koebe representation by touching circles on the sphere; we may assume that the circle representing the node s is the exterior of D. So all the other circles are inside D, and we obtain a mapping of the set of gridpoints in K into D. By letting the grid become arbitrarily fine, it was shown by Rodin and Sullivan [RS] that in the limit we get a conformal mapping of K onto D.

So the Cage theorem is indeed a beautiful bridge between discrete mathematics (graph theory) and continuous mathematics (complex analysis). But where is the Laplacian? We'll see one connection in the next section. Another one occurs in the work of Spielmann and Teng [SpT], who use the Cage theorem to show that the eigenvalue gap of the Laplacian of a planar graph is at most $1/n$. This implies, among others, that the mixing time of the random walk on a planar graph is at least linear in the number of nodes.

5.3 Colin de Verdière's invariant. In 1990, Colin de Verdière [Co] introduced a parameter $\mu(G)$ for any undirected graph G. Research concerning this parameter involves an interesting mixture of ideas from graph theory, linear algebra, and analysis.

The exact definition goes beyond the limitations of this article; roughly speaking, $\mu(G)$ is the multiplicity of the smallest positive eigenvalue of the Laplacian of G, where the edges of G are weighted so as to maximize this multiplicity.

The parameter was motivated by the study of the maximum multiplicity of the second eigenvalue of certain Laplacian-type differential operators, defined on Riemann surfaces. He approximated the surface by a sufficiently densely embedded graph G, and showed that the multiplicity of the second eigenvalue of the operator can be bounded by this value $\mu(G)$ depending only on the graph.

Colin de Verdière's invariant created much interest among graph theorists, because of its surprisingly nice graph-theoretic properties. Among others, it is minor-monotone, so that the Robertson–Seymour graph minor theory applies to it. Moreover, planarity of graphs can be characterized by this invariant: $\mu(G) \leq 3$ *if and only if G is planar.*

Colin de Verdière's original proof of the "if" part of this fact was most unusual in graph theory: basically, reversing the above procedure, he showed how to reconstruct a sphere and a positive elliptic partial differential operator P on it so that $\mu(G)$ is bounded by the dimension of the null space of P, and then invoked a theorem of Cheng [C] asserting that this dimension is at most 3.

Later van der Holst [Ho] found a combinatorial proof of this fact. While this may seem as a step backward (after all, it eliminated the necessity of the only application of partial differential equations in graph theory I know of), it did open up the possibility of characterizing the next case. Verifying a conjecture of Robertson, Seymour, and Thomas, it was shown by Lovász and Schrijver [LoS2] that $\mu(G) \leq 4$ *if and only if G is linklessly embedable in $\mathbf{R}^3$.*

Can one go back to the original motivation from $\mu(G)$ and find a "continuous" version of this result? In what sense does a linklessly embedded graph approximate the space? These questions appear very difficult.

It turns out that graphs with large values of μ are also quite interesting. For example, for a graph G on n nodes with $\mu(G) \geq n-4$, the complement of G is planar, up to introducing "twin" points; and the converse of this assertion also holds under reasonably general conditions. The proof of the latter fact uses the Koebe–Andre'ev representation of graphs.

So the graph invariant $\mu(G)$ is related at one end of the scale to elliptic partial differential equations (Cheng's theorem); on the other, to Riemann's theorem on conformal mappings.

References

[A]   T. VAN AARDENNE-EHRENFEST, On the impossibility of a just distribution, Nederl. Akad. Wetensch. Proc. 52 (1949), 734–739; Indagationes Math. 11 (1949) 264–269.

[Al]   F. ALIZADEH, Interior point methods in semidefinite programming with applications to combinatorial optimization, SIAM J. on Optimization 5 (1995), 13–51.

[AloS] N. ALON, J. SPENCER, The Probabilistic Method. With an appendix by Paul Erdős, Wiley, New York, 1992.

[An]   E. ANDRE'EV, On convex polyhedra in Lobachevsky spaces, Mat. Sbornik, Nov. Ser. 81 (1970), 445–478.

[ArLMSS] S. ARORA, C. LUND, R. MOTWANI, M. SUDAN, M. SZEGEDY, Proof verification and hardness of approximation problems, Proc. 33rd FOCS

(1992), 14–23.

[B]   I. BÁRÁNY, A short proof of Kneser's conjecture, J. Combin. Theory A 25 (1978), 325–326.

[Be] J. BECK, Roth's estimate of the discrepancy of integer sequences is nearly sharp, Combinatorica 1 (1981), 319–325.

[BeC] J. BECK, W. CHEN, Irregularities of Distribution, Cambridge Univ. Press (1987).

[BeS] J. BECK, V.T. SÓS, Discrepancy Theory, Chapter 26, in "Handbook of Combinatorics" (R.L. Graham, M. Grötschel, L. Lovász, eds.), North-Holland, Amsterdam (1995).

[Bj] A. BJORNER, Topological methods, in "Handbook of Combinatorics" (R.L. Graham, L. Lovász, M. Grötschel, eds.), Elsevier, Amsterdam, 1995, 1819–1872.

[Bo] J. BOURGAIN, On Lipschitz embedding of finite metric spaces in Hilbert space, Isr. J. Math. 52 (1985), 46–52.

[C]   S.Y. CHENG, Eigenfunctions and nodal sets, Commentarii Mathematici Helvetici 51 (1976), 43–55.

[Ch] A.J. CHORIN, Vorticity and Turbulence, Springer, New York, 1994.

[Co] Y. COLIN DE VERDIÈRE, Sur un nouvel invariant des graphes et un critère de planarité, Journal of Combinatorial Theory Series B 50 (1990), 11–21.

[Chu] F.R.K. CHUNG, Spectral Graph Theory, CBMS Reg. Conf. Series 92, Amer. Math. Soc., 1997.

[DP] C. DELORME, S. POLJAK, Combinatorial properties and the complexity of max-cut approximations, Europ. J. Combin. 14 (1993), 313–333.

[DeL] M. DEZA, M. LAURENT, Geometry of Cuts and Metrics, Springer Verlag, 1997.

[Du] R.M. DUDLEY, Distances of probability measures and random variables, Ann. Math. Stat. 39 (1968), 1563–1572.

[DyF] M. DYER, A. FRIEZE, Computing the volume of convex bodies: a case where randomness provably helps, in "Probabilistic Combinatorics and Its Applications" (Béla Bollobás, ed.), Proceedings of Symposia in Applied Mathematics, Vol. 44 (1992), 123–170.

[DyFK] M. DYER, A. FRIEZE, R. KANNAN, A random polynomial time algorithm for approximating the volume of convex bodies, Journal of the ACM 38 (1991), 1–17.

[GW] M.X. GOEMANS, D.P. WILLIAMSON, .878-Approximation algorithms for MAX CUT and MAX 2SAT, Proc. 26th ACM Symp. on Theory of Computing (1994), 422-431.

[GrLS] M. GRÖTSCHEL, L. LOVÁSZ, A. SCHRIJVER, Geometric Algorithms and Combinatorial Optimization, Springer, 1988.

[H]   J. HÅSTAD, Some optimal in-approximability results, Proc. 29th ACM Symp. on Theory of Comp. (1997), 1–10.

[Ho]   H. VAN DER HOLST, A short proof of the planarity characterization of Colin de Verdière, Journal of Combinatorial Theory, Series B 65 (1995), 269–272.

[JS]   M. JERRUM, A. SINCLAIR, Approximating the permanent, SIAM J. Computing 18 (1989), 1149–1178.

[K]   M. KNESER, Aufgabe No. 360, Jber. Deutsch. Math. Ver. 58 (1955).

[Ko]   P. KOEBE, Kontaktprobleme der konformen Abbildung, Berichte uber die Verhandlungen d. Sächs. Akad. d. Wiss., Math.–Phys. Klasse 88 (1936), 141–164.

[LR]   F.T. LEIGHTON, S. RAO, An approximate max-flow min-cut theorem for uniform multicommodity flow problems with applications to approximation algorithms, Proc. 29th Annual Symp. on Found. of Computer Science, IEEE Computer Soc. (1988), 422-431.

[LiLR]   N. LINIAL, E. LONDON, Y. RABINOVICH, The geometry of graphs and some of its algebraic applications, Combinatorica 15 (1995), 215–245.

[Lo1]   L. LOVÁSZ, Kneser's conjecture, chromatic number, and homotopy, J. Comb. Theory A 25 (1978), 319-324.

[Lo2]   L. LOVÁSZ, On the Shannon capacity of graphs, IEEE Trans. Inform. Theory 25 (1979), 1–7.

[LoM]   L. LOVÁSZ, P. MAJOR, A note on a paper of Dudley, Studia Sci. Math. Hung. 8 (1973), 151–152.

[LoP]   L. LOVÁSZ, M.D. PLUMMER, Matching Theory, Akadémiai Kiadó - North Holland, Budapest, 1986.

[LoS1]   L. LOVÁSZ, A. SCHRIJVER, Cones of matrices and set-functions, and 0-1 optimization, SIAM J. on Optimization 1 (1990), 166–190.

[LoS2]   L. LOVÁSZ, A. SCHRIJVER, A Borsuk theorem for antipodal links and a spectral characterization of linklessly embeddable graphs, Proceedings of the AMS 126 (1998), 1275–1285.

[LoSi]   L. LOVÁSZ, M. SIMONOVITS, Random walks in a convex body and an improved volume algorithm, Random Structures and Alg. 4 (1993), 359–412.

[M]   J. MATOUŠEK, Geometric Discrepancy, Springer Verlag, 1999.

[MS]   J. MATOUŠEK, J. SPENCER, Discrepancy in arithmetic progressions, J. Amer. Math. Soc. 9 (1996), 195–204.

[O]   D. ORNSTEIN, Bernoulli shifts with the same entropy are isomorphic, Advances in Math. 4 (1970), 337–352.

[PW]   L.E. PAYNE, H.F. WEINBERGER, An optimal Poincaré inequality for convex domains, Arch. Rat. mech. Anal. 5 (1960), 286–292.

[RS]   B. RODIN, D. SULLIVAN, The convergence of circle packings to the Riemann mapping, J. Differential Geom. 26 (1987), 349–360.

[RoH]   G.-C. ROTA, L.H. HARPER, Matching theory, an introduction, in "Advances in Probability Theory and Related Topics" (P. Ney, ed.) Vol I,

Marcel Dekker, New York (1971) 169–215.

[Rot]   K.F. ROTH, Remark concerning integer sequences, Acta Arith. 9 (1964), 257–260.

[S]   W. SCHMIDT, Irregularities of distribution, Quart. J. Math. Oxford 19 (1968), 181–191.

[SpT]   D.A. SPIELMAN, S.-H. TENG, Spectral partitioning works: planar graphs and finite element meshes, Proc. 37th Ann. Symp. Found. of Comp. Sci., IEEE (1996), 96–105.

[St]   V. STRASSEN, The existence of probability measures with given marginals, Ann. Math. Statist 36 (1965), 423–439.

[T]   W. THURSTON, The Geometry and Topology of Three-manifolds, Princeton Lecture Notes, Chapter 13, Princeton, 1985.

[W]   H. WOLKOWITZ, Some applications of optimization in matrix theory, Linear Algebra and its Applications 40 (1981), 101–118.

LÁSZLÓ LOVÁSZ, Microsoft Research, One Microsoft Way, Redmond, WA 98052, USA   lovasz@microsoft.com

GAFA, Geom. funct. anal.
Special Volume – GAFA2000, 383 – 405
1016-443X/00/S10383-23 $ 1.50+0.20/0

© Birkhäuser Verlag, Basel 2000

GAFA Geometric And Functional Analysis

PYTHAGOREAN AND PLATONIC CONCEPTIONS IN XXTH CENTURY PHYSICS

YUVAL NE'EMAN

1 Introduction: Why Start in Greece?

Pythagoreanism: the syncretistic philosophy, expounded by Pythagoras, chiefly distinguished by its description of reality in terms of *arithmetical relationships*. ['American Heritage' dictionary of the English language]

.."Plato alleges that God *forever geometrizes*" [Plutarch, *Quaest. Plat.*]

In a reevaluation of the mathematical mold adopted by Fundamental Physics at the end of the XXth Century, we find it instructive to start our retrospective with a search for "roots" in Greek culture, in the VI-th Century BC – then make a leap forward and concentrate on the evolution of the mathematical formulation throughout the XXth Century itself. In what follows, we motivate this choice.

Science consists of representations of sectors of reality, each based upon a small number of postulates, from which all observable qualitative and quantitative features of that sector should be derivable by logical inference. Euclid's formatting of *Geometry*, in terms of axioms, theorems and corollaries, has provided us ever since with a model of what any science should be made to look like.

Historically, this view replaced religious or mythological interpretations of reality. It thus had to pre-suppose that *any impact upon physical reality – if there be one – originating in hypothetical religious or ethical dimensions*, is already included within the same framework constraining reality and thus *does not disturb the requirement of a good fit with observations, as the key criterion for a theory's validity*. In other words – *for science to be observationally verifiable, "miracles", if any, also have to occur within the laws of nature*.

Another assumption is that the patchy nature of the scientific coverage of reality is a temporary feature, with a gradual, step by step unification process coming ever closer to the ideal of a fully united and complete scientific description of existence. Even areas such as *Life, Consciousness, or Creation* which, at the beginning of the XXth Century, to many, still appeared to lie beyond the reach of science, have since yielded (or at least have begun to yield) to the scientific treatment, in the course of this century.

But why start in Greece? True, science-oriented activities had indeed started in prehistory and developed (sometimes independently) in most early cultures – Egypt, Sumer and Akkad, India, China – and later in those of the American continent. This generally included, for example, the observation of the heavens and the separation of the objects observed in two categories – the rotating tapestry of thousands of "fixed" stars – and seven objects moving with respect to that frame (objects still honored by the seven-day week and having its days named after them).

And yet, *it is only in Greece that we can point to science in the sense we have defined.* Everywhere else, it was either an improvement on magic, or at best, an effort to achieve a better understanding as to the way the Gods were running 'their' world, without the search for completeness and without the idea that it could all be described and explained by a logical inference network. Among the Semitic cultures of the Middle-East, both Eastern (Akkad) and Western (Canaanite, Hebrew) Codes of Law evolved during the Second Millenium BC (Hammurabi's in Akkad, Moses' in Israel). When this movement spread to Greece at the end of the VIIth Century, producing Solon's reforms and his code of laws, its impact included something that had not happened elsewhere, namely the idea that "the world" might also obey a code of laws [1] – beyond and above the attempts to explain nature in terms of anthropomorphic deities.

Greek science flourished between 600BC and 200AD, reaching remarkable peaks. Decay started when the intellectual elite began to be drawn away from science by the spread of *Judeo-Christian ethics* with its emphasis on humanism and social justice, in addition to its raising millenary expectations. The rise of the Christian Establishment – in whose eyes the scientific component of Greek neo-Platonist philosophy was (unjustly) considered inseparable from Idolatry – then completed the eradication of all research centers, until the Mediterranean countries where Greek science had grown and prospered were swept clean of any trace of it (the Academy in Athens was all that was left by 420AD and was finally also closed in 529AD). So much for the history. Let us now view the contents.

2   Greek "Meta-science" as Distinguished from Greek Science

In retrospect, I suggest that it is useful, within the Greek contribution, to treat separately, on the one hand, the clear-cut scientific achievements – and on the other, some ideas which, though they too must have been triggered by some contemporary phenomenology, nevertheless could not be realized in those early stages of the development of science, and yet were profound and pregnant with conceptual advances, finally achieving full vindication in modern times. We therefore classify these as achievements in *meta-science* rather than in science, in *Metaphysics* rather than in Physics; note that I use the prefix *meta* in the sense of *foundations of*, not in the sense of either *pseudo* or *spiritual* which is now often the intended meaning of *metaphysics*.

Examples of *scientific* achievements: in Mathematics, the Pythagoreans' proof of the existence of irrational numbers (∼550BC); Euclid's (Greek Euklides') geometry[1] (∼300BC), or in Physics, Archimedes' law of buoyancy or his conception of angular momentum (to use the modern definition) in his study of the lever (∼250BC); in Physical Geography, Erathostenes' 0.5% precision measurement of the radius of the Earth (∼230BC); in Astronomy, Hipparchos' 1% precision measurement of the distance to the Moon (∼140BC), Heron's invention of the steam engine (60AD), etc. There were also some failures, such as the completely wrong result for the distance to the Sun.

Now, as a first example of what I described as *metaphysical* concepts, we take the idea of *atomism*, as launched by Demokritos of Abdera, around 450BC, influenced by the Pythagorean Philolaos of Croton and incorporating some ideas absorbed by Demokritos from the teachings of his mentor, Leucippos. True, Demokritos also included in his presentation a premature *realization* of his idea, based upon the poor state of Chemistry in his time, which was lagging much behind the advances in Physics or Mathematics. This led him to suggest his four-element theory (soil, water, air and fire). What was required and missing was the work of such as Joseph Priestley or of Antoine Lavoisier in the Eighteenth Century, achieving a better understanding of the simplest chemical reactions (e.g. the discovery of oxygen and the rejection of the *phlogiston/caloric* approach) – which made it then possible for John Dalton to conceive the first serious scientific realization of

---

[1]I have taken the liberty of often reverting to the Greek form for Greek names, instead of the usual Latinized or Anglicized forms, e.g. Euklides instead of Euclid, Demokritos rather than Democritus, Platon instead of Plato, Aristoteles rather than Aristotle, etc.

Demokritos' metaphysical idea. We shall return to the four-element model in our discussion of Geometry, since Demokritos, in his model, related the four elements to four of the five "perfect" polyhedrons, while relating the fifth to the Cosmos. Note that *the linkage between atoms and the Cosmos is by itself yet another fruitful metaphysical idea which has had its successes lately.*

I have recently enumerated five important metaphysical contributions, profound ideas about the science of Physics; and yet, ideas which, could truly not be realized in antiquity [2], whatever the phenomena and circumstances which initially triggered their conception – but have all since been vindicated (the last two only in the XXth Century, though in a very impressive way). Following each of the five notions, I list the key rungs on the logical ladder linking them to present physics. The remaining sections of this article then review the last two notions, as also implied by our title.

(1) **Variational Principles** were introduced by Aristoteles [3] and applied by Heron and fifteen centuries later, by Fermat – with Leibniz then developing both the physical idea and the mathematical tools. The idea was then taken up by the brothers Jean and Jacques Bernoulli and treated by them and by their student Leonhard Euler, then by Joseph Lagrange, with Pierre-Louis de Maupertuis finally arriving at *the Principle of Least Action.* Later additions are due to W.R. Hamilton and K.G. Jacobi, also to Felix Klein's and Sophus Lie's introduction of algebraic variations, with their "Erlangen program" [4], to Emmy Noether with her two theorems [5] relating algebraic invariance to physical conservation laws, and to Richard Feynman, with his version of Path Integrals and applications of Functional Analysis. Our work in refs. [2,6] has covered the evolution of the Variational Principles.

(2) **The Vacuum**, an abstract notion, conceived by Platon, denied by Aristoteles' *effective* approach, revived by Galileo and an essential element for his definition of *inertia* – with Torricelli then providing a physical realization. The concept was essential to Newton's postulation of his laws of mechanics – this is the classical story [6]. The concept acquires a major physical role in the XXth Century, with Quantum Mechanics discovering *vacuum energy,* General Relativity bringing in *the Cosmological Constant* and *Inflationary Cosmogony* unifying the two pictures in the story of *Creation* [7,8,9]. The Platon-to-Newton phase of the "story of the vacuum" is discussed in ref. [6].

(3) **Atomism**: from Demokritos, a big leap to the times of John Dalton. Since then, J.J. Thomson, E. Rutherford, and J. Chadwick discover the constituents of Dalton's atom, followed by the discovery (first in Cosmic Rays and then in particle accelerators) of hundreds of *hadrons*. This lead us to the discovery of the *quark* layer in the onion of matter, and the story has not ended [10]. Note also that the observations by Thomson, Rutherford, etc,, at the beginning of the XXth Century, also disposed of the *positivist* "as-if"ic version (as espoused by E. Mach and others) regarding physical reality, namely that atoms do not "really" exist, they just represent a "model" for the description of phenomena, i.e. things behave "as if" there are atoms. We shall not review the story of Atomism here.

(4) **The Geometrization** of physics (Platon), unexpected and yet fully realized in the XXth Century, first in the works of Einstein (1905-1915) and then again (1955-1975) by the emergence of the "Standard Model". This will be reviewed here, though it was discussed in some detail in refs. [6,11]. As to Platon's belief in Geometry as a vehicle for Physics – as described by Plutarch in the passage we quote, the usual guess (based upon Platon's presentation in the *Timmeos* dialogue of the *geometrical features* of Demokritos' model, inspired by Philolaos) is that this is indeed what he had in mind. We shall see in our discussion of present trends that something of that nature has indeed come to play a role in the candidate future theory at the end of the XXth Century!

(5) **Pythagoreanisms** (or "tetracytes") – see the dictionary quotation above. This music inspired and arithmetically oriented prescription also materialized in the XXth Century exclusively, for reasons I shall explain. I have treated this issue before and am thus reviewing it here in some detail.

3   Pythagoreanisms, as induced by Quantum Mechanics

We are definitely uncertain about the origins or justification of Platon's confidence with respect to the geometrization of Physics, but there is no such doubt in the matter of the Pythagorean claim that *nature should be describable by simple arithmetical relations (e.g. "tetracytes" 1:2:3:4)*. Here, it is the role played by Music in the eyes of Pythagoras and of his School, which caused the conceptual "phase-transition".

The emergence of dimensionless simple ratios in Music [12] is caused by the *wave nature of Sound*, as realized by *standing waves in musical instrumentation*. The arithmetic emerges through an *artificial* "quantization" process. Pythagoras and his School uncovered all these facts watching a monocord; they also noticed the emergence of harmonics, embellishing each basic note – and realized that the originating mechanism in that case is the division in 2, 3, 4, of the vibrating string, in a violin or a lyre, or a similar division of the air-column, in a flute or in any other wind-instrument. *One thus faces a discrete system of frequencies or of wavelengths – or, in other words, a Spectroscopy (though artificial, in the case of music).* For Pythagoras' generalization to be realized, i.e. beyond music, two elements were therefore necessary: a field in which the phenomena include a spectroscopy (natural, this time) and the mathematical tools – i.e. *Group and Group Representation Theory*, with perhaps the Plancherel formula replacing the simple count of the Fourier expansion.

There were one or two false starts. Johannes Kepler once thought he had understood the systematics of the planetary orbits, having (wrongly) guessed that they correspond to the sequential babushka-doll like embedding of the five perfect solids. In the XXth Century, Sir Arthur Eddington, Astronomer Royal, and great proponent of General Relativity, was also a sworn Pythagorean and after a few minor cases, tried in vain, to explain the "137" of the dimensionless $\alpha_{EM} = e^2/(4\pi\hbar c) = 1/137$.

The main mathematical tools, namely *Group Theory* and *Group Representation Theory*, were launched around 1828 by two young men in their teens, Niels Henryk Abel (1802-1829) and Evariste Galois (1811-1832) with further developments by Sophus Lie (1842-1899), Elie Cartan (1869-1951) and others. Phenomenologically, it was at the end of the XIXth Century that atomic spectroscopy started being used and revealed systems of lines which were organized and systematized. Writing λ for the wavelength, n and k for the Lower and the Upper Quantum Numbers, respectively, and R for the Rydberg constant, we have the Rydberg formula, $\lambda^{-1} = R[(n^{-2} - (k)^{-2}]$ from phenomenology. In Hydrogen, we have the series

		n	k
Lyman (1916)	[UV]	1	> 1
Balmer (1885)	[visible,UV]	2	> 2
Paschen (1908)	[IR]	3	> 3
Brackett (1922)	[IR]	4	> 4
Pfund (1924)	[IR]	5	> 5

for which Niels Bohr found an ad hoc derivation and Quantum Mechanics an exact one [13].

The same occurred in the Physics of atomic nuclei, where Maria Goeppert-Mayer and H. Jensen identified the *shell-model* structure [14], with one of the main clues supplied by the observation of the so-called *magic numbers* (doesn't this sound Pythagorean?!). Further work revealed more structure (the *"collective model"*) [15] and some features corresponding to shape-pulsations (quadrupole excitations) [16] and finally features leading to a more unified picture, namely the *Interacting Boson Model* [17] due to A. Arima and F. Iachello.

The "Third Spectroscopy" is a name given to the spectrum of the hadrons, which we identified in 1961 as given by the zero-triality unitary representations of $SU(3)$ [18], or in other words, by the unitary irreducible representations (*"unirreps"*) of the quotient-group $SU(3)/Z(3)$ [19]. Here too, structural deciphering (the "Quark Model" [20]) and dynamical understanding (*Quantum Chromodynamics* [21]) followed, but as this sequel is cast into a *geometrical* mold (of the *local* type), it will be discussed under that category.[2]

What better illustration of Pythagoreanisms than the results obtained from the *static approximation* consisting of the combination of the above *hadron unitary symmetry* $SU(3)_h$ with *spin* in $SU(6)_{static}$, yielding for example for the ratio between the magnetic moments of the proton and neutron $\frac{\mu(p)}{\mu(n)} = -3/2$, or for the ratio between the asymptotic total cross-sections for a meson projectile to that of a nucleon projectile, both being hurled at the same target X, is $\frac{\sigma(\pi,X)}{\sigma(N,X)} = 2/3$, etc.

Note that there are now new questions relating to such "music of the spheres". The Standard Model does not explain the pattern of masses at the level of the fundamental matter fields at this stage, the quark and lepton fields. Why 3 "generations" [(u, d / c, s / t, b) for the quarks (each in 3 *colors*) and (ν_e, e/ν_μ, μ/ν_τ, τ) for the leptons]. Moreover, what is the meaning of the mass (or energy) spectrum they display? This question has become especially "hot" after the discovery of the *top* quark, with a mass of some $170 GeV$, whereas the 5 other quarks and all leptons have masses smaller than $5 GeV$? Will the pattern displayed by this *fourth spectroscopy* prove to be as readable by algebraic means as the preceding ones?

A remark regarding the Pythagoreans, who are generally treated as a "cult". If this classification is due to their bizarre universal, namely a belief

---

[2]In 1963, I experienced a personalized illustration of the similarities between the discovery of the regularities in atomic and hadronic spectroscopies. I had just published a non-technical article about the discovery of the $SU(3)$ charges and currents, when I received a letter from a lady in Sweden who was Rydberg's daughter and gave me an eyewitness description of the similarities between the two stories.

in the possibility of describing nature in terms of arithmetical relations, they should instead be considered as the most scientific and nonreligious intellectual grouping of all. In their interest in music, they had understood and reconstructed a limited sector in the workings of nature – and their *credo* amounted to a feeling that this sector was typical. We now know that it had no analog in macroscopic physics, but with their guess they hit the jackpot once we had crossed into the quantum world.

4 The First Geometrization (Relativity): Global (1905) and Local (1915) Phases

In writing his June 1905 first Relativity paper, "On the Electrodynamics of Moving Bodies" [22], Einstein had no idea he was touching anything having to do with geometry. It was his former teacher at the ETH, Hermann Minkowski, now a professor at Göttingen, who realized and pointed out that Einstein's Principle of Relativity, with its postulated invariance of the velocity of light, was in fact a statement about the 4-dimensional geometry of *spacetime* being pseudo-euclidean $M^{1,3}$, i.e. with metric $g_{\mu\nu} = dia(-1,1,1,1)$, $x^0 = ct$, instead of the Galilean $R^1 \times R^3$. Minkowski explained this result at the National Congress of German Physicians and Nature-researchers in Cologne in 1908 [23]. No doubt that this interpretation was useful in simplifying the application of Lorentz and Poincaré group transformations, but this would not have been crucial, had Einstein not been working on his program of harmonizing Newtonian Physics (invariant under the Galileo group) with Maxwell's Electrodynamics and Special Relativity (invariant under the Poincaré and Lorentz groups). Using $E = mc^2$ which he had meanwhile derived in his second Special Relativity paper [24], and assuming gravity to be coupled to energy, through its $m \equiv E/c^2$ inverse, he evaluated the effects of a gravitational field on a beam of light and got the deflection (half the one resulting from General Relativity) [25]. Gone was the invariance of the speed of light! It was at this point that Einstein was forced to look for a more comprehensive approach – and found the Geometrical line unique, in being able to accommodate changes in the metric – provided it would go back to Minkowski's on the tangent, at any point of spacetime. This was Riemannian Geometry, Einstein learned from Marcel Grossmann, his friend and former schoolmate at the ETH. Einstein, working first with Grossmann [26], and then pursuing on his own, arrived at the "General Theory of Relativity" in 1915 [27].

(Note: the story of Hilbert's getting there first has recently been finally put to rest [28].) The geometry of General Relativity possesses two *local* invariances: *passively, invariance under the local diffeomorphisms of R^4*, while *actively it is invariant under the Lorentz group's action on a local frame, an anholonomic feature.* Such local invariances require the presence of *connection* fields $\Gamma_\mu^{\rho\sigma}(x)$ for the holonomic diffeomorphisms and $\Gamma_{\mu\ b]}^{[a}$, necessary for the anholonomic local Lorentz invariance. Connection fields enter into the realization of *parallel transport* through a *covariant derivative*, $D_\mu = \partial_\mu - \Gamma_\mu(x)$ where the anholonomical indices in the connection field have been contracted with these corresponding matrices in the Lie algebra. Aside from this, one has to use the frame's (algebraic) vector (or tangent) fields and their cotangent inverses (1-forms) in defining the overall covariant derivative $D_a = e_a^\mu D_\mu$

5   Extending the Geometry so as to Include Local "Internal" Symmetries (Creation of the Means, in an Early Attempt at Unification)

Much has been said and written about Einstein's Unification Program, launched in 1918, in which the aim was to unite Gravity with Electromagnetism. Note that the existence of other interactions only became clear around 1932, with the discovery of the neutron – whose strong binding in nuclei is 10^{40} times stronger than gravity. From the strength of the binding between two protons within an alpha particle (=Helium nucleus) one can also estimate that it is several hundred times stronger than the Coulomb repulsion between these protons. Anyhow, before 1932 the only contemplated unification was with Maxwell's field. Three methods got furthest in this program:

(1) Hermann Weyl arrived at a Fiber Bundle Geometry – known in Physics as a *local gauge theory*. His first attempt tried to identify the abelian group in the fiber as the group of scale transformations. This was wrong, and was replaced, ten years later, by $U(1)$ as the Fiber group [29] acting on the complex phase of the electron wave function in Quantum Mechanics. In 1953, C.N. Yang and R.L. Mills constructed the generalization of this geometry for the case of a non-abelian Lie group [30]. *The Principal Fiber Bundle $B(M, G, \pi,)$* is given by the base space M (mostly spacetime), laid quasi-horizontally, the (vertical) fiber (here the Lie group $U(1)$ or $SU(3)$, etc.), the (vertical) projection π from any point $P(x)$ onto

x, and the dot-product is a map from the abstract Lie algebra of G onto operators lying within the tangent of the Bundle manifold as such [30]. Preempting our coming description of its development, we note here that the entire Standard Model [31], as completed in 1975, is a Yang-Mills FB with spacetime $M^{1,3}$ as base space and $G = S[U(3)_{color} \times U(2)_{weak}]$.

(2) Albert Einstein himself tried to extend the geometry over dimensions with an antisymmetric metric, unsuccessfully. After the CPT theorem and what we have learned about quantum field theory, it is clear that the subspace with the antisymmetrical coordinates (i.e. symplectic metric) should have involved *spinorial coordinates with half-integer spin*. This is *superspace* and the natural *superalgebra* of isometry is the Ortho-symplectic superalgebra. However, spacetime with its Poincaré isometry algebra can support a smaller superalgebra, namely *supersymmetry* as introduced by Golfand and Likhtman in 1971 and by J. Wess and B. Zumino in 1973 [32]. Work on the *Hierarchy* issue has shown that supersymmetry is essential and should be observed in the hadron spectrum under 10 TeV. *Supersymmetry also allows (almost uniquely) for the unification of spacetime and internal symmetry groups – and that of gravity with other gauge groups within supergravity.*

(3) Theodor Kaluza and Oskar Klein [33] devised a different merger with spacetime, in which one assumes additional dimensions, generally of angular type, the new angles providing spacetime with new gauge symmetries and the entire embedding manifold obeying the Riemannian constraints. In recent years, the issue has been revived for different and rather more compelling reasons, in two different contexts:

(a) "maximal" supergravity [34] which is constructed with $N = 1$ (a single supersymmetry) in 11 dimensions (yielding, under dimensional reduction; $N = 8$ supersymmetries in "3+1" dimensions, the maximum allowed by the axioms). An example of *spontaneous compactification of the added dimensions* has been worked out [35].

(b) superstrings only exist in $(2 + 8n)$ dimensions (for the cancellation of the dilational Quantum anomaly) and in particular in 10 dimensions (for the cancellation of a tachyon in its excitations' spectrum [36]. Superstring theory (and its newest version, "M-theory" [37]) have been developed as candidate post-Planck-energy physical theories unifying Gravity and the Standard Model, all quantized (i.e. reproducing a finite or renormalizable Quantum Gravity theory [38] .

Summing up, we note that the three main routes to unification, drawn in 1918-1926 when Maxwell's electrodynamics was the only candidate for unification with gravity, are also the main routes now utilized, both in constructing the Standard Model (i.e. preparing the analog to Maxwell's theory, in a Weyl-like geometrized mold) and in unifying it with gravity – and also in going beyond it, whether in a Quantum Field Theory formulation, or within Superstrings theory and its Supermembrane extensions, [37,39], etc.

6   The Second Geometrization (Gauge Theories): Global and Quasi-local Phase (1955-1966)

The thirty years 1925-1955, starting with the birth of Quantum Mechanics, were extremely non-geometrical in their fundamental physics output. This non-platonic era can be considered to have ended around 1955-57 as a result of experiments involving the *hadrons*, i.e. particles partaking in the Strong Interactions, but probed here through both the Weak and the Electromagnetic Interactions. Already in 1935, Enrico Fermi had shown that the *effective* Weak Interactions Hamiltonian could be written as a product of two currents, somewhat like the interaction between two conductors in electromagnetism. For instance, nuclear beta-decay $n^0 \to p^+ + e^- + \bar{\nu}_e^0$ can be described as the interaction between a nucleon current $n^0 \to p^+$ and a lepton current in the "pair creation" mode $\bar{\nu}_e(+)e^-$ – just as we would have described a proton current emitting an electron-positron pair. The 1955-56 Weak Interactions experiments, mostly hailed at the time as proof of the *non-conservation of Spatial Reflection Symmetry ("Parity")*, consisted of probing for the Lorentz behavior of the Fermi currents, i.e. identifying experimentally which amongst the five allowed Dirac spinor bilinears (Scalar, Pseudo-scalar, Vector, Axial-vector, anti-symmetric Tensor) is involved, and the answer was "V-A", or $j_\mu = \bar{\psi}(x)\gamma_\mu[(1 + \gamma_5)/2]\psi(x)$. In a massless *Dirac spinor* (4 complex components), this projects out a ("left-chiral" – with 2 complex components) *Weyl spinor* – a separation which exists (for a Minkowski metric) in $d = 4n$ dimensions (n an integer) – or in $d = 4n + 2$ as a 2-real components *Majorana-Weyl spinor*. The *weak* chiral (electrically charged) currents (no electrically neutral *weak* transitions had been observed at that time) would be precisely conserved in the limit of massless (fermion) matter fields – and thus fit into a Yang-Mills-type interaction, with hypothetical vector-potentials (or induced fields) ensuring parallel transport in the fiber-bundle geometry and – physically – mediating the interaction between two currents. The Weak Interaction appeared to take a turn towards geometry, though the precise group structure wasn't clear yet.

Meanwhile, Robert Hofstadter was probing the electromagnetic structure of the nucleon. His 1955-56 results were explained by Y. Nambu and others, who showed that they indicated that the nucleons' primordial surrounding meson cloud is generated by isospin (an abstract $SU(2)$ transformation in Hilbert space, turning a proton into a neutron) $I = 0$ and $I = 1$, spin-parity $J^P = 1^-$, (massive, i.e. short-ranged) vector-potentials – rather than being constituted of the $I = 1$, $J^P = 0^-$ Yukawa pions, as previously thought. Some 10^{-23} seconds after their emission, these vector-mesons decay into 3 and 2 pions respectively, facts which were soon verified directly with inertial mass plots, in bubble-chamber experiments. J.J. Sakurai then convincingly argued (1960) that the Strong Interactions should be described by a Yang-Mills-type Hamiltonian, generated by Local Invariance under

$$\frac{[SU(2)_I \otimes U(1)_Y]}{Z(2)} \otimes U(1)_B = U(2)_{I,Y} \otimes U(1)_B \qquad (1)$$

involving the local conservation of isospin I, hypercharge Y and baryon-number B. The inclusion of this last $U(1)_B$ charge, corresponding to *baryon number* or *Atomic Mass Number*, followed Edward Teller's original suggestion as a possible mechanism explaining the observed *hard core* limiting the compressibility of nuclear matter. The field emitted by the baryonic charge would produce a Coulomb-like repulsion among any two nucleons. Sakurai now pointed to corroborating experimental evidence.

It is at this stage that I became personally involved in this story, in my search for an algebraic description of the hadron spectrum (see last paragraph of section 3). This was a new approach, considered as "abstract", because it was not pre-supposing any dynamical or structural model – whereas most researchers in Particle Physics were trying to guess at some hypothetical initial conditions, either mechanical (e.g. a *spinning top*) or structural (such as the Sakata model, in which all hadrons are made of p, n, λ^0, 3 "more fundamental" particles). In the *abstract* approach, one was searching for a *simple group, of rank* $r = 2$ (so that it would contain the isospin-hypercharge $U(2)$. The rank 2 semi-simple Lie algebras in Cartan's classification are A_2, B_2, C_2, D_2, G_2 and various experimental observations selected A_2, the generator algebra for $SU(3)$ or $SL(3, R)$. The requirements of finite multiplets corresponding to unitary irreducible representations then selects $SU(3)$. The inclusion $SU(3) \supset U(2)_{IY}$ gave sensible results, including the addition of 4 new vector-mesons ("K^*" and their antiparticles) which were observed experimentally. Baryon number enters via an extension to the non-simple $U(3)$. The result was thus both a classification of all hadrons, a "law-of-force" represented by the Clebsch-Gordan coefficient in the invariant coupling between multiplets, 8 conserved currents (or 9, counting baryon number) by Noether's first theorem, approximate *universal coupling of these currents* – i.e. equal to the matrix-elements of the corresponding charges between the relevant states – to the $8 \oplus 1$ vector-mesons – thus an approximate Yang-Mills Interaction for $U(3)$.

The next step was structural. With Haim Goldberg, I suggested a *structural model* [20]. Normalizing the $U(1)$ of the baryon number to $B = 1/3$ meant that all *baryons* should be in representations appearing in the triple direct self-product of the defining representation of $SU(3)$, namely $3 \otimes 3 \otimes 3 = 1 \oplus 8 \oplus 8 \oplus 10$, whereas mesons should consist in the product $3^* \otimes 3 = 1 \oplus 8$. This algebraic model was further developed by M. Gell-Mann (who coined the term *quarks* for the basic triplet, and by George Zweig. Note that quarks have fractional electrical charges $u^{2/3}, d^{-1/3}, s^{-1/3}$. It was suggested – and later verified experimentally – that quarks have spin $J = 1/2$.

We saw that the Weak Interaction involves quasi-conserved currents – more precisely a fully conserved vector half and a partially conserved axial vector half. The latter's "leak" $\partial_\mu a^\mu = C\phi_\pi^I$ is compensated by the pion field. Conservation laws indicate the presence of symmetry – and such a partial-conservation corresponds to a *spontaneous symmetry breakdown*, generated by a *Goldstone boson*. This amounts to relegating the symmetry-breaking agent to the boundary conditions (here the Hilbert space of quantum states, with *a non-trivial assignment* of the vacuum state – and the Goldstone states constituting the remaining states in this vacuum-containing multiplet, thus also forcing their masses to vanish in a limit in which the compensation to the broken conservation law becomes precise. With the pion and other members of the $J^P = 0^-$, $SU(3)$ octet plus singlet mesons no more taken as fundamental, the expression used has changed to *pseudo-Goldstone* states.

7   The Second Geometrization's Local Phase: (a) QCD (1973)

Meanwhile, the mystery of the paradoxical non-observation of states belonging in unirreps with non-zero triality deepened. The *Triality* [19] of a representation of $SU(3)$ is the number of triplet factors needed to reach that representation in a direct product $3 \otimes 3 \otimes 3 \otimes ...mod\ 3$. The representations of $SU(3)/Z(3)$ are the T=0 representations of $SU(3)$, as $Z(3)$, the *center* of $SU(3)$ is a group of three 3 *by* 3 matrices $a1$, with a the three complex cubic roots of the identity – so that in $T = 0$ unirreps, they are all represented by the identity. Also, the anti-quark unirrep 3* has triality $T = 2 \equiv -1$ as can be seen in the reduction of the product $3 \otimes 3 = 6 \oplus 3^*$ into its symmetric and antisymmetric pieces. The observed unirreps can be reached either by a product of 3 quarks, or by multiplying a quark by an anti-quark $3 \otimes 3^* = 8 \oplus 1$, i.e. in either case, $T = 0$. Moreover, even very high energy scattering never resulted in a quark "breaking off"; any extra energy just brought about the emission of mesons, thus preserving the triality of the original hadron. This feature came to be known as quark *confinement.*

Working with quarks in a static approximation induces *coupled spin-unitary spin invariance*, i.e. an invariance under the exchange of the six basic quark helicity states, or an $SU(6)_{static}$ symmetry [40]. It yielded several hundred verifiable results – which were indeed validated (see our discussion of Pythagoreanisms). However, using $SU(6)$ further complicates the situation, because all good results confirm an assignment which *puts the ground state hadrons in unirreps symmetrical in the entire system of quark Quantum Numbers*, thus appearing to *violate the spin-statistics theorem.* The only possible resolution of the paradox is to assume the existence of an additional charge (or Quantum Number), carried by quarks and in which these matter multiplets are antisymmetrical. Since all these multiplets are 3-quark states, the new QN should have a totally antisymmetrical state in a triple product – and is therefore most probably yet another $SU(3)$! Moreover, to guarantee that all "ground state" hadrons do coincide with the antisymmetrical assignment for the new QN and vice versa, it has to be *dynamical*, i.e. the agent involved in gluing quarks into hadrons. There was a first version of the new $SU(3)_{color}$, due to Han and Nambu, in which the quarks do not have fractional charges (and these appear as an average of three types of integer charge quarks), but experimental results rejected this rather elegant model and definitely pointed to a straightforward solution, in which the generator algebras of the two $SU(3)$ groups [41] commute, $[su(3)_{hadron}, su(3)_{color}] = 0$. Since it has to be dynamical, the $SU(3)$ invariance is local – i.e. a Yang-Mills model. This is *Quantum Chromodynamics ("QCD")*. It was postulated immediately after the publication of a mathematical proof of *asymptotic freedom*, a feature almost unique to Yang-Mills theories and consisting of a *gradual and monotonic weakening of the theory's universal coupling at ever shorter distances*. This explains yet another mystery, namely *the detectability of the quarks inside a hadron – in what is supposed to be a very Strong Interaction*. It also raised hopes of a dynamical derivation of *color confinement*: with the coupling weakening at short distances, it must become stronger with increasing distance. Full confinement would imply that at a certain finite distance, the value of the coupling goes to infinity, which is why quarks cannot escape. Note that as of the year 2,000 – all existing mathematical proofs of color confinement either use the approximation of replacing spacetime by a finite-interval lattice – or involve additional mechanisms, which we do not have in a "pure" Yang-Mills theory, such as QCD. Mostly, models apply the Higgs mechanism. In the most elegant of such proofs [42], there is an added assumption of *supersymmetry*, which brings in scalar fields, accompanying the matter field in its supermultiplets. One of these scalars becomes the Higgs field. In that case, confinement is just the effect of a short ranged penetration, due to the chromodynamical field becoming massive. As it is in fact the "chromo-electic" component which has to acquire mass, we have a "dual" to the picture in superconductivity, where the Meissner effect relates to the magnetic field.

The geometrical nature of the QCD fiber bundle has been exhibited in the application of the theory's topological solutions, such as the *instantons* [43], which can be seen to play an important role in the dynamics. Another "strongly geometrical" feature was revealed by J. Thierry-Mieg [44] in his indentification of the unitarity-preserving *ghost fields* with the "vertical" components of the connection, with the "BRST" equations corresponding to Cartan's structural equation for bundle, stating the "horizontality" of the curvature. Moreover, the same solutions have been exploited by the geometricians [45] and have produced the first serious advance in many years in the *exploration and classification of 4-dimensional manifolds*. This study has produced entirely new types and "exotic" manifolds [46]. Further progress has been achieved in this exploration by exploiting the methods developed in [42].

8   The Second Geometrization's Local Phase: (b) the Electroweak Theory (1967-1982)

We have seen that – up to a form of spontaneous symmetry breakdown – the charged weak currents are basically *conserved* and coupled *universally*, indicating the presence of a *local gauge symmetry*, i.e. a Yang-Mills fiber-bundle geometry – except that the group was not yet identified. There was, for a while, some confusion caused by the coexistence of related transitions, as in the beta decays of the neutron (I) $n \rightarrow p + e + \bar{\nu}_e$ and of the lambda hyperon (II) $\Lambda^0 \rightarrow p + e + \bar{\nu}_e$ as against muon decay (III) $\mu^- \rightarrow e^- + \nu_\mu + \bar{\nu}_e$ where the measured couplings appear in the proportion 96 : 4 : 100 when normalized to (III). The problem was solved by N. Cabibbo [47], who showed that the answer lay in the definition of the decaying state, which should be $n' = [n \cos \theta + \Lambda^0 \sin \theta] \rightarrow p + e + \bar{\nu}_e$ with Cabibbo angle $\theta = 15^0$. Applying the quark model means replacing $d \rightarrow n$, $s \rightarrow \Lambda^0$, $d' \rightarrow n'$. From the point of view of Relativistic Quantum Field Theory, it is the state d' seen by the Weak Interaction which corresponds to the original and as yet "unmixed" *matter fields*, here the *primed "down" quark field* or *2nd current-quark state*, and it is the Strong Interaction which mixes d' and the orthogonal s', producing our low-energy SI virtual *constituent quark* eigenstates d and s. More correctly, the SI results in n and Λ^0 hadrons, which are then seen as compounds and used to define the constituent quarks. Note that we now have *six* such *quark flavors* with further such SI mixing, parametrized in the Cabibbo–Kobayashi–Maskawa matrix [47], generalizing the angle $\theta_{cabibbo}$ – and ample experimental confirmation for this picture.

Once the transitions were correctly understood and seeing that all there was consisted of current densities, w^+ for the charge-raising and w^- for the charge-lowering operators within an $su(2)_{weak}$ generator algebra, one could add the missing (neutral-transitions) third component w_3 and some additional hypothetical piece w_y. Such a piece is called for in the Electroweak conjecture by S. Weinberg and A. Salam [48], in which one assumes that the electromagnetic current arises as a linear combination of $aw^3 + bw_y = w_{em}$ [48], i.e. an $su(2) \times u(1)$ originally suggested by Glashow [49]. The orthogonal combination $(-b)w^3 + aw_y$ is then a prediction of the existence of *electrically neutral weak currents*. These new currents were observed experimentally in 1973. The gauge group is thus $SU(2) \times U(1)$, except that we still have to account for the *breaking of this symmetry group*, as exhibited by the masslessness of the electromagnetic field, versus the very massive mediator bosons (the masses were predicted and indeed found at the predicted value).

The answer to the symmetry-breaking here has been given by P. Higgs [50], T. Kibble and others, and is inspired by the Landau-Ginzburg method for phase transitions. It represents yet another type of *spontaneous symme-*

try breakdown, this time for a local symmetry (it also *destroys asymptotic freedom*). The ensuing "Higgs Model" involves a scalar field $\Phi(x)$ with a "Mexican Hat" potential $V = -\mu^2 + \lambda\Phi^4$, whose minimization produces non-zero values $\Phi_{min} = \pm[\mu^2/2\lambda]^{-1/2}$. This has led to correct predictions with respect to the masses of the vector-mesons (connections) and many other features. The spontaneously broken Yang-Mills theory has topological solutions, the main one being the *monopole* [51], appearing when the residual invariant subgroup is $U(1)$, taking the second homotopy group as a mapping unto the asymptotic regions (and the vacuum) and applying the reduction theorem $\Pi_2(G/H) = \Pi_1(H)$.

More recently, D. Quillen's *superconnection* [52] and A. Connes' *noncommutative geometry* [53] (NCG), two related new branches of geometry, have been applied in lieu of the Higgs model of Spontaneous Symmetry Breakdown (SSB). Connes and Lott [54] have reproduced the Weinberg-Salam theory as a local symmetry in NCG, and a model developed back in 1979 by me [55] and by D.B. Fairlie [56] independently and involving the supergroup $SU(2/1)$ is an example of a superconnection [57] and was also shown to be derivable from NCG [58]. I have recently applied a similar technique in *reproducing Einsteinian Gravity from SSB of an Affine Geometry* [59] in the form of a superconnection with $P(4, R)$ as NCG local symmetry.

9   The Standard Model and Beyond: Towards a Third Geometrization?

In 1975, the above gauge theories were consolidated in *the Standard Model* [31]. This is again a fiber-bundle, with Gauge group $S[U(3) \otimes U(2)]$. The "S" (for "special") correlates the two original systems: the leptons have integer Y_w and null (i.e. even triality) $SU(3)_{color}$ whereas the quarks have fractional Y_w and odd triality. Within the $U(2)$, the half-integer $SU(2)_w$ (=left-chiral) leptons have odd eigenvalues of Y_w while (left-chiral) quarks have an odd number of thirds of the unit for Y_w, whereas the right-chiral leptons and quarks (i.e. $SU(2)_w$ scalars) have an even number of Y_w units for leptons and of thirds of a unit for the quarks.

In the SM, some constraints are "naturally" fulfilled, e.g. *the vanishing of the dilational anomalies occurs because the sum of the electrical charges of the 12 quarks and 3 leptons in each generation vanishes*. This is most probably an indication that there is a larger algebraic structure (a GUT –

gauge-unified theory) in which this is a basic postulate. Another feature which seems to point to a larger symmetry results from *the renormalization group calculations, showing that all three couplings (Strong – and the two Electroweak) "meet" somewhere around* $10^{17} GeV$. Moreover, this value is rather close to the Planck energy, where gravity dominates – which might hint at further unification.

The next step is indeed expected to bridge between the two geometrized theories we mentioned, with gravity first undergoing a few adaptations. In such theories, *gravity first has to be quantized* so as to be match-able with the SM, a quantized theory, formulated as a RQFT. Note that even though everything up to $.5TeV$ is described by either GR or the SM, there are indications of possible new interactions. For instance, the bare masses of leptons and quarks are considered as input parameters in the SM, though they may be "the tip of the iceberg" of new interactions, etc.

Two sets of models have been considered as candidate frameworks for *further unification*, supergravity [60] and a formalism going beyond RQFT, namely theories of *quantized extended bodies*. The latter entered the scene with the *string* and its supersymmetric version, the *superstring* [36] and have since been generalized into *membranes* and *supermembranes* and *p-dimensional "p-branes" and "super p-branes"* [37,39,61,62], a joke I tried to fight in a purist spirit by introducing instead *d-dimensional "extendons" with their "superextendons"*, but failed badly. In these theories, the *relativistic quantum fields correspond to a low-energy truncation of the super-extendon and the particle states represent resonating frequencies of these geometrical elements*. Isn't this in the spirit of Philolaos, Demokritos and Platon's discussion [63]? Note that this theory is fully quantized and provides a perturbative solution to the Quantum Gravity riddle.

Supergravity theories, as relativistic quantum field theories (RQFT) of gravity, have yet to be quantized – though this negative status might not yet represent the final verdict. The superextendon theories, however, are fully quantized and contain gravity (thus already in quantized form), embedded in some supergravity framework (also quantized). The presently popular superextendon model, known as *"M theory"* [37,39] contains GR embedded within the most comprehensive supergravity model [64]. Thus, replacing the RQFT by an "extended structure limited localization" provides a dimensional cut-off by replacing a point by a region. In the opposite direction, the divergences enter through the truncation process.

Should this extended direction prove to win, as the next theoretical consolidation and advance, we would have come close to the spirit of the ideas which, according to that conjecture, triggered and motivated Platon's faith in geometry as a key to physics, namely Demokritos' model incorporating Philolaos' idea of *the various atoms as manifestations of four of the five perfect polyhedra, i.e. matter itself born out of geometry*, somewhat like Boticelli's Venus emerging from sea-wave ripples. In the "post-M" theory, the basic objects are also geometrical, except that they all live in a super-manifold with $D = 11$ (bosonic) dimensions – and are distinguished each by its *dimensionality*, namely $d = 1$ for the string, $d = 2$ for the membrane, etc., and topological features (such as "open" and "closed" for the string, the genus for the membrane, etc.). There is, in addition, an overlap with non-commutative geometry. Altogether, *a third geometrization appears to be in the making, very close in spirit to the ideas launched by Philolaos, and developed by Demokritos and Platon.* The formulation of the theory is as yet rather ad hoc; however, some new and fundamental invariance postulate may be missing.

Meanwhile, there is an independent effort which also appears to have achieved a *formulation [65] of Quantum Gravity, based on Dirac's idea [66] of Hamiltonian quantization, a non-perturbative mode* which is therefore harder to juxtapose with the SM theories. The key (gauge-invariant) variable is the *Wilson (loop) integral* (already present in H. Weyl's work) which serves as a transform which "inscribes" the connection $A(x)$ in spacetime: $T_A(\gamma) = N^{-1}Tr(\exp \int_\gamma A)$. This brings in the diffeomorphism-invariance of the loop, so that only *knot theory* can provide the classification of such QG states [68]. Note the resemblance to strings (is it fortuitous?). And yet, this approach has also been shown to lead to the *Quantization of Geometry – areas and volumes* [67,68]. Here too, there is an overlap with non-commutative geometry and there is also some resemblance with the extendons of the superstring's successors, M-theory and the like.

Returning to the elements of a Third Geometrization, we note that these already appear to provide a view beyond "Planck dimensions", both as geometry and as physics. With classical GR and "non-relativistic" QM we already get a limit on localizability when the Schwarzschild radius equals the Compton wavelength, namely at Planck dimensions. This is further clarified [69] when applying the *decoherence* interpretation for the measurement process in QM [70]. With superextendons in M-theory, one achieves a further insight. The theory has a set of *duality* symmetry transformations, similar to both Hodge duality (in the Grassmann algebra, as in electric-magnetic inversions) and *bootstrap duality* (between the dynamical variables and the solitonic solutions). With magnetic monopoles fulfilling Dirac's condition $eg = n$, we observe that a strengthening of e (the electric charge) weakens the magnetic charge g and vice versa. This implies that the *effective* couplings behave mutually according to $e \equiv (1/r) \leftrightarrow g \equiv r$ and vice versa. Thus, going to smaller dimensions than r_{Pl} will take us back into larger dimensions in the dual system. This is just a first glimpse at the third geometrization; two other recent successes have consisted of (1) the recovery of the macroscopically-derived Bekenstein formula [71] for black hole entropy – from a statistical analysis of the density of states of the gravitational field in this Quantum Gravity model [72]; and (2) a calculation of QCD color-confinement parameters from supergravity in 5-dimensional anti-De Sitter spacetime [73] (by applying a transformation related to the Wigner-Inönu group-contraction). One may guess that much further geometrical bliss awaits the XXIst Century's physicists.

Acknowledgement. I would like to thank my friend Dr. Maurice Cohen for his outstanding support of a study of some geometrical features of present gauge-theories, both "internal" and "external".

References

[1] C. PREAUX, in "Connaissance Scientifique et Philosophie", Academie Royale de Belgique (Proc. 1973 Colloquium, commemorating 200th Ann. of Royal Acad. of Belgium), Bruxelles (1975), 39–100.

[2] Y. NE'EMAN, Insight and foresight in physics: from Plato and Aristotle to Einstein and the standard model, in "Nuclear Matter, Hot and Cold (Proc. School of Physics and Astronomy, Tel Aviv University, April 1999) (D. Ashery, J. Alister, eds.), to appear.

[3] W. YOURGRAU, S. MANDELSTAM, Variational Principles in Dynamics and Quantum Theory, Dover pub., New York, 1968, 201pp.

[4] H.A. KASTRUP, in "Symmetries in Physics (1600-1980)" (M.G. Doncel, et al., eds.), University of Barcelona Pub., Barcelona (1987), 113–164.

[5] Y. NE'EMAN, The impact of Emmy Noether's theorems on XXIst Century physics, in "Israel Mathematics Conf. Proc. 12 (1999), 83–101.

[6] Y. NE'EMAN, De l'autogeometrisation de la physique, in "Les Relations entre les Mathematiques et la Physique Theorique", Festschrift for the 40th Anniversary of the IHES (1998), 145–152.

[7] Y. NE'EMAN, Inflation, the top and all that (3 lectures at Erice, NATO Advanced Study Institute, May 1994), in "Currents in High Energy Astrophysics", (1995), 255–280.

[8] Y. NE'EMAN, The Eternal Inflationary Universe (in Hebrew), Weizmann Institute, Youth Activities Publications, 1995, 63pp.

[9] Y. NE'EMAN, Whereto, Lambda, paper read on 22 April 1999 at the Symposium in Astrophysics and Cosmology, held on the occasion of the 75th Anniversary of Wolfgang Priester, at the Bonn Radioastronomical Observatory.

[10] Y. NE'EMAN, Y. KIRSCH, The Particle Hunters, 2nd edition, Cambridge University Press, 1997, 272pp. See Chap. 11.

[11] Y. NE'EMAN, Plato and the self-geometrization of physics in the twentieth century, Functional Differential Equations 5:3-4 (1998), 19–34.

[12] E.E. HELM, Sc. Am. 217 (Dec. 1967), 92–103.

[13] E. SCHROEDINGER, Ann. d. Physik 79 (1926), 361; W. PAULI, Zeit. Physik 36 (1926), 336.

[14] M.G. MAYER, Phys. Rev. 75 (1949), 1969; O. HAXEL, J.H.D. JENSEN, H.E. SUESS, Phys. Rev. 75 (1949), 1766.

[15] J.P. ELLIOTT, Proc. Roy. Soc. London A245 (1958), 128.

[16] Y. DOTHAN, M. GELL-MANN, Y. NE'EMAN, Phys. Lett. 17 (1965), 148–151; Y. NE'EMAN, DJ. ŠIJAČKI, Phys. Lett. 157B (1985), 267–274.

[17] A. ARIMA, F. IACHELLO, Ann. Rev. Nucl. Sc. 31 (1981), 75.

[18] Y. NE'EMAN, Nucl. Phys. 26 (1961), 222; M. GELL-MANN, Caltech preprint CTSL-20 (1961), unpub.

[19] G.E. BAIRD, L.C. BIEDENHARN, Proc. 1st (1964) Coral Gables Conf. on Symmetry Principles at High Energy (B. Kursunuglu, et al., eds.), W.H. Freeman Pub., 58.

[20] H. GOLDBERG, Y. NE'EMAN, Nuovo Cim. 27 (1953), 1; M. GELL-MANN, Phys. Lett. 8 (1964), 214; G. ZWEIG, CERN rep. TH 401, unpub.

[21] S. WEINBERG, Phys. Rev. Lett. 31 (1973), 494; H. FRITZSCH, M. GELL-MANN, in "Proc. XVIth ICHEP" 2 (Chicago, 1972) 135.

[22] A. EINSTEIN, Ann. d. Physik 17 (1905), 891.

[23] H. MINKOWSKI, Lecture at 80th Assembly of German researchers in Natural Science and physicians, Cologne (1908).

[24] A. EINSTEIN, Ann. d. Physik 18 (1905), 639; and 23 (1907), 371.

[25] A. EINSTEIN, Jahrbuch d. Radioactivitaet und Electronik 4 (1907), 411.

[26] A. EINSTEIN, M. GROSSMANN, Zeit. f. Math. u. Physik 62 (1913), 225.

[27] A. EINSTEIN, Ann. d. Physik 49 (1916), 769.

[28] L. CORRY, J. RENN, J. STACHEL, Science 278 (1997), 1270.

[29] H. WEYL, Zeit. F. Physik 56 (1929).

[30] C.N. YANG, R.L. MILLS, Phys. Rev. 96 (1954), 191.

[31] S. WEINBERG, Phys. Rev. D11 (1975), 3583.

[32] Y.A. GOLFAND, E.P. LIKHTMAN, JETP Letters 13 (1971), 323; J. WESS, B. ZUMINO, Nucl. Phys. B70 (1974), 39.

[33] TH. KALUZA, Sitz. Preu. Akad. Wiss. (1921), 966; O. KLEIN, Nature 118 (1926), 516.

[34] E. CREMMER, B. JULIA, Phys. Lett. B80 (1978), 48; and Nucl. Phys. B159 (1979), 141.

[35] P.G.O. FREUND, M.A. RUBIN, Phys. Lett. B97 (1980), 233.

[36] J.H. SCHWARZ, ED., Superstring Theory (2 vols.) World Scientific Pub., Singapore (1985), 1145pp.; M.B. GREEN, J.H. SCHWARZ, E. WITTEN, Superstring Theory (2 vols.), Cambridge U. P. (1987), 920pp.

[37] P.K. TOWNSEND, Phys. Lett. B350 (1995), 184.

[38] Y. NE'EMAN, Quantizing gravity and spacetime: Where do we stand?, in Annal der Physik 7 (1998), 1–15.

[39] Y. NE'EMAN, E. EIZENBERG, Membranes and Other Extendons ('p-branes'), World Scientific, Singapore, 1995, 178pp.

[40] F. GÜRSEY, L. RADICATI, Phys. Rev. Lett. 13 (1964), 175; see also F. DYSON, Symmetry Groups in Nuclear and Particle Physics, W.A. Benjamin pub., New York, 1966, 320pp.

[41] Y. NE'EMAN, Phys. Rev. B134 (1964), 1355; and The original fifth interaction, in "Elementary Particles and the Universe, Essays in honor of Murray Gell-Mann" (Proc. Aniv. Conf., Pasdena, Jan. 1989) (J.H. Schwartz, ed.), 47–59.

[42] N. SEIBERG, E. WITTEN, Nucl. Phys. B426 (1994), 19; idem B431 (1994), 484.

[43] A.M. POLYAKOV, Phys. Lett. B59 (1975), 82; G. 'T HOOFT, Phys. Rev. Lett. 37 (1976), 8.

[44] J. THIERRY-MIEG, J. Math. Phys. 21 (1980), 2834.

[45] S. DONALDSON, J. Diff. Geom. 18 (1983), 279; M. FRIEDMAN, J. Diff. Geom. 17 (1982), 357.

[46] R.F. GOMPF, J. Diff. Geom. 18 (1983), 317.

[47] N. CABIBBO, Phys. Rev. Lett. 10 (1963), 531; M. KOBAYASHI, K. MASKAWA, Prog. Theor. Phys. 49 (1972), 282.

[48] S. WEINBERG, Phys. Rev. Lett. 19 (1967), 1264; A. SALAM, in "Elementary Particle Theory" (N. Svartholm, ed.), Almquist Vorlag AB, Stockholm (1968).

[49] S.L. GLASHOW, Nucl. Phys. 22 (1961), 579.

[50] P. HIGGS, Phys. Lett. 12 (1964), 132; F. ENGLERT, R. BROUT, Phys. Rev. Lett. 13 (1964), 2386.

[51] G. 't Hooft, Nucl. Phys. B79 (1974), 276; A.M. POLYAKOV, JETP Lett. 20 (1974), 430.

[52] D. QUILLEN, Topology 24 (1985), 89.

[53] A. CONNES, Noncommutative Geometry, Academic Press, NY, 1994, 661pp.

[54] A. CONNES, J. LOTT, Nucl. Phys. B18, Proc. Suppl. (1990), 29.

[55] Y. NE'EMAN, Phys. Lett. 81B (1979), 190.

[56] D. FAIRLIE, Phys. Lett. B82 (1979), 97.

[57] S. STERNBERG, Y. NE'EMAN, Superconnections and internal supersymmetry dynamics, Proc. Nat. Acad. Sci. USA 87 (1990), 7875–7877.

[58] R. COQUEREAUX, R. HÄUSSLING, N.A. PAPADOPOULOS, F. SCHECK, Int. J. Mod. Phys. A7 (1992), 2809; R. COQUEREAUX, G. ESPOSITO-FARÉSE, F. SCHECK, Int. J. Mod. Phys. A7 (1992), 6555.

[59] Y. NE'EMAN, Phys. Lett. B247 (1998), 19–25.

[60] D.Z. FREEDMAN, P.V. NIEUWENHUIZEN, S. FERRARA, Phys. Rev. D13 (1976), 3214; S. DESER, B. ZUMINO, Phys. Lett. B62 (1976), 335.

[61] P.A. COLLINS, P.W. TUCKER, Nucl. Phys. B112 (1975), 150.

[62] Y. NE'EMAN, D. ŠIJAČKI, Phys. Lett. 4 (1986), 165–170; Y. NE'EMAN, D. ŠIJAČKI, B174 (1986), 171–175.

[63] PLATO, Timmeos 32, 35, 36.

[64] E. CREMMER, B. JULIA, Nucl. Phys. B159 (1979), 141.

[65] A. ASHTEKAR, Phys. Rev. Lett. 57 (1986), 2244.

[66] P.A.M. DIRAC, Can. J. Math. 2 (1950), 129; Proc. Royal Soc. A246 (1958), 326, 333.

[67] J. BAEZ, ED., Knots and Quantum Gravity, Oxford Science pub., Clarendon Press, Oxford, 1994.

[68] C. ROVELLI, L. SMOLIN, gr-qc/9411005

[69] Y. NE'EMAN, Phys. Lett. A186 (1994), 5.

[70] H.D. ZEH, Found. Phys. 1 (1970), 69.

[71] J. BEKENSTEIN, Lett. Nuov. Cimento 4 (1972), 737.

[72] A. STROMINGER, C. VAFA, Phys. Lett. B379 (1996), 99.

[73] J. MALDACENA, The large N limit of superconformal field theories and supergravity (hep-th 97/11200); Wilson loops in large-N field theories (hep-th 98/03002).

YUVAL NE'EMAN, Sackler Faculty of Exact Sciences, Physics, Tel Aviv University, Tel Aviv 69978, Israel
and
Center for Particle Physics, University of Texas, Austin, TX 76712, USA

GAFA, Geom. funct. anal.
Special Volume – GAFA2000, 406 – 424
1016-443X/00/S10406-19 $ 1.50+0.20/0

© Birkhäuser Verlag, Basel 2000

I GAFA Geometric And Functional Analysis

1. CLASSICAL AND MODERN TOPOLOGY
2. TOPOLOGICAL PHENOMENA IN REAL WORLD PHYSICS

Sergey P. Novikov

According to the Ancient Greeks, the famous real and mythical founders of Mathematics and Natural Philosophy, such as Pythagoras, Aristotle and others, in fact, borrowed them from the Egyptian and Middle Eastern civilizations. However, what had been told before in the hidden mysteries Greek scientists transformed into written information acceptable to everybody. Immediately after that the development of science in the modern sense started and had already reached a very high level 2000 years ago. Therefore you may say that the free exchange of information and making it clear to people were the Greeks' most important discoveries. I would say it is the basis of our science now. As you will see, any violation of this fundamental rule does serious harm to our science and inevitably leads to its decay.

1 Classical and Modern Topology

Prehistory: The first fifty years of Topology. The first important topological ideas were observed by famous mathematicians and physicists such as Euler, Gauss, Kelvin, Maxwell and their pupils, during the XVIIIth and XIXth Centuries. As everybody knows, it was Poincaré who really started Topology as a branch of Mathematics in the late XIXth Century. Many top class mathematicians participated in the development of Topology in the first half of our century. A huge number of mutually connected fundamental notions were invented: degree of maps and singularities of vector fields, homotopy and homology groups, differential forms and smooth manifolds, the fundamental idea of transversality, the simplicial/cell(CW) and singular complexes as tools for studying topological invariants, braids, knot invariants and 3-manifolds, coverings, fibre bundles and characteristic classes and many others. Deep connections with Qualitative Analysis, Calculus of Variations, Complex Geometry and Dynamical Systems were

This work is supported by NSF Grant DMS9704613.

established in this period. Combinatorial Group Theory and Homological Calculus started from topological sources. A great new field of topological objects unknown to the classical mathematics of the XIXth Century appeared finally in the 1940s. At that time this new area was known to a few mathematicians only. However there was a very high density of really outstanding scientists among them.

1950s and 60s: The Golden Age of Classical Topology. The fundamental set of algebraic ideas unifying all these branches of mathematics appeared in the 40s; a new era started about 1950. Spectral sequences of fibre bundles, sheaves, highly developed homological algebra of the groups, algebras and modules, Hopf algebras and coalgebras, were invented and heavily used for the calculation of topological invariants needed for the solution of the fundamental problems of topology. Let me point out that, in many cases, it was a completely new type of calculations based on the deep combination of the very general "categorial properties" of these quantities with very concrete geometric, algebraic or analytical study of a completely new type. In the previous period, people had not even dreamt about how they could be calculated. Regular methods were built to calculate homotopy groups, for example. It was one of the most difficult problems of topology. A lot of them were computed completely or partially including the homotopy groups of spheres, Lie groups and homogeneous spaces. The topologically important cobordism rings were computed and used in many topological investigations. The famous signature formula for differentiable manifolds was discovered. It has innumerable of applications in the topology of manifolds. Besides that, this formula played a key role in the proof of the so-called Riemann-Roch theorem in Algebraic Geometry and later in the study index, the famous homotopy invariant of Fredholm operators.

The mutual influence of Topology and Algebraic Geometry during that period led to the broad extension of the ideas of homology: the extraordinary (co)homology theories like K-theory and cobordisms appeared. They brought a new type of technic to topology with many applications. Representation theory and complex geometry of manifolds deeply unified with homological algebra and Hopf algebras. The technic of formal groups appeared here. It has been applied in particular for the improvement of the calculations of stable homotopy groups of spheres. As everybody knows, during this period topology solved the most fundamental problems in the theory of multidimensional smooth manifolds:

Nontrivial differentiable structures on multidimensional spheres were discovered on the basis of the results of algebraic topology combined with a new understanding of the geometry of manifolds and bundles. The multi-dimensional analogue of the Poincaré Conjecture and H-cobordism theorem were proved. Counterexamples to the so-called "Hauptvermutung der Topologie" were found. A classification theory for the multidimensional smooth (and for PL-manifolds as well) was completely constructed. The role of the fundamental group in this theory led to the development of a new branch of algebra: the algebraic K-theory. Topological invariance of the most fundamental characteristic classes was finally proved. The so-called "Annulus Conjecture" was proved. No matter how elementary these results can be formulated, nobody has succeeded to avoid the use of a whole bunch of results and tools of algebraic and differential topology in the proof. The classification theory for the immersions of manifolds was constructed. The theory of multidimensional knots was constructed. Several classical problems of the theory of 3-manifolds also were solved during that period: the so-called Dehn's program was finished after a 50 year break; the algorithm for recognizing the trivial knot in three-space has been theoretically constructed as a part of the deep understanding of the structure of 3-manifolds and the surfaces in them. As a by-product of topology, the fundamental breakthrough in the topological understanding of generic dynamical systems was reached. A new great period started in this area. Qualitative theory of foliations has been constructed with especially deep results for 3-manifolds.

As a summary, I would like to add one more very important characteristic of the topological community in the golden age of classical topology:

All important works have been carefully checked. If some theorem had not been proved, it immediately became known to everybody.

So you can find a full set of proofs in the literature. Unfortunately, a full set of textbooks covering all these developments (1950-1970) has not been written yet. Many modern textbooks are written in a very abstract way. Even if they cover some pieces formally, it is more difficult to read them than the original papers. Let me recommend to you the Encyclopedia article [1] written exactly for the exposition of these ideas.

1970s: Period of decay. In my opinion, the period of the 1970s can be characterized as a period of decay for classical topology. There are many indications for that. Several leading scientists left topology for new areas, such as algebra and number theory, Riemannian and symplectic geometry, dynamical systems and complexity theory, functional analysis and representations, PDEs, and different branches of mathematical/theoretical physics.... It is certainly a good characterization of the community if it could generate such a flux of scientists in many different areas and bringing to them completely new ideas. Anyway, this community dispersed.

What can we say about the topological community after that?

First of all, some important new ideas appeared in the 70s (e.g. localization technique in homotopy topology, the nicely organized theory of the rational homotopy type, hyperbolic topology of 3-manifolds were developed by Sullivan and Thurston among others). However, **a huge informational mess was created in the 1970s.** Let me point out that a series of fundamental results of that period was not written, with full proof, until now. Let me give you a list:

Sullivan's Hauptvermutung theorem was announced first in early 1967. After the careful analysis made by Bill Browder and myself in Princeton of the first version in May 1967 (before publication), his theorem was corrected: a necessary restriction on the 2-torsion of the group $H_3(M, Z)$ was missing. This gap was found and restriction was added. Full proof of this theory has never been written and published. Indeed, nobody knows whether it has been finished or not. Who knows whether it is complete or not? This question is not clarified properly in the literature. Many pieces of this theory were developed by other topologists later (they sometimes used different ideas). Nobody has unified them until now. Indeed, these results were used by many others later. In particular, the final Kirby-Siebenmann classification of topological multidimensional manifolds therefore is not proved yet in the literature.

The second story is the theory of Lipschitz structures on manifolds. In the mid-seventies Sullivan distributed a preprint containing the idea of how to prove existence and uniqueness of such structures on the manifold $N^n, n \neq 4$. This idea obviously included (for uniqueness) the direct use of the Annulus Conjecture (and therefore of all ideas and techniques needed in the proof of topological invariance of the rational Pontryagin Classes inside). Proof of the Lipschitz theory has never been published. Indeed, many years later, already in the 1990s, some brilliant younger scientists developed a very nice theory of Fredholm (elliptic) operators on Lipschitz manifolds. As a corollary, they claimed that a new proof of topological invariance of rational Pontryagin classes has been obtained from Analysis (this was a problem posed by Singer in the 60s). Young scientists made a "logical circle" believing in the classical results. Nobody told them that the corresponding theorems had never been proved. How could it happen? This funny story shows the modern state of information in the topological community.

Another informational mess has been created in 3D hyperbolic topology. This beautiful area was started by Thurston in the mid-70s. For many years people could not find out what was proved here. In this area the situation has been finally resolved: it has been acknowledged that these methods lead to the proof of the original claim (the so-called Geometrization Conjecture) only for the special class of Haken manifolds. The Geometrization Conjecture means more or less that (in the case of closed 3-manifolds) the fundamental group can be realized as a discrete subgroup acting in the 3D hyperbolic space if trivial necessary conditions are satisfied: all its abelian subgroups are cyclic and $\pi_2 = 0$. However, it is difficult to find out who actually proved this theorem? It seems for me that the younger mathematicians who managed to finish this program did not receive proper credit.

I would like to mention that this kind of informational mess has happened since 1970 not only in topology. For example, the famous results of KAM in the three-body problem known since the early 60s were found recently unproved. It was announced for the first time at the Berlin Congress last year (see the plenary talks of M. Hermann and J. Moser. Kolmogorov and Moser never claimed solution of this specific degenerate problem in their works.). In this case, some works supposedly containing full proof were published in the first half of the 60s. Does this mean that nobody actually read them for at least 30 years?

Do you think that algebra is better? Let me tell you as a curious remark that all works of the Steklov Institute (i.e. Shafarevich's) school in algebraic number theory, algebraic geometry and theory of finite p-groups awarded by the highest (Lenin and State) prizes in the former Soviet Union since 1959, did not contain full proof. The gaps in the proofs were found many years later. Not all these gaps were really deep. However, some of these authors knew their mistakes many years before they became publicly known and could not correct them. They managed to fill gaps after many years, using much later technical achievements made by other people. Does it mean that in the corresponding time, despite many public presentations, nobody in fact read these great works? Can we say that all proofs are known now in all these cases?

There are much worse cases in modern algebra indeed. How many of you know that the so-called classification of simple finite groups did not exist as a mathematical theorem until now? In this case we can even say that in fact (as a few real experts have known since 1980) not one work existed claiming that this problem was finished in this work. All public opinion has been based only on the "New York Times Theorem" for the past 20 years.

1980s and 90s: Period of recovery. The role of Quantum Field Theory. It became clear already in the late 70s that modern quantum field theory started to generate new ideas in topology. It gave several new alternative ways to construct topological invariants: Path integral for metric-independent actions on manifolds was used for the first time. The famous self-duality equation appeared first in the works of physicists. It was applied in the 80s for the solution of fundamental topological problems in the theory of 4-manifolds. Quantum string theory brought in the early 80s new deep results in the theory of the classical Fuchsian groups and moduli spaces. At first physicists (like 'tHooft and Polyakov) were not very interested in such by-products of their activity. They always said that they were doing physics of the real world, not pure mathematics. However the next wave of brilliant physicists (like Witten, Vafa and others) started to solve problems of pure mathematics. Such purely topological subjects, as Morse theory and cobordism theory associated with action of compact groups on manifolds, were developed in the 80s from the completely new point of view. Symplectic Topology reached a very high level in the late 80s. We are facing now impressive development of Contact Topology.

Certainly Quantum Theory brought new beautiful ideas. Besides that, the fundamental new invariants of knots were discovered in the 80s by the topologists who came from functional analysis and theory of C^* algebras. These invariants also received quantum treatment in the late 80s. The beautiful connection of the specific Feinmann diagrams with surfaces was borrowed from physics literature. It became a very effective tool for the solution of several topological problems. Unfortunately, only a few mathematicians learned this technique and started to apply it in topology. I know only Singer, Kontsevich and a very small number of others. Even if you will add here the names of pure mathematicians who learned this with the intention to do real physics, this list will increase inessentially. I do not count here people who were trained originally in the physics community. A large number of them moved into pure mathematics with the intention to prove rigorous theorems about the models serving (in their opinion) as an idealization of theoretical physics. They call this area Mathematical Physics, but not everybody agrees with such a definition of mathematical physics. This community does not do topology.

I would like to make a remark here concerning a beautiful work of Kontsevich calculating certain Chern numbers on the punctured moduli spaces of Riemann Surfaces through the special solution to the KdV hierarchy. This formula has been known as a Witten Conjecture. You have to specify for this some compactification of the moduli spaces of punctured Riemann surfaces, otherwise it makes no sense. Kontsevich actually proved this formula for one specific ("Strobel-Penner") compactification in 1991. What about the standard Deligne-Mumford compactification? Kontsevich claimed in 1992 in his work in Inventiones that it is true. However, no proof has been presented until now. So this problem is open. There was a mistakable statement about this at the Berlin Congress.

Let me point out that the physics community did not create any informational mess in topology. According to their training tradition, theoretical work produces Conjectures which should be proved only by some kind of experiment. Starting to do beautiful nonrigorous mathematics, they do not claim that they "proved" something. They are saying that they "predicted this fact". In the case of pure mathematics, the final proof done by pure mathematicians these people may treat as an "experimental confirmation". In the past ten years several deep results have been obtained in the 4D topology. We cannot say this about 3D topology: quantum invariants here created some sort of "invariantology": a lot of people are constructing topological invariants but no one new topological result has been obtained for almost 10 years. Indeed, these ideas look beautiful in some cases. In my opinion, new deep results will appear after better understanding of the relationship of new invariants with classical topology.

2   Topological Phenomena in Real World Physics

Topological ideas in physics in the period of the early 80s.   I spent about 10 years learning different parts of Modern Theoretical Physics in the 60s and 70s. After joining the physics community (i.e. Landau school) in the early 70s I found out that most physicists did not know at all the new areas of mathematics like topology, dynamical systems and algebraic geometry, including analysis on Riemann surfaces. The quantum people knew some extracts from group theory and representations because they needed it in Solid State Physics as well as in Elementary Particles Theory since the 1960s. A lot of them knew something about Riemannian Geometry because of the Einsteinian General Relativity. However, these people had already heard something about the new mathematics of the XXth century and badly wanted to find its realization in physics. You have to take into account that among them there was a great number of extremely talented people at that time with very good training in practical mathematics. In some cases I was able to help physicists (like Polyakov, Dzyaloshinski, Khalatnikov and some others, especially at the Landau Institute) to learn and to use topology in the 70s. I worked this period in General Relativity (Homogeneous Cosmological Models) and Periodic KdV Theory with my pupils and collaborators. We found completely nonstandard applications of Dynamical Systems and Algebraic Geometry in these areas. However, until the late 70s I did not produce any new topological ideas. My very first topological work in physics was made in 1980 (see [2]). I started to use in the spectral theory of the Schrödinger operators in periodic lattice and magnetic field the idea of transversality applied to the families of Hermitian matrices or elliptic operators on the torus. This idea led to the discovery of the series of topological invariants, Chern Numbers of Dispersion Relations. They are well-defined for the generic operators only. The classical Spectral Theory in mathematics never considered such quantities because they are not defined for every operator with prescribed analytical properties of coefficients. The ideology of transversality is important here. This work was not understood by my colleagues-physicists at that time (the vice-editor of JETP did not want to publish it as "nonphysical", so I published it in the math literature). People thought that the important integer-valued observable quantities in Solid State Physics may come from symmetry groups only. Indeed, the Integral Quantum Hall phenomenon was discovered soon. Some famous theoretical physicists rediscovered my mathematical idea after that. It is certainly a sum of the Chern classes of dispersion relations below the Fermi level.

My next topological discovery was made in the joint work with student I. Schmelzer in 1981, dedicated to the very special problem of classical mechanics and hydrodynamics (see [3]). I immediately realized its value for modern theoretical physics, as well as for mathematics, and developed this idea in several directions in the same year ([4]). The series of works in the Theory of Normal Metals which I am going to discuss today, is also one of by-products of that discovery. Doing the Hamiltonian factorization procedure for the top systems on the phase spaces like $T^*(SO_3)$ by the action of S^1, you are coming to the systems mathematically equivalent to the motion of the charge particle on the 2-sphere. This sphere is equipped by some nontrivial Riemannian metric. What is important and has been missed by the good experts in analytical mechanics like Kozlov and Kharlamov is that the effective magnetic field like Dirac monopole appears here for the nonzero values of the "area integral" associated with S^1-action. This means precisely that the magnetic flux along the sphere is nonzero. The reason for this is that the symplectic (Poisson) structure after factorization is topologically nontrivial. In terms of modern symplectic geometry, the magnetic field is equivalent to the correction of the symplectic structure. This fact is not widely known in the geometric community even now. The appearance of the topologically nontrivial symplectic structures after S^1-factorization of symplectic manifolds was independently discovered and formulated in geometric, nonphysical terminology in 1982 in the beautiful work [6] for different goals (calculating of integrals).

It has been realized in [3, 4] that the action functional for such systems is in fact a closed 1-form on the spaces of loops. These functionals have been immediately generalized for higher dimensions, to the spaces of mappings F of q-manifolds in some target space M where a closed $q+1$-form is given instead of magnetic field. We are coming finally to the action functional well-defined as a closed 1-form on the mapping spaces F. The topological quantization condition for such actions was formulated in 1981 [4, 5] as a condition that this closed 1-form should define an integral cohomology class in $H^1(F, Z)$. It is necessary and sufficient for the Feinmann amplitude to be well-defined as a circle-map

$$\exp\{iS/h\} : F \to S^1 .$$

For the case $q = 1$ the original Dirac requirement was based on a different idea: the magnetic field should be a Chern class for the line bundle whose space of sections should serve as a Hilbert space of states for our Quantum Mechanics. Therefore it should be integral in $H^2(M, Z)$.

In pure topology and in the Calculus of Variations these ideas led to the construction of the Morse-type theory for the closed 1-forms on the finite- and infinite-dimensional manifolds. Let me refer to the last publication of the present author (with P. Grinevich) in this direction [7] where the survey of results and problems is discussed. I would like to point out that for the compact symplectic manifolds the action functional for any nontrivial Hamiltonian system is multivalued. The cohomological class of symplectic form cannot be trivial here. I do not know of such cases in real physics where the symplectic manifold is compact. However, even in the community of symplectic geometers nobody paid attention to such properties of action functional until the 90s.

After that I started to think about different aspects of the Hamiltonian Theory where the class of one-valued functions naturally can be extended to the class of all closed 1-forms. For every symplectic (Poisson) manifold M with $H^1(M) \neq 0$ we may consider Hamiltonian Systems generated by the closed 1-form dH where the function H is multivalued. Instead of energy levels $H = const$ we have to consider nontrivial codimension 1 foliation $dH = 0$ with Morse (or Morse-Bott) singularities. We are coming to the topological problems of studying such foliations. It has been posed in [5]. Several participants of my seminar (A. Zorich, Le Tu Thang, L .Alania) have made very important contributions to the study of this subject. Interesting quasiperiodic structure appears here. It is not revealed fully in my opinion (see references and discussions in the article [9]).

Multivalued Hamiltonians in real physics. I started to look around in 1982 asking the following question: can you find such systems in real physics where Hamiltonian or some other important integral of motion is multivalued (i.e. dH is well-defined as a closed 1-form)? Much later people realized that in the theory of the so-called Landau-Lifshitz equation (which is a well-known physical integrable system with zero-curvature representation elliptic in the spectral parameter) the momentum is a multivalued functional.

At that time (1982) I found only one such system describing motion of the quantum ("Bloch") electron in the single crystal D-dimensional normal metal (D=1,2,3) under the influence of the homogeneous magnetic field B. We are working here with one-particle approximation for the system of Fermi particles whose temperature is low enough. For the zero temperature our electrons fill in all one-particle quantum Bloch states ψ_p below the so-called "Fermi Level" $\epsilon \leq \epsilon_F$. Its value depends on the number of electrons in the system. It is the intrinsic characteristic of our metal. The index p here may be considered finally as a point in the torus T^D defined by the reciprocal lattice dual to the crystallographic one

$$p \in T^D, \quad T^D = R^3/\Gamma^*.$$

There is a Morse function $\epsilon(p) : T^D \to R$ (dispersion relation) such that the domain $\epsilon \leq \epsilon_F$ in the torus T^D is filled in by Bloch electrons.

Its boundary $\epsilon = \epsilon_F$ is a closed surface $M_F \subset T^D$ for $D = 3$. We call it a **Fermi Surface**. It is homologous to zero in the group $H_2(T^3, Z)$. For finite but very small temperature all essential events are happening nearby the Fermi Surface.

Add now a homogeneous magnetic field to our system (i.e. put metal in the magnetic field B). Nobody succeeded in constructing a suitable well-founded theory for the exact description of electrons in the magnetic field and lattice. Irrational phenomena appear in the spectral theory of Schrödinger operators and destroy the whole geometric picture. However, since the late 50s physicists have used some sort of adiabatic approximation which they call "semiclassical". Let me warn you that this approximation has nothing to do with the standard understanding of semiclassical approximation. We take dispersion relation $\epsilon(p)$ as a function on the torus T^3 extracted from the exact solution of the one-particle Schrödinger operator in the lattice without magnetic field. We consider a phase space $T^3 \times R^3$ with coordinates $p_i, x^j, i, j = 1, 2, 3$ and Poisson bracket of the form,

$$\{x^j, x^k\} = 0, \quad \{x^j, p_k\} = \delta_k^j, \quad \{p_j, p_k\} = B_{jk},$$

where $B_{jk}(x)$ are components of the magnetic field B treated as a 2-form. Our space R^3 is Euclidean, so we can treat the magnetic field as a vector B with components B^j.

Take now the function $\epsilon(p)$ as a Hamiltonian. It generates through the Poisson structure above a Hamiltonian system in the phase space $T^3 \times R^3$. For the homogeneous (i.e. constant) magnetic field we can see that our phase space projects on the torus T^3 with Poisson bracket $\{p_j, p_k\} = B_{jk}$. This Poisson bracket has a Casimir (Annihilator) $C_B(p) = \epsilon^{ijk} p_i B_{jk} = B^i p_i$. This Casimir is multivalued: it is defined by the closed 1-form $\omega_B = \sum_i B^i dp_i$ on the torus. As you will see, this is the main reason for the appearance of nontrivial topological phenomena in this problem.

Our Hamiltonian $H = \epsilon(p)$ depends on the variable p only. Therefore all important information can be extracted from the Hamiltonian system on the 3-torus with Poisson bracket defined by the magnetic field. The electron trajectories for the low temperature can be described as a curves in this torus such that

$$\epsilon(p) = \epsilon_F, C_B(p) = const.$$

However, the levels of the Casimir on a Fermi surface are in fact leaves of foliation given by the closed 1-form restricted on the Fermi surface

$$\omega_B|_{M_F} = \sum_i B^i dp_i|_{M_F} = 0 \,.$$

Some people in ergodic theory studied in fact the most generic ergodic properties of "foliations with transversal measure" on the Riemann surfaces. In a sense, our situation is a partial case of that. However, our picture in 3-torus is nongeneric in that sense. We cannot apply any results of that theory. We have to work with foliations obtained in the 3-torus by this special procedure only. Our use of word "generic" here is restricted by that requirement. As we shall see, ergodicity is a nongeneric property within this physically realizable subclass of foliations 2-surfaces given by the closed 1-form. What is interesting is that ergodic examples exist in our picture but they occupy a measure zero subset on the sphere of directions of the magnetic fields (if generic Fermi surface is fixed).

As I realized in 1982 (see [5]), this picture leads to nontrivial 3-dimensional topology, and I posed it as a purely topological problem to my students. The first beautiful topological observation was made by A. Zorich [8] for the magnetic fields closed to the rational one. After new discussion and reconsidering all conjectures (see [9]), I. Dynnikov made a decisive breakthrough in the topological understanding of this problem for the generic directions of magnetic fields (see [10]). S. Tsarev constructed in 1992 the first nontrivial ergodic examples, later improved by Dynnikov (see [14]).

However, several years passed before some physical results were obtained (see the first remark about the possibility of that in my article [11]). We made a series of joint works with A. Maltsev (see [12, 13]) dedicated to physical applications. Essentially, we borrowed topological results from the works of Zorich and Dynnikov. However, the needs of applications required that we not apply their theorems directly, but extract the key points from the proofs and reformulate them. So the modern topological formulations of these results are by-products of these works with applications (see the most modern survey in [14]). Let me formulate here our main physical results and after that explain the topological background and generalizations.

This picture has been extensively used in solid state physics since the late 50s. The leading theoretical school in that area has been the Kharkov–Moscow school of I. Lifshitz and his pupils, like M. Azbel, M. Kaganov, V. Peschanski, A. Sludskin and others. You may find all proper quotations to physics literature in the survey article [13]. The following fundamental **Geometric Strong Magnetic Field Limit** was formulated by that school (and fully accepted later by the physics community):

All essential phenomena in the conductivity of normal metals in a strong magnetic field should follow from the geometry of the dynamical system described above.

How to understand this principle? You have to take into account that this picture certainly will be destroyed by the "very strong" magnetic field where quantum phenomena (of the magnetic origin) are important. It should happen for such magnetic fields that magnetic flux through the elementary lattice cell is comparable with quantum unit. However, the lattice cell in solid state physics is so small that you need for that magnetic field the order of magnitude $B \sim 10^8 Gauss$ or $B \sim 10^4 t$ where $1t$ is equal to $10^4 Gauss$.

Therefore we are coming to the conclusion that even for the "real strong" magnetic fields like $10^2 t$ this picture still works well.

For our goal we need to consider such metals that a Fermi surface is topologically nontrivial. It means precisely that the imbedding homomorphism of fundamental groups

$$\pi_1(M_F) \to \pi_1(T^3) = Z^3$$

is onto. As people have known already for many years, the noble metals like copper, gold, platinum and others satisfy this requirement. Probably the very first time this property was found was by Pippard in 1956 for copper. Many other materials with really complicated Fermi surfaces are known now.

By definition, the electron orbit is **compact** if it is periodic and homotopic to zero in T^3. Therefore it remains compact on the covering surface in R^3, where R^3 is a universal covering space over the torus T^3. All other types of trajectories will be called **noncompact**.

Normally all pictures in physics literature are drawn in R^3, but everybody knows that quasimomentum vectors $p_1, p_2 \in R^3$, such that $p_1 - p_2$ belongs to the reciprocal lattice Γ^*, are physically identical.

The Lifshitz group started to study this dynamical system about 1960 and made the first important progress. For example, Lifshitz and Peschanski found some nontrivial examples of noncompact orbits stable under the variation of the direction of magnetic field. It looks like nobody could understand them properly in the physics community at that time. It was several decades before this community started to understand the geometry of dynamical systems. The Lifshitz group was ahead of its time. They made some mistakes leading to wrong conclusions and investigations were stopped. You may find the detailed discussion in our survey article [13]. Their mistakes have been found only now because they contradicted our final results describing the conductivity tensor.

Our main results. Consider projection of the conductivity tensor on the direction orthogonal to magnetic field. This is a 2×2 tensor σ_B. Applying any weak electric field E orthogonal to B, we get current j. Its projection $\sigma_B(E)$ orthogonal to B is only what is interesting for us now. We claim that for the strong magnetic field $|B| \to \infty$ of the generic direction in S^2 only two types of asymptotics are possible.

Topologically Trivial Type.

$$\sigma_B \to 0, \quad |B| \to \infty.$$

More exactly, we have $\sigma_B = O(|B|^{-1})$ for the topologically trivial type. All directions with trivial type occupy a set U_0 of measure equal to μ_0 on the two-sphere $U_0 \subset S^2$.

Topologically Nontrivial Type.

$$\sigma_B \to \sigma_B^0 + O(|B|^{-1}).$$

Here 2×2 tensor σ_B^0 is a nontrivial limit for the conductivity tensor. We claim that it has only one nonzero eigenvalue on the plane orthogonal to B. Let us describe the topological properties of this limiting conductivity tensor. It has exactly one eigen-direction $\eta = \eta_B$ with eigenvalue equal to zero. Consider any small variation B' of the magnetic field B. For the new field B' we have an analogous picture if perturbation is small enough. We have a new 2×2 tensor $\sigma_{B'}$ with one zero eigen-direction $\eta_{B'} = \eta'$. Our statement is that the plane $a\eta + b\eta', a, b \in R$, generated by this pair of directions, is locally stable under the variations of magnetic field. This plane is integral (i.e. generated by two reciprocal lattice vectors). It contains zero eigen-directions $\eta_{B''}$ for all small variations of the magnetic field B. It can be characterized by 3 relatively prime integer numbers $m = (m_1, m_2 m_3)$. This triple of integer numbers is a measurable topological invariant of the conductivity tensor. An open set of directions $U_m \subset S^2$ with measure μ_m corresponds to this type. The total measure of all these types is full:

$$\mu_0 + \sum_{m \in Z^3} \mu_m = 4\pi.$$

We started to look in the old experimental data obtained in the Kapitza Institute in the 60-s by Gaidukov and others (see references in [13]). They

measured resistance for the single crystal gold samples in the magnetic field
about 2t-4t following the suggestion of Lifshitz. Confirming the ideas of the
Lifshitz group, several domains with nonisotropic behavior of conductivity
were found and many suspicious "black" dots (maybe domains of small
size) on the sphere S^2. It is not hard to see even now that several larger
domains in these data with nonisotropic conductivity should correspond to
the simplest stable topological types like $(\pm 1, 0, 0), (\pm 1, \pm 1, 0), (\pm 1, \pm 1, \pm 1)$
up to permutation in the natural basis of this cubic lattice. However,
for good checking it would be nice to increase magnetic field to 20t-40t
for a more decisive conclusion. The black dots either correspond to the
smaller domains with larger values of the topological integers or to some
ergodic regimes occupying measure zero set on the sphere. For the final
decision these experiments should be repeated and increased about 10 times
magnetic field and smaller temperature like $10^{-2}K$.

Let me explain now the topological background of these results. Consider the generic Morse function $\epsilon : T^3 \to R$ and its generic nonsingular
level $M_F \subset T^3, \epsilon = \epsilon_F$ in the torus and in the covering space $M' \subset R^3$. We
call the surface $M' \subset R^3$ **a periodic surface**. Apply now generic magnetic
field B and make the following construction:

Remove all nonsingular compact trajectories (NCT) from the periodic
surface M' and its image M_F in the torus. The remaining part is exactly
some surface with boundary if it is nonempty:

$$M_F \backslash (NCT) = \bigcup M_i$$

(i.e. Fermi surface minus all NCT is equal to the union of surfaces with
boundary). We call these surfaces M_i and their closure below **the Carriers of Open Trajectories**. All boundary curves are the separatrix type
trajectories homotopic to zero in T^3. They bound 2-discs in the corresponding planes orthogonal to magnetic field B. Let us fill them by these discs in
the planes. We get closed piecewise-smooth surfaces $\bar{M}_i$. We denote their
homological classes by $z_i \in H_2(T^3, Z)$.

We use the following extract from the proofs of the main theorems of
Zorich and Dynnikov (see [8, 10]; their theorems have not been formulated
in that way, but you may extract these key points from the proofs):

**In the generic case all these homology classes are nontrivial
and equal to each other up to sign $0 \neq z_i = \pm z \in H_2(T^3, Z)$ where
z is some indivisible class in this group. All these closed surfaces
have a genus equal to 1.**

As you may see, this statement means in fact some kind of the "Topological Complete Integrability" of our systems on the Fermi surfaces for the generic magnetic field.

For obtaining our final result on the conductivity tensor, we need to use the Kinetic Equation for the quasi-particles based on Bloch waves nearby the Fermi level. This equation has been used a lot by solid state physicists for the past 30 years. For the small (but nonzero) temperature, strong magnetic field and appropriate general assumptions on the impurities, the motion of quasi-particles concentrates along the electron trajectories above. This fact leads to our conclusions. Despite the fact that this theory is considered a well established one already for many years in the physics community, any attempt to prove such things as the rigorous mathematical theorems would be a huge mess. As we see, our final conclusion is separated from all theorems by some gap which cannot be eliminated. Let me point out that it is always so. "Rigorous proofs" in mathematical physics never prove anything in real world physics.

What about nongeneric trajectories? Tsarev and Dynnikov constructed very interesting examples where genus of carriers of the open trajectories is larger than 1 (see[14]). We call such cases **stochastic**. Sometimes we call them **ergodic**. There were some attempts to extract from their properties highly nontrivial asymptotics of the conductivity tensor in the strong magnetic field [15]. However, these attempts need a better understanding of the properties of such trajectories. We have to answer the following questions:

1. How many directions of the magnetic field on the sphere S^2 admit ergodic trajectories?

According to my conjecture, for the generic Fermi surface, this set of directions has a Hausdorf dimension not greater than some number $a < 1$ on the sphere S^2. For the special Fermi surfaces $\epsilon = 0$ of the even functions like $\cos p_1 + \cos p_2 + \cos p_3 = 0$, we expect to have ergodic trajectories for the set of directions with Hausdorf dimension like $1 < a < 2$. Dynnikov started to investigate this example in his Thesis and proved several general properties. Recently R. Deleo investigated such examples more carefully and performed more detailed calculations ([16]). His results confirm our conjectures. However, the Hausdorf dimension of this set has been unknown in this example until now.

2. Which geometric properties does a "typical" ergodic trajectory have?

According to the conjecture of Maltsev, these trajectories are typically the "asymptotically self-similar" plane curves in the natural sense. His idea (if it is true) leads to the interesting unusual properties of the asymptotic conductivity tensor. Anyway, this problem is very interesting.

Dynnikov investigated also the dependence of these invariants on the level ϵ_F of the dispersion relation (see [14]). These results are useful for the right understanding of our conjectures.

Multidimensional generalizations. Consider the following **problem:** What can be said about topology of the levels $f(x, y) = const$ of the quasiperiodic functions with m periods on the plane x, y?

For the case $m = 3$ this problem exactly coincides with our subject above: By definition, a quasiperiodic function on the plane is a restriction on the plane $R^2 \subset R^m$ of the m-periodic function. Our space R^3 was a space of quasi-momenta (more precisely, its universal covering). Our plane was orthogonal to the magnetic field. Can this theory be generalized to the case $m > 3$? According to my conjecture, it can be generalized to the case $m = 4$. I think that for small perturbations of the rational directions this theory can be generalized to any value of m. We consider now any 4-periodic function $f : R^4 \to T^4 \to R$ and pair of the rational directions l_1^0, l_2^0 corresponding to some lattice Z^4 in R^4.

Let me formulate the following theorem.

Theorem. *There exist two nonempty open sets U_1, U_2 on the sphere S_3 containing the rational directions l_1^0, l_2^0 correspondingly such that:*

For every plane $R_l^2 \subset R^4$ from the family given by 2 equations $l_1 = const, l_2 = const$, the quasiperiodic functions f_l have only the following two types of connectivity components of the levels $f_l = const$ on the plane R_l^2.

1. *The connectivity component of the level is a compact closed curve on the plane.*

2. *The connectivity component of the level is an open curve lying in the strip of finite width between 2 parallel straight lines with the common direction η. This situation is stable in the following sense. After any small variations of the directions $l_1 \in U_1, l_2 \in U_2$, of function f on T^4 or the level we still have such open component with direction η'. For all possible perturbations this set of directions $\eta, \eta', \ldots$ belong to some integral 3-hyperplane in R^4.*

This property can be formulated in terms of the integral homology class in the group $H_3(T^4, Z)$ and of the torical topology of the carriers of the open trajectories. The idea of the proof was recently published by the author in [17].

We may reformulate this problem in terms of Hamiltonian systems. Let the constant Poisson Bracket B_{ij} be given on the torus T^m whose rank is equal to 2. Any Hamiltonian f generates such systems whose trajectories are equal to the levels of f on the planes. Our theorem means that in these cases this Hamiltonian system is Completely Integrable in the specific topological sense described above.

References

[1] S. Novikov, Topology-I. Encyclopedia of Mathematical Sciences, vol. 12, Springer-Verlag, Berlin-Heidelberg-New-York.

[2] S. Novikov, Bloch functions in a magnetic field and vector bundles. Typical dispersion relations and their quantum numbers, Doklady AN SSSR 27:3 (1981), 538–543.

[3] S. Novikov, I. Schmelzer, Periodic solution of the Kirchhoff equations, Functional Analysis Appl. 15:3 (1981), 54–66.

[4] S. Novikov, Multivalued functions and functionals, Doklady AN SSSR 260:1 (1981), 31–35.

[5] S. Novikov, The Hamiltonian formalism and multivalued analog of the Morse theory, Russian Math Surveys 37:5 (1982), 3–49.

[6] J. Duistermaat, G. Heckmann, On the variation in the cohomology of the symplectic form of the reduced phase space, Inventiones Math. 69 (1982), 259–269.

[7] P. Grinevich, S. Novikov, Nonselfintersecting magnetic orbits on the plane. Proof of the overthrowing of the cycles principle, American Math Society Translations Ser 2, 170 (1995) (Topics in Topology and Math Physics, edited by S. Novikov) 59–82.

[8] A. Zorich, Novikov's problem on the semiclassical electron in the homogeneous magnetic field, Russian Math Surveys 39:5 (1984), 235–236.

[9] S. Novikov, Quasiperiodic structures in topology. Proceedings of the Conference "Topological Methods in Modern Mathematics-Stonybrook,June 1991", Stonybrook University, 1993.

[10] I. Dynnikov, Proof of Novikov conjecture on the semiclassical electron, Math. Zametki 53:5 (1993), 57–68.

[11] S. Novikov The semiclassical electron in a magnetic field and lattice. Some problems of low-dimensional "periodic topology", Geometric And Functional Analysis 5:2 (1995), 433–444.

[12] S. Novikov, A. Maltsev, Topological quantum characteristics observed of the conductivity in normal metals, Letters of JETP 63:10 (1996), 809–813; Translated into English by the American Institute of Physics.

[13] S. Novikov, A. Maltsev, Topological phenomena in the normal metals, Uspekhi Phys. Nauk 168:3 (1998), 249–258; translated in English by the

American Institute of Physics.

[14] I. DYNNIKOV, Geometry of stability zones in the Novikov problem on the semiclassical motion of an electron, Russian Math Surveys 54:1 (1998), 21–60.

[15] I. DYNNIKOV, A. MALTSEV, JETP-Journ. Experimental and Theoretical Physics 112 (1997), 371–376.

[16] R. DELEO, Existence and measure of ergodic leaves in Novikov's problem on the semiclassical motion of an electron, Russian Math Surveys 54:6 (1999).

[17] S. NOVIKOV, The levels of Quasiperiodic functions on the plane and topology, Russian Math Surveys 54:5 (1999).

Sergey P. Novikov, Math Department and IPST, University of Maryland, College Park, MD 20742-2431, USA
and
Landau Institute for Theoretical Physics, Moscow 117940, Kosygina 2;

GAFA, Geom. funct. anal.
Special Volume – GAFA2000, 425 – 433
1016-443X/00/S10425-9 $ 1.50+0.20/0

I GAFA Geometric And Functional Analysis

SOME PROBLEMS IN THE THEORY OF DYNAMICAL SYSTEMS AND MATHEMATICAL PHYSICS

Ya.G. Sinai

The problems considered below are not new. They have been discussed before in many publications and important results have been obtained. However, it seems that an adequate understanding is still lacking and serious progress is expected which could be even more important than the initial goals.

A brief look at the history of the theory of dynamical systems over the last fifty years shows that it developed in bursts around the great discoveries, when new ideas, methods and phenomena appeared. Let me mention these discoveries explicitly (in more or less chronological order):

1. KAM-theory.
2. The works of Smale, Anosov and others on the structural stability and hyperbolicity of dynamical systems, strange attractors and the Lorenz model, and hyperbolic billiards.
3. Entropy theory which started with the classical work by Kolmogorov and ended with the famous works of Ornstein and his coauthors.
4. Feigenbaum universality in period-doubling bifurcations.

Remarkable results were obtained in other directions such as non-local bifurcations, holomorphic dynamics, flows on surfaces, ergodic theory on homogeneous spaces, billiards but it seems that 1–4 were dominating events.

Fifty years ago it was absolutely impossible to anticipate any of these four breakthroughs. This unpredictability is one of the attractive features of the theory of dynamical systems. For this reason, on the occasion of this Conference one can only hope to indicate some directions where new and serious progress looks possible. The text below presents some suggestions.

1 Generically Quasi-periodic and Hyperbolic Dynamics are the Only Structurally Stable Types of Dynamics

The problem goes back to Kolmogorov. In the case of Hamiltonian dynamics a positive answer would mean that the phase space can be decomposed onto subsets of positive measure consisting of invariant tori with quasi-periodic motions and stochastic layers where at least some Lyapunov exponents are non-zero. The statement of this type is important for applications in differential geometry (the structure of geodesic flows), astronomy, statistical mechanics and other fields using Hamiltonian dynamics.

For dissipative systems some version of this problem was formulated recently as the whole program by J. Palis (see [P])

There are some results which can be considered as related to this problem.

First of all I mean one-dimensional dynamics where due to the important results of Jakobson [J], Benedicks and Carleson [BC], Lyubich [Ly], Graczyk and Swiatek [GrS], we know that in analytic families of one-dimensional maps the space of parameters can be decomposed onto a subset where the dynamics has a stable periodic orbit, a subset where the dynamics has an absolutely continuous invariant measure and is chaotic and a negligible subset of measure zero.

KAM-theory gives the conditions under which the invariant tori with quasi-periodic dynamics persist in Hamiltonian systems. In spite of a recent activity around the construction of these tori it is hard to expect new discoveries based on this technique even in its modernized form. On the other hand, many problems which grew out of KAM-theory remain completely open. The most notable one is the problem of classification of KAM-islands. It is highly non-trivial even within the class of two-dimensional twist maps, i.e. symplectic maps of the two-dimensional cylinder $T(z, \varphi) = (z', \varphi')$, $z' = z + kV'(\varphi)$, $\varphi' = \varphi + z'$. Here V is a smooth periodic function of φ, k is a parameter. It is widely believed that the families of twist maps are the next candidates for deep study after the families of one-dimensional maps. One of the non-direct consequences of KAM-theory is the existence for generic twist maps of KAM-islands, i.e. invariant or periodic domains bounded by invariant or periodic curves. Near the islands one can find similar islands of a smaller scale and so on (see Figure 1). We clearly see a hierarchical structure of the islands with some scale invariance. The basic problems here are the following:

1) to establish the existence of this hierarchical structure;
2) to describe rotation numbers for the induced maps on the boundaries of the islands;
3) to estimate the global measure of KAM-islands as $k \to \infty$.

Apparently, 1) can be studied with the help of renormalization group technique because in typical situations asymptotically the structure of the

Figure 1

corresponding power of the map near any small island depends only very weakly on the structure of the initial map and belongs to some class of universality. These ideas were proposed in the works of Lichtenberg [L], Zaslavsky [Z], Melnikov [Me] and others. The renormalization group approach to the construction of KAM-tori can be found in [AK], [KS] (in [AK] one can find other related references).

Problem 2) is connected with another important question. Consider an invariant curve Γ given by the equation $z = f_\Gamma(\varphi)$. If Γ is a KAM-curve then it is not isolated and is in a natural sense a density point of other KAM-curves. Therefore boundary curves which separate stochastic layers from KAM-curves should be non-smooth. On the other hand, they are "last" curves with given rotation numbers and this is their characteristic feature (see Figure 2.) With the change of parameter they bifurcate into Aubry-Mather sets. Thus another problem along these lines is

   4) to construct or to prove the existence of non-smooth invariant curves of twist maps.

At the moment the only hope to attack this problem is again the renormalization group theory. Chirikov [Ch] stresses that rotation numbers of induced maps on these curves should be special. In this connection one can formulate the next problem:

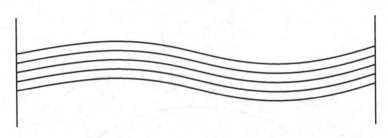

Figure 2

5) Why the rotation number of the last invariant curve is a golden mean.

It is easy to show that for large enough k twist maps have no invariant curves Γ described by the equations $z = f_\Gamma(\varphi)$ (see for example, [Si]). Therefore there exists a critical value k_{cr} such that for $k > k_{cr}$ there are no such curves. It was a numerical discovery by J. Greene [Gre] that for some twist maps the rotation number on the last curve is a golden mean. R. MacKay [M] proposed a renormalization group technique to study these curves.

A satisfactory answer to the last problem can explain why the golden mean appears so often in mathematics and not only in mathematics (architecture, construction, art,...).

Problem 3) is connected with the following well-known hypothesis:

Hypothesis 1. *For any $k \neq 0$ the twist map has positive metric entropy.*

Hypothesis 2. *For generic V the set of k for which T has no islands at all and is ergodic has density 1.*

This hypothesis is motivated by analogy with the Jakobson theorem in one-dimensional dynamics and Benedicks–Carleson results for Henon map. It means that, for k in Hypothesis 2, the map T is hyperbolic but the hyperbolicity should be very weak. In particular, stable and unstable manifolds are not transversal everywhere and can have tangency up to some order. It would be interesting to show that typically the tangency is of finite order. However, even the precise formulation of an exact statement can be non-trivial. One should mention the result by Duarte [D] who showed that the set of k for which T has KAM-islands is open and everywhere dense.

The extension of these problems to the multidimensional setting is certainly very important but apparently very difficult.

2   Qualitative Theory of PDE and Infinite-Dimensional Systems

It seems that this theory will be different from its finite-dimensional counterpart started by Poincaré, Denjoy, Lyapunov and others. So far the concept of PDE as a vector field on a topological manifold was not so fruitful, probably, because the topology of the phase space did not play an essential role. On the other hand, except general questions of existence, uniqueness and smoothness of solutions the existing theory is much more concrete. There are relatively few PDE which are important for applications: the Navier–Stokes and other hydrodynamical systems, the Korteweg–de-Vries equation, the KP-equation, the Ginzburg–Landau equation, the Kuramoto–Sivashinski equation, reaction-diffusion equations, the equation of general relativity, mean curvature flows and a few others. Basic qualitative and numerical results are concentrated around the construction of various types of space patterns with different geometry (rolls, dislocations and others). The transition from one type of pattern to another seems to be a new type of bifurcation. The basic problem can be formulated as follows:

1) to classify spatial patterns which are displayed by PDE and to describe possible types of their bifurcations. Some results in this direction can be found in the book by Golubitsky and Schaeffer [GS].

   In previous years there were many examples of computer-assisted proofs in the theory of dynamical systems. The most notable one was the proof by Lanford of the existence of a Feigenbaum–Collet–Tresser fixed point in the theory of period-doubling bifurcations. Recently Tucker gave a computer-assisted proof of the existence of Lorenz attractor (see [T]). However in general this direction did not attract much attention. People usually do not want to provide detailed mathematical arguments if numerical evidence is convincing enough. It is quite possible that in the case of PDE rigorous results will go together with serious numerical studies. It is already clear that complete proofs require in many cases the most powerful computers.

   Navier–Stokes equations deserve special attention. The existence problem of strong solutions for a 3D-Navier–Stokes system still remains an open outstanding problem in spite of the efforts of many researchers having different opinions about the answer. If we adopt the point of view that blow-ups in finite time are possible even in the presence of viscosity, then we can take the next step and ask which flows with finite energy and infinite enstrophy can be limits of local solutions of a 3D-Navier–Stokes system. Probably this formulation of the basic problem is more amenable to direct attacks since it is local. The problem of turbulence will be in the center of interest of many mathematicians, physicists, and engineers. For mathematicians one of the main problems is to formalize notions which are widely used by physicists, engineers, etc. Here is one example.

2) To propose definitions of solutions displaying the flow of energy over the spectrum and to prove the existence of such solutions.

It is believed that one of the mathematical models of turbulence is Navier–Stokes system with random forcing. Some results concerning the existence of invariant measure for this model were obtained by Flandoli and Maslowski (see [FM]) and others. However, basic questions remain unanswered. The main problem can be formulated as follows.

3) To construct Markov random fields for Navier–Stokes systems where forcing is a white in time random field and to study the properties of their correlation functions in the limit of zeroth viscosity.

Progress in this direction can lead to the mathematical justification of Kolmogorov's theory of turbulence.

3 Quantum Chaos and Anderson Localization

Quantum chaos as a part of physics appeared about twenty years ago. For mathematicians its main purpose was to describe properties of quantum systems which depend on their classical limits. Problems of quantum chaos are discussed also by P. Sarnak (at this conference). The first mathematical result in quantum chaos was proven by Shnirelman (see [S]) who showed that if a geodesic flow on a smooth compact Riemannian manifold is ergodic then the squares of corresponding eigenfunctions of Laplacian are in some sense uniformly distributed.

Actually problems of quantum chaos in mathematics started even earlier. Assume that Q is a compact smooth d-dimensional Riemannian manifold and $N(\lambda) = \sharp\{\lambda_i \leq \lambda\}$ where λ_i are eigenvalues of the Laplacian. Weyl's famous asymptotics tells us that

$$N(\lambda) = c\lambda^{d/2} + n(\lambda)$$

where c a dimension-dependent constant and $n(\lambda)$ is a remainder.

It is widely believed that the behavior of $n(\lambda)$ is non-universal and is determined by ergodic properties of the geodesic flow on the unit tangent bundle of Q. It follows basically from the works of Colin de Verdiere (see [C], the two-dimensional case in [KoMS]) that in the integrable case the analysis of $n(\lambda)$ can be reduced to number-theoretic problems and the result essentially has the form $n(\lambda) = \lambda^{\frac{d-1}{4}} Q(\lambda) + n_1(\lambda)$ where $Q(\lambda)$ is a quasi-periodic function and $n_1(\lambda)$ has a smaller order of magnitude.

In the case of manifolds of negative curvature where the geodesic flow is strongly mixing numerical and qualitative results indicate that $n(\lambda)$ grows only as $(\log \lambda)^\gamma$ for some γ. This resembles the asymptotics of the number of zeros of the Riemann ζ-function. Other results suggest that the limiting probability distribution of spacings between the nearest eigenvalues behaves similarly to the distribution of spacings in some ensembles of random matrices of growing dimension. Let me mention in this connection results by Hejhal and Rackner (see [HR]) who convincingly showed that the probability distributions of eigenfunctions of Laplacians on such manifolds are Gaussian in the limit $\lambda \to \infty$. All these results lead to the following problem.

1) Why Laplacians on compact manifolds of negative curvature behave as elements of Wigner ensembles of random matrices.
   So far asymptotic expressions for eigenfunctions of Laplacians are known for the above mentioned cases of Q with integrable geodesic flows. Schnirelman's theorem shows that in the case of mixing flows they are much more random.
   Random eigenfunctions appear in a quite different part of mathematical physics, the so-called Anderson localization (AL). AL is a property of Schroedinger operators with random potentials and means that with probability 1 all eigenfunctions decay exponentially at infinity. AL is based on quantum resonances, i.e. of the appearance or disappearance of approximate eigenfunctions with close eigenvalues. One of the main open problems here is

2) to construct for $d \geq 3$ a non-localized eigenfunction belonging to the continuous spectrum.
   It would be interesting to connect the theory of AL and quantum chaos. It looks plausible that the asymptotic of eigenfunctions of Laplacians in mixing cases has something common with AL.

References

[AK]   J. ABAD, H. KOCH, Renormalization and periodic orbits for Hamilton flows, Comm. Math. Physics, submitted.

[BC]   M. BENEDICKS, L. CARLESON, The dynamics of the Menon map, Ann. Math. 133:1 (1991), 73–170.

[C]   B. CHIRIKOV, Private communication.

[Co]   Y. COLIN DE VERDIÉRE, Spectre conjoint d'operateurs pseudo-differentiels qui commutent, Le cas integrable, Math. Zeitschrift 171 (1980), 51–73.

[D]   P. DUARTE, Plenty of elliptic islands for the standard family, Ann. Inst. Henri Poincaré, Phys. Theor. 11 (1994), 359–409.

[FM]   F. FLANDOLI, B. MASLOWSKI, Ergodicity of the 2D-Navier–Stokes equation under random perturbation, Comm. Math. Phys. 171 (1995), 119–141.

[GS]   M. GOLUBITSKY, D. SCHAEFFER, Singularities and Groups in Bifurcation Theory, Vols. 1,2, Springer-Verlag, New York 1985, 1988.

[GrS] J. GRACZYK, G. SWIATEK, Generic hyperbolicity in the logistic family, Ann. of Math. 146 (1997), 1–52.

[Gre] J. GREENE, A method for determining a stochastic transition, J. Math. Phys. 20 (1979), 1183–1201.

[HR]   D. HEJHAL, B. RACKNER, On the topography of Maass waveforms for PSL (2 , Z), Experiment. Math. 1 (1992), 275–305.

[J]   M. JACOBSON, Absolutely continuous invariant measure for certain maps of an interval, Comm. Math. 81 (1981), 17–51.

[KS]   K. KHANIN, YA. SINAI, Renormalization Group Method and the KAM Theory, in "Nonlinear Phenomena in Plasma Physics and Hydrodynamics" (R.Z. Sagdeev, ed.), Moscow, MiR, 1986.

[KoMS] D. KOSYGIN, A. MINASOV, YA. SINAI, Statistical properties of spectre of Laplace-Beltrami operators on Lionville surfaces, Russian Math. Surveys 48:4 (1993), 1–142.

[L]   A.J. LICHTENBERG, Universality from resonance renormalization of Hamiltonian maps, Physica D 14 (1985), 387–394.

[Ly]   M. LYUBICH, Regular and stochastic dynamics in the real quadratic family, Proc. National Academy of Sciences, U.S.A. 95 (1998), 1405–1427.

[M]   R. MACKAY, Renormalization in Area Preserving Maps, World Scientific, 1993.

[Me]   V. MELNIKOV, in "Transport, Chaos and Plasma Physics" 2nd ed. (S. Benkadda, F. Doveil, Y. Elskens, eds.), World Scientific, 1996.

[P]   J. PALIS, A global view of dynamics and a conjecture on the denseness of finitude of attractors, preprint, IMPA, Brazil, (1998).

[S]   A. SHNIRELMAN, On the asymptotic properties of eigen-functions in the regions of chaotic motion, in "KAM theory and semi-classical approximations to Eigen-functions" (V.F. Lazutkin, ed.), Springer-Verlag, Berlin, 1993.

[Si]   YA. SINAI, Topics in Ergodic Theory, Princeton University Press, 1994.

[T]   W. TUCKER, The Lorenz attractor exists, preprint, Department of Mathematics, Uppsala University, S-75106, Uppsala, Sweden.

[Z]   G. ZASLAVSKY, Physics of Chaos in Hamiltonian Systems, Imperial College Press, 1998.

YA.G. SINAI, Department of Mathematics, Princeton University, Princeton, NJ 08544, USA
and
Landau Institute of Theoretical Physics

GAFA, Geom. funct. anal.
Special Volume – GAFA2000, 434 – 453
1016-443X/00/S10434-20 $ 1.50+0.20/0

I GAFA Geometric And Functional Analysis

SOME GEOMETRICAL CONCEPTS ARISING IN HARMONIC ANALYSIS

Elias M. Stein

That geometric considerations enter in a decisive way in many questions of harmonic analysis is by now a well-known fact. In explicit form such ideas arose first in the estimation of the Fourier transform of surface-carried measure; they have since played a key role in averages over lower-dimensional varieties, restriction theorems, in connection with the study of oscillatory integrals and Fourier integral operators, and in application to linear and non-linear dispersive equations.

What we want to consider in this paper, however, is another fundamental occurrence of geometric concepts in this subject. Here our focus will be on singular integrals and the geometry that describes the singularities of the kernels of these operators, and which controls the relevant estimates that are made.

The kind of geometry that arises is "local" in nature and is based, in the first instance, on a "distance" $= d(x,y)$. This metric controls what happens when y is near x; however, the exact size of $d(x,y)$ is not crucial, but what matters is the order of magnitude of d as $y \to x$.

According to past experience, for each kind of singular integral that arose, one could find related to it such a metric (more precisely, an equivalence class of such metrics), which was used in describing the underlying framework and expressing all the estimates that arose. This was a very satisfactory state of affairs. However, more recently we have begun to realize (in view of some interesting examples which occur in sub-elliptic p.d.e. and several complex variables) that matters are not always that simple. In fact it may well happen that there appear two or more distinct and conflicting metrics, each arising naturally from the context at hand, and each describing different aspects of the singular integral in question. Such singular integral operators are by their nature more complex than those hitherto studied. In particular, the Calderón–Zygmund paradigm used to prove L^p estimates is no longer applicable. Thus the problem that needs to be addressed is how we can reformulate our earlier views about singular integrals to take these more general circumstances into account.

In its most general form, a singular integral operator is given as a linear mapping, $f \mapsto T(f)$ of the kind

$$T(f)(x) = \int_M K(x,y)f(y)dy. \qquad (0.1)$$

Here M is a manifold and the measure dy has a smooth sensity. The kernel K is a distribution on $M \times M$ which is "singular" at least on the diagonal (and possibly on other sub-varieties of $M \times M$), but is otherwise smooth. While the study of such operators represents an extremely broad and diverse subject, our attention here will be more narrowly focused on the problems alluded to above. The issues we want to deal with are connected with the following questions:

(I) How do such operators arise in applications?
(II) What is the underlying "geometry"?
(III) How are these operators (and their kernels) described by this geometry?
(IV) How does one prove L^2 (or L^p) regularity for these operators?

We shall organize our exposition by describing three different stages of development of these questions: First, a review of some past results; next, our present point of view in terms of several examples of interest where we shall describe some joint work with A. Nagel and F. Ricci which is in progress; and finally we will indulge in speculation about some future directions of the subject.

1   "Classical Theory"

This theory can be thought as originating about a half-century ago with work of Mihlin, Calderón, and Zygmund. It has since undergone numerous developments. Of particular note are the further ideas introduced to study the Cauchy integral (on Lipschitz curves), and the subsequent general formulation referred to as the "$T(1)$ theorem."[1] The essence of that theorem can be stated in a somewhat simplified form as follows. We consider an operator T as in (0.1), where $M = \mathbb{R}^n$, $dy =$ Lebesgue measure, and the distribution K is C^∞ away from the diagonal.[2] We make two assumptions
(a) *Differential inequalities:*

$$\left| \partial_x^\alpha \partial_y^\beta K(x,y) \right| \lesssim |x-y|^{-n-|\alpha|-|\beta|}; \qquad (1.1)$$

---

[1] See the expositions in Meyer [Me], Stein [S]; also David, Journé and Semmes [DJS].
[2] C^∞ regularity can be relaxed to C^1 or even less.

(b) *Cancellation conditions*

$$\|T(\varphi)\|_{L^2} \lesssim r^{n/2} \quad \text{and} \quad \|T^*(\varphi)\|_{L^2} \lesssim r^{n/2}. \tag{1.2}$$

This is to hold for any φ which is a "normalized bump function" associated to a ball $B = \{x \in \mathbb{R}^n : |x - x_0| < r\}$ of radius r; i.e. we require only that φ is supported in B, and $\sup_B |\varphi| + r \sup_B |\nabla \varphi| \leqslant 1$

Note that here the underlying geometry is the obvious one – it is given by the Euclidean distance $d(x, y) = |x - y|$.

The main result for such operators is their L^2 (and L^p, $1 < p < \infty$) boundedness. This is proved, as is well known, in two steps. First one deals with the L^2 regularity. In the translation-invariant setting (i.e. when $K(x, y) = K(x - y)$), the Fourier transform is the natural tool to use. This is also the case when one considers "variable coefficient" versions of these operators (i.e. pseudo-differential operators). However for the more general situation envisaged above other tools must be used, in particular arguments involving "almost orthogonality."

Once the L^2 boundedness is established, the L^p regularity is then proved in a series of steps which is commonly referred to as the *Calderón–Zygmund paradigm*. This approach involves a covering argument for disjoint balls, and a decomposition of a given function into a "good" part and "bad" part supported on the union of these balls.[3]

These results can be localized and extended to the situation where M is a (compact) manifold, and the geometry is that given by a suitable Riemannian metric. This extension is particularly relevant for applications. We shall briefly recall two of the most fundamental examples:

(a) *The Cauchy integral in* $\mathbb{C}^1$. Here $M = \gamma$ is a suitably smooth simple closed curve in $\mathbb{C}$, which divides the plane into two regions: Ω_+ the interior, and Ω_- the exterior. For f given on γ, we obtain a pair of holomorphic functions $F_\pm$ defined in $\Omega_\pm$ respectively by $F(z) = \frac{1}{2\pi i} \int_\gamma \frac{f(t)}{t-z} dt$. Taking the boundary radius of $F_\pm$ we obtain a pair of singular integral operators, $f \to T_\pm(f)$. Recall that if γ is the real axis (with Ω_+ the upper half-space), then $T_+ = \frac{(I+iH)}{2}$, where H is the Hilbert transform.

(b) *Elliptic p.d.e.* We limit ourselves to the simplest situations. First: "interior estimates" for a second-order elliptic operator L, given by

$$L(u) = \sum_{i \leqslant i,j \leqslant n} a_{ij}(x) \frac{\partial^2 u}{\partial x_i \partial x_j}$$

---

[3] A general formulation of this kind of argument is in Coifman and Weiss [CoW], and Stein [S, Chap. 1].

with $\{a_{ij}(x)\}$ a smooth real symmetric positive-definite matrix. Using a fundamental solution operator K (i.e. where $KL = I + E$, with E infinitely smoothing), one can then show that the operators $T = \frac{\partial^2}{\partial x_i \partial x_j} K$ are singular integrals of the above kind.

Singular integrals also arise when studying *coercive* boundary value problems for L. We mention here a very special instance because a variant will arise later in §5. We take $M = \partial\Omega$, where Ω is a bounded domain in $\mathbb{R}^n$ with smooth boundary. We then consider the "Dirichlet to Neumann" operator $F \mapsto D(f)$ defined as follows: $D(f) = \frac{\partial u}{\partial \nu}\big|_M$, for u satisfying $L(u) = 0$, with $u|_M = f$ (where $\partial/\partial\nu$ is the normal derivative at the boundary). The operator D can be inverted modulo an infinitely smoothing operator. Writing D^{-1} for such a relative inverse, then the operator XD^{-1}, is a singular integral (with conditions analogous to (1.1) and (1.2)), whenever X is a smooth vector field on M.

2   Extensions; The Setting

Starting in the late 1960's an increasing number of suggestive examples came to light which lead to a recasting of the theory so as not to be bound by the "elliptic" (or Riemannian) framework described above. From our particular point of view, this development can be schematized as follows:

$\mathbb{R}^n \to$ nilpotent groups with dilations $\to$ geometry associated to a collection of vector fields.

The heading $\mathbb{R}^n$ signifies the (translation-invariant) theory in $\mathbb{R}^n$ alluded to in §1. The nilpotent groups are Lie groups equipped with dilations (one-parameter group of automorphisms); these groups were originally referred to as "homogeneous" or "graded."[4] The collection of vector fields are given in some neighborhood in $\mathbb{R}^n$, and the Lie algebra they generate is assumed to span. The resulting geometry is then the "control" geometry we shall describe below in §3.

We now list a number of examples of the associated singular integrals:

(a) *Intertwining operators.* We let G be a semisimple Lie group of non-compact type with Iwasawa decomposition $G = KAN$ and associated symmetric space G/K. A distinguished part of its boundary can be identified with the nilpotent group N. Dilations on N are of the form $n \mapsto a_t n a_t^{-1}$, where $\{a_t\}$ is an appropriate one-parameter sub-group of A. The intertwining operators of the principal series of representations of G can be realized

---

[4]More recently they have been dubbed "Carnot–Caratheodory groups."

as "singular integrals" on N, and in this setting their L^2 study came to the fore.[5] However the interest of the operators was not limited to representation theory, but was also relevant in complex analysis. In fact when $G = SU(n+1,1)$ the corresponding symmetric space is the complex unit ball in $\mathbb{C}^{n+1}$, and N is the Heisenberg group $\mathbb{H}^n$, which can be naturally identified with the boundary of the ball (with the exception of one point which corresponds to the point of infinity for $\mathbb{H}^n$). It then turns out that the Cauchy–Szegö projection operator is a particular example of such an intertwining operator. Thus the L^p regularity of this operator could be subsumed in the more general theory of singular integrals.[6]

(b) *Analysis on strongly pseudo-convex domains.* The insights regarding the complex ball and the Heisenberg group provide important hints for the study in the more general case of strongly pseudo-convex domains Ω in $\mathbb{C}^{n+1}$. Here one takes $M = \partial\Omega$, and the relevant geometry is the control geometry determined (locally) by vector fields $\{X_j, Y_j\}_{j=1}^n$, so that $\overline{Z}_j = X_j + iY_j$, $1 \leqslant j \leqslant n$, are a basis of the tangential Cauchy–Riemann vector fields on $\partial\Omega$. The focus of attention is then the $\overline{\partial}_b$ complex, the Cauchy–Szegö projection operator C, and the inverse of the Kohn Laplacian $\Box_b = \overline{\partial}_b\overline{\partial}_b^* + \overline{\partial}_b^*\overline{\partial}_b$.[7]

A result of that analysis[8] is that the operator C, as well as $Q(X,Y)\Box_b^{-1}$ are singular integrals attached to the control metric (here $Q(X,Y)$ is a quadratic polynomial in $X_1, \ldots, X_n, Y_1, \ldots, Y_n$).

(c) *The sub-Laplacian.* $\mathcal{L} = \sum_{j=1}^k X_j^2$. The Hörmander operator is determined by real vector fields $X_1, X_2, \ldots, X_k$, whose commutators are assumed to span. As a result $\mathcal{L}$ is hypoelliptic.[9] The further optimal estimates for the regularity of $\mathcal{L}$ are consequences of the following: One can construct a parametrix S (an approximate inverse) for $\mathcal{L}$; (essentially $S\mathcal{L}$ and $\mathcal{L}S$ differ from I by a smoothing operator). Then the sharp estimates follow from the fact that the operators X_iX_jS are singular integrals associated to the control metric determined by $\{X_1, X_2, \ldots, X_k\}$.[10]

We now turn to a description of this control geometry, and the charac-

---

[5]See Knapp and Stein [KS].

[6]See Korányi–Vágy [KoV].

[7]Each $\Box_b$ is an operator on $(0,q)$ forms, and is invertible if $1 \leqslant q \leqslant n-1$, the case to which we limit ourselves.

[8]See Fefferman [F], Folland and Stein [FSt]

[9]See Hörmander [H].

[10]See Rothschild and Stein [RS], Nagel, Stein and Wainger [NSW2], and Fefferman and Sanchez–Calle [FS].

terization of the attached singular integrals.

3   Control Metric and Associated Singular Integrals

Suppose $X_1, \ldots, X_k$ are k given real and smooth vector fields defined in some neighborhood M in $\mathbb{R}^n$, whose commutators span. We then define the control distance, d, as $d(x, y) = \inf T$, so that there is a curve $\gamma(t)$, with $\gamma(0) = x$, $\gamma(T) = y$, and $\dot{\gamma}(t) = \sum_{j=1}^{k} a_j(t) X_j(\gamma(t))$, where $\sum_{j=1}^{k} (a_j(t))^2 \leqslant 1$. The important fact first realized in special cases, but proved in Chow [Ch], is that any two nearby points x and y could be joined by a curve whose tangent points in a direction spanned by these vector fields. That is,

$$d(x, y) < \infty, \quad \text{for } y \text{ near } x. \tag{3.1}$$

We next define $B(x, \delta)$ to be the ball in this metric of radius δ, centered at x,

$$B(x, \delta) = \{y \colon d(x, y) < \delta\}. \tag{3.2}$$

Application of the Calderón–Zygmund paradigm requires the following properties of this metric, stated in terms of those balls:

(a) The engulfing property: $B(x, \delta) \cap B(y, \delta) \neq \phi \Rightarrow B(y, \delta) \subset B(x, 1 < \delta)$. (This follows immediately with $c = 3$, since d is a metric.)

(b) The doubling property:[11] $|B(x, 2\delta)| \leqslant C|B(x, \delta)|$.

To prove (b) and for the further analysis below, one needs an analytic description of these balls.[12] To this we now turn. With $X_1, \ldots, X_k$ our given vector fields, we define a commutator of degree r to be the vector field $X_I = [X_{i_1}[X_{i_2}[\cdots X_{i_r}]]]$, where $I = (i_1, i_2, \ldots, i_r)$, $1 \leqslant i_j \leqslant k$, and degree $(I) = |I| = r$. Our assumption of the spanning of the commutators is then the existence of an integer m, so that the collection $\{X_I\}$, $|I| \leqslant m$, span.

Next, a *frame* $\mathbf{f}$ is defined to be any n-tuple of the X_I, $\mathbf{f} = \{X_{I_1}, X_{I_2}, \ldots, X_{I_n}\}$. We define degree $(\mathbf{f}) = |\mathbf{f}| = |I_1| + |I_2| + \cdots + |I_n|$, and $\det \mathbf{f} = \det(X_{I_1}, X_{I_2}, \ldots, X_{I_n})(x)$. With these we form the basic "invariant", $\Lambda_x(\delta)$, which for each $x \in M$ is a polynomial in δ of degree $\leqslant m$,

---

[11] It has been shown by Nazarov, Treil, and Volberg [NaTV] that in many circumstances one does not need to assume this property. However, because of the nature of our framework (in particular (3.6) below), this property still arises in an essential way in what follows.

[12] For details of the fact given here see Nagel, Stein, and Wainger [NSW1,2]. A different analysis, having interesting connections with asymptotics of eigen values is in Fefferman and Phong [FP], and Fefferman and Sanchez–Calle [FS].

given by

$$\Lambda_x(\delta) = \sum_{|\mathbf{f}| \leqslant m} |\det(\mathbf{f})(x)| \delta^{|\mathbf{f}|} . \tag{3.3}$$

For each x, the behavior of $\Lambda_x(\delta)$ as $\delta \to 0$, determines the local geometry near x. In fact choose for each (x, δ) a frame $\mathbf{f}$ so that $|\det(\mathbf{f})| \delta^{|\mathbf{f}|} \geqslant c\Lambda_x(\delta)$, for some (small) but fixed constant. This can be done by choosing a largest term in (3.3). Then one has:

PROPOSITION.

$$B(x, \delta) \approx \left\{ \exp\left(\sum_{I_j \in \mathbf{f}} a_j X_{I_j} \right)(x) , \quad |a_j| \leqslant \delta^{|I_j|} \right\} \tag{3.4}$$

$$|B(x, \delta)| \approx \Lambda_x(\delta) . \tag{3.5}$$

Conclusion (3.5) immediately implies the doubling property. The assertion (3.4) states that the ball $B(x, \delta)$ has axis length $\delta^{|I_j|}$ in the direction of X_{I_j}. It is important to note that even if x is fixed, the coordinate system (i.e. choice of frame) used to describe the ball may depend on δ; in fact, finitely many different choices of coordinate systems may be needed.[13] These circumstances are illustrated by the following simple, but suggestive, example.

Here $M = \mathbb{R}^2 = \{(x, y)\}$, and we take two vector fields $X = \frac{\partial}{\partial x}$, $Y = x\frac{\partial}{\partial y}$, The only commutation relation is $Z = [X, Y] = \frac{\partial}{\partial y}$. It follows that $\Lambda_x(\delta) = |x|\delta^2 + \delta^3$, and hence

$$B(x, \delta) \approx \begin{cases} \{\exp(aX + bY)(x, y), \ |a| \leqslant \delta, \ |b| \leqslant \delta\}, & \text{if } |x| \geqslant \delta \\ \{\exp(aX + cZ)(x, y), \ |a| \leqslant \delta, \ |c| \leqslant \delta^2\}, & \text{if } |x| \leqslant \delta. \end{cases}$$

We now turn to the analytic consequences of this geometric framework: with M endowed with the metric d arising as above, we consider the operators of the form

$$T(f)(x) = \int_M K(x, y) f(y) dy .$$

The kernel K will be assumed to be smooth away from the diagonal and the following additional properties, with r fixed (the "non-isotropic smoothing" order of T):

$$|X_x^\alpha X_y^\beta K(x, y)| \lesssim \frac{d(x, y)^{r - |\alpha| - |\beta|}}{V(x, y)} . \tag{3.6}$$

---

[13]This fundamental fact seems to have been misunderstood by some later writers. For instance Gromov [Gr, p. 98], mistates (3.4), requiring only one frame; the proof of the doubling property p. 125, is therefore incorrect. Bellaïche [B] deals with the example below, but does not allude to the fact that there two coordinate systems are needed.

Here $V(x,y)$ is the volume of the ball $B(x,\delta)$, with $\delta = d(x,y)$ (so $V(x,y) \approx \Lambda_x(\delta)$. Also $X^\alpha = X_1^{\alpha_1} X_2^{\alpha_2} \cdots X_k^{\alpha_k}$), $|\alpha| = \alpha_1 + \alpha_2 + \cdots + \alpha_k$. One also assumes,

$$\|T(\varphi)\|_{L^2} + \|T^*(\varphi)\|_{L^2} \lesssim \delta^r |B(x,\delta)|^{1/2}. \qquad (3.7)$$

The function φ is any *normalized bump function* for the ball $B(x,\delta)$: i.e. φ is supported in $B(x,\delta)$ and satisfies

$$\sup_B \left\{ |\varphi| + \delta \sum_{j=1}^k |X_j(\varphi)| \right\} \leqslant 1.$$

On the basis of (3.6) and (3.7) one can prove the following.

PROPOSITION. (a) *If $r = 0$, the operator T is bounded on L^p, $1 < p < \infty$.*

(b) *The parametrix S (for the sub-Laplacian $\mathcal{L}$ discussed in §2) is an operator of the above type with $r = 2$.*

(c) *The operators $X_i X_j S$ are of the above type with $r = 0$.*

It is useful to remark that in all the above the metric d can be replaced by an *equivalent metric d'* in the formulations and conclusions stated. The equivalence of d and d' is the assertion that $c_1 \leqslant \frac{d(x,y)}{d'(x,y)} \leqslant c_2$ for a pair of positive constants c_1 and c_2. One can even extend this to a wider form of equivalence, in which $c_1 \leqslant \frac{\mu(d)}{d'} \leqslant c_2$, where μ is a suitable strictly monotone function. Then for each x, the family of balls $\{B(x,\delta)\}$ and $\{B'(x,\delta)\}$ are essentially the same; they differ only insofar that $\{B'(x,\delta)\}$ arises essentially from $B(x,\delta)\}$ by reparametrizing the radius δ.[14]

4   The Conflict of Metrics; First Example

The situation described above – where in a given analytic context there arises a natural metric which determines the relevant class of singular integrals – represents a very satisfactory state of affairs. Unfortunately there are now an increasing number of examples which do not fall under the scope of these ideal circumstances; there may in fact be two different and conflicting metrics, each inherent in the particular situation, and each describing different features of the relevant singular integrals that arise. We shall give two examples of this.

In the first instance we take Ω to be a smooth and bounded domain in $\mathbb{C}^{n+1}$, and $M = \partial\Omega$. Under the conditions on Ω below we are lead to two different metrics:

---

[14]Of course then, the conditions (3.6) and (3.7) must be appropriately reinterpreted.

(i) the "control metric;"

(ii) the "Szegö metric."

For (i) we assume that Ω is of finite-type in the sense that the Lie algebra generated by the tangential Cauchy–Riemann vector fields and their conjugates span the tangent space of M. Then we define the control metric, d_C, as in §3, where $X_1, \ldots, X_{2n}$ are the real and imaginary of the Cauchy–Riemann vector fields.

To define (ii) we assume that Ω is convex, and of finite-type in the sense that any complex line has at most a finite-order contact with $M = \partial\Omega$. The balls of this metric, d_S, are determined as follows: For each $x \in \partial\Omega$, we let $\nu(x)$ denote the unit inward normal at x. Then $B(x, \delta)$ is the "cap" on $\partial\Omega$ "cut out" by inserting a maximal polydisc in Ω, centered at $x + \delta\nu(x)$.[15] It turns out that in this finite-type convex situation the Szegö metric is the one that controls the Cauchy–Szegö projection as a singular integral.[16]

We now make the following remarks:

OBSERVATION 1.   When Ω is strongly pseudo-convex (e.g. the unit ball in $\mathbb{C}^{n+1}$), then $d_c \approx (d_S)^{1/2}$, so that these metrics are comparable in the wide sense described in §3, and therefore give essentially the same family of balls.

OBSERVATION 2.   In general, however, this is not the case. Take for example $\Omega \subset \mathbb{C}^3 = \{(z_1, z_2, w)\}$, with $\Omega = \{Im(w) > |z_1|^{2n} + |z_2|^{2m}\}$, where n and m are integers and $1 \leqslant n \leqslant m$, and $m > 1$. If $t = Re(w)$, the points of $M = \partial\Omega$ can be parametrized by (z_1, z_2, t), and a natural pair of tangential Cauchy–Riemann radius on M is given by

$$\overline{Z}_1 = \tfrac{\partial}{\partial \bar{z}_1} - in|z_1|^{2n-2}z_1\tfrac{\partial}{\partial t} = X_1 + iY_2$$
$$\overline{Z}_2 = \tfrac{\partial}{\partial \bar{z}_2} - im|z_2|^{2m-2}z_2\tfrac{\partial}{\partial t} = X_2 + iY_2 \,.$$

In this case the distances of a point (z_1, z_2, t) to the origin are given essentially by

$$d_C(z_1, z_2, t, 0) \approx |t|^{1/2n} + |z_1| + |z_2|,$$
$$d_S(z_1, z_2, t, 0) \approx |t| + |z_1|^{2n} + |z_2|^{2m}\,.$$

A comparison of these formulae highlights the different natures of the metrics d_C and d_S. Namely, the control metric d_C weighs the z_1 and z_2

---

[15]These balls are described in McNeal [Mc]; there is an earlier version in the real case, considered in Bruna, Nagel, and Wainger [BrNW]. It is remarkable that in the latter situation, the balls are intimately connected with the decay of the Fourier transform of smooth measures carried on $\partial\Omega$.

[16]For this see McNeal and Stein [McS].

directions equally, and weighs the t direction according to the least number of commutators needed to span that direction. However, the Szegö metric, d_S, assigns different weights to z_1 and z_2, calibrated in terms of their differing relation to the t-direction.

For the domain Ω and its boundary M, there are three basic and related operators of interest:

$$C = \text{Cauchy–Szegö projection}$$
$$-\Box_b = Z_1\overline{Z}_1 + \overline{Z}_2 Z_2 \qquad {}^{17}$$
$$\mathcal{L} = X_1^2 + Y_1^2 + X_2^2 + Y_2^2 \,.$$

Now C is controlled by the metric d_S as noted above. Also the inverse of $\mathcal{L}$ is controlled by d_C; see §3. However $-\Box_b = \mathcal{L}+$ first-order terms, and would seem to reflect both d_C and d_S Added to this is the remarkable fact pointed out in Rothschild [R], that as opposed to the strongly pseudo-convex situation, the operator $\Box_b$ is *not* maximally sub-elliptic. A question that arises is:

PROBLEM. *What is the nature of* $\Box_b^{-1}$, *the resulting singular integrals, and their regularity?*

Some of these questions in the case where $n = 1$, $m > 1$, have been considered by Machedon [M]. In that case, besides the usual control metric, he constructs a second "control" metric, based on different vector fields, which happens to be equivalent with the Szegö metric defined above. He also shows that the kernel for $\Box_b^{-1}$ satisfies certain differential inequalities involving both metrics. However, the construction for the second metric and resulting analysis, seems to be applicable only in the case $n = 1$.

5 The Conflict of Metrics; Second Example

This example arises when we consider the $\bar{\partial}$-Neumann problem. We present a modified and simplied version of the problem.[18]

We take Ω to be a bounded smooth domain in $\mathbb{C}^{n+1}(=\mathbb{R}^{2n+2})$, which we also assume is of finite-type. Consider the inhomogeneous Laplace equation

$$\Delta u = f \quad \text{in } \Omega, \tag{5.1}$$

---

[17]Strictly speaking, this represents only one component at $\Box_b$ on $(0, 1)$ forms, but is nevertheless typical.

[18]In this version, the problem is for a scalar-valued functions, instead of a corresponding system.

taken with the complex Neumann-like boundary condition

$$\left(\tfrac{\partial}{\partial\nu} + i\tfrac{\partial}{\partial\tau}\right)(u) + a\cdot u\big|_{\partial\Omega} = 0. \tag{5.2}$$

Here $\partial/\partial\nu$ denotes the inward unit normal, $\partial/\partial\tau = J(\partial/\partial\tau)$, where J is the involution of the complex structure of $\mathbb{C}^{n+1}$ (thus $\tfrac{\partial}{\partial\nu} + i\tfrac{\partial}{\partial\tau}$ is the restriction to the boundary of a transverse Cauchy–Riemann vector field); a is an appropriate complex-valued function.

Solution of this problem can be reduced to the inversion of a corresponding boundary operator, a "Dirichlet to complex Neumann operator", $F \to \not{D}(F)$, defined as follows. For F given on $\partial\Omega$ let U be the solution of the Dirichlet problem, $\Delta U = 0$, $U|_{\partial\Omega} = F$. Then $\not{D}(F)$ is defined to be $= \left(\tfrac{\partial}{\partial\nu} + i\tfrac{\partial}{\partial\tau} + a\right)U\big|_{\partial\Omega}$.

Now $\not{D}$ is a first-order pseudo-differential operator. As far as its invertibility and the nature of this inverse, not much is known in general. However, it is very reasonable to expect that two different metrics must come into play in these questions: There is first the restriction of the Euclidean metric (of $\mathbb{R}^{2n+2}$) to the boundary (we call this metric d_E); it reflects the Euclidean structure inherent in the Laplacian Δ. This interior influence competes with the control metric, d_C, which reflects the interplay of the boundary with the complex structure of $\mathbb{C}^{n+1}$.

Besides these general observations, we do not know a great deal about the inverse of $\not{D}$, except when either Ω is strongly pseudo-context, or when $n = 1$ and Ω is pseudo-convex of finite type.[19] That situation has been studied in detail and it is known that $\not{D}$ is invertible when the range of a excludes certain forbidden values. In the related special case corresponding to the unit ball, when the boundary is realized as $\mathbb{H}^n$, the inverse operator is a convolution operator $f \to T(f) = f \star K$, whose kernel can be described rather explicitly. Remarkably it can be written $K = \sum_{k,\ell} K^{(k,\ell)}$ where each summand $K^{(k,\ell)}$ is itself the product of homogeneous functions with respect to the two dilations, with $\delta > 0$,

$$(z,t) \longrightarrow (\delta z, \delta t) \tag{5.3}$$

$$(z,t) \longrightarrow (\delta z, \delta^2 t). \tag{5.4}$$

The first is the (non-automorphic) isotropic dilations connected with the "Euclidean" metric d_E. The second are the automorphic dilations connected with the control metric d_C on $\mathbb{H}^n$. More precisely,

$$K^{(k,\ell)}(z,t) = E_k(z,t)H_\ell(z,t)$$

---

[19]See Greiner and Stein [GS], Treves [T], and Chang, Nagel, and Stein [CNS].

where E_k is homogeneous of degree $-k$ with respect to (5.3), and H_ℓ is homogeneous of degree $-\ell$ with respect to (5.4), and both are smooth away from the origin.[20]

The most interesting "critical case" of such operators arises when $k = 2n$, $\ell = 2$. A typical example is

$$K(z,t) = \frac{w(z,t)}{(|z|^2 + t^2)^n} \cdot \frac{1}{|z|^2 + it} \tag{5.5}$$

where ω is homogeneous of degree 0, with respect to (5.3), is smooth when $|z|^2 + t^2 = 1$, and has vanishing mean-value on the sphere $|z| = 1, t = 0$.

It is clear that the kernel (5.5) displays features of both metrics: the first factor is controlled by d_E; the second is controlled by d_C. The question then becomes:

PROBLEM. *How are we to understand the class of convolution operators whose kernels are typified by (5.5)? How does one prove their L^2 and L^p regularity?*

6   A New Viewpoint

What we want to do in the rest of this essay are three things:

(a) Present an initial frame-work in terms of which we can hope to deal with these further classes of singular integrals.

(b) Explain more concretely how this point of view allows one to resolve the questions raised in §4 and §5.

(c) Speculate about some of the features of a general theory that might encompass these ideas.

Turning to the first point, we shall try to explain matters in terms of what might be called the "two-factor case." The picture that should be kept in mind is the following

$$M_1 \times M_2$$

$$\Big\downarrow \pi \tag{6.1}$$

$$M$$

Here M is our underlying space – which is the world we actually see. Above it, what we don't see but is merely a construct, is a product of spaces

---

[20]See Phong and Stein [PS].

$M_1 \times M_2$. What is going on will be accounted for by this product and the projection π. Besides "products" and "projections" the other key-words attached to this view-point are "liftings" and "flags."

We will illustrate this first in terms of the operator arising on the Heisenberg group in §5, whose kernel is given by

$$K(z,t) = \frac{\omega(z,t)}{(|z|^2 + t^2)^n} \cdot \frac{1}{|z|^2 + it}.$$

One proceeds[21] by introducing an additional *virtual* singularity in K; that is we deal with it as if it were the kernel $K'(z,t)$ given by

$$K'(z,t) = \frac{\omega(z,0)}{|z|^{2n}} \cdot \frac{1}{|z|^2 + it}. \tag{6.2}$$

The singularities of K' are now represented by a flag structure

$$\{0\} \subset S \subset \mathbb{H}^n, \tag{6.3}$$

where S is the central sub-group $\{(0,t)\}$. The kernel K' has its heaviest singularity at the origin; it has also a singularity on S – but to a lesser degree; on the complement of S it is regular. The differential inequalities satisfies by K' (and a fortiori by K) are

$$\left| \partial_z^\alpha \partial_t^\beta K'(z,t) \right| \lesssim |z|^{-2n-|\alpha|} \left(|z| + |t|^{1/2} \right)^{-2-2\beta} .eqno(6.4)$$

The kernels also satisfy appropriate cancellation conditions involving separate integrations in the z and t variables.

We now outline how we can deal with singular integrals which have such kernels (e.g. K or K')

- The picture (6.1) becomes in this case

$$\mathbb{H}^n \times R$$

$$\downarrow \pi$$

$$\mathbb{H}^n$$

with $\pi((z,t),u) = (z, t+u)$. This realizes $\mathbb{H}^n$ as $(\mathbb{H}^n \times \mathbb{R})/\Gamma$, where Γ is the sub-group $= \{(0,u,-u)\}$.

- We next lift the kernel K on $\mathbb{H}^n$ to a kernel $\widetilde{K}$ on $\mathbb{H}^n \times \mathbb{R}$, so that $\widetilde{K}$ is a "product kernel." Among other things this means $\widetilde{K}$ should satisfy

---

[21]Here we describe the treatment given in Müller, Ricci, and Stein [MüRS].

the differential inequalities

$$\left|\partial_z^\alpha \partial_t^\beta \partial_u^\gamma \widetilde{K}(z,t,u)\right| \lesssim \cdot\left(|z| + |t|^{1/2}\right)^{-2n-2-|\alpha|-\beta} \cdot |u|^{-1-\gamma} \qquad (6.5)$$

and appropriate cancellation conditions (involving separate integrations in the (z,t) and u variables).

- The lifted kernel $\widetilde{K}$ is constructed so (and this is the main point) that it "projects" to K, i.e.

$$K(z,t) = \int_{-\infty}^{\infty} \widetilde{K}(z, t - u, u)\,du\,, \qquad (6.6)$$

taken in the sense of distributions.

- Operators $\widetilde{T}$ corresponding to product kernels $\widetilde{K}$ can ultimately be dealt with in terms of tensor products of operators acting each variable separately, and in this way one can obtain their L^p boundedness.[22]
- From the L^p regularity of $\widetilde{T}$ we obtain that of T by transference, using the projection π (i.e. (6.6)).

7   A New Viewpoint, II

The first step in a general theory is a proof that the results for the Heisenberg group extend to any nilpotent Lie group with dilations.[23]

We assume N is such a group, which we identify with its Lie algebra $\mathfrak{n}$. A flag for N will be an increasing set of homogeneous sub-groups.

$$\{0\} \subset S_1 \subset S_2 \cdots \subset S_\ell = N\,. \qquad (7.1)$$

We identify S_j with its Lie algebra $\mathfrak{s}_j$, obtaining a filtration of sub-algebras

$$\{0\} \subset \mathfrak{s}_1 \subset \mathfrak{s}_2 \cdots \subset \mathfrak{s}_\ell = \mathfrak{n}\,.$$

Our assumption about the flag (7.1) is that there is a gradation $\mathfrak{s}_j = \mathfrak{s}_{j-1} \oplus w_j$ of subspaces w_j, $j = 1, \ldots, k$, so that $[w_j, w_k] = 0$, if $j \neq k$.

This is equivalent with the existence of another group $\widetilde{N}$ which is the direct product $\widetilde{N} = N_1 \times N_2 \cdots \times N_\ell$, where each N_j is homogeneous, and a homomorphism π of $\widetilde{N}$ onto N, so that

$$S_j = \pi(N_1 \times N_2 \cdots \times N_j \times \{0\} \cdots \{0\})\,, \qquad 1 \leqslant j \leqslant \ell\,.$$

In analogy with the above we can define flag kernels, with respect to the flag (7.1). As in the case of $\mathbb{H}^n$, each such kernel can be lifted to a product

---

[22]This procedure which uses square functions and Littlewood–Paley theory, was used in the context of $\mathbb{R}^n$ in Fefferman and Stein [FSt].

[23]This is the subject of a forthcoming paper of Nagel, Ricci and Stein [NRS].

kernel on $\widetilde{N}$. The properties of the convolution operator on N are then consequences of the corresponding properties on $\widetilde{N}$.

The theory just described has a number of further applications. In the case where $N = \mathbb{R}^n$ it can be used to obtain an explicit description of the singularities (and L^p regularity) of the Cauchy–Szegö projection for tube domains over polygonal cones. Also the case when N is a two-step nilpotent group allows one to generalize this to the Cauchy–Szegö projections associated to CR sub-manifolds of $\mathbb{C}^N$ which are given in terms of simultaneously diagonalizable quadratic forms. The singular integrals for the corresponding flags make it possible to study the (relative) regularity of $\Box_b$ for these sub-manifolds.[24]

The general point of view described here also applies to the example in §4.[25] We sketch very briefly the set-up.

Starting with $\Omega \subset \mathbb{C}^3$ given by $\Omega = \{Im(w) > \Phi_1(z_1) + \Phi_2(z_2)\}$, $(\Phi_1(z_1) = |z_1|^{2n}, \Phi_2(z_1) = |z_2|^{2m})$. We let

$$M = \partial\Omega = \{Im(w) = \Phi_1(z_1) + \Phi_2(z_2)\}$$

and set $M_j = \partial\Omega_j = \{Im(w_j) = \Phi_j(z_j)\} \subset \mathbb{C}^2$, $j = 1, 2$. Here $\pi : M_1 \times M_2 \to M$ is given by $\pi(w_1, z_1; w_2, z_2) = (w_1 + w_2, z_1, z_2)$.

One again studies flag-type operators. Here there are two alternative flag structures, each corresponding to a variable $x \in M$:

$$\{x\} \subset V_x^{(i)} \subset M, \quad i = 1, 2. \tag{7.2}$$

Here $V_x^{(1)} = \pi(M_1 \times \{\bar{x}_2\})$, where $(\bar{x}_1, \bar{x}_2)$ is any point so that $\pi(\bar{x}_1, \bar{x}_2) = x$. There is a similar definition of the alternate manifold $V_x^{(2)}$. It then turns out that the operator $\Box_b^{-1}$ can be expressed in terms of appropriate kernels attached to these flags. Their analysis leads to sharp estimates and also accounts for the restricted sub-ellipticity mentioned in §4.

We now take up again the issue of the occurrence of a multiplicity of metrics on M, which motivated our considerations in §4 and §5. We readily see that this phenomenon can be seen as arising from the above frame-work. The starting point is the premise that each factor M_j carries its own natural metric. These individual metrics can be combined in a variety of different ways to give a metric on the product $M_1 \times M_2$. Using the projection $\pi \colon M_1 \times M_2 \to M$, we then obtain a number of different metrics on M.

---

[24]For further details see Nagel, Ricci, and Stein [NRS].

[25]This is the subject of a forthcoming paper of Nagel and Stein [NS].

In the example where $M = \mathbb{H}^n$, $M_1 = \mathbb{H}^n$, $M_2 = R$, the natural metric on $\mathbb{H}^n$ and $\mathbb{R}$ are (respectively) given by $d_1 = |z| + |t|^{1/2}$, $d_2 = |u|$; (here d_1 and d_2 denote the distance of $(z, t) \in \mathbb{H}^n$, and $u \in \mathbb{R}$ from the origin). Consider the metrics $\tilde{d}_E = d_1 + d_2$ and $\tilde{d}_C = d_1 + d_2^{1/2}$ on $M_1 \times M_2$. Then under π, $\tilde{d}_E$ and $\tilde{d}_C$ project to metrics equivalent with d_E and d_C in any bounded neighborhood of the origin.

8 Some Aspects of a General Theory

The various examples we have dealt with, and the results already obtained for homogeneous groups, raise the question of finding a general formulation for singular integrals encompassing these ideas. At this stage we can have no clear picture, but we can nevertheless speculate about the possible shape of the outcome. In trying to do this we can envisage a framework based on certain geometric data which we now describe:

First, we suppose we are given for each $x \in M$ a flag of submanifolds centered at x, of increasing dimensions,

$$\{x\} \subset V_x^{(1)} \subset V_x^{(2)} \cdots \subset V_x^{(\ell)} = M, \quad x \in M. \tag{8.1}$$

Here each $V_x^{(j)}$ is a sub-manifold, depending smoothly on x.

Second, connected with this there is assumed to be an increasing collection of vector fields,

$$\Gamma_1 \subset \Gamma_2 \cdots \subset \Gamma_\ell, \tag{8.2}$$

where Γ_ℓ consists of vector fields $X_1, X_2, \ldots, X_k$ and their commutators up to a certain order. It would also be supposed that the vectors in Γ_j span the tangent space of $V_x^{(j)}$ at x.

To such an underlying geometric structure we should like to associate singular integrals,

$$T(f)(x) = \int_M K(x, y) f(y) dy. \tag{8.3}$$

The connection with the data (8.1) and (8.2) we have in mind that the distribution $K(x, y)$ would be allowed to be singular when $y \in V_x^{(j)}$, $j < \ell$, and that the "degree" of that singularity (measured in terms of the vector fields in Γ) would decrease as j increases.

This rather vague description must of course be made precise. To do this a number of issues would have to be addressed and resolved:

(1) What are the further geometric properties of the $V^{(j)}$ and Γ_j that need to be assumed?

(2) What are the differential inequalities and cancellation conditions that should be imposed on K?

(3) How does one prove the L^2 and L^p boundedness of these operators?

(4) Do these operators form an algebra?

To answer these questions fully looks at present to be quite a daunting task. There is, however, a further example that might provide some useful hints: the "singular Radon transforms."[26]

In terms of the formulation sketched above, these operators arise when we are dealing with a two-step flag

$$\{x\} \subset V_x \subset M$$

(where each V_x is k-dimensional, with $0 < k < n$), and when we restrict ourselves to the special case when the kernel K in (8.3) is entirely supported on $\{y \in V_x\}$.

Let us try to make this connection a little more precise. One realizes V_x in parametrized form, $V_x = \{\gamma_t(x), t \in U\}$ with $(x,t) \mapsto \gamma_t(x)$ a smooth mapping of $M \times U \to M$ (where U is an open neighborhood of the origin in $\mathbb{R}^k$); we also suppose $\gamma_0(x) \equiv x$. Such mappings γ_t have an asymptotic expansion

$$\gamma_t(x) \sim \exp\left(\sum_\alpha X_\alpha t^\alpha/\alpha!\right)(x), \quad \text{as } t \to 0 \qquad (8.4)$$

for appropriate vector fields $\{X_\alpha\}$. It is convenient to set $X_J = [X_{\alpha_1}[X_{\alpha_2}[\cdots X_{\alpha_j}]]]$, and assign weight $|\alpha_1| + |\alpha_2| + \cdots + |\alpha_j|$ to X_J, where $J = (\alpha_1, \alpha_2, \ldots, \alpha_j)$. The assumption that is made on γ_t is that the $\{X_J\}$ span the tangent space of M. A singular Radon transform is then of the form

$$R(f)(x) = \int_U k_x(t) f(\gamma_t(x)) dt \qquad (8.5)$$

where for each x, $k_x(t)$ is a Calderón–Zygmund kernel on $\mathbb{R}^k$, depending smoothly on x, with compact support in M.

We now indicate a class of operators of the class (8.1)–(8.3) that would seem to subsume the singular Radon transform (8.5). For this, consider the X_j, $1 \leqslant j < k$, which are coefficients of the linear monomials appearing in (8.4). Choose also $X_{J_1}, X_{J_2}, \ldots, X_{J_{n-k}}$ so that $X_1, X_2, \ldots, X_k, X_{J_1}, X_{J_2}, \ldots, X_{J_{n-k}}$ is a basis of the tangent space of M. Define $\tilde\gamma(t, s)$

---

[26]For a general treatment of these operators see Christ, Nagel, Stein, and Wainger [ChrNSW].

by

$$\tilde{\gamma}(t,s)(x) = \exp\left(\sum_{j=1}^{n-k} s_j X_{J_j}\right) \cdot \gamma_t(x)$$

where (t,s) lies in a neighborhood of the origin in $\mathbb{R}^k \times \mathbb{R}^{n-k}$. We then take $k_x(t,s)$ (for each $x \in M$) to be a flag kernel on $\mathbb{R}^n$, relative to the flag

$$\{0\} \subset \mathbb{R}^k \subset \mathbb{R}^n, \quad \mathbb{R}^k = \{(t,0)\},$$

and homogeneity $\delta: (t,s) \rightarrow (\delta t_1, \delta t_2, \ldots, \delta t_k, \delta^{a_1} s_1, \ldots, \delta^{a_{n-k}} s_{n-k})$. Here a_j is the weight of X_{J_j}; we also suppose that $x \rightarrow k_x$ is smooth and has compact support. Then the kind of operator we wish to consider is

$$T(f)(x) = \int k_x(t,s) f(\tilde{\gamma}(t,s)(x)) dt\, ds,$$

which is of the form (8.3) with

$$K(x,y) = k_x(t,s) J^{-1}(x,y), \quad y = \tilde{\gamma}(t,s)(x)$$

and where $J(x,y)$ is the Jacobian of the mapping $(t,s) \rightarrow y = \tilde{\gamma}(t,s)x$ for each fixed x. We note that in the special case $k_x(t,s) = k_x(t)\delta(s)$, where δ is the Dirac delta function at the origin in $\mathbb{R}^{n-k}$, our operator T becomes essentially the operator R in (8.5).

References

[B]   A. BELLAÏCHE, The tangent space in sub-Riemannian geometry, in "Sub-Riemannian Geometry", Progress in Math. 144, Birkhauser (1996), 1–78.

[BrNW] J. BRUNA, A. NAGEL, S. WAINGER, Convex hypersurfaces and Fourier transforms, Ann. of Math. 127 (1988), 333–365.

[CNS]   D.C. CHANG, A. NAGEL, E. STEIN, Estimates for the $\bar{\partial}$-Neumann problem in pseudoconvex domains of finite-type in $\mathbb{C}^2$, Acta Math. 169 (1992), 153–228.

[Ch]   W. CHOW, Uber systeme von linearer parteillen differentialgleichungen enrter ordnung, Math. Ann. 117 (1940), 98–115.

[ChrNSW] M. CHRIST, A. NAGEL, E. STEIN, S. WAINGER, Singular and maximal Radon transforms: analysis and geometry, Ann. of Math., to appear.

[CoW]   R. COIFMAN, G. WEISS, Analyse harmonique non-commutative sur certains éspaces homogenes, Springer Lecture Notes in Math 242 (1971).

[DJS]   G. DAVID, J.L. JOURNÉ, S. SEMMES, Opérateurs de Calderón–Zygmund, fonctions para-accrétives et interpolation, Rev. Math. Iberoamer 1 (1985), 1–56.

[F]   C. FEFFERMAN, The Bergman kernel and biholomorphic mappings of pseudo-convex domains, Invent. Math. 26 (1974), 1–65.

[FP]   C. FEFFERMAN, D. PHONG, Subelliptic eigenvalue problems, in "Confer-
        ence on Harmonic Analysis in honor of Antoni Zygmund" (Beckner, et.
        al.), Wadsworth (1983), 590–606.

[FS]   C. FEFFERMAN, A. SANCHEZ–CALLE, Fundamental solutions of second
        order elliptic operators, Ann. of Math. 124 (1986), 247–272.

[FSt]   R. FEFFERMAN, E. STEIN, Singular integrals on product spaces, Adv. in
        Math. 45 (1982), 117–143.

[FoS]   G. FOLLAND, E. STEIN, Estimates for the $\bar{\partial}_b$ complex and analysis on the
        Heisenberg group, Comm. Pure and Appl. Math. 27 (1974), 429–522.

[GS]   P. GREINER, E. STEIN, Estimates for the $\bar{\partial}$-Neumann problem, Math.
        Notes 19, Princeton University Press (1977).

[Gr]   M. GROMOV, Carnot–Caratheodory spaces seen from within, in "Sub-
        Riemannian Geometry", Progress in Math. 144, Birkhauser (1996), 79–
        323.

[H]   L. HÖRMANDER, Hypoelliptic second-order differential equations, Acta
        Math. 119 (1967), 147–171.

[KS]   A. KNAPP, E. STEIN, Interwining operators for semi-simple groups, Ann.
        of Math. 93 (1971), 489–578.

[KoV]   A. KORÁNYI, S. VÁGY, Singular integrals in homogeneous spaces and
        some problems in classical analysis, Ann. Scuola Norm. Sup. Pisa 25
        (1971), 575–648.

[M]   M. MACHEDON, Estimates for the parametrix of the Kohn–Laplacian on
        certain domains, Inv. Math. 91 (1988), 339–364.

[Mc]   J. MCNEAL, Estimates on the Bergman kernels of convex domains, Adv.
        Math. 109 (1994), 108–139.

[McS]   J. MCNEAL, E. STEIN, The Szegö projection on convex domains, Math.
        Zeit. 224 (1997), 519–553.

[Me]   Y. MEYER, Ondelettes et Opérateurs, II, Hermann, 1990.

[MüRS] D. MÜLLER, F. RICCI, E.M. STEIN, Marcinkiewicz multipliers and
        multi-parameter structure on Heisenberg-type groups I, Inv. Math. 119
        (1995), 199–233.

[NS]   A. NAGEL, E. STEIN, in preparation.

[NRS]   A. NAGEL, F. RICI, E. STEIN, Flag singularities and analysis on certain
        quadratic CR manifolds in $\mathbb{C}^N$, in preparation.

[NSW1] A. NAGEL, E. STEIN, S. WAINGER, Boundary behavior of functions
        holomorphic in domains of finite type, Nat. Acad. Sci. U.S.A. 78 (1981),
        6596–6599.

[NSW2] A. NAGEL, E. STEIN, S. WAINGER, Balls and metrics defined by vector
        fields, Acta Math. 155 (1985), 103–147.

[NaTV] F. NAZAROV, S. TREIL, A. VOLBERG, Weak type estimates and Cotlar
        inequalities for Calderón–Zygmund operators in non-homogeneous spaces,
        Int. Math. Res. Notices 1 (1998), 463–487.

[PS]   D. PHONG, E. STEIN, Some further classes of pseudodifferential and singular integral operators arising in boundary value problems, Amer. Jour. Math. 104 (1982), 141–172.

[R]   L. ROTHSCHILD, Nonexistence of optimal L^2 estimates for boundary Laplacian operator on certain weakly pseudoconvex domains, Comm. Part. Diff. Eq. 5 (1980), 897–912.

[RS]   L. ROTHSCHILD, E. STEIN, Hypoelliptic differential operators and nilpotent groups, Acta Math. 137 (1976), 247–320.

[S]   E. STEIN, Harmonic Analysis: Real-variable Methods, Orthongonality, and Oscillatory Integrals, Princeton University Press, 1993.

[T]   F. TREVES, Introduction to Pseudo-Differential and Fourier Integral Operators, I, Plenum Press, 1982.

ELIAS M. STEIN, Department of Mathematics, Princeton University, Princeton, NJ 08544-1000, USA